Also from WILEY...

A FIRST LAB IN CIRCUITS AND ELECTRONICS

by Yannis Tsividis, Columbia University

Written by an award-winning educator and researcher, the 16 experiments in this book have been extensively tested and fine-tuned. This lab manual, like no other,

➤ provides an exciting, *active exploration* of concepts and measurements.

➤ *connects circuits to real applications* which students can relate to, in order to motivate them and convince them that what they learn is for real.

➤ pays special attention to *strengthening intuition.*

➤ encourages students to *tinker and experiment* on their own.

➤ *includes self-contained background* for all electronics experiments, so that it can be run concurrently with any circuits or electronics course.

Your students will not only learn the principles correctly, but will connect the circuits they build to the real world through light and temperature sensors, microphones, loudspeakers, and antennas. Examples include:

➤ In the frequency response experiment, students use their **RC** circuits to process music signals and listen to them.

➤ Students combine their knowledge from the rectifier, **LC** circuit, and amplifier experiments to make a simple, but complete, radio receiver.

As a result, the material is not only made interesting, but helps motivate further study in circuits, electronics, communications, and semiconductor devices.

A lab based on this book can be easily put together, thanks to *an extensive "Putting the Lab Together" manual.* This manual, along with a detailed table of contents and a description of the supplements available, can be found on the book's Web site:

www.wiley.com/college/tsividis

Fundamentals of Electronic Circuit Design

Fundamentals of Electronic Circuit Design

David Comer
Donald Comer

BRIGHAM YOUNG UNIVERSITY

 John Wiley & Sons, Inc.

http://www.wiley.com/college/comer

EXECUTIVE EDITOR	Bill Zobrist
MARKETING MANAGER	Katherine Hepburn
SENIOR PRODUCTION EDITOR	Christine Cervoni
SENIOR DESIGNER	Dawn Stanley
ILLUSTRATION EDITOR	Gene Aiello
COVER DESIGNER	David Levy
COVER IMAGE	Triptych by Frank Lloyd Wright/© Superstock. Reproduced with permission of Artists Rights Society, New York.

This book was set in Times, Helvetica and Trade Gothic by TechBooks and printed and bound by Donnelley/Willard. The cover was printed by Brady Palmer.

This book is printed on acid-free paper. ∞

Library of Congress Cataloging in Publication Data:

ISBN 0–471–41016–0

Printed in the United States of America

10   9   8   7   6   5   4   3   2   1

Preface

The university curriculum in electrical engineering and the field of electronic circuits have changed drastically in the last two decades. Many programs have incorporated fundamental courses in digital signal processing and expanded the list of required courses in computer engineering. This expansion has led to a reduction in the number of mandated electronic circuits courses. In some programs, electrical engineering students are required to take only one electronic circuits course.

At the same time, the metal-oxide-semiconductor field-effect transistor (MOSFET) has replaced the bipolar junction transistor (BJT) as the workhorse of the electronics industry. Not only is the MOSFET used in all personal computers, pushing clock speeds into the gigahertz range, but also more analog circuits are now designed with the MOSFET than ever before. The BJT is still popular in discrete circuit design and in high-speed heterojunction processes, but the volume of integrated circuits fabricated in a CMOS process overwhelms all competition. Integrated circuit design using MOSFETs is an absolute necessity for the broad education of the electrical engineer, regardless of the chosen option within the field.

A Modern Approach

This textbook accounts for these trends. It concentrates on the MOSFET and the BJT without attempting to cover the less well-used devices. The fundamental ideas of circuit design can then be presented in a textbook of reasonable length. The length and topics are designed for approximately six semester hours. The text can be used for two three-credit classes with a small amount of required supplemental material for the instructor. If used for a four-credit class, it contains extra topics to allow some latitude in choice of topics by the instructor.

The importance of the MOSFET is reflected by the fact that this amplifying device is introduced prior to the introduction of the BJT. Additionally, in the chapters on integrated circuit design, the MOSFET is discussed first as the dominant electronic device.

The significance of the integrated circuit cannot be neglected in the initial electronics course. Almost all industrial companies now have access to a semiconductor foundry, whether their own or an independent foundry. For this reason, Chapters 8, 9, and 10 are devoted exclusively to integrated circuit design.

Circuit simulation is an absolute necessity in integrated circuit design. This topic is generally introduced in the previous course that covers fundamental concepts and theorems of circuit analysis. Several simulation examples are included in this textbook, especially in the integrated circuit chapters.

V

Organization

The textbook attempts to motivate the student in several ways. Chapter 1 discusses the importance of the field and the history of electronics. By becoming familiar with the individuals who have made significant contributions to this field, students can understand the potential for their own future contributions. Chapter 2 overviews several important subfields of electronic circuits to give the student an appreciation for the need to learn more about devices and circuits. Several theorems, important to circuit design, are reviewed in this chapter showing applications of Thévenin's theorem and the Miller effect. Chapter 3 begins with a discussion of a general amplifier model. Gain elements and frequency-dependent factors are introduced here so that later discussions of the MOSFET and BJT can be related directly to this general amplifier model. We hope that this overview will enable the student to understand the specific considerations required in the amplifier design of later chapters.

Chapter 4 treats the first amplifying element, the op amp. This device is treated first because it represents a near-ideal model that allows the topic of modeling to be introduced. Although op amps are typically covered in the preceding circuit analysis class, this chapter adds the frequency response limitations that must be considered in amplifiers that contain op amps. Chapter 5 treats the basic ideas of semiconductor physics and the *pn* junction. The motivation for modeling nonlinear devices is also discussed here. Applications of the diode and breakdown diode are considered at the end of this chapter.

The basic operation of the MOSFET along with simple MOSFET amplifiers are the topics of Chapter 6. Chapter 7 parallels Chapter 6 using the BJT as the amplifying element. Not every possible configuration of these devices is included in this coverage. These chapters are more concerned with teaching the basic principles of amplifiers than with being an exhaustive coverage of all possibilities.

Chapters 8, 9, and 10 build on the earlier chapters in treating integrated circuit design. Chapter 8 is a detailed discussion of the differences between discrete and integrated circuit design. Basic layout of integrated circuits is considered. Material on Spice simulation is also included here. Chapter 9 covers basic building blocks or single amplifying stages often used in MOSFET integrated circuit amplifier design, and Chapter 10 covers similar material for the BJT integrated circuit amplifier.

Chapter 11 discusses differential input stages and op amp design from an integrated circuit standpoint. Stability and compensation of the op amp are also discussed. The topic of Chapter 12 is feedback amplifiers. The general theory of feedback is followed by specific examples of practical stages. Again, not every possible configuration of feedback amplifier is covered. The examples use only voltage feedback stages to demonstrate the principles involved in feedback analysis.

Chapters 13 and 14 are directed toward digital circuits. The first of these chapters discusses fundamental ideas of switching circuits followed by simple discrete multivibrator circuits and integrated circuit timing configurations. Chapter 14 is devoted to CMOS combinational logic circuits. Basic operation of the inverter, NOR gate, and NAND gate is emphasized followed by a discussion of general combinational logic function realization with CMOS circuits. More advanced digital CMOS circuits are deferred to succeeding digital design courses.

Pedagogical Features

To help students develop their understanding of circuit design, we have built several features into the textbook.

➤ **Chapter Outline.** Each chapter opens with an outline of the major headings to inform the student of the subject and scope of the topics to be covered.

➤ **Important Concepts.** Each section opens with a summary of the key points included in that section. This helps the student focus on these significant topics as the material is studied.

➤ **Demonstration Problem.** Most chapters introduce the subsequent material by providing a typical circuit design Demonstration Problem at the beginning of the chapter. The principles required to complete this design problem are then developed throughout the chapter. At the end of the chapter, these principles are applied to the solution of the Demonstration Problem.

➤ **Practical Considerations.** Scattered liberally throughout the text, the Practical Considerations are sections that relate the developed theory to the actual practice of circuit design. A more thorough understanding of practical implications of the theory can be obtained from these sections.

➤ **Practice Problems.** Practice problems with accompanying answers are included to allow the student to test comprehension before moving to succeeding material.

➤ **Summary.** Each chapter ends with a summary of key concepts developed in the chapter.

➤ **End-of-Chapter Problems.** These problems are segmented by section headings to help the instructor identify appropriate problems to assign and to help students identify concepts needed for solution. The numbers of the more difficult problems are preceded by an asterisk. Design problem numbers are prefaced by the letter *D*.

➤ **Worked Examples.** Many examples that apply the previously developed theory are provided to solidify concepts for the student.

➤ **Spice Examples.** While Spice examples are used throughout the book, the three chapters on integrated circuit design, Chapters 8, 9, and 10, provide a strong focus on Spice simulation.

In demonstrating the use of Spice, a netlist file is included as well as the schematic for each circuit. Although basic circuit simulation is popularly done using a schematic capture program, it is important that the student also understand the notation used in a netlist. For more advanced circuit layout tools such as layout versus schematic (LVS) programs, the output is generated in the format of a netlist. It is essential that the designer be capable of using this netlist information appropriately. Consequently, the netlist file is included for all simulated circuits. It is our intent to provide additional practice in simulation using Web-based supplements for this textbook. A Spice tutorial and the results of several simulations, appearing as interactive problems, are available at the Web address: www.ee.byu.edu/circuit.

The prerequisites for a study of this text are an understanding of math through differential equations, although this topic is used sparingly in the textbook, and an understanding of basic circuit analysis. The typical prerequisite circuits course covers the fundamental analysis methods such as Kirchhoff's node and loop methods, Thévenin's and Norton's theorem, and some notions of frequency response.

This text also supplies engineering definitions of commonly used electronic terms, providing the student an opportunity to learn the jargon of circuit design. Finally, this text attempts to include the information that will be essential to any graduating electrical or computer engineer, assuming that the curriculum provides no further study of electronic design.

Acknowledgments

We would like to thank the following colleagues who reviewed the manuscript for their thoughtful comments and suggestions.

Robert Bernick, *California State Polytechnic University, Pomona*

Dan Chen, *Virginia Polytechnic and State University*

Yogendra P. Kakad, *University of North Carolina—Charlotte*

Robert J. Krueger, *University of Wisconsin—Milwaukee*

Jeff Schowalter, *University of Wisconsin—Madison*

Guru Subramanyam, *University of Dayton*

Trevor J. Thornton, *Arizona State University*

Douglas Tougaw, *Valparaiso University*

Joseph Tront, *Virginia Polytechnic and State University*

It has been a pleasure to work on this project with Bill Zobrist, Acquisitions Editor. His encouragement and suggestions have influenced the content and appearance of the textbook. Barbara Heaney, Development Editor, Christine Cervoni, Senior Production Editor, and the production team are also to be thanked for their excellent work.

David J. Comer

Donald T. Comer

Brief Contents

Contents

5 ...The Semiconductor Diode and Nonlinear Modeling *124*

12 ...Feedback Amplifiers *369*

13 ...Large-Signal Circuits *404*

Introduction to Electronics

In the United States and other parts of the civilized world, few people realize the tremendous influence of electronics on world affairs. In Section 1.1 we emphasize the effects of this field, both in world events and in the daily life of each individual. Electronics is a field in which an individual can have a lasting impact on society, as we show in succeeding sections of this chapter.

1.1 The Significance of Electronics

The cliche, "We live in an electronic age," implies the tremendous significance of the electronics field. Some feel that electronics is one of the most important fields in existence today. The next few paragraphs will provide the rationale for this opinion.

The list of Table 1.1 includes several important areas or subfields of electronics that can lead to an appreciation for the effect this field has had on humanity.

Just as the electronics field has a great impact on humanity, the electrical engineer has a great impact on the electronics field. The engineer is involved in the invention, design, and development of electronic devices and systems. In recent years, electrical engineers have led the way in the development of integrated circuits, compact disks (CDs), high-definition television (HDTV), the microprocessor, and personal computers. The engineer plays an important part in the advancement of the challenging and exciting field of electronics that will yet produce many unique new developments.

1.1.1 COMMUNICATION SYSTEMS

It is easy to take electronics for granted because we have had the telephone, radio, and television with us for so many years. Even these traditional systems continue to evolve as advancements in the radio and television areas lead to improved quality of performance. High-definition television will replace present systems in the near future, providing much improved resolution. The old, traditional area of AM radio has also been improved in quality. The bandwidth of transmission of certain AM channels has been increased, stereo is available, and other improvements have taken place in recent years.

Newer devices have become prominent in the communications field in the last several decades. Satellite communication systems that offer instantaneous transmission of television signals from any part of the world have been available since the 1960s. Facsimile printers that transmit printed matter over phone lines became popular in the 1980s. Cellular phones that allow communication from automobiles and other areas that are not wired directly to

1

Table 1.1 Important Areas of Application of Electronics

Area	Examples
Communications and entertainment	Radios, telephones, televisions, CDs, communication satellites, stereos, military communications, VCRs, DVDs
Computers and calculators	Digital circuits, microprocessors, personal computers, mainframe computers, supercomputers, hand calculators
Automatic control systems	Space vehicles, guided missiles, airplane controls, numerically controlled lathes and milling machines, robotics, disk controllers, auto manufacturing, integrated circuit fabrication
Instrumentation	Electronic instruments such as meters and oscilloscopes, medical instruments such as CAT scan, MRI, and X-ray equipment
Automotive electronics	Electronic ignition systems, antiskid braking systems, automatic suspension adjustment, performance optimization, computation and display of performance parameters
Power generation and distribution	Control and optimization of generators, control of distribution paths, monitoring of system performance
Radar	Air traffic control, security systems, military systems, law enforcement
Integrated circuits	Design and fabrication of microelectronic circuits

the telephone lines also saw a tremendous surge in usage in the 1980s. Lasers and fiber optic communication links represent another area of great development in recent years.

The Internet with all its data transmission channels is now used by a large number of subscribers that continues to grow each day. This capability is changing the way we do business, the way we communicate with each other, and the way we obtain information.

Emergency mobile services for police, health personnel, and fire departments are vital to the efficient operation of those groups. Air traffic control is done over communication channels as are ship-to-shore voice links.

1.1.2 COMPUTERS AND CALCULATORS

If not for the development of the transistor, efficient computers could not have been developed. The early vacuum tube versions developed in the 1950s were too inefficient and unreliable to make the computer a useful industrial system. The transistor led to significant advancements in the computer and allowed this important system to streamline the operation of many companies.

The integrated circuit further advanced the digital electronics area by lowering the cost of the computer and introducing the hand calculator. The minicomputers of the late 1960s were affordable by smaller professional offices and research groups that previously could not afford a computer. The microprocessor ushered in the era of the personal computer, as this device further decreased the cost of a highly capable computer. The electrical engineer contributes to the computer area in several ways. The integrated circuit designer provides circuits that can be used in the computer. The computer architect designs the actual computer. The systems engineer often specifies how a computer is to be used to control some complex system. All of these functions are generally performed by the electrical engineer.

It would be difficult to describe the importance of the computer in today's world. This system has been important for years in industrial applications, but now extends its influence to personal applications. A huge impact on society has been the attraction of nontechnical users to the computer. From the 1950s to the 1980s, the major computer users were technically trained. The year 2000 found millions of everyday people utilizing the advantages offered by the computer. Small businessmen, elementary school children, and people of all ages have embraced the personal computer.

1.1.3 AUTOMATIC CONTROL SYSTEMS

The historic flight of the Apollo space vehicle to the moon in 1969 was accomplished under the guidance of an electronic system. The guidance and control systems for the Challenger vehicles as well as all other space vehicles are based on electronic principles. Only an automatic control system can make the precise and intricate maneuvers required in space travel.

All military missiles are controlled by electronic guidance systems operating under the control of a computer. During the Gulf war of 1990, the world watched SCUD missiles with an electronic guidance system being shot down by Patriot missiles with an electronic guidance system. Fighter pilots use electronic displays and other instruments to "zero in" on targets that can be destroyed by electronically guided missiles such as the heat-seeking Sidewinder missile.

1.1.4 INSTRUMENTATION

Electrical engineers design systems to aid in measuring performance of other circuits or systems. This field is called instrumentation. Voltmeters, oscilloscopes, signal generators, and spectrum analyzers are examples of typical instruments. In addition to these rather conventional examples, electronics plays a large role in many medical instruments. The magnetic resonance imaging (MRI) system, computer-aided tomography (CAT scan) systems, and digital heart monitoring systems rely heavily on electronics to function effectively.

1.1.5 AUTOMOTIVE ELECTRONICS

The automobile has increased its use of electronics in the last decade. We now have antiskid braking mechanisms, digital speedometers, and monitoring systems that indicate the instantaneous miles per gallon achieved, the number of miles until the gas tank is empty, and other information. Microprocessors now control the fuel–air mixture and timing of combustion in the cylinders to optimize performance. These kinds of recent applications add to the traditional automotive uses of electronics in AM/FM radios and tape players and electronic ignition systems.

Some luxury automobiles have memories that electronically control the settings preferred by each person that will drive the car. Seat position, mirror position, and steering wheel position can be set under the control of the memory for each driver, once that person's identity is indicated.

1.1.6 POWER GENERATION AND DISTRIBUTION

For years, the major application of electronics to the power area was in monitoring the performance of the power generating and transmitting equipment. In the last few decades, computers have become common to this field as they automatically control generators and automatically switch distribution paths to avoid overloads. Much of the switching is done with thyristors controlled with electronic circuits rather than the relays of past years. The continual availability of electrical power in the United States has helped maintain a very high standard of living that would not be possible without the field of electronics.

1.1.7 RADAR

Radar is indispensable for the worldwide airline industry. In recent years, air traffic has increased to the point that it would be impossible to handle without radar. Air traffic controllers must be able to view all nearby planes to allow planes to leave from and arrive at the airport. Of course, flight in limited visibility situations is also aided by radar.

Radar was originally developed for military applications and continues to be important in that area also. Security systems and monitoring of speed by law enforcement officers represent other uses of radar.

1.1.8 INTEGRATED CIRCUITS

An integrated circuit is a silicon chip that contains resistors, capacitors, and transistors inside or on the chip. These components are formed by diffusing "impurity" atoms into the silicon or depositing material on the surface of the chip. Using photolithographic techniques, millions of components can now be formed within a single silicon chip. The tremendous reduction in size over discrete electronic circuits has resulted in corresponding reductions in cost of electronic systems.

Integrated circuits are now used in almost all types of electronic systems such as computers, calculators, television sets, audio components, and guidance systems. Two areas in which integrated circuits are somewhat limited are those of high-frequency and high-power circuits.

1.2 Electronic Circuit Design

Before we begin a formal study of the field of electronic circuit design we should become familiar with a few definitions. We must clearly define the words electronic devices and electronic circuits in order to appreciate the scope and significance of this field.

One dictionary defines electronics as

The physics of electrons and their utilization.

Although this definition is accurate, it is also quite general, which renders it vague and almost useless to one who is initiating a study of the field of electronic circuits. The word electronics takes on more meaning when used with the word device. Consequently, we will offer a less formal definition of an *electronic device.*

An electronic device is a component that utilizes some form of energy such as voltage or light to control the flow of electronic current.

Some examples of electronic devices are bipolar transistors, field-effect transistors, diodes, light-emitting diodes, silicon-controlled rectifiers, and thyristors. Resistors, capacitors, and inductors are also electronic devices, although these devices are generally called *components*—a custom we will adhere to in this textbook.

An electronic circuit uses several electronic devices or components to control current flow in a meaningful way. The current may be used to drive a loudspeaker to produce audio sounds. It may be used to drive an antenna to produce electromagnetic waves that carry television signals through the atmosphere. It may be used to drive semiconductor memories that store digital information. It may be used to detect the energy carried by a laser light beam in a fiber optic light pipe.

Resistors and capacitors are typically required by electronic devices to create a useful circuit. Prior to 1960 a circuit was produced by connecting *discrete components* to *discrete devices.* Each component or device was packaged separately, hence the name discrete, and connected by means of conducting wire. In the 1960s the *integrated circuit* (IC) was developed. The IC creates an entire electronic circuit on or in a silicon wafer. The devices are interconnected through the silicon material or by very small conducting metal strips, deposited on the wafer. Thousands of devices and components can be created on a small wafer to produce very complex circuits. Although the IC is technically a circuit composed of several devices and components, it is often referred to as an electronic device. In view of the definition of an electronic device, it may be appropriate to designate the IC as such; however, the difference in complexity and capability between the IC and the discrete device is enormous.

Electronic circuit design refers to the selection and interconnection of electronic devices and components in such a way that the circuit efficiently performs its intended function. This can be accomplished in several possible ways. In discrete circuits, the design is completed using physically separate devices and components linked together with conducting wire or metal strips. Other circuit designs may utilize standard integrated circuit chips to create a system. Each chip is connected to others using wires or metal strips. Many types of integrated circuits are available from various manufacturers. Discrete devices may also be included with standard integrated circuit chips, if necessary, to complete this type of design. Typically, the implementation of these designs would place the devices or chips on a printed circuit board.

Another type of design is accomplished by creating an integrated circuit to perform the desired function or functions. The design and fabrication of integrated circuits consist of selecting transistors, resistors, and capacitors that will be formed within the silicon chip by *ion implantation* or by the *diffusion* of vapor impurities and by the deposition of metal conductors or thin films on the chip. Devices or components are connected either by placing them physically adjacent within the chip or by depositing metal conductors appropriately on the chip. This deposition is accomplished by vaporizing metal and allowing it to be deposited in selected areas.

Obviously, electronic circuit design using discrete elements or standard integrated circuit chips is quite different from electronic circuit design involving the fabrication of integrated circuit chips. A common requirement for both forms of design is an understanding of the behavior of electronic devices.

1.3 Brief History of Electronics

The field of electronics has greatly influenced world affairs since the early 1900s. This field is entirely responsible for the development of such traditionally important areas as communications, computers, instrumentation, aviation, and space travel. Without electronic devices we could not have the telephone, radio, television, high-performance computer, or precision control circuit. Life as we now know it would be drastically different were it not for this significant field. In this section we briefly describe some of the milestones of the electronics field.

It is interesting to note that all of the following significant developments took place after 1900 when a U.S. Patent Office employee stated, "Everything that can be invented has already been invented."

1.3.1 THE VACUUM TUBE AND RADIO

The general-purpose vacuum tube is rarely used in the electronics field, although a few special-purpose tubes are used in the communications area and some musicians still prefer vacuum tube amplifiers. The invention of this device is credited with ushering in the era of electronics. The development of the vacuum tube triode [Davis, 1983] in 1906 is attributed to Lee de Forest (1873–1961). Dr. de Forest received the patent on the triode, then called the audion, in 1908. This device could amplify small signals such as those produced by a microphone. Using several triode stages, the signal could be amplified sufficiently to drive the speaker of a public address system. John Fleming, who had patented a two-element bulb or tube in 1904 [Finn, 1966], fought de Forest in the courts for years over the rights to the triode. It was not until 1943, two years before Fleming's death, that the U.S. Supreme Court ruled in de Forest's favor.

Lee de Forest was surrounded by controversy in his professional career. Edwin Armstrong was involved in a long legal battle with de Forest over the rights to the regenerative receiver, an important radio component. Armstrong had received the patent in 1914 for this circuit, which used a triode. In 1924, after a long legal battle, the Supreme Court ruled in favor of de Forest, although most knowledgeable observers and the lower courts felt that Armstrong deserved credit for this development [Lewis, 1991]. The Institute of Radio Engineers (IRE), the forerunner of the IEEE, had awarded Armstrong the Medal of Honor for this development in 1918. In 1934, Armstrong attempted to return the medal to the IRE. He indicated that the courts represent a world in which men substitute words for truth and then talk about or argue over the words rather than the truth. Since the courts had ruled in favor of de Forest, Armstrong felt the medal should be returned. The president and the board of directors of the IRE refused to accept the return of the medal on the grounds that they knew the regenerative receiver had, in spite of the court decision, been developed by Armstrong. He had developed the circuit, wrote technical papers that explained its operation, and built the first regenerative receivers. Lee de Forest did not even seem to understand its operation when questioned in the lower courts [Lewis, 1991]; yet dè Forest ultimately was awarded the patent.

AM radio flourished for decades using the basic idea of the regenerative receiver. However, neither Armstrong nor de Forest realized significant income from this invention. RCA had purchased the rights to this circuit from Armstrong, who consulted for them in his earlier career. They later paid for any claim de Forest had by paying the company that had obtained de Forest's patent rights. Thus, RCA profited greatly, but the inventors received minimal payment. This situation, in which a company benefits greatly by an invention while the inventor receives little reward, is quite common in the electronics field.

The process of amplification was very important to the electrical industry, especially in the communications area. After a demonstration of the triode amplifier by de Forest in 1912, AT&T secured patent rights to the triode from the inventor for the then rather large sum of $50,000 [Davis, 1983]. AT&T later paid de Forest $90,000 for rights to an improved version of the triode to be used as a telephone amplifier and a radio amplifier.

What may have been the first audio wireless transmission from a vacuum tube transmitter occurred in 1912 in the lab of Frederick Lowenstein of New York City. Using an antenna, voice could be transmitted for several blocks. The construction of crystal sets became the hobby of thousands of people this same year [Davis, 1983].

On April 14, 1912, a coded distress message was received by an operator of the Marconi Radio Station in New York from the S.S. *Olympic*, a ship that was about 1400 miles away from New York in the Atlantic Ocean. The message stated, "S.S. *Titanic* ran into iceberg— sinking fast." The *Titanic* sank with a resulting loss of 1517 lives. Mainly as a result of the wireless message, rescuers saved 705 lives. It was immediately obvious that ships equipped with wireless systems could reduce the number of lost lives in such tragic accidents. Congress then passed an act requiring all ships with 50 or more passengers to include wireless systems on board.

As a note of interest, the wreckage of the *Titanic* was finally discovered at the bottom of the Atlantic in 1985. The discovery and subsequent exploration of this historic vessel was done by a remote-controlled underwater autonomous vehicle that transmitted video images back to the surface command vessel.

The next few years saw a great development of interest in radio. Several stations were constructed by the U.S. Navy, private firms, and the government. In 1914 the Federal Telegraph Company used 608-foot antenna towers in San Francisco and Heeia, Hawaii to transmit sound over a 2400-mile distance.

The necessity of good communications during World War I served as the impetus for rapid development of vacuum tubes and radio circuits. The tetrode (1919) and pentode (1926) improved on the capability of the triode. These devices were indispensable in the tremendous growth of the field from the 1920s to the early 1960s.

Much of the fundamental development in the commercial radio and communications areas took place between 1920 and 1940, although John Carson [Davis, 1983] of AT&T applied for a patent on single sideband transmission in 1915. The patent was granted in 1923. This basic idea was used for many years to simultaneously transmit several long-distance phone conversations over a single channel. It was not until the 1960s that pulse code modulation methods began to replace single sideband methods of telephone transmission.

During World War I, President Woodrow Wilson ordered the dismantling of all amateur radio equipment in the United States. In April 1917, station 9XM at the University of Wisconsin began transmitting voice and music, but was shut down because of the war. Later that year, President Wilson placed all radio facilities in the United States under government control. The war ended on November 11, 1918, but it was not until October 1, 1919 that the restrictions on amateur radio were removed by the Secretary of the Navy.

Several stations returned to the air in 1919. Regular radio broadcasts using amplitude modulation (AM) began from Detroit and Pittsburgh in 1920 [Davis, 1983]. For many years, the AM broadcast ruled the airways. Although the concept of frequency modulation (FM) was conceived around 1912, this form of transmission was felt to be impractical by many well-known engineers such as John Carson. Carson expressed the opinion in 1922 that FM would experience low signal-to-noise ratios compared to AM signals [Carson, 1922]. His conclusion was based on the use of a very small deviation in frequency from the carrier frequency value.

Edwin H. Armstrong (1890–1954) began pursuing FM in 1925. He received a patent for FM in 1933. In a paper presented by Armstrong in 1936 [Armstrong, 1936], he indicated that he had demonstrated FM transmission using larger frequency deviations with very successful results. Commercially, however, FM transmission did not become popular until the 1950s. Armstrong demonstrated the advantages of FM over AM by setting up a receiver of each type in a side-by-side arrangement. The same information was then transmitted over FM and AM transmitters. When a static noise source was added, the AM signal could hardly be discerned but the FM signal remained almost crystal clear. Although Armstrong demonstrated to RCA that FM was far less sensitive to interference than AM, RCA apparently did not want to convert to the better system. It was the dominant radio company and had sold millions of AM sets to the American public [Lewis, 1991]. In fact, RCA was opposed to further development of FM, and Armstrong's relationship with the company deteriorated rapidly from this point.

After Armstrong failed to interest RCA in his FM system in the mid-1930s, he began promoting this transmission method on his own. He was quite successful in this promotion and in 1941 there were over a dozen licensed FM transmitters, 500 applications for FM transmitters pending before the Federal Communications Commission, and 25 FM receiver manufacturers [Finn, 1966]. Armstrong's experience is only one of many that demonstrate the need for public relations or promotional skills in addition to technical creativity. A great technical idea may not have much impact unless someone can sell it. Generally, that someone is the inventor. Once the idea starts generating profit, numerous companies become willing to become involved.

At about this time, Armstrong rejected an offer of over one million dollars for the rights to the system from RCA, which was now quite interested in FM [Brittain, 1977]. RCA developed its own version of FM, but it apparently used many of Armstrong's concepts. After World War II, Armstrong accused RCA of infringement on his FM patents. The case began in 1949, but dragged on in courts for years. In February 1954, Armstrong wrote a note to his wife and then jumped from the 13th floor of his apartment house. Although there were other problems in his life, the legal battles he had fought had certainly contributed to Armstrong's state of mind at the time of his death.

Before the end of that same year, a settlement was reached between RCA and Armstrong's estate. Just over one million dollars was to be paid by RCA to the estate, with the first payment of $125,000 to be made on Christmas Eve 1954. This settlement was perhaps smaller than Armstrong would have exacted, but the court costs to RCA probably reached two to three times that amount.

In addition, this settlement money was used by Armstrong's wife, Marion, to pursue damages from many other radio companies including Philco, Motorola, Bendix, Admiral, DuMont, Sylvania, Packard Bell, and Raytheon. Most settled by 1957, but Emerson Radio held out until 1959 and Motorola fought the settlement until 1967 when the Supreme Court refused to rehear the case. Marion Armstrong lived until 1979 in quiet luxury from the various settlements with these radio companies [Lewis, 1991]. Perhaps her greatest joy was not the money but the establishment of her husband as the unquestioned developer of the FM system.

The remainder of the FM saga is known to anyone who listens to the radio. This type of modulation became more prominent in the 1950s and 1960s and overtook the AM system in total listeners in the late 1970s. Although the carrier frequencies assigned to FM limit the distance of transmission compared to AM distances, the larger bandwidth used makes the quality of sound much higher. FM is used for music whereas traditional AM is now used primarily for news and informational programs. Only in the late 1980s did some improvement in AM quality occur, which allowed serious competition to take place between the FM system and the improved AM systems that are used to transmit by some approved stations.

1.3.2 TELEVISION

Television was being developed as early as the 1870s [Shiers, 1970]. Although a mechanical/electrical system was patented in 1885 by Paul Nipkow [Shiers, 1970], this system was not reduced to practice. Vladimir K. Zworykin applied for patents on the iconoscope in 1923 and the kinescope in 1927 while working for the Westinghouse Electric Company [Brittain, 1977]. The iconoscope was not reduced to practice and, in 1934, a U.S. Patent Board gave priority to Philo T. Farnsworth on the basis that Farnsworth had developed a working model of the image dissector, the first electrical scanning TV camera.

Farnsworth had conceived the idea of using electrons to create a means for transmitting pictures as a 15-year-old farm boy in Idaho. He later discussed these ideas with his high school chemistry teacher. In 1934, this same teacher gave testimony that was valuable in determining that Farnsworth's work had priority over Zworykin's work.

Farnsworth was largely self-educated, having spent only one year at Brigham Young University before embarking on his technical career. He produced the first working model of an all-electric system in 1928 [Abramson, 1987]. This system was observed by employees of the Pacific Telephone Company and was reported in the *San Francisco Chronicle*.

Farnsworth's pioneering efforts have evidently been forgotten by more recent historians because they credit RCA with the dominant developments in TV. In fact, this then very powerful company did more to hinder the development of TV than to foster it. Even the well-publicized image orthicon tube was unpatentable by RCA due to a prior patent held by Farnsworth [Farnsworth, 1989]. After years of attempting to develop its own system, RCA finally agreed to pay a fee to use many of Farnsworth's basic patents in 1939. Several patents considered imperative to the operation of TV systems were held by Farnsworth [Abramson, 1987]. The significance of his pioneering work is underscored by the fact that the agreement with RCA represented the first time this powerful firm had agreed to pay continuing royalties for the use of a patent. Farnsworth had already signed licensing agreements with Philco, Baird (England), Fernseh (Germany), and AT&T [Abramson, 1987]; thus his work was now recognized by the leading electronics firms of the day. In addition to his pioneering efforts in American television, Farnsworth's systems were used in England and other parts of Europe in the mid-1930s. Farnsworth never received the recognition or the wealth commensurate with his contributions to this field, due first to the government's refusal to set TV standards and then to the occurrence of World War II. After the war, many of Farnsworth's patents had expired or were near the end of the 17-year expiration time for U.S. patents.

Throughout the 1930s, RCA had lobbied to keep the government from developing broadcast standards for TV since the RCA system was inferior to Farnsworth's and others. When RCA finally purchased Farnsworth's patent rights and was ready to enter the TV scene, World War II was imminent, further delaying the development of TV standards. Although six companies produced TV receivers for the New York market prior to 1940—American Television Corp., Andrea Radio Corp., Allen B. DuMont Labs, General Electric, Philco, and RCA— the field became significant only after the end of World War II in the late 1940s.

The transmission of radio and TV signals along with the reception of these signals were based on electronic circuits using vacuum tubes as the device of primary importance. The conversion of TV to transistors took place in the 1960s. Up to this time, many stores including supermarkets had vacuum tube testers available to check the tubes of a TV set. These stores also sold vacuum tubes to replace defective devices.

Color television was first demonstrated to the public in 1951 in the United States [Dummer, 1983]. Private industries had been working on color schemes since the mid-1940s, but different companies used different approaches to the production of colored images.

In the United States, changes or additions to public communication procedures must be approved by the Federal Communications Commission (FCC). CBS was one of the first to petition the FCC for acceptance of its wideband scheme for color transmission. This scheme required a 15-MHz bandwidth rather than the 6-MHz required by the black-and-white scheme [*Electronics*, 1981]. In March 1947, the FCC denied this petition. In July 1949, CBS demonstrated a second system to the American Medical Association convention in Atlantic City, New Jersey. This system could transmit color on a 6-MHz bandwidth. Although the resolution was not as good as that on existing black-and-white sets, the FCC adopted the CBS standard in 1950. The methods used by this color TV were not compatible with the black-and-white receivers in use at the time. All other TV firms were now forced by the FCC/CBS standard to adopt this method of transmission or attempt to get the standard changed.

RCA, which had been involved in its own development of a color scheme, opposed the FCC's acceptance of the CBS standard, taking the matter to court. In May 1951, the U.S. Supreme Court upheld the FCC order. On June 25, 1951 regular color telecasts to the public began in New York using the CBS system.

RCA continued work on its scheme, which was compatible with black-and-white receivers. Ultimately, the National Television Standards Committee (NTSC) recommended a standard that was closely related to the RCA system. In December 1953, the FCC reversed itself and adopted the NTSC recommendation as the standard, rescinding the CBS standard. This reversal by the FCC was probably a wise decision, but cost those manufacturing companies using the CBS system dearly, and some were forced to file for bankruptcy.

The NTSC standard has been used with little change for over four decades. This standard is not the one observed by the entire world. Only Japan and South America adopted a similar standard, and many other countries developed their own.

A similar problem to that experienced with color TV arose in conjunction with the high-definition television (HDTV) schemes proposed. This system uses digital image-processing methods and greatly improves the resolution and capabilities of the analog TV system. Although the FCC adopted its necessary "wait-and-see" posture to determine whether one method would ultimately prove to be the best before setting a standard in the United States, it made one significant decision. In 1993, the FCC encouraged the competing companies to form a group referred to as "the Grand Alliance." This group set several standards that define the field in the United States. Ultimately, the companies that meet the standards with the highest quality and reasonable complexity will prosper.

1.3.3 THE TRANSISTOR AND THE INTEGRATED CIRCUIT

The transistor was invented in 1947 by John Bardeen, Walter H. Brattain, and William Shockley, all of Bell Telephone Laboratories [Brittain, 1977]. In 1956, these three men were awarded a Nobel Prize in physics for their invention.

The discovery of the transistor is an outstanding example of an important scientific development. It did not take place due to chance or serendipity as do many developments. The background information for the use of germanium and silicon as semiconductors was obtained during World War II by several researchers. The general research program leading to the transistor was initiated and directed by Shockley at Bell Labs [Bardeen, 1949]. A combination of theoretical and experimental work led to the discovery of transistor action by Bardeen and Brattain in 1947. This development was based on the point-contact transistor that used gold wire and germanium material to form the device. Shockley then led the team in the development of the junction transistor, which now forms the basis of the integrated circuit.

Due to the heavy usage of vacuum tube circuits at this time, the transistor was slow to displace the vacuum tube in electronic circuits. An area that first took advantage of the small size and low power requirements of the transistor was that of the AM radio receiver. This area grew rapidly during the late 1950s as hand-held radios that could fit into a shirt pocket underscored the size advantage of transistors over the bulkier vacuum tube.

Another area that soon began utilizing the transistor was the computer industry, which we will consider in succeeding paragraphs of this section.

In the early stages of transistor development, one very significant possibility was overlooked. The potential for fabricating microcircuits within a small chip of material was unknown until long after the discovery of the transistor. The field of integrated circuit fabrication could not exist without the use of the transistor as a basic building block. This field creates thousands and even millions of transistors, resistors, and capacitors in a small silicon chip. Without integrated circuits we would not have the low-cost personal computer, the hand calculator, or the small video cassette recorders of today. Thus, the transistor was required before the integrated circuit field could begin to develop.

Jack Kilby of Texas Instruments (TI) developed the first integrated circuit in 1959 when he demonstrated a germanium oscillator circuit. Robert Noyce at Fairchild Semiconductors had also given serious thought to the concept of an integrated circuit, and this company contributed greatly to the early development of the device. Fairchild led the development of the planar process that provided isolation between each of the many components on a silicon chip and a means of interconnection [Comer, 1994].

Although Fairchild was the major player in integrated circuit development, Motorola Semiconductor Corporation and Texas Instruments capitalized more quickly on the technology. Motorola introduced the MECL logic family that was used for decades, and TI developed the TTL family that has been improved on by many companies over the years.

It was rumored that Fairchild's problems were related to organization and management. Top management resided on the East Coast near Fairchild's corporate headquarters and pursued a conservative management style. The West Coast management team needed to function in an aggressive style to beat other companies to the market. Delays in project approvals and support caused major problems in competing with other semiconductor firms. Many of the top engineers left Fairchild to start other companies in this fast-growing area. More will be said about this in the next section on digital computers.

Because of low yields in the production processes, early integrated circuits might cost $100 in small quantities. As production processes improved, increasing the yield and producing more components on a chip, the price dropped dramatically. In 1965, flip-flop circuits on a chip cost a few dollars while logic gate prices dropped below $1. By 1995, the cost of these integrated circuits included in larger systems was less than one-thousandth of a cent.

The continual decrease in price of integrated circuits has led to a corresponding decrease in many electronic systems, the most notable being electronic computers and calculators. Over a three-decade period from the 1960s to the 1990s, the computer has decreased in price by a factor exceeding 1000.

The transistor has changed the face of society leading to space travel, worldwide satellite communications, computers, guided missiles, and many other significant accomplishments. This device, along with its descendant, the integrated circuit, may prove to be the most significant technical development in the history of mankind.

1.3.4 THE DIGITAL COMPUTER

One of the first electronic digital computers developed in the United States was the ENIAC, an acronym for Electronic Numerical Integrator And Computer. Vacuum tubes were used in this computer, which was constructed between 1943 and 1946 under the supervision of John G. Brainard, John W. Mauchly, J. P. Eckert, and Thomas K. Sharpless [Brainard, 1948]. The construction of this computer took place at the Moore School of Engineering at the University of Pennsylvania. This work was supported by the Army and the Ballistics Laboratory. As in many other new developments, several people may have contributed to the basic concepts. Mauchly had visited John V. Atanasoff at Iowa State University during the summer of 1941 and discussed some ideas that might be useful in electronic computing. Atanasoff later testified in a patent infringement case involving a patent held by Eckert and Mauchly. Atanasoff evidently contributed to Mauchly's original work [*Electronics*, 1981].

The ENIAC was vastly different from today's sleek computer models, containing over 18,000 vacuum tubes, weighing 30 tons, and occupying a 30×50 foot room [*Electronics*, 1981]. This 60,000-pound computer consumed 150 kilowatts of power, in contrast to today's 10-pound lap-top computer that consumes a few watts of power.

The ENIAC was designed to solve equations of motion of ballistic missiles, but was in fact a general-purpose computer. Unfortunately, the mean time between failures for the ENIAC was very low, due primarily to the continual failure of the vacuum tubes. During 1952, for example, approximately 19,000 tubes were replaced to keep the computer functional. In addition, the program could only be changed by rewiring the circuitry, which was a very time-consuming operation, often requiring several days.

A smaller cryptanalysis (code-breaking) computer with perhaps the first stored program that could be changed electronically was developed by the British in 1943. Some observers of the field believe that this machine, called the Colossus, is the first true electronic digital computer [*Electronics*, 1981]. This system used only 1500 vacuum tubes.

It was difficult to predict the future impact of the digital computer. Thomas Watson Sr., the president of International Business Machines, is reputed to have said, "The world market for computers is about six." Little did he know that the size and price of the computer would be reduced by a factor exceeding 1000! Even before these reductions, thousands of large companies paid millions of dollars to IBM and other companies for the mainframe computers of the 1960s. In the early 1980s, when size and price were reduced sufficiently, the personal computer became available to the world.

A few improved computers were built with vacuum tubes after ENIAC, but it was obvious that the transistor's smaller size, smaller power consumption, and rugged construction would lead to more efficient computer design. Just prior to 1960, computer systems designed exclusively with transistors began to appear on the market. These computers were quite large and expensive and were used primarily by large companies that could afford this very expensive system. By 1965, the IC had become available and computer manufacturers quickly took advantage of the size and price advantages over discrete transistor models to provide more capability in the large mainframe computer.

It was also in 1965 that Digital Equipment Corporation introduced the PDP-8, generally considered the world's first minicomputer. Other companies followed quickly with a variety of minicomputer types. These systems offered a smaller and cheaper alternative to the large mainframe computers. Instead of prices that could approach $1,000,000, the costs of these systems ranged from $20,000 to $200,000. The minicomputer was based on the concept that a computer with less precision and speed than the large mainframe models could still be valuable. As the fabrication processes for integrated circuits continued to improve, lower prices for the minicomputer made this system available to many small scientific labs.

One of the significant developments in the integrated circuit industry was the movement of a small group of brilliant engineers and scientists that chose to move to a 20-mile strip of land between Palo Alto and San Jose, California [Comer, 1994], the area now known as Silicon Valley.

A little over a decade after the invention of the transistor, Shockley decided to start a company in his home town of Palo Alto. He brought some of his colleagues from Bell Labs to assist in this venture. The initial staff of Shockley Laboratories included Robert Noyce, Gordon Moore, Andrew Grove, and Jean Hoerni. This company focused on developing a 4-layer semiconductor device that never gained broad acceptance. Several members of the staff decided to leave Shockley in 1957 and formed the Fairchild Semiconductor Corporation. This firm was financed by Sherman Fairchild of Fairchild Instrument and Camera Corporation.

The previous section alluded to a problem in management style within the Fairchild organization; this problem is reported to have led to a small group of key people leaving Fairchild. In 1968, Robert Noyce and Gordon Moore left Fairchild and founded Intel (INTegrated ELectronics) Corporation. Andrew Grove joined this company within a short time. Others also left Fairchild at about that time. Charles Sporck, who had been hired as a Fairchild production manager, left to become president of National Semiconductor. Jerry Sanders, a marketing manager at Fairchild, ultimately founded Advanced Micro Devices. Jean Hoerni also started several companies, one the forerunner of Signetics. Many of the Silicon Valley semiconductor companies in the 1970s were and still are traceable as descendants of the Fairchild Semiconductor Corporation.

Shortly after Intel was formed, a Japanese firm named Busicom asked Intel to develop high-performance calculator chips. From 1969 to 1971, work continued at Intel on this project. The result was the first commercial microprocessor chip, the Intel 4004. Although this device is very primitive by today's standards, it contained the central processing unit or CPU of a computer on a single silicon chip.

Although the microprocessor was introduced by Intel Corporation, a 1990 court decision upheld a patent for the basic idea of the microprocessor filed in 1970 by Gilbert Hyatt. Intel's first model was based on a small 4-bit digital word, but succeeding models expanded the capability to 8-bit, then 16-bit, and finally 32-bit words. The results of this development are manifest in the intelligent electronic instrumentation and the low-cost personal computers now available. If not for the microprocessor chip, the personal computer and the more powerful workstation could not exist.

The field of electronic systems is one of the few that continually offers increased capability for lower cost. This advantage is due to the development of integrated circuit design methods that result in chips with higher densities of devices and higher yields. Most university students now possess more computing power than that of the initial million-dollar transistorized mainframe computers of the 1960s.

In the mid-1980s, the capability of including one million transistors on a silicon chip was achieved. In the mid-1990s, over ten million transistors could be fabricated on a chip. These very large scale integrated circuits are applied primarily to the digital circuit area. In the near future, systems with one billion devices on a single chip are anticipated.

Computers or digital memories require large numbers of the same type of circuit and can effectively utilize this capability of placing many devices on a single chip. Analog integrated circuits are not able to pack as many devices into a small volume, but still offer a tremendous volume reduction when compared to the discrete version of the circuit.

Other important advancements in the electronics field are the development of handheld calculators, lasers and electro-optic devices, the development of high-power switching devices such as the silicon-controlled rectifier, the continual creation and improvement of electronic instruments such as oscilloscopes and medical equipment, and the development of automatic control systems to guide space vehicles and missiles. How different the world would be without the development of the transistor and the integrated circuit.

In this section many individuals have been mentioned in connection with various important developments. This should underscore the impact that individual men and women can have on society as a whole. Note, however, that although a single person may generate a new idea, a team of engineers, an entire company, or several companies may be involved in its development.

1.4 Electronics Education

The field of electronics is obviously very broad and changes rapidly as new electronic devices are developed. These facts make it difficult to produce a comprehensive, up-to-date textbook.

We are now in the era of the integrated circuit, yet the discrete circuit is still used in several applications. The diode, the bipolar transistor, and the junction field-effect transistor are all useful elements with relative importances that will change over the next few years as integrated circuit technology advances.

In an attempt to make this textbook a topical and useful work, the following approach is taken. The fundamentals of active circuit analysis are treated, and then design methods are applied both to discrete circuits and to integrated circuits. Integrated circuit designs will be emphasized for those areas in which the integrated circuit (IC) is almost universal. An example of this is the low- and intermediate-frequency amplifier using the operational amplifier chip. Another is the digital circuit area where discrete circuits are rarely needed.

While the same basic circuit design principles apply to either discrete or integrated circuit design, the philosophy of design is somewhat different. The ultimate goal in engineering is often to minimize the cost/performance ratio of a circuit or system. Although other factors such as size, reliability, or ease of use may become more important in some applications, the cost for a specified performance is always significant in a profit-oriented firm.

The most expensive element in discrete circuits is often the transistor or capacitor, while the resistor is the least expensive. Consequently, resistors are used liberally in an effort to minimize transistors and capacitors in discrete circuit design. In designing integrated circuits, one attempts to minimize the total volume of silicon used. Transistors occupy the least amount of space, while capacitors and resistors occupy a great amount of silicon chip volume. In designing integrated circuits, transistors often replace resistors to minimize required volume and cost. Thus, configurations change as we move from a discrete circuit to an integrated circuit that performs a similar function.

It is very difficult to integrate a high-quality inductor, and it is difficult to integrate a very large value of capacitance. Whereas discrete circuits may use such elements, the integrated circuit may need to be much more complex to perform an equivalent function without the use of these elements. Typically, many additional transistors are utilized in the integrated circuit compared to the equivalent discrete circuit. The transistor is easy to integrate and, if extra transistors can lead to the elimination of inductors or large capacitors, a circuit that was not integrable can now be fabricated on an IC chip. The previous considerations lead to differences in design, and these differences will be treated in the following chapters.

One last point relating to integrated circuit design is the constraints imposed by fabrication and debugging. With discrete circuits, a design can easily be implemented and subsequently checked to see whether specifications are met. Element values can be modified or perhaps configuration may be changed to lead to satisfactory performance. For simpler discrete circuits, design, implementation, and modification may take place in a single day.

For integrated circuits, even a simple design may take weeks to implement. The process is complex and requires many steps to create the electronic devices within the silicon chip through photolithography and diffusion of impurities into selected areas of the chip. If an undetected error is made in the initial design stage, it may not be discovered until the finished chip is tested. At this point, the design must be modified and several more weeks will be required for implementation of a second chip. It is imperative that the initial design be accurate to minimize total time for realization of an acceptable circuit.

1.4.1 CIRCUIT SIMULATION

There are two approaches to minimize design time for integrated circuits. One is to check the paper design by computer simulation of the circuit. Many good simulation programs are now available that allow circuit performance to be predicted by a computer. The second approach is to implement the paper design with a discrete element circuit. In this approach, separate transistors, resistors, diodes, and capacitors that have been fabricated by integrated circuit techniques are used to construct and test the circuit. Once this discrete circuit is debugged and meets the specifications, work proceeds to implement the entire circuit on a single chip. This approach requires both discrete and integrated circuit design abilities and is used by many analog integrated circuit firms.

Perhaps the most useful simulation programs are those based on Spice, such as PSpice©, hSpice©, SmartSpice©, or AllSpice©. Spice is an acronym derived from the words Simulation Program with Integrated Circuit Emphasis. This program was developed in the late 1960s at the University of California (Berkeley) and steadily improved over the years. Free Spice programs, with no technical support, can be obtained from the University of California. The commercial simulation programs provide full technical support and now can be run on personal computers as well as larger mainframe systems.

Spice and its derivatives allow the configuration or connections of a circuit to be specified to the computer along with element values. An input signal can be specified and the program automatically calculates the output signal. Very complex circuits can be simulated with this program.

This powerful capability to analyze circuit behavior does not eliminate the need to understand and utilize sound design principles. Circuit concepts must be applied to select the proper configuration and to modify parameters to achieve acceptable performance. Simulation programs assist this process and thus become an important design aid.

Although Spice programs are very important to the field of electronic circuit design, this text will not cover the basic concepts of Spice operation. It is assumed that the student has been exposed to such simulation programs in an earlier electronics course. The simulations included in this textbook will use netlists as the input file. As explained in the Preface, an understanding of netlists is important in the application of more advanced computer-aided layout tools. A schematic for each simulated circuit will also be shown for those that use a schematic capture program for data entry. A tutorial on PSpice© is available on the Internet at www.ee.byu.edu under the ECEn313 class page.

1.5 Level of the Textbook

This textbook is designed for use in the third or fourth year of a standard electrical engineering curriculum, which implies that the student has completed calculus and differential equations. Although this level of math is not required in every chapter, some derivations require integral calculus or differential equations. The student should also be familiar with basic methods of network analysis. A familiarity with Ohm's law, Kirchhoff's laws, Thévenin's theorem, loop and node analysis techniques, in both dc and ac circuits, is assumed. Laplace transform methods and Fourier series analysis are not required.

1.5.1 DESIGN VS. ANALYSIS

The purpose of this textbook is to teach sound principles that apply to the design of electronic circuits. It must be recognized that most principles of analysis also apply to design; thus the earlier chapters include more analysis than the later chapters. In establishing the behavior or models of electronic devices, design methods are not required. On the other hand, when using these devices in electronic circuits, the design approach will always be emphasized.

The difference between design and analysis could be said to lie in the starting material. A design starts from a concept or a set of specifications rather than a physical entity. A mechanical engineer designs a new automobile to meet a set of specifications. When these specifications are given to the engineer, the automobile does not yet exist. During the design process, the engineer invents or creates the configuration of components that will comprise the actual car. Once the car is produced and sold, a mechanic can analyze the operation of the car. The mechanic analyzes an existing physical entity whereas the engineer both designs and analyzes as the car is created.

This same difference applies in the electrical engineering field. The engineer may invent a new type of television or design a unit to satisfy a set of specifications. The new High-Definition Television (HDTV) represents an example of a design problem. The engineer creates the physical television set to satisfy the specifications. Once it is created, constructed, tested, and sold, the TV technician can service it by using analysis methods.

The analysis of electronic circuits is considerably easier than the design of these circuits. Furthermore, exact analysis procedures can be developed, whereas design methods are much less procedural. It is the purpose of this book to develop an ability to both analyze and design electronic circuits in an orderly and scientific manner.

REFERENCES

Abramson, A. *The History of Television, 1880 to 1941.* Jefferson, N.C.: McFarland, 1987.

Armstrong, E. H. "Method of reducing disturbances in radio signalling by a system of frequency modulation," *Proc. I. R. E.*, Vol. 24 (May 1936).

Bardeen, J. and Brattain, W. H. "Physical principles involved in transistor action," *Bell System Technical Journal*, Vol. 28 (April 1949), pp. 239–247.

Brainard, J. G., and Sharpless, T. K. "The ENIAC," *Electrical Engineering*, Vol. 67 (Feb. 1948), pp. 163–172.

Brittain, J. E. (Ed.). *Turning Points in American Electrical History.* New York: IEEE Press, 1977.

Carson, J. "Notes on the theory of modulation," *Proc. I. R. E.*, Vol. 10 (Feb. 1922), pp. 57–82.

Comer, Donald T., *Introduction to Mixed Signal VLSI.* Provo, Utah: Array Publishing, 1994.

Davis, H. B. O. *Electrical and Electronic Technologies.* Metuchen, N.J., and London: Scarecrow Press, 1983.

Dummer, G. W. A. *Electronic Inventions and Discoveries*, 3rd edition. Elmsford, N.Y.: Pergamon Press, 1983.

Electronics (Ed.). *An Age of Innovation: The World of Electronics 1930–2000.* New York: McGraw-Hill, 1981.

Farnsworth, E. G. *Distant Vision.* Salt Lake City, Utah: Pemberlykent Publishers, 1989.

Finn, B. S. "Controversy and the growth of the electrical art," *IEEE Spectrum*, Vol. 3, No. 1 (Jan. 1966), pp. 52–56.

Lewis, T. *Empire of the Air.* New York: Harper Perennial, 1991.

Shiers, G. "Early schemes for television," *IEEE Spectrum*, Vol. 7, No. 5 (May 1970), pp. 24–34.

Applications of Electronic Circuits

2

The expression, "I can't see the forest for the trees," may be appropriate to the area of electronics education. This expression could be translated to the electronics area as, "We are taught too much 'bottom-up' design rather than 'top-down' design."

We should not teach students how transistors, resistors, capacitors, and inductors work before they are taught how these elements are used to construct electronic systems. Bottom-up design starts at the lowest, most detailed level and works up to the more complex. In circuit design, the low level is the device and element level, and the highest level is the finished electronic system. The bottom-up method does not provide good motivation for the student because the student does not see for what purposes the transistors and elements are to be used. The trees must not be carefully studied before the forest is seen.

Top-down design begins at the complex level and works down to the simpler level. The top-down design of a system begins at the system level. The overall system is divided into simpler modules, which when interconnected properly will perform the desired function. The modules are then further broken down to a level that can be implemented with transistors and other elements.

Although both approaches are useful in design methods, the top-down approach is quite popular and provides more motivation to the student. When the student understands how the system is to operate, it is easier to select the components that implement that operation. In this case, the student first looks at the forest, then makes a close inspection of the trees.

This chapter looks at the forest by discussing several popular types of electronic systems. We will also consider the characterization of performance of these systems. By discussing the overall system operation and performance, we provide a framework on which to base a study of the components required to design such a system. Thus, this chapter will introduce the forest of electronic systems and subsequent chapters will introduce the trees. With this background, top-down design can then be applied.

2.1 Amplifiers

IMPORTANT Concepts

1. Most signals developed by sensors, antennas, or other transducers are very small (in the microvolt or millivolt range) and must be amplified by an amplifier circuit to become useful.

17

2. Midband voltage gain, upper 3-dB frequency, and lower 3-dB frequency are important specifications that characterize an amplifier.

3. Input and output impedances are also significant in determining the "in-circuit" or "loaded" gain of the amplifier.

2.1.1 APPLICATIONS OF AMPLIFIERS

A very significant use of electronic circuits is in the amplification of signals. A large stereo speaker may require a signal that reaches a peak value of 10 volts (V) with an accompanying current of 10 amps (A). The head that reads from a magnetic tape may produce a signal that has a 50-millivolt (mV) peak value with a current of 40 microamps (μA). If this small magnetic head signal were applied directly to the speaker, it would not produce an audible sound. This signal must be amplified; that is, both the current and voltage must be increased to the levels required by the speaker before the appropriate sound results. An amplifier is used to enlarge the voltage and current to the desired levels.

To demonstrate the need for the amplifier, we will consider the models for a microphone and a loudspeaker to be used in a public address system. An electrical model for some physical device consists of electrical devices connected in such a way as to exhibit the same electrical characteristics as does the physical device to be modeled. One possible microphone model is shown in Fig. 2.1 along with an electrical model of a speaker.

This particular microphone generates peak voltages of 50 mV when a person speaks into the device with normal volume. The output speaker requires peak currents of 500 mA, corresponding to peak voltages of 8 V. The voltage amplification between the microphone and the speaker must be equal to at least

$$A_v = \frac{8}{0.050} = 160 \text{ V/V}$$

An amplifier with the proper voltage and current gains can be designed and inserted between the microphone and the loudspeaker to achieve the desired amplification.

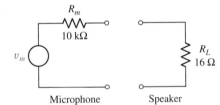

Figure 2.1
Electrical models of a microphone and a speaker.

Microphone   Speaker

PRACTICAL Considerations

A high-quality microphone is listed in an audio catalog as having a 10-kΩ output impedance. For normal input speech, the peak value of voltage generated by the microphone is 50 mV. If the output of this device were shorted, the total voltage would drop across its output impedance and the current would have a peak value of

$$i = \frac{v}{R_m} = \frac{50 \text{ mV}}{10 \text{ k}\Omega} = 5 \text{ }\mu\text{A}$$

A cone speaker is also listed in the catalog as a 15-W, 8-Ω speaker. If a single sinusoidal signal were applied to this speaker, a peak voltage of 15.5 V and a peak current of 1.94 A would result in maximum output power. This is calculated from

$$P_{max} = \frac{V_{rms}^2}{R}$$

Substituting 15 W for P_{max} and 8 Ω for R leads to

$$V_{rms}^2 = 15 \times 8 = 120$$

Solving for V_{rms} gives

$$V_{rms} = 10.95 \text{ V}$$

A sinusoid has a peak value of $V_p = \sqrt{2}V_{rms}$ leading to a 15.5-V peak value. Dividing this voltage by 8 Ω leads to a peak current of 1.94 A.

We note the discrepancy of output signal from the microphone and required signal for maximum power to the speaker. A 50-mV signal with a 10-kΩ output impedance cannot supply a 15.5-V, 1.94-A signal to the speaker. If this microphone is to drive the speaker, an amplifier that boosts both voltage and current must be inserted between the microphone and the speaker. The voltage must be amplified by a factor of about $15.5/.05 = 310$, and the current amplification must exceed a factor of $1.94/5 \times 10^{-6} = 388,000$.

PRACTICE Problem

2.1 If the microphone is connected directly to the speaker in Fig. 2.1, calculate the resulting peak loudspeaker current for a peak voltage of 50 mV generated by the microphone. *Ans:* $I = 4.99 \ \mu A$.

Amplifiers are also needed to receive the very low energy radio signals received by radio and television antennas. These signals are in the microvolt range and again must be amplified by a factor of several thousand before they can be used to produce the audio and/or video signals necessary to drive the speaker or the picture tube.

Another application of amplifiers occurs in the digital computer area. A single logic circuit may be required to drive several other logic circuits simultaneously. The driving circuit may produce enough current to drive three or four following circuits, but it may be required to drive 40 circuits. In such a case, the output current of the driving circuit must be amplified by a current amplifier, then applied to the following 40 circuits.

The mouthpiece of a telephone is similar to a microphone in that the speech pressure waves are converted to corresponding voltages by this device. The output voltage and current levels are too small to transmit directly over the phone lines and drive the earpiece, which is similar to a speaker. An amplifier is used to increase the signal strength of the mouthpiece output before transmitting this signal over the lines.

A typical amplifier produces an output signal that is a magnified version of the input signal. To do this requires that the circuit amplify all frequencies contained in the signal by the same factor, generally called the *gain factor* or *gain* of the amplifier. For example, a public address system designed to amplify speech must amplify all frequency components in speech by the same factor. Speech contains frequencies ranging from approximately 100 Hz up to 8 kHz. The amplifier may have a voltage gain of 200 V/V, implying that all frequency components in the input signal that have a frequency between 100 Hz and 8 kHz will be amplified by a factor of 200.

Some amplifiers, such as those used in conjunction with the telephone mouthpiece, may discriminate against a certain frequency range of the input signal. Although speech sounds contain frequencies above 3400 Hz and below 300 Hz, some of the circuits involved in transmitting and receiving the phone signal may only amplify frequencies in this range. This allows the phone company to use a smaller frequency range or bandwidth to transmit the signal from one point to another. The quality of the signal is degraded, but the resulting voice produced by the earpiece is intelligible.

Another amplifier, the high fidelity amplifier, is designed to modify or enhance the sound by boosting certain frequency ranges more than others. The bass or treble controls allow

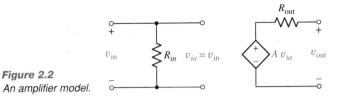

Figure 2.2
An amplifier model.

the modification of the gain factor in a lower range of frequencies, perhaps the 50 Hz to 500 Hz range (bass), and in a higher range, perhaps 2 kHz to 15 kHz range (treble). A gain in the bass range that is higher than the gain in the mid-range of frequencies emphasizes the bass sounds. A gain in the treble range that is higher than the gain of the mid-range of frequencies emphasizes the treble sounds.

Although we have mentioned only a few applications of the amplifier, this electronic circuit is very important in the field of electronics.

2.1.2 CHARACTERIZATION OF AMPLIFIERS

Some parameters that describe amplifier performance are midband gain, input resistance, output resistance, upper corner frequency, and lower corner frequency.

Midband Gain: Although current gain or power gain is sometimes of interest, the most frequently used gain parameter is that of midband voltage gain. This parameter is measured by the ratio of the magnitude of the output voltage to the magnitude of input voltage. The measurement must be taken at a midband frequency that lies considerably below the upper corner frequency, yet considerably above the lower corner frequency of the amplifier. There are no units for voltage gain, since it is a ratio of volts to volts; however, many textbooks use units of volts per volt or V/V. A useful block diagram for an amplifier is shown in Fig. 2.2. The voltage reaching the amplifier input is designated v_{ia}.

Input and Output Resistances: The input resistance is measured across the input terminals of the amplifier. The output resistance is measured across the output terminals. These resistances reduce the overall or in-circuit gain of the amplifier configuration, which is demonstrated in Fig. 2.3.

The resistance R_g is the resistance of the signal generator, and R_L is the resistance of the load to be driven. Typically, we are interested in overall voltage gain given by

$$A_o = \frac{v_{out}}{v_g} \tag{2.1}$$

This value of gain is more meaningful than the amplifier gain, since it will be used in specifying the amplifier system. For example, if the input is a microphone that outputs a 50-mV peak value and the load is a speaker that requires a 5-V peak signal, the overall gain must be 100. This value will be less than the amplifier gain due to the attenuation of v_g by the resistances R_g and R_{in} and the attenuation of $A v_{ia}$ by R_{out} and R_L.

Figure 2.3
A practical amplifier configuration.

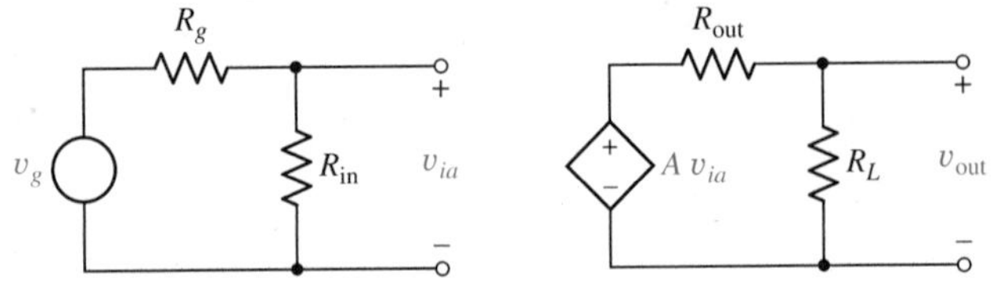

In terms of the circuit of Fig. 2.3, the overall gain can be written as

$$A_o = \frac{v_{out}}{v_g} = \frac{v_{ia}}{v_g} \times \frac{Av_{ia}}{v_{ia}} \times \frac{v_{out}}{Av_{ia}} \tag{2.2}$$

Expressing the gain in this manner allows each factor in Eq. (2.2) to be easily calculated. The three factors are

$$\frac{v_{ia}}{v_g} = \frac{R_{in}}{R_{in} + R_g}$$

$$\frac{Av_{ia}}{v_{ia}} = A$$

$$\frac{v_{out}}{Av_{ia}} = \frac{R_L}{R_L + R_{out}}$$

This allows the overall gain to be written as

$$A_o = \frac{R_{in}}{R_{in} + R_g} \times A \times \frac{R_L}{R_L + R_{out}} \tag{2.3}$$

Both resistive ratios in Eq. (2.3) will be less than unity; thus the overall gain is less than the amplifier gain. The amount of gain reduction is dependent on all four resistive values. If R_{in} is much greater than R_g and R_{out} is much smaller than R_L, the value of A_o is close to A. If these relationships are not true, the resistive ratios can be considerably less than unity. This attenuation is referred to as *loading*. The amplifier is said to load the microphone if R_{in} is not large compared to R_g. The driven load is said to load the amplifier if R_L is not large compared to R_{out}. The value of A is often called the *unloaded gain* of the amplifier, since it is the overall gain that would be measured if $R_g = 0$ and $R_L = \infty$. The overall gain could also be called the *loaded gain* of the amplifier.

EXAMPLE 2.1

A microphone generates a 50-mV peak signal and has an output resistance of 10 kΩ. This microphone output is to be coupled through an amplifier to a 100-Ω speaker that requires a 5-V peak signal. If the input resistance of the amplifier is 50 kΩ and the output resistance is 80 Ω, what value of amplifier gain is required?

SOLUTION The overall gain of this amplifier system must be

$$A_o = \frac{v_{out}}{v_g} = \frac{5}{0.050} = 100 \text{ V/V}$$

Substituting this value into Eq. (2.3) gives

$$100 = \frac{50}{50 + 10} \times A \times \frac{100}{100 + 80} = 0.463\,A$$

Solving for A results in

$$A = 216 \text{ V/V}$$

The resistive loading effects require an unloaded amplifier gain of 216 in order to achieve an overall gain of 100. This effect must always be considered as an amplifier is designed. Although this problem is rather simple, we will use it to review the concept of circuit simulation, which is a valuable tool in designing more complex circuits, especially integrated circuits on a silicon chip.

There are commercial versions of Spice designed for use with Windows©. Some versions allow two methods of data entry: schematic capture or a connectivity list that is called a *netlist*. Details of schematic capture can be found in numerous instruction manuals. Using this entry method, the circuit schematic along with element values are drawn on the computer screen. The netlist method specifies connectivity and element values by way of a specified format.

Most students use the schematic capture approach to input the data to a simulator. This approach is convenient and gives the designer a visual representation of the circuit. However, it is also important to understand the netlist format for presenting circuit information. In integrated circuit design, the final circuit must be implemented in silicon. This implementation is based on a photolithographic mask layout that specifies the areas in the chip that will become transistors, resistors, capacitors, or metal conductors. Once this layout is done, the chip can be fabricated. Because fabrication is a very expensive step, it is necessary to ensure that the layout will actually result in the circuit that has been simulated. One step in the entire procedure to perform this verification is the LVS or layout versus schematic procedure that must be done before fabrication proceeds. This step uses the layout information as input data and forms the netlist information that would result if this layout were implemented. This layout netlist is then compared to that for the original circuit schematic for accuracy. Errors or problems are indicated in terms of the netlist format. This format must be understood by the designer to make the necessary corrections. Consequently, we will specify each circuit to be simulated in terms of both the schematic and the netlist even though the preferred method in a basic circuits course is generally the schematic capture approach.

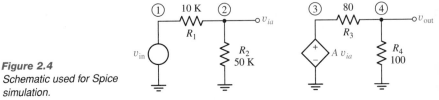

Figure 2.4
Schematic used for Spice simulation.

The two input methods are demonstrated in the netlist that follows. The schematic of Fig. 2.4 shows node numbers and element numbers that normally do not appear in the schematic. These are shown only to relate them to the information in the netlist.

```
EX2-1.CIR
*MICROPHONE AMPLIFIER
*WITH LOADING

R1 1 2 10K
R2 2 0 50K
R3 3 4 80
R4 4 0 100
VIN 1 0 AC 0.05         50mV AC input signal
EAMP 3 0 VALUE = 216*V(2)   Amplifier gain = 216
.AC DEC 1 1000 1000     Single frequency only
.PRINT AC V(4)
.END
```

The netlist file is named in the first line. The second and third lines are comment lines indicated by asterisks. Comments may be used liberally in complex circuits to allow clarification of steps.

Components such as resistors, capacitors, and inductors are identified and numbered in the first field. A space or tab separates fields from each other. The next two fields specify the nodes to which a component connects. The fourth field specifies the value of the component.

The independent input source is specified next as VIN. The initial letter of an independent voltage source is V. The next two fields again list the two nodes to which this source is connected. The fourth field indicates that this is an AC source, and the fifth field specifies the strength of the source. For an AC source, the specified voltage is taken as an rms value.

The dependent voltage source is named EAMP, observing the requirement that a dependent voltage source name must begin with the letter E. The nodal connection information is given next followed by a field that specifies the value of the source. In this case, the value is given as 216 times the voltage at node 2, the voltage at the amplifier input.

A frequency of 1000 Hz is used for this analysis, as specified by the .AC statement. The first field in this instruction indicates that AC analysis is to be performed. The next field specifies that decades will be used to vary the frequency. However, the fourth and fifth fields indicate the starting and ending frequencies over which the analysis should be done. Since both are given as 1000 Hz, the analysis is done at only one frequency. The third field tells how many points should be calculated in each decade, a moot point in this case.

The output voltage, V(4), was calculated and printed out for this simulation. Simulations were done for various values of gain. A few values of gain are shown in the following table.

Iterations	
Gain	**Output, V(4)**
100	2.31 V
200	4.63 V
216	5.00 V

Upper and Lower Corner Frequencies: Figure 2.5 shows a typical frequency response of a wideband amplifier. The magnitude of amplifier gain falls off at high frequencies due to transistor frequency limitations or undesired capacitances. At low frequencies, the falloff is due to capacitors added to the amplifier for other purposes.

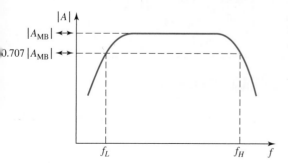

Figure 2.5
Frequency response of an amplifier.

Table 2.1 Some Values of dB for Key Values of A_o

A_o	A_{dB}
1.0	0
0.707	−3
1.414	3
2.0	6
10	20
100	40
1000	60

PRACTICE Problems

2.2 A voltage source with a resistance of 2 kΩ produces a peak voltage of 10 mV. This signal is to be amplified to drive a load of 1 kΩ to a peak voltage of at least 2 V. Which of the three following amplifiers will satisfy the design?
(a) $R_{in} = 1$ kΩ $R_{out} = 1$ kΩ
A = 1000
(b) $R_{in} = 10$ kΩ
$R_{out} = 1$ kΩ A = 500
(c) $R_{in} = 2$ kΩ
$R_{out} = 100$ Ω A = 500
Ans: (b) and (c).
2.3 Calculate A_{dB} for
$A_o = 394$. *Ans:* 51.9 dB.
2.4 Calculate A_{dB} for
$A_o = -628$. *Ans:* 56 dB.
2.5 Calculate A_{dB} for
$A_o = 0.674$. *Ans:* −3.43 dB.
2.6 Calculate A_{dB} for $A_o = -0.879$. *Ans:* −1.12 dB.
2.7 Use Table 2.1 to approximate A_{dB} for
$A_o = 3000$. *Ans:* 69–70 dB.

The magnitude of the gain may be constant at some midband value over a large range of frequency. This value is called the midband voltage gain. Those frequencies at which the magnitude of the voltage gain falls to 0.707 of the midband value are called the upper and lower corner frequencies. These points are also called the upper 3-dB frequency and the lower 3-dB frequency for reasons to be discussed later. Input signals with frequencies above the upper corner frequency or below the lower corner frequency are not amplified to the extent of the midband signals. For a high-fidelity audio amplifier, the upper corner frequency is often 20 kHz and the lower corner frequency is 20 Hz. When an amplifier is designed, the required corner frequencies must be known.

One very important electronic component called the operational amplifier or op amp extends the midband gain down to zero frequency or dc. There is no lower corner frequency in this situation, and the only 3-dB corner frequency of the voltage gain is the upper corner frequency. A later chapter will discuss the op amp in greater detail.

Decibels: The acronym dB stands for decibels. This unit is a relative unit rather than an absolute one such as volts or amps. Decibels are defined by the equation

$$A_{dB} = 20 \log_{10} |A_o| \tag{2.4}$$

It is not obvious from Eq. (2.4) that the dB is a relative unit until the gain A_o is written as

$$A_o = \frac{v_{out}}{v_{in}}$$

If the absolute value of A_o is 124, the dB value is

$$A_{dB} = 20 \log_{10} 124 = 20 \times 2.093 = 41.9$$

The quantity v_{out} is said to be 41.9 dB above v_{in}. Some values of dB for key values of $|A_o|$ are tabulated in Table 2.1.

The value of A_{dB} is zero for unity gain and becomes negative for gains below unity.

An important property of dB is the additive nature of this unit. For example, a gain of 200 corresponds to a dB gain of 46 dB. This may be found from

$$A_{dB} = 20 \log_{10} 200 = 20 \log_{10} (2 \times 100)$$
$$= 20 \log_{10} 2 + 20 \log_{10} 100 = 6 + 40 = 46 \text{ dB}$$

Because the decibel is defined by a logarithm, the product of two factors in absolute gain results in the sum of these factors expressed in dB.

2.2 Digital Circuits

IMPORTANT Concepts

1. Digital circuits are used to construct computers or microprocessor chips and other digital systems.
2. These circuits are driven with either a high voltage level or a low voltage level and generate either a high or a low voltage level at the output.

2.2.1 APPLICATIONS OF DIGITAL CIRCUITS

Digital circuits and systems have had a great impact on society in the last three to four decades. The digital computer grew to become very large and powerful in the 1960s, then began to move toward smaller systems near the end of that decade. The minicomputer became available in various forms between 1965 and 1968. These systems used the same type of logic or digital circuits as the mainframe computer, but took advantage of the rapid developments in the IC area to decrease cost and size. Several of the presently available logic families were developed in these formative years of the 1960s. The TTL (transistor-transistor logic) family and the ECL (emitter-coupled logic) family are examples of digital circuits that have changed little for almost 30 years. MOS (metal-oxide semiconductor) logic circuits have contributed greatly to the development of portable digital equipment as well as to the computer field. The popular CMOS (complementary MOS) digital circuit is used in any system that requires minimal energy drain on the energy source such as the hand calculator. Most high-performance microprocessor chips are now fabricated with CMOS circuits.

The 1970s experienced the development of the microprocessor and the later use of this device for the low-cost personal computer. It is interesting to note that the first microprocessors were intended for use in "smart" electronic instruments or in "smart" computer peripherals. They did not exhibit enough capability on which to base a computer design. The sophistication of the microprocessor long ago reached the level necessary for use in a computer, and the areas of electronic instrumentation and computer peripherals continue to apply the microprocessor. The microcomputer has been one of the great developments of the century. It has brought the power of a digital computer to the home and the office.

2.2.2 CHARACTERIZATION OF DIGITAL CIRCUITS

An individual digital logic circuit is so designed that the output is either at a high voltage level or at a low voltage level. The inputs to the circuit are driven by either high or low voltages. An example of these two levels is a low level of 0 V and a high level of 4 V. In logic circuits one of these levels is associated with binary 1, and the other is associated with binary 0.

Current and Voltage Definitions   The next few paragraphs define the voltage and current parameters that characterize digital logic circuits.

V_{IHmin} = minimum input voltage that the logic circuit is guaranteed to interpret as the high logic level

V_{ILmax} = maximum input voltage that the logic circuit is guaranteed to interpret as the low logic level

V_{OHmin} = minimum high logic-level voltage appearing at the output terminal of the logic circuit

V_{OLmax} = maximum low logic-level voltage appearing at the output terminal of the logic circuit

I_{IHmax} = maximum current flowing into an input when a specified high logic-level voltage is applied

I_{ILmax} = maximum current flowing into an input when a specified low logic-level voltage is applied

I_{OH} = current flowing into the output when a specified high-level output voltage is present

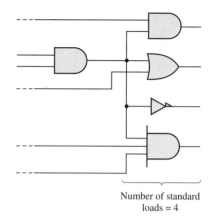

Figure 2.6
Partial schematic demonstrating
loading.

Number of standard
loads = 4

I_{OL} = current flowing into an output when a specified low-level output voltage is present

I_{OS} = current flowing into an output when the output is shorted and input conditions are such to establish a high logic-level output. (Current flowing out of a terminal has a negative value.)

Fan-Out In a digital system a given gate may drive the inputs to several other gates. The designer must be certain that the driving gate can meet the current requirements of the driven stages at both high and low voltage levels. The number of inputs that can be driven by the gate is referred to as the *fan-out* of the circuit. This figure is expressed in terms of the number of standard inputs that can be driven. Most circuits of a family will require the same input current, but a few may require more. If so, the specs for such a circuit will indicate that the input is equivalent to some multiple of standard loads. For example, a circuit may present an equivalent input of two standard loads. If fan-out of a gate is specified as 10, only five of these circuits could be safely driven. Figure 2.6 shows an AND gate loaded with four inputs, assuming each circuit presents one standard load to the AND gate output.

In several handbooks, the current requirements are given and fan-out can be calculated. For example, one logic gate has the following current specs:

$$I_{IH} = 40 \ \mu A \qquad I_{IL} = -1.6 \ mA$$

$$I_{OH} = -400 \ \mu A \quad I_{OL} = 16 \ mA$$

If this gate is to drive several other similar gates, we see that the output current capability of the stage is 10 times that required by the input. We note that the output stage can drive 400 μA into the following stages at the high level and sink 16 mA at the low level. The fan-out of this gate is 10.

Noise Margin Although current requirement is the major factor in determining fan-out, input capacitance or *noise margin* may further influence this figure. Noise margin specifies the maximum amplitude noise pulse that will not change the state of the driven stage, assuming that the driving stage presents a worst-case logic level to the driven stage. Noise margin can be evaluated from a consideration of the voltage levels $V_{IH\min}$, $V_{IL\max}$, $V_{OH\min}$, and $V_{OL\max}$. Figure 2.7 shows two logic circuits that are cascaded.

Figure 2.7
Cascaded stages used to
calculate noise margin.

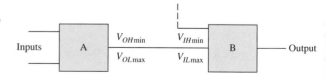

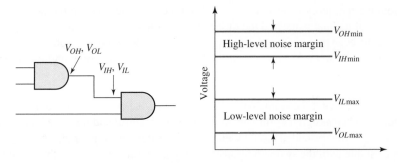

Figure 2.8
Noise margin definitions.

If we assume that $V_{IL\max} = 0.8$ V for circuit B, this means that the input must be less than 0.8 V to guarantee that circuit B interprets this value as a low level. If circuit A has a value of $V_{OL\max} = 0.4$ V, a noise spike of less than the difference $0.8 - 0.4$ V cannot lead to a level misinterpretation by circuit B. The difference

$$NM_L = V_{IL\max} - V_{OL\max} \qquad (2.5)$$

is called the low-level noise margin.

Assuming that $V_{IH\min} = 2$ V for circuit B and $V_{OH\min} = 2.7$ V for circuit A, the high-level margin is $2.7 - 2.0 = 0.7$ V. This high-level noise margin is found from

$$NM_H = V_{OH\min} - V_{IH\min} \qquad (2.6)$$

Since the minimum voltage developed by circuit A at the high level is 2.7 V, while circuit B requires only 2.0 V to interpret the signal as a high level, a negative noise spike of -0.7 V or less will not result in an error. Both low- and high-level noise margins are demonstrated in Fig. 2.8.

As we consider the noise margin we recognize that the values calculated in Eqs. (2.5) and (2.6) are worst-case values. A particular circuit could have actual noise margins better than those calculated.

As more gate inputs are connected to a given output, the voltages generated at both high and low levels are affected as a result of increased current flow. Thus, fan-out is influenced by noise margin.

Switching Times Another quantity that is used to characterize switching circuits is the speed with which the device responds to input changes. For switching circuits, the graph of Fig. 2.9 is useful in defining delay times. This figure assumes an inverting gate. There is a finite delay between the application of the input pulse and the output response.

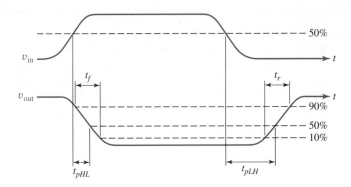

Figure 2.9
Definition of switching times.

PRACTICE Problems

2.8 If the output of gate A drives an input to gate B, how should $V_{OHmin}(A)$ relate to $V_{IHmin}(B)$? How should $V_{OLmax}(A)$ relate to $V_{ILmax}(B)$? *Ans:* $V_{OHmin} > V_{IHmin}$, $V_{OLmax} < V_{ILmax}$.

2.9 If $I_{IH} = 25\ \mu A$, $I_{IL} = -1.2$ mA, and the fan-out of the circuit is 8, what are the minimum values of I_{OH} and I_{OL}? *Ans:* $I_{OH} = -200\ \mu A$, $I_{OL} = 9.6$ mA.

2.10 If $V_{ILmax} = 0.7$ V and the low-level noise margin is 0.5 V, what is V_{OLmax}? *Ans:* $V_{OLmax} = 0.2$ V.

2.11 If two inverting gates are connected in series, each having a propagation delay time of 12 ns, what is the total propagation delay between circuit input and output? Is propagation delay time additive with respect to number of series circuits? Explain. *Ans:* 24 ns.

A quantitative measure of this delay is the difference in time between the point where v_{in} rises to 50% of its final value and the time when v_{out} falls to its 50% point. This quantity is called leading-edge propagation delay t_{pHL}. The trailing-edge propagation delay t_{pLH} is the time difference between 50% points of the trailing edges of the input and output signals. Fall and rise times are defined by 10 to 90% values as the output voltage swings between lower and upper voltage levels.

The speed of switching is important in determining the operating speed of a computer. For example, when a digital arithmetic unit is used to perform addition, the input operands are presented as binary-coded digital voltages at the arithmetic unit input. The output result will not be stable until the switching transients of the circuit within the arithmetic unit have been completed. These switching times lead to a limit on the clock speed of a computer. Clock frequencies for modern personal computers now exceed 1 GHz.

PRACTICAL Considerations

The statement, "Time is money," applies directly to computer operation. In earlier days, when digital computers cost millions of dollars per unit, faster computers allowed the solution of more complex problems. Problems that could not be solved with slower computers became workable as computer speed increased. More programs were submitted to these high-speed computers and more revenue was generated.

With personal computers, higher speeds are desirable for complex graphic programs and games as well as number crunching. In spite of the fact that earlier personal computers may still be useful as they run programs for which they were originally designed, newer programs may not run on the computer. Thus, many users buy newer, faster computers to upgrade these older models. Computer manufacturers with the fastest models typically sell the most computers.

2.3 Electronic Instrumentation

Instruments that measure electrical quantities are necessary to the advancement of the electronics field. The oscilloscope that displays voltage waveforms on a cathode ray tube (CRT) is one of the most important electronic instruments. This system includes amplifiers, power supplies, filters, waveform generators, and other electronic circuits. Voltmeters, signal generators, ammeters, counters, frequency meters, and curve tracers are examples of instruments that require electronic circuits.

2.4 Modulation Circuits

Modulation circuits are used primarily in communication systems. The popular commercial AM and FM broadcast bands utilize different types of modulators to transmit audio information through the atmosphere. The AM station uses an amplitude modulator and the FM station uses a frequency modulator. These modulation schemes allow several radio stations to transmit energy into the atmosphere simultaneously. The receiver is able to select the signal of a single station and recover the audio information on this station even though several other signals are present at the receiver input.

Television transmitters use a combination of amplitude modulation (video information) and frequency modulation (audio information) to transmit information to the receivers. In the last few decades, two significant developments that greatly improve the overall capability of television took place. One is the development of satellite transmission networks, and the other is the development of cable TV systems. Satellite transmission allows signals from one part of the world to be transmitted instantaneously to any other point. The second significant development uses cable TV systems to distribute signals to remote or hard-to-reach geographic areas and provide a large number of channels to these areas. Sophisticated modulation methods are used in both satellite and cable systems.

The HDTV systems present several noteworthy improvements over the older TV system. The aspect ratio (ratio of width to height) of the TV screen has been changed from the present 4:3 to 16:9, which allows wide screen movies to be presented with no loss of information on the video screen. The sound is digital rather than analog and is on a par with the compact disk (CD) for quality. The number of frames per second has been doubled and the number of picture elements has been increased by as much as six times that of the analog TV, which offers considerably higher resolution in the video image. Using digital methods also decreases the interference due to other noise sources. The introduction of the HDTV represents a large advancement in the television industry.

Transmission of digital signals over phone lines applies modulation techniques. A remote terminal connected to a computer several miles away uses a *modem* to communicate. The modem is a device that provides the proper signals to the phone line to allow the exchange of information between the computer and terminal. Modems are also used in Internet communications and are important in that regard.

Scientific data transmitted from space exploration vehicles are transmitted back to earth using various types of modulation, usually some form of pulse modulation. Long distance telephone conversations are generally converted to a pulse-code modulated signal before being transmitted over the phone lines, which allows several phone conversations to be transmitted over a single pair of lines and separated at the receiving station. Several conversations with the same city of destination are detected in the receiving station for that city. The individual conversations are then recovered and distributed locally to the proper recipients.

There are many other applications of modulation, with those listed above being some of the more significant ones.

2.5 Filters

IMPORTANT Concepts

1. Filters are used to pass the frequencies contained in a desired signal and reject all other frequencies.
2. Common frequency responses for filters are narrowband, wideband, low-pass, high-pass, and band-reject.

2.5.1 APPLICATION OF FILTERS

Almost every application of electronic circuits requires filters to improve performance. In many cases, unwanted electrical noise appears with the desired signal. Filters are used to minimize this noise problem. Filters are also used to change the quality of an audio signal. In high-fidelity stereo systems, bass boost, treble boost, and equalization are accomplished with filter circuits.

Radio and TV receivers use filters to separate the desired channel signal from the undesired channel signals that are present at the receiver input. A TV receiver must not only separate one channel from all others, it must also separate the sound or audio signal from the video signal of the channel to which it is tuned. Again, filter circuits are important in this application.

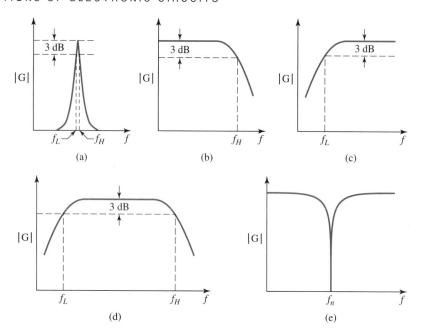

Figure 2.10
Filter responses:
(a) narrowband response,
(b) low-pass response,
(c) high-pass response,
(d) wideband response,
(e) band-reject or notch
response.

Figure 2.10 demonstrates the frequency responses of several important types of filters. These figures show the magnitude of the filter transfer function, $|v_{out}/v_{in}|$, as a function of frequency.

The narrowband filter is used in communication circuits to pass a desired channel while rejecting other channels. The low-pass filter passes low-frequency signals while attenuating high-frequency signals above a specified 3-dB frequency. The high-pass filter passes high-frequency signals and attenuates low-frequency signals. The wide-band response attenuates low-frequency and high-frequency signals while passing a broad midband range of signals. The notch filter drastically attenuates signals of a particular frequency or small range of frequencies.

2.5.2 CHARACTERIZATION OF FILTER PERFORMANCE

Narrowband filters are specified by resonant frequency and 3-dB bandwidth, which is given by $f_H - f_L$. The Q of the filter is given by

$$Q = \frac{f_o}{f_H - f_L} \tag{2.7}$$

where f_o is called the resonant frequency and corresponds to the frequency at which the magnitude of the transfer function is maximum. A high-Q circuit is very selective, passing signals over a small range of frequencies.

The low-pass and high-pass filters are specified by the 3-dB corner frequencies and the asymptotic falloff in the magnitude of the transfer function at frequencies far removed from the corner frequencies. For example, a filter may fall at -12 dB/octave which is equivalent to -40 dB/decade. An octave represents a change of frequency by a factor of 2. A decade represents a change in frequency by a factor of 10.

If a reference frequency is 246 Hz, an octave above this frequency is 492 Hz and an octave below this frequency is 123 Hz. A decade above this frequency is 2460 Hz and a decade below is 24.6 Hz. Mathematically, the octave and decade can be expressed from the equations

$$f_a = 2^n f_b \tag{2.8}$$

PRACTICE Problems

2.12 The asymptotic value of the magnitude of the transfer function for a low-pass filter is $|v_{out}/v_{in}|$ of 0.1 at a frequency of 280 Hz. At a frequency of 620 Hz, this value has fallen to 0.045. How many dB/octave is this filter response falling? *Ans:* 6 dB/oct.

2.13 Express the answer to Practice Problem 2.12 in dB/decade. *Ans:* 20 dB/dec.

2.14 Repeat Practice Problem 2.12 if the magnitude has fallen to 0.0092 at a frequency of 620 Hz. *Ans:* 18 dB/oct.

2.15 Express the answer to Practice Problem 2.14 in dB/decade. *Ans:* 60 dB/dec.

and

$$f_a = 10^m f_b \qquad\qquad (2.9)$$

where n is the number of octaves relating f_a to f_b and m is the number of decades relating these frequencies.

2.6 Power Electronics

The computer area applies a large number of low-power devices and components to collect and process information. The transistors used, generally in integrated circuit form, conduct currents in the nanoamp to milliamp range. There are several other important areas of electronics that require much larger currents. These areas are generally included in the power electronics field. Examples include the control of a 4-horsepower electric motor, the control of the intensity of a lighting system, and the production of dc voltages and currents for power supply circuits. Typically, the currents involved in such applications might range from 1 to 1000 A.

Power bipolar junction transistors, power field effect transistors, and silicon controlled rectifiers are common elements used in power electronics circuits. This list corresponds to only a few significant applications of electronic circuits. There are many others; however, it is hoped that the applications included might provide some motivation for further study of electronic circuits.

2.7 Review of Thévenin's Theorem for Electronic Circuit Design

IMPORTANT Concepts

1. Use of the Thévenin equivalent circuit is imperative in the analysis of electronic circuits.
2. The Thévenin equivalent circuit can be used to replace complex subcircuits, reducing the number of nodes in the circuit. Manual analysis of the circuit can be simplified considerably with this tool.

Many beginning students in electronic circuit design are amazed by the ability of an instructor to simplify complex circuit problems to a point where the solution becomes a trivial exercise. They later realize that the simplifications were all logical applications of valid circuit theorems, but the steps were done in the instructor's head and did not follow the mathematical rigor that the student had learned in earlier network analysis classes.

An electronic circuit design class generally assumes a math and basic circuits background for those taking the class. The student has typically studied dc circuits, ac circuits, and transient analysis before embarking on a study of electronic design. Ohm's Law, Kirchhoff's voltage and current laws, and Thévenin's and Norton's theorems should be familiar to the student.

All circuit theory remains valid in electronic circuit design. It is the method used to apply that theory that is often different. Theory is regularly applied informally or in a piece-wise manner to achieve a desired result. A portion of an electronic circuit may be analyzed using

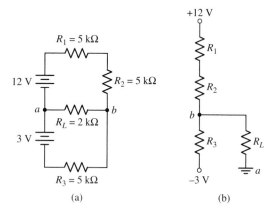

Figure 2.11
(a) Conventional circuit diagram.
(b) Electronic circuit version.

one method while a second part of the circuit may be analyzed in a different way. The choice of methods used is often related to the level of experience of the designer.

An experienced engineer chooses an appropriate circuit structure to accomplish the required signal processing function. The engineer's understanding of analysis can then allow a trial-and-error optimization. In such cases the design equation reduces to

$$\text{experience} + \text{analysis} + \text{trial} = \text{design}$$

It is the purpose of this text to provide the methods of analysis of electronic circuits and demonstrate how these methods can apply to the design of such circuits. The next few subsections will consider how methods learned in a basic circuits class might be applied to electronic circuits. We will also consider an explanation of some common differences.

2.7.1 CIRCUIT DIAGRAMS
Rather than drawing dc voltage sources as done in Fig. 2.11(a), electronic circuits often use a shorthand notation such as that of Fig. 2.11(b). Whereas the circuit diagram of Fig. 2.11(b) shows no loops, the two loops of part (a) are implied in part (b). The dc voltages (usually called power supply voltages in a practical circuit) may be designated by labels as shown in Fig. 2.11(b). The other side of the battery is assumed to be connected to ground to complete the circuit as in Fig. 2.11(a). The circuit diagrams of Figs. 2.11(a) and 2.11(b) are elementary forms of what is referred to as *circuit schematics*.

In most electronic circuit schematics, convention dictates that the positive supply voltage is drawn at the top of the schematic while negative voltages are drawn at the bottom. Thus, Fig. 2.11(b) is drawn in correct schematic form and perhaps provides a better basis for visualizing circuit simplifications.

Another point worth noting is that the resistor circuit values are given in "kilohms" or kΩ. Many network analysis textbooks deal in ideal or normalized units, and it is not unusual to work problems where resistor values are 1 or 2 Ω. In electronic practice, most resistors are in the kΩ range, and milliamperes (mA) or microamperes (μA) are the most popular units for currents. In a network analysis class, a correct answer for a current I might be 0.000375 amps. A circuit design engineer would most likely specify the answer as 0.375 mA or 375 μA. The product of current in mA and resistance in kΩ results in units of volts.

2.7.2 THE THÉVENIN EQUIVALENT
Although Thévenin's theorem is discussed in detail in every basic circuits class, the full significance of this theorem is rarely appreciated until electronic circuits are studied. In this

field, it is important to know such things as the input impedance or the output impedance of an amplifier. The input impedance is nothing more than the Thévenin equivalent impedance at the input terminals of the amplifier. The output impedance is the Thévenin equivalent impedance looking into the output terminals of the amplifier. In hand analysis, this theorem is indispensable.

An approximate statement of Thévenin's theorem follows.

Any linear two-terminal network consisting of current or voltage sources and impedances can be replaced by an equivalent circuit containing a single voltage source in series with a single impedance.

The mechanics of applying this theorem are simple.

1. To find the Thévenin equivalent voltage at a pair of terminals, the load is first removed, leaving an open circuit. The open-circuit voltage across this terminal pair is then calculated. This voltage is the Thévenin equivalent voltage.

2. The equivalent resistance is found by replacing each independent voltage source with a short circuit (zeroing the voltage source), replacing each independent current source with an open circuit (zeroing the current source), and calculating the resistance between the terminals of interest. Dependent sources are not replaced and can have an effect on the value of equivalent resistance.

A practical voltage source consists of a voltage in series with a small resistance. The ideal voltage source has a zero source resistance. The ideal source is zeroed by replacing it with a short circuit. A practical current source consists of a current in parallel with a large resistance. The ideal current source has infinite resistance. The ideal source is zeroed by replacing it with an open circuit.

The actual calculation of the Thévenin equivalent resistance for a circuit can be done in several ways. The simplest method, when there are no dependent sources, is to combine all resistors between the terminals of interest into a single equivalent resistance. In some instances, especially when dependent sources are present, it may be necessary to calculate the ratio of voltage across the terminal to the entering current. This can be done by assuming an applied voltage across the terminals and finding the resulting current. In some cases, it may be more convenient to assume a current is entering the terminal and then calculate the resulting voltage. The following examples will clarify these approaches.

EXAMPLE 2.2

Calculate the current through the load resistance, R_L, in the circuit of Fig. 2.11(b) by taking a Thévenin equivalent circuit of the remaining network.

SOLUTION The Thévenin equivalent voltage at terminals a-b is found from the network of Fig. 2.12(a). The load resistance has been removed from the circuit of Fig. 2.11(b) for this calculation.

The voltage at terminal b is determined after finding the current through the three series resistors. This value is

$$I = \frac{V_1 - V_2}{R_1 + R_2 + R_3} = \frac{12 - (-3)}{15} = 1 \text{ mA} \qquad \textbf{(2.10)}$$

The Thévenin equivalent voltage is then

$$V_{\text{Th}} = V_1 - I(R_1 + R_2) = 12 - 1 \times 10 = 2 \text{ V} \qquad \textbf{(2.11)}$$

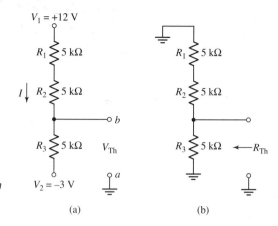

Figure 2.12
*Circuits used to find Thévenin
equivalents: (a) voltage,
(b) resistance.*

(a)   (b)

The equivalent resistance is calculated from the circuit of Fig. 2.12(b), which has replaced the independent voltage sources with short circuits. There are two paths from terminal b to terminal a: one through R_1 and R_2, the other through R_3. These paths are in parallel, resulting in an equivalent resistance of

$$R_{Th} = 10 \parallel 5 = 3.33 \text{ k}\Omega \tag{2.12}$$

The Thévenin equivalent circuit is shown in Fig. 2.13(a). The load resistance is added in Fig. 2.13(b), allowing the load current to be written as

$$I_L = \frac{V_{Th}}{R_{Th} + R_L} = \frac{2}{3.33 + 2} = 0.375 \text{ mA} \tag{2.13}$$

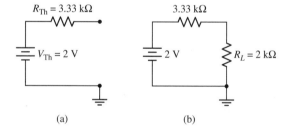

Figure 2.13
*(a) Thévenin equivalent circuit.
(b) External load added.*

(a)   (b)

The resistance between any two terminals is often referred to as the "resistance seen looking into" those terminals. For example, the effective resistances between different terminals of the circuit of Fig. 2.14 are determined as follows:

$$R_{20} = R_A \parallel R_C \parallel (R_B + R_L) \tag{2.14}$$

$$R_{10} = 0 \tag{2.15}$$

$$R_{30} = R_L \parallel (R_B + R_C \parallel R_A) \tag{2.16}$$

$$R_{23} = R_B \parallel (R_A \parallel R_C + R_L) \tag{2.17}$$

The effective or equivalent resistances between the various terminals were determined after replacing the voltage source with a short circuit and the current source by an open circuit.

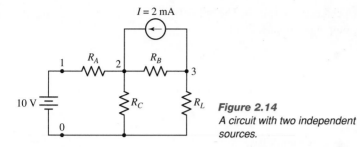

Figure 2.14
A circuit with two independent
sources.

A second method of calculating the Thévenin equivalent resistance is demonstrated in Example 2.3.

EXAMPLE 2.3

Find the output resistance of the circuit in Fig. 2.15.

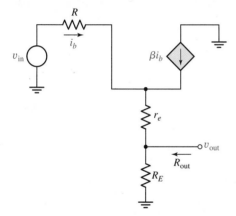

Figure 2.15
Circuit for Example 2.3.

SOLUTION The term "output resistance" is synonymous with the Thévenin equivalent resistance between the output terminal and ground. To obtain this value, we use the equivalent circuit of Fig. 2.16 that replaces the input source with a short circuit.

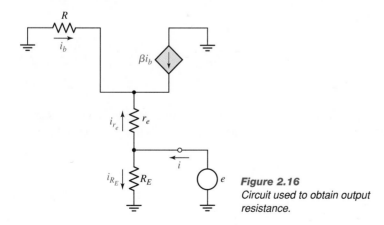

Figure 2.16
Circuit used to obtain output
resistance.

We note that the dependent current source of value βi_b remains in the circuit. We assume that a voltage of e is applied to the output terminal, and we will calculate the resulting value of current, i, that enters this terminal. The ratio of these values will be the Thévenin equivalent resistance or output resistance; that is,

$$R_{out} = \frac{e}{i}$$

The current, i, consists of the sum of the currents through R_E and through r_e, giving

$$i = i_{R_E} + i_{r_e}$$

The current i_{R_E} is, by Ohm's law,

$$i_{R_E} = \frac{e}{R_E}$$

Summing currents into the node above r_e, we can write

$$i_{r_e} = -i_b - \beta i_b = -(\beta + 1)i_b$$

The current i_b is found by summing voltages between the input and output terminals, resulting in

$$Ri_b - r_e i_{r_e} + e = 0$$

Substituting for i_{r_e} leads to

$$Ri_b + r_e(\beta + 1)i_b + e = 0$$

and i_b is determined to be

$$i_b = \frac{-e}{R + (\beta + 1)r_e}$$

The value of i_{r_e} is

$$i_{r_e} = -(\beta + 1)i_b = \frac{(\beta + 1)e}{R + (\beta + 1)r_e}$$

The total current can be written as

$$i = i_{R_E} + i_{r_e} = \frac{e}{R_E} + \frac{(\beta + 1)e}{R + (\beta + 1)r_e}$$

The output resistance is

$$R_{out} = \frac{e}{i} = \frac{1}{\frac{1}{R_E} + \frac{1}{\frac{R}{\beta+1}+r_e}} = R_E \parallel \left(r_e + \frac{R}{\beta + 1} \right)$$

The final example of this section will demonstrate the method of resistance calculation that assumes a current into the terminal and then finds the resulting voltage.

EXAMPLE 2.4

Find the input resistance of the circuit shown in Fig. 2.17.

SOLUTION The input resistance is synonymous with the Thévenin equivalent resistance looking into the input terminals. The circuit of Fig. 2.18 can be used to find this resistance.

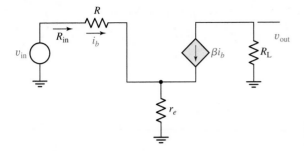

Figure 2.17
Circuit for Example 2.4.

We assume a current input of i. We will then calculate the voltage, e, that results between the input terminal and ground. The current i also equals i_b; hence we can write

$$e = Ri_b + r_e i_{r_e} = Ri_b + r_e(\beta + 1)i_b = Ri + r_e(\beta + 1)i$$

The input resistance is then

$$R_{in} = \frac{e}{i} = \frac{Ri + r_e(\beta + 1)i}{i} = R + (\beta + 1)r_e$$

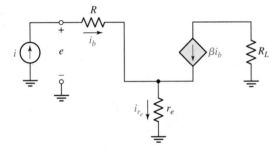

Figure 2.18
Calculation of R_{in} for Example 2.4.

PRACTICAL Considerations

A thorough understanding of the concept of the Thévenin equivalent circuit and its relation to electronic circuits is imperative in the area of circuit design. For this reason, we summarize some important points here.

1. An ideal voltage source of zero value is a short circuit.
2. An ideal current of zero value is an open circuit.
3. Unlike independent sources, dependent voltage or current sources are not automatically zeroed.
4. The input impedance of a circuit equals the Thévenin equivalent impedance at the input terminals.
5. The output impedance of a circuit equals the Thévenin equivalent impedance at the output terminals.

There are occasions when it is important to know how much resistance appears in series with a pair of terminals. The time constant of a capacitor may be found as the product of this resistance and the capacitance.

Figure 2.19(a) shows a simple transistor amplifier circuit employing resistors, capacitors, and a bipolar junction transistor or BJT as an active device. The object of our problem is

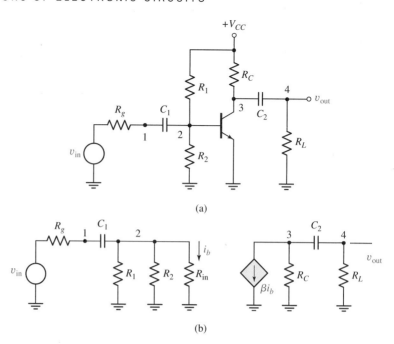

Figure 2.19
A circuit with a dependent source.

to determine the *effective resistance* as seen from each of the capacitors' terminals. For purposes of simplification, we will here simply state that the circuit of Fig. 2.19(a) can be represented by the circuit shown in Fig. 2.19(b). That is, the circuit of Fig. 2.19(b) is an *equivalent circuit* for that of Fig. 2.19(a).

Using Fig. 2.19(b), we can evaluate the resistance seen by the capacitor C_1 by finding the Thévenin resistance at terminals 1-2. This is

$$R_{12} = R_g + R_1 \parallel R_2 \parallel R_{in} \tag{2.18}$$

The series resistance seen by capacitor C_2 is

$$R_{34} = R_C + R_L \tag{2.19}$$

Three important points should be noted in this circuit. The first is that the calculations are made for incremental voltages or currents. The input signal may typically be a small sinusoid. The second point is that a dc voltage source is a short circuit for incremental signals, which can be seen by considering that a changing current, ΔI, is impressed on the dc source. The voltage across a dc source will remain constant at its dc value. Therefore, ΔV will be zero. The equivalent resistance of the source for incremental signals is then

$$r_{inc} = \frac{\Delta V}{\Delta I} = \frac{0}{\Delta I} = 0$$

When incremental signals are considered, a dc voltage source can always be replaced with a short circuit.

The third point is that the dependent current source of Fig. 2.19(b) has a zero value when R_{34} is found, since the input voltage source is shorted to evaluate the Thévenin equivalent resistance. The current i_b is zero and, consequently, so is βi_b. A current source of zero value is replaced by an open circuit.

2.7.3 SIGNAL GENERATOR IMPEDANCE

The effective resistance of a signal generator is sometimes referred to as its *source impedance* or *generator impedance* even though it may be purely resistive. For example, many signal generators used for bench work have a 600-Ω generator impedance.

Figure 2.20 shows the equivalent circuit of such a signal generator. The effects of a 600-Ω generator impedance may be observed by noting that if the signal generator output is adjusted to 1 volt with no load connected to its output terminals, then the output will be reduced when the load is connected.

From earlier studies many students are under the impression that load impedances should always be set equal to generator impedances when connecting a load to a source. This is not generally true, and often it is advantageous to make the load resistance as large as possible compared to the generator resistance. Doing so will ensure that the loading effect (attenuation due to R_L) is minimized. For example, from Fig. 2.20

$$v_L = v_g \frac{R_L}{R_g + R_L}$$

If $R_L \gg R_g$, then $v_L \approx v_g$, which implies that the entire source voltage reaches the load resistance.

If $R_L = R_g$, then $v_L = v_g/2$. A loss of 50% of the signal will result from setting $R_g = R_L$. This is usually not significant when coupling to a signal generator, since the loaded output can be readjusted to any desired level within the range of the signal generator output. However, when coupling low-level signals from most transducers, it is very seldom that the loss of signal can be afforded without a serious reduction in system performance as a result of the high output impedance of many transducers. This impedance becomes the generator impedance in the preceding loading equation.

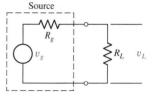

Figure 2.20
Signal generator connected to a load.

PRACTICE Problems

2.16 Calculate the output impedance for the circuit of Example 2.3 if $\beta = 100$, $R = 1 \text{ k}\Omega$, $r_e = 26 \ \Omega$, and $R_E = 500 \ \Omega$. *Ans:* 33.5 Ω.
2.17 Calculate the input impedance for the circuit of Example 2.4 if $\beta = 100$, $R = 1 \text{ k}\Omega$, and $r_e = 26 \ \Omega$. *Ans:* 3.63 kΩ.

2.8 The Miller Effect

IMPORTANT Concepts

1. An impedance that is connected between the input and output nodes of an inverting amplifier can drastically affect the input impedance of the amplifier. This effect is referred to as the Miller effect.
2. The current drawn by the bridging impedance when an input voltage is applied is $(1 + |A|)$ times the current that would be drawn if this impedance were connected between input and ground. The amplifier gain A can be a large number, leading to a large input current.
3. Calculations can be simplified by replacing the bridging impedance with a smaller impedance between the amplifier input terminal and ground. The value of this impedance equals that of the bridging impedance divided by the factor $(1 + |A|)$.

The Miller effect was named after its discoverer, John M. Miller, who did this pioneering work on the vacuum tube triode over 50 years ago. His efforts were directed toward defining the multiplicative effect of an inverting amplifier on the capacitance that linked input and output loops. A more general derivation considers the effect of any impedance that links the input and output of an inverting amplifier.

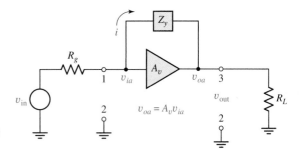

Figure 2.21
An amplifier with impedance bridging.

The Miller effect relates closely to the Thévenin equivalent impedance seen at a terminal pair. Figure 2.21 displays an amplifier with an impedance that bridges the input and output nodes.

The circuit designer is often interested in the loading effects of the amplifier. The input current required by the amplifying circuit will cause a voltage drop across the generator resistance, diminishing the voltage that reaches the amplifier input. Output current produced by the amplifier and intended to flow through the load resistance will be diminished by the current that flows to the impedance Z_y.

In order to evaluate amplifier loading effects, the input and output impedances, due to the impedance Z_y, can be found. In this evaluation, we will assume an ideal amplifier block with infinite input and zero output impedances. Of course, the input impedance to the amplifying stage is the Thévenin equivalent impedance looking into terminals 1-2 toward the amplifier. The output impedance is the Thévenin equivalent impedance looking into terminals 3-2 toward the amplifier.

2.8.1 MILLER EFFECT ON INPUT LOOP

The amplifier produces an output voltage of

$$v_{oa} = A_v v_{ia} \tag{2.20}$$

where A_v is the voltage gain and v_{ia} is the input voltage of the amplifier.

The input impedance due to Z_y can be calculated as

$$Z_{inR} = \frac{v_{ia}}{i}$$

where i is the current flowing into Z_y. This current can be found using Ohm's law as

$$i = \frac{v_{ia} - v_{oa}}{Z_y} = \frac{v_{ia} - A_v v_{ia}}{Z_y} = \frac{(1 - A_v)v_{ia}}{Z_y} \tag{2.21}$$

The Thévenin equivalent input impedance due to Z_y is then

$$Z_{inR} = \frac{v_{ia}}{(1 - A_v)v_{ia}/Z_y} = \frac{Z_y}{1 - A_v} \tag{2.22}$$

Equation (2.22) is an interesting equation. If the voltage gain of the amplifier is real and a large positive number, the input impedance becomes negative. In effect, this means that when v_{ia} is positive, current flows out of Z_y toward the input voltage source. This is due to the fact that the voltage on the output side of the circuit is much larger than the voltage on the input side, forcing current to flow from right to left through Z_y. Although negative impedances are important in some circuits, the most common situation in amplifier design occurs when A_v has a large negative value.

If $A_v = -A$, where A is a constant, Eq. (2.22) leads to an input impedance of

$$Z_{inR} = \frac{Z_y}{1 + A} \qquad (2.23)$$

In the case of an inverting amplifier, the input impedance due to Z_y is reduced by the factor $(1 + A)$. This factor can be significant if A is a large number.

The Miller effect tells us that, if an impedance bridges the input and output nodes of an inverting amplifier, the equivalent input impedance due to this bridging element will be given by Eq. (2.23). We often refer to this Thévenin equivalent input impedance as the impedance reflected to the input. If the impedance Z_y is a resistor of value R_y, the reflected impedance is also resistive and has a value of

$$R_{inR} = \frac{R_y}{1 + A} \qquad (2.24)$$

This reflected resistance, R_{inR}, is typically much smaller than the bridging resistance, R_y.

If the impedance Z_y is produced by a capacitor of value C_y, the value of reflected capacitance is

$$C_x = (1 + A)C_y \qquad (2.25)$$

The value of the reflected capacitance is larger than the original value because the impedance of a capacitor is an inverse function of the capacitor value. For the reflected impedance to be smaller than the original impedance, the reflected capacitance must be larger than the original value.

If the amplifier input impedance, Z_{ia}, is finite rather than infinite, the total input impedance to the amplifier circuit is

$$Z_{in} = Z_{inR} \parallel Z_{ia} \qquad (2.26)$$

If the amplifier output impedance is finite rather than zero, the reflected impedance is

$$Z_{inR} = \frac{Z_y + Z_{oa}}{1 + A} \qquad (2.27)$$

where Z_{oa} is the output impedance of the amplifier. This equation reflects the total impedance appearing between the input node of the amplifier and the dependent voltage source.

A configuration that often occurs in practice is shown in Fig. 2.22. The effect of R_L on the impedance reflected to the input is readily found by taking a Thévenin equivalent circuit of the dependent source, R_{oa}, and R_L. This leads to the circuits of Fig. 2.23.

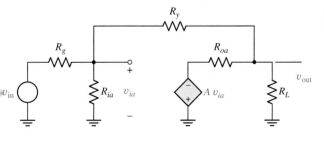

Figure 2.22
An amplifier stage with finite input and output impedances.

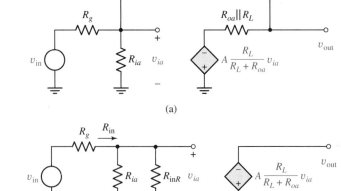

(a)

Figure 2.23
(a) An equivalent circuit.
(b) Impedance reflected to input.

(b)

After taking the equivalent circuit of Fig. 2.23(a), the reflected resistance is found as

$$R_{\text{in}R} = \frac{R_y + R_{oa} \parallel R_L}{1 + \frac{A R_L}{R_L + R_{oa}}} \tag{2.28}$$

We note that the total resistance between input node and the dependent voltage source is $R_y + R_{oa} \parallel R_L$. The strength of the dependent source is reduced by the loading of R_L on R_{oa}. The total input resistance seen by the signal generator is found from Eq. (2.26) to be

$$R_{\text{in}} = R_{ia} \parallel R_{\text{in}R}$$

2.8.2 MILLER EFFECT ON OUTPUT LOOP

The effect of the bridging element on the output loop can also be approximated rather easily. We will confine this discussion to the situation of a large negative amplifier gain given by $A_v = -A$. When a voltage of v_{ia} is present at the amplifier input, the output voltage is $v_{oa} = A_v v_{ia} = -A v_{ia}$. The Thévenin equivalent impedance at the output, due to Z_y, is

$$Z_{\text{out}R} = \frac{v_{oa}}{i}$$

where i is the current flowing from the output toward Z_y as a result of the voltage v_{oa}. This current is

$$i = \frac{v_{oa} - v_{ia}}{Z_y} = \frac{v_{oa} + \frac{v_{oa}}{A}}{Z_y} = \frac{\left(1 + \frac{1}{A}\right) v_{oa}}{Z_y} \tag{2.29}$$

The output impedance is then

$$Z_{\text{out}R} = \frac{Z_y}{\left(1 + \frac{1}{A}\right)} \tag{2.30}$$

The term $1/A$ is often much smaller than unity and can be neglected in such a case. We then get a reflected output impedance of

$$Z_{\text{out}R} = Z_y \tag{2.31}$$

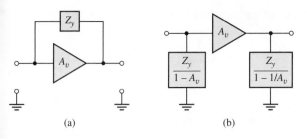
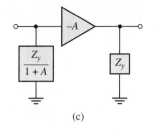

Figure 2.24
(a) Bridging circuit. (b) Miller equivalent circuit. (c) Miller equivalent circuit for large negative amplifier gain.

The total output impedance of the amplifier circuit is

$$Z_{\text{out}} = Z_{\text{out}R} \parallel Z_{oa} \tag{2.32}$$

where Z_{oa} is the amplifier output impedance. This impedance is generally very small, due to the feedback through the bridging impedance, resulting in a value of Z_{out} that approaches zero.

Note that, in calculating the effective output impedance, we ignored the effect of the voltage fed back from output to input through the bridging impedance. This voltage gets amplified and modifies the total current at the output. For applications of the Miller effect to single-stage BJT and MOSFET circuits, this modification of output current and reflected impedance is generally negligible. Thus, we will neglect it in future discussions, realizing that in some circuits a larger error may be introduced by so doing.

2.8.3 MILLER EFFECT EQUIVALENT CIRCUIT

Based on the discussion of this section, the equivalent circuit of an amplifier with a bridging impedance should appear as in Fig. 2.24. The circuit of Fig. 2.24(c) is the most useful for amplifier design. If the amplifier has a finite input impedance, the overall input impedance to the amplifier circuit is the Miller effect input impedance in parallel with the amplifier input impedance.

A finite amplifier output impedance will affect the amplifier gain, and this effect must be considered to obtain the appropriate gain to calculate both reflected impedances. An example will clarify these points.

EXAMPLE 2.5

Assume that the amplifier of Fig. 2.25 has infinite input impedance, zero output impedance, and a frequency-independent voltage gain of $A_v = -800$ V/V.

(a) Choose a bridging capacitor, C, to lead to an input time constant of 100 ms.

(b) Find the value of time constant for the output loop of the circuit.

(c) Repeat part (a) if the output resistance of the amplifier is $R_{oa} = 300 \ \Omega$.

(d) Repeat part (b) using the output resistance of part (c).

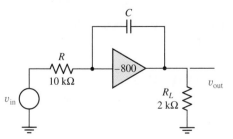

Figure 2.25
Amplifier circuit for Example 2.5.

SOLUTION The Miller equivalent circuit for parts (a) and (b) is shown in Fig. 2.26.

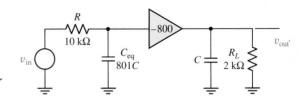

Figure 2.26
The Miller equivalent circuit for Example 2.5.

(a) The input time constant is $\tau_i = RC_{eq} = 10^4 \times 801C$. This value is equated to 100 ms, and C is found to be

$$C = \frac{\tau_i}{801R} = \frac{0.1}{8.01 \times 10^6} = 12.5 \text{ nF}$$

The bridging capacitor is 12.5 nF, as also is the reflected output capacitance.

(b) The output loop time constant is zero since the output resistance of the amplifier, which drives the output capacitance, is zero.

(c) For this case, the Miller equivalent circuit is shown in Fig. 2.27. The gain from the input side of C to the output side is now reduced by the loading of R_L on R_{oa}. This reduced value of gain is

$$A_L = -800 \times \frac{2000}{2000 + 300} = -696 \text{ V/V}$$

This calculation assumes that the bridging capacitor does not load the output circuit. The equivalent input capacitance is now $697C$, resulting in a time constant of $\tau_i = RC_{eq} = 10^4 \times 697C$. Solving for C gives

$$C = \frac{0.1}{6.97 \times 10^6} = 14.3 \text{ nF}$$

(d) The output time constant is now

$$\tau_o = R_L \parallel R_{oa}C = 2000 \parallel 300 \times 14.3 \times 10^{-9} = 3.73 \ \mu s$$

This time constant is a factor of several thousands less than the input time constant and will have negligible effect on the overall time constant of the circuit.

Figure 2.27
Miller equivalent circuit with amplifier output impedance.

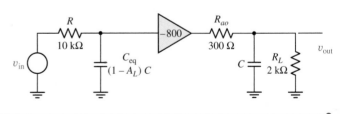

An amplifier output impedance can lower the overall gain of the amplifying element and decrease the Miller multiplication factor.

2.8.4 IMPORTANCE OF THE MILLER EFFECT
The humorous statement, "I have some good news, and I have some bad news," applies to the Miller effect. It can be used to advantage in certain special cases, but it often leads to degrading effects in electronic amplifier design.

The good news is that this effect can be used to multiply the apparent value of a capacitance. For over 50 years, circuits that can take the integral of an input voltage have been based on this effect. These types of circuits are called Miller integrators and require very large capacitance values for high accuracy. A second need for capacitance multiplication became apparent after the development of the integrated circuit. Compensation capacitors to stabilize the operation of high-gain op amps and capacitors used in IC filters are often rather large. Capacitors larger than tens of pF require prohibitive amounts of space on the IC chip. These larger element values are obtained from small capacitors multiplied by the Miller effect.

PRACTICAL Considerations

The popular 741 op amp chip uses a 30-pF capacitor for compensation of the amplifier, as will be seen in Chapter 11. This capacitor is small enough to be fabricated on the chip, but far too small to compensate properly. An amplifier stage with a voltage gain of about -200 V/V is used to multiply the apparent value of this capacitance to $201 \times 30 = 6030$ pF. The amplifier is then compensated with this large Miller effect capacitance that requires very little actual chip space or "real estate."

PRACTICE Problems

2.18 Repeat part (a) of Example 2.5 if the input time constant is to be 240 μs. *Ans:* 30 nF.
2.19 Calculate the input time constant for the circuit of Example 2.5 if $C = 20$ pF. *Ans:* 160 ms.

The bad news or disadvantage of the Miller effect is that the parasitic capacitances that result in the manufacture of transistors, both bipolar junction and metal-oxide semiconductor devices, often appear larger than the physical value of capacitance. Certain useful amplifier configurations of these devices lead to poor high-frequency gain due to the multiplied capacitance. We will consider this effect in more detail in later chapters.

2.9 Transient Waveforms

IMPORTANT Concepts

1. Single time constant circuits lead to exponential responses when excited by a step change of input.
2. A general equation for an exponential response can be used to express the time variation of the response.
3. The Thévenin equivalent circuit can often be used for a complex circuit to simplify the calculations of time constant, initial voltage, and final voltage.

Many waveforms in electronic circuits result from an abrupt change of signal level applied to a single time constant circuit. The single time constant circuit often consists of a capacitor being charged or discharged through some equivalent resistance. Thévenin's theorem is generally applied to such transient circuits to evaluate the charging/discharging resistance and the initial or final capacitor voltage.

In the circuit of Fig. 2.28, the switch is in the first position long enough to charge the capacitor to an initial voltage, v_i. The switch then changes position and the capacitor begins to charge toward a target voltage, v_t. The voltage across the capacitor changes exponentially between the initial voltage and the target voltage. An important general equation that describes this exponential voltage change is

$$v(t) = v_i + (v_t - v_i)(1 - e^{-t/\tau}) \qquad (2.33)$$

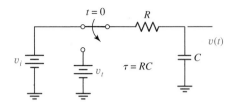

Figure 2.28
A single time constant circuit.

If v_i is -2 V, v_t is 8 V, and $\tau = 10\ \mu$s for the circuit of Fig. 2.28, the time taken for the capacitor voltage to reach 0 V from the time the switch changes position can be calculated from Eq. (2.33). If this time is designated t_1, the equation can be solved to give

$$t_1 = \tau \ln \frac{v_t - v_i}{v_t - v(t_1)} \tag{2.34}$$

Using the values given with $v(t_1) = 0$ V results in $t_1 = 2.23\ \mu$s. These equations are very useful in time calculations for single time constant circuits.

A large number of electronic circuits behave as single time constant circuits. One difficulty in dealing with such circuits is that, mathematically, the exponential waveform never reaches its final value. For example, if we use Eq. (2.34) with $v(t_1) = v_t$, the denominator of the logarithm becomes zero and the log term becomes infinite. In real circuits, the voltage gets close enough to the final value in some finite time. In a time equal to 3τ, the transition is about 95% complete; in a time equal to 4τ, the transition is 98% complete; and in a time equal to 5τ, the transition is 99% complete.

In an effort to mathematically define a more accurate transition time, the 10% to 90% rise time was proposed. As the circuit makes an exponential transition, the time required to move 10% of the total transition is designated t_{10}. The time required to move 90% of the total transition is designated t_{90}. The rise time, t_r, can then be found to be

$$t_r = t_{90} - t_{10} = 2.2\tau \tag{2.35}$$

The proof of this equation is left to the student.

Although we have only briefly discussed the 3-dB bandwidth of a circuit, we will here state a relationship between the time constant and the circuit bandwidth in rad/s for a single time constant circuit. This relationship is

$$\omega_{3dB} = \frac{1}{\tau} \tag{2.36}$$

If the time constant of the circuit is known when a step change of input occurs, the bandwidth can easily be calculated from Eq. (2.36) for a sinusoidal input signal.

We now give a demonstration of the application of Thévenin's theorem to a transient circuit. Figure 2.29 shows a switching circuit with an input that switches instantaneously from 2 V to 12 V at $t = 0$. We are to sketch the waveform of the output voltage transient.

Figure 2.29
A switching circuit.

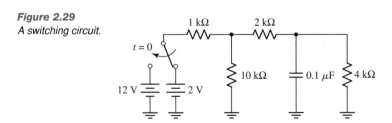

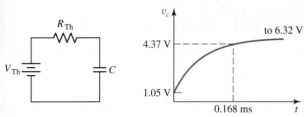

Figure 2.30
Equivalent circuit and output waveform.

The solution to this problem requires that the initial voltage, the target voltage, and the time constant be found. Taking a Thévenin equivalent circuit will aid in finding each of these quantities. This equivalent circuit should be taken at the capacitor terminals to see what voltage and resistance are presented to the capacitor. The equivalent resistance is

$$R_{Th} = 4 \parallel (2 + 1 \parallel 10) = 1.68 \text{ k}\Omega$$

The equivalent voltage is given by

$$V_{Th} = V_{in} \frac{10 \parallel 6}{10 \parallel 6 + 1} \times \frac{4}{6} = 0.526 V_{in}$$

The initial voltage of $v_i = 1.05$ V results when $V_{in} = 2$ V. A final voltage or target voltage of $v_t = 6.32$ V results when $V_{in} = 12$ V. The time constant is $\tau = 1680 \times 10^{-7} = 0.168$ ms. The equivalent circuit and a sketch of the output waveform are shown in Fig. 2.30.

PRACTICE Problem

2.20 If the switch in Fig. 2.29 has been connected to 12 V for a long time, then switches to connect to the 2-V source at $t = 0$, how long will it take for the capacitor to discharge to 2 V?
Ans: 0.288 ms.

SUMMARY

➤ Because most electronic signals of interest are small, these signals must be amplified to become useful. The amplifier becomes an important circuit for this reason.

➤ The appropriateness of an amplifier for a given application depends on such parameters as the gain, input impedance, output impedance, and bandwidth of the amplifier.

➤ Digital circuits are used extensively in the construction of computer systems. These circuits are characterized by such parameters as noise margin, fan-out, and switching times.

➤ Other important areas of electronics are instrumentation, communications, filters, and power transmission and distribution.

➤ The Thévenin equivalent circuit is indispensable in the analysis of many electronic circuits. The mechanics of calculating the equivalent voltage and equivalent resistance must be understood to successfully perform circuit analysis and design.

➤ The Miller effect is encountered in high-frequency circuits as well as high-gain circuits such as op amps. This concept must also be understood to proceed to more advanced circuit analysis and design.

➤ Digital circuits often switch between voltage levels with an exponential variation. A general equation for this variation can be used to calculate switching times.

PROBLEMS

SECTION 2.1 AMPLIFIERS

2.1 A transducer generates a 10-mV peak signal and has a 2-kΩ resistance. This signal must be amplified to a 6-V peak level to drive a 4-kΩ load. Assume the input resistance to the amplifier is infinite while the output resistance is zero.

(a) Calculate the required voltage gain.

(b) Calculate the required load current.

2.2 An ideal amplifier has an infinite input resistance and a zero output resistance. This amplifier is used to couple a 20-mV peak voltage, 10-kΩ source to a 200-Ω load.

(a) If the overall voltage gain is to be 200 V/V, calculate the necessary amplifier voltage gain.

(b) Repeat part (a) if the amplifier resistances are $R_{in} = 100$ kΩ and $R_{out} = 20$ Ω.

(c) Repeat part (a) if the amplifier resistances are $R_{in} = 10$ kΩ and $R_{out} = 1$ kΩ.

(d) Repeat part (a) if the amplifier resistances are $R_{in} = 1$ kΩ and $R_{out} = 1$ kΩ.

If efficient voltage transfer is desired, what can you conclude about R_{in} and R_{out} of the amplifier?

2.3 An amplifier is used to couple a 20-mV peak voltage, 10-kΩ source to a 900-Ω load. The overall voltage gain should be 120 V/V. If the amplifier has a 10-kΩ input resistance and a 1-kΩ output resistance, calculate the necessary value of amplifier gain A.

2.4 Repeat Problem 2.3 if the source resistance is 1 kΩ and the load resistance is 4 kΩ.

2.5 Calculate A_{dB} for $A_o = 1260$ V/V.

2.6 Two stages are cascaded. The first has a voltage gain of 28 V/V and the second has a voltage gain of 36 V/V. Both gains have accounted for loading effects. Calculate the overall gain in V/V and in dB.

☆ **2.7** Two stages are cascaded to result in an overall gain of 680 V/V. The loaded gain of the first stage is 36 dB. What must the gain of the second stage be in dB to satisfy this specification? In V/V?

SECTION 2.2 DIGITAL CIRCUITS

2.8 The current specs on a two-input AND gate are $I_{IHmax} = 40$ μA, $I_{ILmax} = -1.6$ mA, $I_{OH} = -800$ μA, and $I_{OL} = 16$ mA. How many similar inputs can one of these gates drive?

2.9 Repeat Problem 2.8 for a logic gate buffer with the following specs: $I_{IHmax} = 40$ μA, $I_{ILmax} = -1.6$ mA, $I_{OH} = -2400$ μA, and $I_{OL} = 48$ mA.

2.10 If a circuit rise time is governed by a single time constant τ, calculate the 10% to 90% rise time in terms of τ.

2.11 If two noninverting stages are cascaded and driven with an ideal step function, is the overall rise time of the output equal to the sum of the individual rise times? Explain.

SECTION 2.5 FILTERS

2.12 How many octaves above 240 Hz is 2860 Hz?

2.13 How many decades above 240 Hz is 2860 Hz?

2.14 Find the frequency that is 3 octaves above 420 Hz.

2.15 Find the frequency that is 3.6 octaves above 420 Hz.

☆ **2.16** If the output of a filter falls with frequency at a rate of -6 dB/octave, how many dB per decade does the output fall?

2.17 What is the necessary Q for a tuned filter with a resonant frequency of 700 kHz and a bandwidth of 14 kHz?

2.18 The Q of a circuit is 20 at a resonant frequency of 620 kHz. Calculate the 3-dB bandwidth of this circuit.

2.19 An AM radio station transmits a carrier frequency of 1160 kHz. If the output signal must be down by 3 dB at frequencies of 1155 kHz and 1165 kHz, find the required Q.

SECTION 2.7 REVIEW OF THÉVENIN'S THEOREM

2.20 Given the circuit shown:

(a) Find the "effective resistance" as seen by the capacitor terminals, R_{12}.

(b) Determine the Thévenin voltage as seen from node 1 to ground, $V_{Th-1,0}$.

(c) Determine the Thévenin voltage as seen from node 2 to ground, $V_{Th-2,0}$.

Figure P2.20

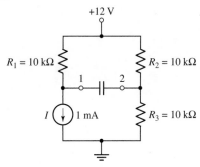

2.21 For the circuit shown, redraw the circuit in design schematic form. Find the current in R_3.

Figure P2.21

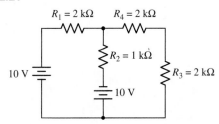

2.22 The circuit shows a signal generator with a 50-Ω generator resistance followed by an attenuator. Calculate the Thévenin equivalent voltage and resistance at terminals 1-2.

Figure P2.22

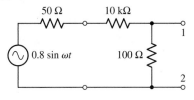

☆ **D 2.23** The signal generated by a source with 50-Ω impedance is to be attenuated by a factor of 10. The output impedance of the source followed by the attenuator should also be 50 Ω. Choose values for the resistive divider.

☆ **2.24** Given the circuit shown:

(a) Find the Thévenin equivalent circuit as seen by R_L.

(b) Using the information of part (a), calculate v_{out}.

(c) Explain why v_{out} is less than V_{Th}.

Figure P2.24

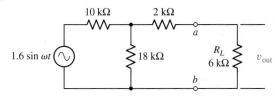

☆ **2.25** Calculate I_B for the circuit shown.

Figure P2.25

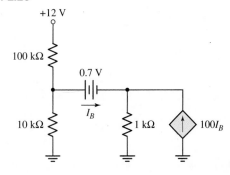

☆ **2.26** Assume that $\alpha = 0.99$ in the circuit shown.

(a) Calculate the voltage gain, $A = v_{out}/v_{in}$.

(b) Calculate the input impedance, R_{in}.

(c) Calculate the output impedance, R_{out}.

(d) A signal generator is now connected to the input of the circuit. This generator can be represented by a voltage source, v_g, in series with a 50-Ω generator impedance. Calculate the overall voltage gain of the circuit, $A_o = v_{out}/v_g$.

(e) In terms of R, r_e, and α, relate A_o to A.

Figure P2.26

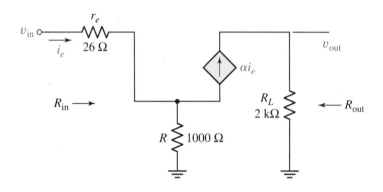

☆ **2.27** Given the circuit shown:

(a) Find the Thévenin equivalent voltage and resistance at terminals a-b.

(b) Using the equivalent circuit of part (a), evaluate the voltage gain, $A_v = v_{out}/v_{in}$.

Figure P2.27

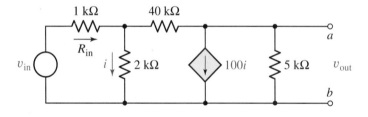

2.28 Find the input impedance to the circuit of Problem 2.27.

☆ **2.29** Given the circuit shown:

(a) Find the Thévenin equivalent circuit at terminals c-d.

(b) From the Thévenin equivalent circuit, evaluate the voltage gain, $A_v = v_{out}/v_{in}$.

Figure P2.29

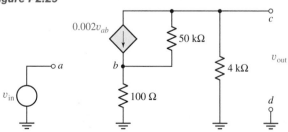

☆ **2.30** Given the circuit shown:

(a) Find the Thévenin equivalent circuit at terminals b-c.

(b) From the Thévenin equivalent circuit, evaluate the voltage gain, $A_v = v_{out}/v_{in}$.

Figure P2.30

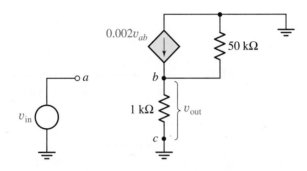

SECTION 2.8 THE MILLER EFFECT

2.31 Calculate the input impedance of the circuit shown if

(a) $k = 0.9$

(b) $k = 4$

(c) $k = -82$

Figure P2.31

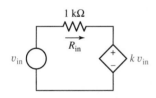

2.32 Assume an ideal amplifier for the circuit shown. Choose A_v to result in an input capacitance of 36 nF.

Figure P2.32

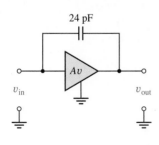

2.33 Find R_{in}, R_{out}, and overall voltage gain $A_o = v_{out}/v_{in}$ for the circuit shown.

Figure P2.33

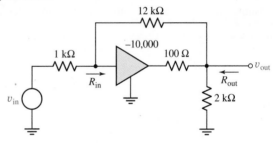

☆ **2.34** Given the circuit shown:

(a) Calculate the value of gain, v_{out}/v_{in}, at low frequencies such that the capacitor can be considered as an open circuit.

(b) Calculate the frequency at which the magnitude of v_{out}/v_{in} equals $1/\sqrt{2}$ times the value in part (a).

Figure P2.34

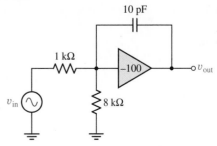

2.35 Calculate the time constant of

(a) the input loop of the circuit shown

(b) the output loop of the circuit

Figure P2.35

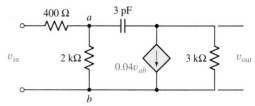

SECTION 2.9 TRANSIENT WAVEFORMS

2.36 If the circuit of Fig. 2.28 has values of $v_i = -4$ V, $v_t = 16$ V, $R = 10$ kΩ, and $C = 5000$ pF, calculate the time for the capacitor voltage to change from -4 V to $+4$ V.

2.37 Calculate the rise time for the circuit of Problem 2.36.

2.38 Calculate the time for the circuit of Problem 2.36 to move to the halfway point of the voltage transition.

2.39 Derive the relationship of Eq. (2.35).

☆ **2.40** If v_{in} switches abruptly from 0 V to 4 V in the circuit shown, sketch v_{out} until equilibrium is reached. Assume the circuit was in equilibrium prior to switching.

Figure P2.40

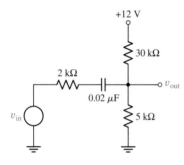

2.41 After the output voltage has become stable in the circuit shown, the input voltage switches from $+4$ V to -6 V.

(a) Calculate the time constant.

(b) Calculate the time taken for the output to reach -3 V.

Figure P2.41

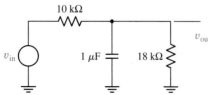

2.42 Repeat Problem 2.41 for the circuit shown.

Figure P2.42

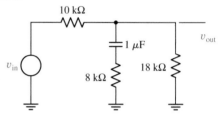

2.43 Plot the capacitor voltage for the circuit shown if the switch moves from terminal a to b at $t = 0$, then returns to terminal a at $t = 10$ μs.

Figure P2.43

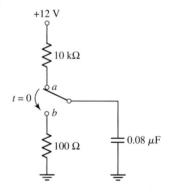

☆ **2.44** Verify Eq. (2.36) for the circuit shown by (a) finding the time constant, then (b) finding the frequency at which the magnitude of v_{out}/v_{in} equals

$$\frac{1}{\sqrt{2}} \times \frac{R_2}{R_1 + R_2}$$

Figure P2.44

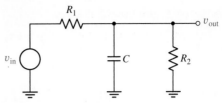

3 | Amplifier Models and Frequency Response

In Chapter 2 we saw that most electronic signals of interest must be amplified or made much larger in order to use them. The output voltages of telephone mouthpieces, microphones, or TV receiving antennas are in the μV or mV range, whereas tens, hundreds, or thousands of volts may be needed to drive telephone lines, speakers, or picture tubes.

A microphone, for example, converts the small pressure variation produced by the voice or by musical instruments into an output voltage that carries this audio information as a voltage. A good microphone may produce a peak voltage of 50 mV for normal voice when the lips of the speaker are placed within an inch of the microphone. This voltage is far too small to perform any useful function such as driving a speaker. The electronic signal or voltage must be amplified by a factor of 100 or more before it can activate a typical speaker. Thus, an amplifier is important to fully utilize the potential of the microphone.

Another example is a radio or television signal transmitted from the station through the atmosphere to our receiver. A receiving antenna may convert the transmitted electromagnetic wave into a signal that is measured in hundreds of microvolts. This signal is completely useless until it is amplified to the point that various circuits in a TV set (or receiver) can respond to the signal. Again, amplifiers are important in the communications field.

An amplifier circuit is used to increase the amplitude of the input signals to the desired level. In applications that call for significant power outputs to be delivered to the load, the amplifier may require two distinct sections: a *preamplifier* that amplifies voltages or currents efficiently and a *power amplifier* that amplifies power efficiently. This chapter will introduce the voltage amplifier.

A given amplifier is based on amplifying devices such as the metal-oxide-semiconductor field-effect transistor (MOSFET) or the bipolar-junction transistor (BJT). These devices can amplify a time-varying signal from a low level to a higher level. A dc power source supplies the energy to accomplish this amplification. Some amplifiers may use a single transistor while others may require several transistors. Each single device, with its associated resistors and capacitors or other components, is called a *stage* of the amplifier. In some cases, such as the differential pair, two devices make up a single stage of amplification.

The amplifiers in this chapter are discussed in a general way rather than considering the specific amplifying elements included in the amplifier. Later chapters will consider the individual stages used in these amplifiers. Here, we are interested in establishing the specifications of amplifiers and considering how these specifications interact with the circuits that either drive the amplifier or use the amplifier's output signal.

The radio receiver system of Fig. 3.1 demonstrates the importance of amplification in this important application. The first subsection of the receiver consists of a tuned

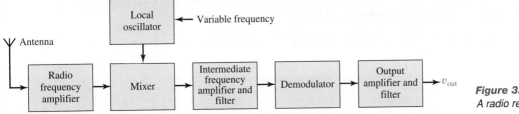

Figure 3.1
A radio receiver system.

amplifier. This special circuit amplifies only signals in a small frequency range, centered around the desired carrier or center frequency of the incoming channel. After this range of frequencies is translated to a new center frequency, the signal is passed through several other amplifying stages in the intermediate frequency amplifier. After the signal is operated on by the demodulator to obtain the transmitted information, this signal is amplified by an audio amplifier. The resulting voltage signal is then used to activate the speaker and convert the voltage signal back to audio pressure waves that can be heard by the ear.

DEMONSTRATION PROBLEM

At the beginning of several chapters, a problem will be introduced to demonstrate the principles on which the chapter will focus. At the end of the chapter, a full discussion of the problem will take place to emphasize the methods that are considered in the chapter.

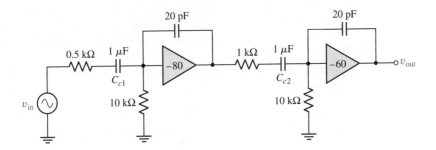

An amplifier.

The demonstration problem for this chapter is the amplifier indicated in the figure. It is desired to calculate the midband voltage gain of this amplifier along with the 3-dB bandwidth of the voltage gain. The bandwidth is equal to the difference of the upper and lower 3-dB frequencies of the voltage gain. An expression for the voltage gain as a function of frequency should also be derived.

The points that must be understood in order to make these calculations are:

1. There is a midband range of frequencies over which the capacitors have no effect on voltage gain.

2. The midband voltage gain calculation must account for impedance loading.

3. Coupling capacitors cause low-frequency falloff of voltage gain.

4. The Miller effect must be used to determine apparent values of capacitors that bridge input and output nodes of an amplifier.

5. Multiple low-frequency and high-frequency corner frequencies must be considered to calculate the actual amplifier bandwidth.

6. An expression for voltage gain as a function of frequency can be related to the upper and lower corner frequencies and the midband gain.

After developing the necessary background material, we will make these calculations at the end of the chapter.

3.1 The General Model of an Amplifying Element

IMPORTANT Concepts

1. Most amplifiers are designed to amplify small variational signals that vary in both directions about zero volts rather than a dc signal.
2. To amplify these variational signals, an amplifying element must be powered by one or more dc voltages. These voltages are called the power supply voltages.
3. The amplifying element must be driven to a proper operating point by applying a dc voltage to the input that creates an appropriate dc output voltage.
4. The variational input voltage is added to the dc input voltage. A larger variational voltage plus a dc voltage appears at the output. Often, the desired variational voltage is passed to a load through a capacitor, which blocks the dc voltage.

The two most important amplifying elements in the electronics field are the BJT and the MOSFET. The underlying principles governing operation of these devices will be discussed in later chapters. This section proposes a block diagram model for an amplifying element that will be invoked repeatedly in these later chapters.

Figure 3.2 shows the ideal general amplifier model. This ideal amplifying element would accept a small incremental input signal and produce a magnified version at the output. The ratio of output voltage to input voltage is called the voltage gain. In Fig. 3.2, this gain is

$$A_v = \frac{v_{\text{out}}}{v_{\text{in}}} = \frac{K v_{\text{in}}}{v_{\text{in}}} = K \tag{3.1}$$

The signal to be amplified is typically a small ac signal that varies incrementally in both directions from zero volts. Figure 3.3 indicates typical output waveforms of a microphone and an FM antenna tuned to a nearby radio station. Most signals to be amplified, including those shown in the figure, contain no dc component. Often, the amplified signal must also contain no dc component.

Unfortunately, the amplifying elements available for amplifiers do not behave ideally. The basic operation of a more practical amplifying stage is demonstrated in Fig. 3.4. Amplification occurs as the input voltage controls the output current. The amplifying element

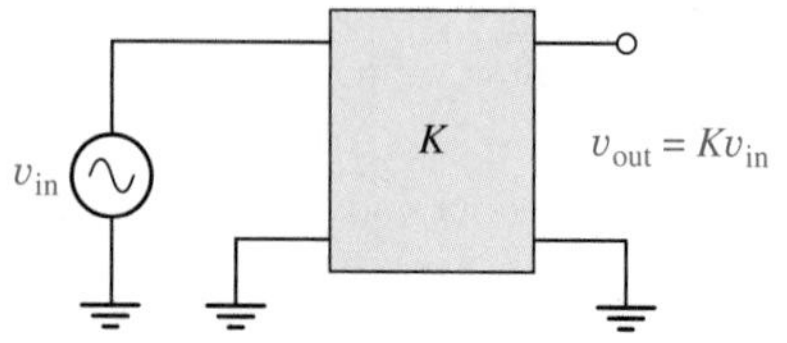

Figure 3.2
An ideal amplifying stage.

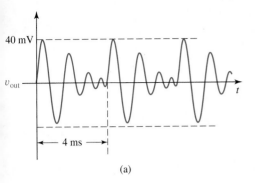

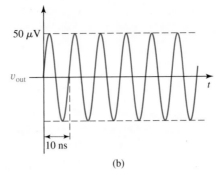

(a)

(b)

Figure 3.3
*Typical signals to be amplified:
(a) microphone output, (b) FM
antenna output.*

requires dc power to function properly. A dc bias current or voltage is added to v_{in} to move the operating point of the amplifying element into its active region. This dc bias is provided by the dc power supply along with components such as resistors.

One result of the bias is that the amplifying element can now produce an output signal that varies in both directions from the dc voltage that is present when no input signal is applied. This dc level is called the *no-signal* or *quiescent* voltage.

The input voltage to the amplifier in Fig. 3.4 consists of the signal to be amplified plus the bias voltage, $V_{IN} = V_{bias} + v_{in}$. The resulting output current is $I = I_Q + kv_{in}$. This output current is channeled through a load resistance to produce an output voltage of

$$V_{OUT} = I_Q R_L + kv_{in} R_L = V_Q + kv_{in} R_L = V_Q + v_{out} \tag{3.2}$$

Note that V_{OUT} is the total output voltage including both dc and ac components. The quantity v_{out} consists of only the ac component.

Since we are not interested in amplifying the dc component, the voltage gain for this situation is

$$A_v = \frac{v_{out}}{v_{in}} = \frac{kv_{in} R_L}{v_{in}} = k R_L \tag{3.3}$$

In many amplifying elements, the load resistor is connected between the output and the dc power supply voltage, V_{DC}, rather than ground. In this case, the current flows from the dc power supply to the output of the amplifier. The quiescent voltage is then

$$V_Q = V_{DC} - I_Q R_L \tag{3.4}$$

The key point is that the output current is controlled by the voltage to be amplified, v_{in}. This current develops an output voltage as it flows through R_L.

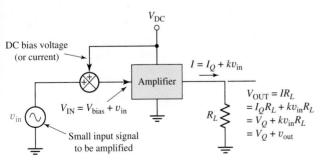

Figure 3.4
A practical amplifying stage.

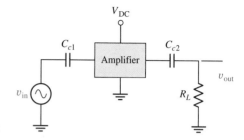

Figure 3.5
Using coupling capacitors in an amplifier.

The use of bias to establish an appropriate quiescent output voltage leads to some problems. For example, the input transducer may not operate properly if dc current passes through it. The same point is also true of the output load. Thus, the input terminal and output terminal of the amplifying element often use series coupling capacitors to block dc current. This arrangement is shown in Fig. 3.5. The amplifying element includes a resistance through which both dc current and output signal current flow. This signal voltage developed across this resistor is coupled through C_{c2} to R_L.

PRACTICAL Considerations

In discrete circuit design, large values of coupling capacitors in the range of 1 to 100 μF are often required. This need is generally satisfied by electrolytic capacitors that pack a large capacitance into a small volume.

These capacitors are *polarized*; that is, one terminal, marked with a plus sign, must always be positive in voltage with respect to the other terminal, marked with a minus sign. If this requirement is not observed, the capacitor will conduct, rather than block, dc current. When connected improperly, the capacitor becomes very leaky and can be modeled as a capacitor in parallel with a resistor.

To connect an electrolytic capacitor properly, the dc voltage levels of the two nodes to be bridged by the capacitor must be considered. In making this consideration we must recognize that an ac signal generator has a zero dc voltage output. If the capacitor connects from the signal generator to the input node of the amplifier, and this node is biased to a positive dc voltage, the plus terminal of the electrolytic capacitor must connect to this input node.

If a node *a* having a dc voltage level of +6 V must be connected through a capacitor to node *b* that is at +2 V, the positive terminal of the capacitor must connect to node *a*. In short, the capacitor terminal marked plus must connect to the most positive of the two nodes.

There are two major aspects in the use of an amplifying element for an amplifier stage: the dc analysis or design and the incremental analysis or design. As mentioned earlier, it is necessary to establish an output quiescent point that will allow the output signal to swing in both directions. For example, if the output voltage of an amplifying element is a linear function of the input voltage only for outputs between 2 V and 6 V, the quiescent output voltage might be chosen to be 4 V. When an input signal is applied, the output voltage can then swing in a positive direction from +4 V to +6 V and in the negative direction from +4 V to +2 V. Larger output swings than ±2 V will cause distortion of the output signal.

Once the dc design or analysis is completed, the incremental considerations can be made. These considerations determine performance parameters such as the voltage gain of the circuit, the input and output impedances, and the 3-dB bandwidth of the voltage gain. The input signal is often a small ac signal and can be considered as an incremental signal.

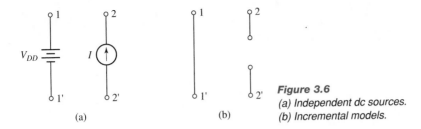

Figure 3.6
(a) Independent dc sources.
(b) Incremental models.

The output signal is much larger, but can also be considered an incremental signal as long as the output is a linear function of the input.

The most popular method of determining incremental parameters is through the use of an incremental equivalent circuit. Models or equivalent circuits will be discussed in more detail in Chapters 4 and 5. Some important points will be mentioned here in preparation for the general incremental models appearing in succeeding sections of this chapter.

3.1.1 INDEPENDENT VOLTAGE AND CURRENT SOURCES

Power is supplied to an electronic circuit by means of a dc voltage source; thus, this element is very important in electronics. The incremental model for an independent dc voltage source is a short circuit. For an independent dc current source, the incremental model is an open circuit. Figure 3.6 indicates these equivalencies.

We can see that a short-circuit model is an accurate representation of the voltage source by assuming that a small, incremental current is forced through the dc source. Since the voltage of the source is constant regardless of the current flowing through it, there is zero incremental voltage developed across it. The incremental resistance is then

$$r_v = \frac{\Delta v}{\Delta i} = \frac{0}{\Delta i} = 0 \,\Omega \tag{3.5}$$

The same reasoning can be used for the dc current source. An incremental voltage change across the source results in no incremental current change, giving

$$r_i = \frac{\Delta v}{\Delta i} = \frac{\Delta v}{0} = \infty \tag{3.6}$$

Because the use of an incremental circuit is the major method applied to manual circuit analysis and design, the transition from an actual to an equivalent circuit will now be discussed. A single-stage amplifier appears in Fig. 3.7. The midband, incremental circuit is found by assuming that the capacitor C is a short circuit as well as the dc voltage source,

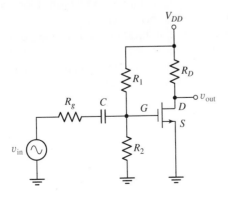

Figure 3.7
A single-stage amplifier.

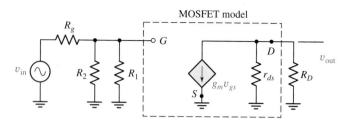

Figure 3.8
The midband equivalent circuit for the amplifier of Fig. 3.7.

V_{DD}. The MOSFET is replaced by its incremental circuit, to be discussed in a later chapter. The resulting circuit is indicated in Fig. 3.8.

We note that all sources in Fig. 3.8, dependent or independent, are incremental sources. No dc sources appear in this equivalent circuit or in any incremental equivalent circuit. This circuit is used to calculate voltage gain, input or output impedance, and other important parameters.

In the remainder of this chapter we will consider only the incremental amplifier model that allows small-signal ac analysis. Later chapters will treat the dc design of amplifying elements.

3.2 Gain Elements

IMPORTANT Concepts

1. An ideal voltage amplifier would have infinite input impedance and zero output impedance. The practical amplifier has a finite value for both input and output impedances. These imperfections always cause the *in-circuit* voltage gain to be less than that of the amplifier by itself.

2. Coupling capacitors are often used to isolate the dc voltages of the amplifier from the input transducer and the output load. These elements introduce low-frequency falloff in overall gain.

3. Parasitic capacitances associated with the amplifying elements cause high-frequency falloff of gain.

The key element in producing amplification is an electronic device that will produce a voltage gain or current gain that is greater than unity. Three of the most well-used amplifying elements over the last 90 years have been the vacuum tube, used from about 1908 to about 1970, the bipolar transistor or BJT, used from about 1950 to the present, and the metal-oxide-semiconductor field-effect transistor or MOSFET, used from the late 1960s to the present. Each of these devices was improved upon throughout their histories, and the BJT and MOSFET continue to reach higher levels of performance.

A single device may not offer enough amplification to meet the needs of a given application. In such a case, several devices or stages may be required. While some amplifiers consist of only one stage, many others are *multistage* amplifiers. In this chapter we will concentrate on single-stage amplifiers with only a short consideration of multistage amplifiers in the last section of the chapter.

3.2.1 AN IDEAL AMPLIFYING ELEMENT

An ideal electronic amplifying element would consist of two sets of terminals. One set, the input terminals, would accept the input signal while the second set, the output terminals, would provide the output signal. A voltage or current would be applied to the input terminals

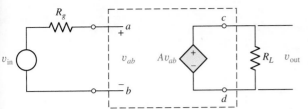

Figure 3.9
An ideal voltage amplifying element.

and a larger voltage or current would result at the output terminals. The magnitude of the output signal would be linearly proportional to the magnitude of the input signal. The model for an ideal voltage amplifying element is shown in Fig. 3.9.

The output voltage is the product of an amplification constant, A, and the input voltage, that is,

$$\frac{v_{out}}{v_{in}} = \frac{v_{cd}}{v_{ab}} = A$$

The ideal element would have an infinite input resistance as indicated by the open circuit between terminals a and b and zero output resistance as indicated by the absence of a resistor in the output terminals. With these extreme values of resistances, the overall voltage gain between a signal source with a generator resistance and a load of finite resistance would equal A. No attenuation of the voltage gain (*loading*) would take place for these ideal resistances.

In addition to the ideal input and output resistances, there would be no parasitic capacitances across the input or output terminals. The resulting voltage gain would be unaffected by frequency. The ideal amplifying element could then result in an amplifying stage with infinite bandwidth.

3.2.2 THE PRACTICAL AMPLIFYING ELEMENT

Actual amplifying elements differ in various ways from the ideal element. In general, there will be a finite input resistance, a nonzero output resistance, and parasitic capacitances across both sets of terminals. In addition, as previously mentioned, amplifying elements require dc voltages and currents even though a pure ac signal is to be amplified. The source and load may not operate properly if dc current flows through these elements; thus the amplifying element may need to be surrounded by coupling capacitors. These components can keep the dc voltages applied to the amplifying element from reaching the source or load while passing the ac signal to be amplified. A model for a nonideal amplifying element for ac signals may appear as in Fig. 3.10.

The capacitors C_{c1} and C_{c2} are generally fairly large, perhaps in the μF range. The capacitors C_{in} and C_{out} are parasitic capacitors with values that are very small, perhaps in the range of fractions of a pF to several pF. The input resistance depends on the type of amplifying element used. For a BJT, it may be in the kΩ range whereas a MOSFET device may have an input impedance in the 100-MΩ range. The output resistance may be in the Ω to 100-kΩ range for either the BJT or the MOSFET.

The model for an amplifying stage is very useful in circuit design, but can often be simplified depending on the particular application. The coupling capacitors typically offer

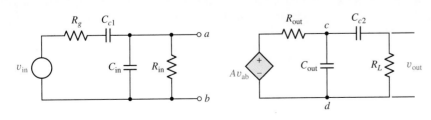

Figure 3.10
A model for a nonideal amplifying stage.

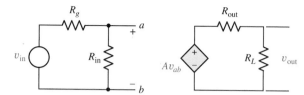

Figure 3.11
The midband model.

little impedance to moderate- or high-frequency signals and can be approximated by short circuits for such frequencies. The parasitic capacitors exhibit very large impedances for moderate- or low-frequency signals and can be approximated as open circuits for these signals.

3.2.3 THE MIDBAND MODEL

Over that range of frequencies for which the coupling capacitors can be considered as short circuits and the parasitic capacitances can be considered as open circuits, the midband model can be used for the amplifier stage. This model is shown in Fig. 3.11.

The midband model was discussed in Chapter 2. The overall voltage gain is given by Eq. (2.3) and is

$$A_{MB} = \frac{v_{out}}{v_{in}} = \frac{R_{in}}{R_{in} + R_g} A \frac{R_L}{R_{out} + R_L} \tag{2.3}$$

This gain is called the *midband voltage gain* and is the most significant parameter of the amplifier stage. The overall midband voltage gain differs from the amplifier voltage gain because of loading. The first factor in Eq. (2.3) accounts for resistive loading or attenuation in the input loop. The third factor accounts for resistive loading in the output loop.

In general, the signal that is applied to an amplifier consists of several frequency components. For example, a high-fidelity music amplifier might amplify all signals between 50 Hz and 18 kHz by the midband gain. Frequencies below 50 Hz or above 18 kHz will be amplified by a smaller factor than the midband voltage gain.

Although a broad spectrum of frequencies may be applied simultaneously to an amplifier, it is customary to characterize the frequency performance in terms of the application of a single frequency signal. A frequency response is created by measuring the gain of the amplifier at a number of individual frequencies. These frequencies are chosen to exceed the spectrum of signals for which the midband gain is valid. The frequency response for the amplifier mentioned in the preceding paragraph might use frequencies ranging from 5 Hz up to 100 kHz to measure the response.

PRACTICE **Problems**

3.1 A microphone generates a 40-mV peak signal for typical inputs. The microphone impedance is 50 kΩ. An amplifier with $R_{in} = 25$ kΩ, $R_{out} = 100$ Ω, and $A = 180$ V/V is used to couple the microphone to a 50-Ω load. Calculate the resulting peak voltage across the load for typical inputs. *Ans:* 800 mV.

3.2 If R_{in} is changed to 250 kΩ and R_{out} to 25 Ω in Problem 3.1, calculate the resulting peak voltage across the load for typical inputs. *Ans:* 4 V.

3.3 Calculate the midband voltage gain of the amplifier represented in Fig. 3.11 if $R_g = 2$ kΩ, $R_{in} = 10$ kΩ, $A = 200$, $R_{out} = 1$ kΩ, and $R_L = 4$ kΩ. *Ans:* 133 V/V.

3.4 For the amplifier of Problem 3.3, what is the minimum value of R_L to result in an overall gain of 92? *Ans:* 1.23 kΩ.

PRACTICAL **Considerations**

From Eq. (2.3) it can be seen that a large value of R_{in} relative to R_g will increase the overall gain of the amplifier. Likewise, a small value of output resistance R_{out} relative to R_L increases the overall gain of the amplifier. In earlier classes, the student is told that in order to maximize power transfer, the load resistance and output resistance should be matched. This statement would also imply that the generator resistance, R_g, and the input resistance should be matched. These results do not apply to the voltage amplifier. In most cases, we are attempting to design an amplifier to result in the best *voltage transfer* from generator to load rather than the best power transfer.

We typically do not have control over R_g and R_L; thus, we control the input and output impedances of the amplifier. For good voltage transfer, we attempt to make R_{in} much larger than R_g and R_{out} much less than R_L.
    The two conditions for efficient voltage transfer are

$$R_{in} \gg R_g$$

and

$$R_{out} \ll R_L$$

PRACTICE Problems

3.5 Explain why coupling capacitors can be considered short circuits at midband frequencies. Demonstrate by calculating the magnitude of the impedance of a 1-μF capacitor at $f = 2\,$kHz. Compare this magnitude to an input impedance of 20 kΩ. *Ans:* 80 Ω, 0.4% of R_{in}.
3.6 Calculate the magnitude of the impedance of a 1-μF capacitor at $f = 30\,$Hz. If $R_{in} = 10\,$kΩ, can this frequency be considered a midband frequency? *Ans:* 5.3 kΩ; no.

3.2.4 THE LOW-FREQUENCY MODEL

As the frequency of the input signal becomes lower, the impedances of the capacitors become larger. The parasitic capacitors are in parallel with resistors and can then be approximated as open circuits. On the other hand, the coupling capacitors are in series with resistors. As the capacitive impedances increase, the voltage drops across these elements increase and less voltage reaches the amplifier input and the load. This leads to low-frequency falloff of the voltage gain. Figure 3.12 shows the low-frequency model that pertains to this situation.

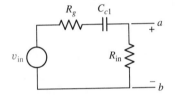
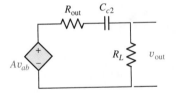

Figure 3.12
The low-frequency model.

3.2.5 THE HIGH-FREQUENCY MODEL

At higher frequencies, the impedances of the coupling capacitors become small and can be approximated as short circuits. The impedances of the parasitic capacitors also decrease. Since these capacitors are in a parallel configuration, as their impedances decrease, more current flows into these components. Less current now flows through R_{in} and R_L compared to the midband situation, and the voltage gain is decreased over the midband value. Figure 3.13 shows the high-frequency model.
    Since the voltage gain at low frequencies and at high frequencies becomes frequency dependent, it is useful to review the types of frequency dependence that is often encountered in amplifiers. We will then return to the specific topic of amplifier frequency response.

PRACTICE Problems

3.7 Show that C_{in} of Fig. 3.13 can be neglected at a frequency of 10 kHz with little error if $C_{in} = 100\,$pF and $R_{in} = 3\,$kΩ. *Ans:* $X_c = 159\,$kΩ, which is much larger than R_{in}.
3.8 Calculate the frequency at which $X_c = R_{in}$ from Problem 1. *Ans:* $f = 531\,$kHz.

Figure 3.13
The high-frequency amplifier model.

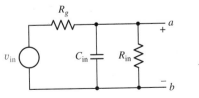
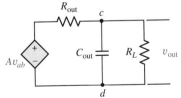

3.3 Frequency-Dependent Factors

IMPORTANT Concepts

1. The decibel or dB is significant in frequency response plots.
2. Gain can be expressed in terms of certain factors that occur often in gain expressions.
3. The dB allows products or quotients of factors to be handled as sums or differences, which allows frequency response plots to be constructed rather simply.

We have expressed the midband gain in terms of a number in the previous chapter. This number is the ratio of output voltage to the source voltage. It is convenient in frequency response work to express this number in terms of decibels. We defined dB in Chapter 2 by

$$A_{dB} = 20 \log_{10} |A_o| \tag{2.4}$$

The relationship between A_{dB} and $|A_o|$ is demonstrated in Table 3.1. This table can be used to express other values of gain magnitude in decibels. We note that

$$20 \log F_1 F_2 = 20 \log F_1 + 20 \log F_2$$

A gain magnitude of 20 can be factored into 2×10, which allows us to write

$$A_{dB}(20) = A_{dB}(2) + A_{dB}(10) = 6\,dB + 20\,dB = 26\,dB$$

A gain magnitude of 16 results in a value of 24 dB since

$$20 \log 16 = 20 \log 2^4 = 4 \times 20 \log 2 = 24\,dB$$

The gain of an actual circuit is often expressed in terms of several frequency-dependent factors. The gain in dB can easily be found by finding the dB value of each individual factor and summing the results. A typical gain expression is

$$A(f) = \frac{-25\left(j\frac{f}{10}\right)}{\left(1 + j\frac{f}{10}\right)\left(1 + j\frac{f}{10^4}\right)} \tag{3.7}$$

The factors involving f are called *zero factors* or *pole factors*. Zero factors appear in the numerator of the gain expression, and pole factors appear in the denominator. A *zero frequency* is the value of frequency that causes a zero factor to equal zero. A *pole frequency* is the value of frequency that causes a pole factor to equal zero. In the preceding expression for $A(f)$, a zero occurs at $f_z = 0$ and pole frequencies occur at $f_{p1} = j10$ and $f_{p2} = j10^4$. If the input frequency equals a zero value, the gain becomes equal to zero. When the input frequency approaches a pole value, the gain approaches infinity since the denominator approaches zero. These zero and pole frequencies are referred to as *critical frequencies*, and their significance will be discussed in more detail in a later chapter. In this chapter, we will deal with real frequencies rather than imaginary values.

The gain in dB is

$$A_{dB} = 20 \log \left| \frac{-25\left(\frac{jf}{10}\right)}{\left(1 + j\frac{f}{10}\right)\left(1 + j\frac{f}{10^4}\right)} \right|$$

which can also be expressed as

$$A_{dB} = 20 \log |-25| + 20 \log \left| j\frac{f}{10} \right| - 20 \log \left| 1 + j\frac{f}{10} \right| - 20 \log \left| 1 + j\frac{f}{10^4} \right|$$

Table 3.1 Comparison of Gain Magnitude to dB

$|A_o|$	A_{dB}, dB
1.0	0
1.414 ($\sqrt{2}$)	3
0.707 ($1/\sqrt{2}$)	−3
2.0	6
0.5 (1/2)	−6
10.0	20
0.1 (1/10)	−20
100.0	40
0.01 (1/100)	−40
1000.0	60
0.001 (1/1000)	−60

In the dB expression for gain, the multiplicative factors become additive factors. Since we can break the overall expression down into factors that can be added or subtracted, we can then consider a plot of each individual factor. The overall plot is simply the sum of all individual positive factors minus all negative factors. Because of the large range of frequencies involved in a wideband amplifier, a logarithmic frequency scale is normally used in frequency response work.

We note that the gain expression above could be manipulated by dividing the numerator and the denominator by the factor $jf/10$. This step results in a gain of

$$A(f) = \frac{-25}{\left(1 - j\frac{10}{f}\right)\left(1 + j\frac{f}{10^4}\right)} \tag{3.8}$$

Thus, the gain in dB could also be written as

$$A_{\text{dB}} = 20\log|-25| - 20\log\left|1 - j\frac{10}{f}\right| - 20\log\left|1 + j\frac{f}{10^4}\right|$$

This alternate expression for gain relates more closely to the physical expression to be derived in succeeding paragraphs. We will discuss factors involved in both forms simply to give a more thorough treatment of frequency-dependent terms.

PRACTICAL Considerations

Although introduced in Chapter 2, it is useful here to consider octaves or decades of frequency. A frequency f_2 that is twice the frequency f_1 is said to be an octave above f_1. A frequency f_3 that is one-half the frequency f_1 is said to be one octave below f_1. The musical scale deals in these same octaves. Middle C on a piano corresponds to a fundamental frequency of 264 Hz, and the next higher C is one octave above middle C or 528 Hz. The C note that is one octave lower has a fundamental frequency of 132 Hz.

We normally associate the prefix *oct* with the number eight. An octagon is an eight-sided figure, and octal numbers refer to numbers expressed with a base of eight. While an octave refers to a doubling or halving of frequency, it was so named to indicate that eight full musical notes are included in one octave.

Another popular term used to express frequency variations is the decade, which refers to an increase by a factor of 10. For example, a frequency of 100 Hz is a decade above a frequency of 10 Hz. A slope of 6 dB/octave is equivalent to a slope of 20 dB/decade.

It is possible to calculate the number of octaves or decades between two frequencies from the equations

$$\text{Number of octaves} = \log_2 \frac{f_2}{f_1} = 3.32 \log_{10} \frac{f_2}{f_1} \tag{3.9}$$

$$\text{Number of decades} = \log_{10} \frac{f_2}{f_1} \tag{3.10}$$

EXAMPLE 3.1

If a function has a magnitude of 230 at a frequency of 1.2 kHz and falls at a rate of -12 dB/octave as frequency increases, calculate the magnitude of the function at $f = 10$ kHz.

SOLUTION The number of octaves between 1.2 kHz and 10 kHz is found from Eq. (3.9). This number is

$$\text{Number of octaves} = \log_2 \frac{10}{1.2} = 3.059$$

Over this interval, the magnitude of the function will fall by

$$3.059 \text{ octaves} \times 12 \text{ dB/octave} = 36.7 \text{ dB}$$

We can now write

$$-36.7 \text{ dB} = 20 \log_{10} \frac{Mag_2}{230}$$

Solving for Mag_2 gives

$$Mag_2 = 3.36$$

A second approach to the solution of this problem is to work directly with the magnitude rather than with dB values. A -6 dB/oct falloff implies that the magnitude varies inversely with frequency. A falloff of -12 dB/oct suggests that the magnitude varies inversely with the square of the frequency. With this in mind, we can write

$$Mag_2 = Mag_1 \left(\frac{f_1}{f_2}\right)^2 = 230 \left(\frac{1.2}{10}\right)^2 = 3.31$$

With this background we can now discuss the gain expression given previously.

3.3.1 THE CONSTANT FACTOR

A constant term results in a constant number of dB at all frequencies. The -25 factor is equivalent to

$$A_{dB}(-25) = 20 \log |-25| = 28 \text{ dB}$$

Because the gain in dB deals with a magnitude, the sign of the factor does not affect the result.

3.3.2 DIRECT VARIATION WITH FREQUENCY

A term such as jf/f_a, where f_a is a constant, results in a plot that has a 6 dB/octave slope. The expression in dB is

$$A_{dB}(jf/f_a) = 20 \log \frac{f}{f_a}$$

This term will result in 0 dB gain when $f = f_a$. Each time the frequency is doubled, the value of A_{dB} increases by 6 dB. This curve increases at the rate of 6 dB/octave or 20 dB/decade. Figure 3.14 shows a plot in dB for the factor $jf/10$. For a frequency term that is of the

Figure 3.14
Magnitude response of jf/10.

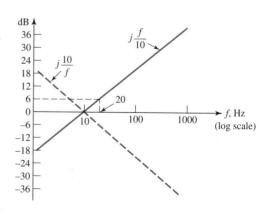

form $j10/f$, the gain gets smaller as frequency increases. The slope is -6 dB/octave or -20 dB/decade, as shown by the dashed line in Fig. 3.14.

3.3.3 THE $\left(1 + j\frac{f}{f_2}\right)$ FACTOR

This factor approaches a constant value when the frequency is much smaller than the constant f_2. The imaginary component is negligible for this case. For frequencies greatly exceeding f_2, the factor varies as jf/f_2, since the real component of unity is now negligible. If this factor appears in the numerator of a gain expression, it has a zero slope at low frequencies and approaches a 6 dB/octave or 20 dB/decade slope at high frequencies. If the factor appears in the denominator, the slope at high frequencies approaches -6 dB/octave.

The actual plot of magnitude as a function of frequency can be made by applying the equation for magnitude of a complex number. This magnitude is found by taking the square root of the sum of the squares of the real part and the imaginary part. For this factor the gain in dB is

$$A_{dB}(1 + jf/f_2) = 20 \log_{10}\left(1 + (f/f_2)^2\right)^{1/2}$$

A key point in frequency occurs when the imaginary component equals the real component. At this frequency, the magnitude of the factor is

$$|1 + j1| = \sqrt{2}$$

The gain in dB is then

$$A_{dB} = 20 \log 1.414 = 3 \text{ dB}$$

Since the gain over the constant portion of the curve is unity or 0 dB, this key frequency is the 3-dB point of the factor. For factors having a real and an imaginary component, the frequency that makes both components equal is a 3-dB point.

This equation for gain is used to generate the plot of Fig. 3.15. We have discussed this factor as a zero factor, occurring in the numerator of a gain expression. If the same factor appears in the denominator, it is a pole factor and equals the reciprocal of the zero factor. We note that if the reciprocal of the $(1 + jf/f_2)$ factor is plotted, it is simply the negative of the plot for $(1 + jf/f_2)$ since

$$\log \frac{1}{x} = -\log x$$

An actual plot of this factor from the equation is somewhat time consuming, and a quicker method of approximation has been developed. As previously mentioned, the curve is 3 dB above the low-frequency value when $f = f_2$. The frequency f_2 is called the *corner*

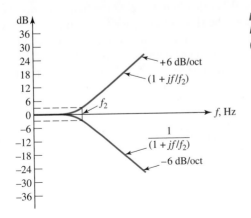

dB

36
30
24 ← +6 dB/oct
18 — $(1 + jf/f_2)$
12
6 — f_2
0
→ f, Hz
-6
-12 — $\dfrac{1}{(1 + jf/f_2)}$
-18
-24 — -6 dB/oct
-30
-36

Figure 3.15
Magnitude response of
$(1 + jf/f_2)$.

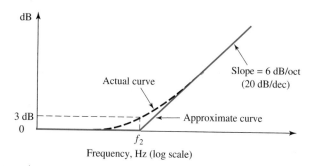

Figure 3.16
Approximate frequency response for $(1 + jf/f_2)$.

frequency or 3-dB point. The approximation to the actual curve consists of two straight-line segments. Below the corner frequency a segment with zero slope is used. Above the corner frequency a line segment with a slope of 6 dB/octave is drawn. Once the corner frequency is known, the two segments can be quickly sketched. The maximum error between the actual and approximate curves is 3 dB, and this error occurs at the corner frequency. An actual and approximate curve is shown in Fig. 3.16.

The approximate curve can be used to sketch the frequency response with corrections made at the corner frequency if more accuracy is needed. The approximate curve is called the asymptotic response since the actual curve approaches the approximate curve asymptotically as the frequency approaches very small or very large values. The approximate curve is also called a Bode diagram in honor of the pioneering work done by Hendrik W. Bode at Bell Labs in the 1930s.

As noted earlier, when this factor appears in the numerator of an expression, it is called a *zero* factor. When it appears in the denominator of an expression, it is called a *pole* factor. A zero factor asymptotically approaches a slope of 6 db/octave at high frequencies considerably above the corner frequency, and a pole factor approaches a slope of −6 dB/octave at high frequencies.

The factor $(1 + jf/f_1)$ varies exactly the same way as the factor previously discussed, except that the corner frequency is f_1. Some factors show a negative sign for the imaginary term. The negative sign in this factor would not affect the magnitude, although it would affect the phase shift versus frequency. Since the magnitude of a complex number is formed by taking the square root of the sum of the squares of the real and imaginary parts of the number, the sign of either real or imaginary component has no effect on the resulting magnitude.

3.3.4 THE $\left(1 - j\frac{f_1}{f}\right)$ FACTOR

This factor approaches a constant value of unity or 0 dB at frequencies considerably above f_1. When $f = f_1$, the real and imaginary components both equal unity and the magnitude of the factor is $\sqrt{2}$ or 3 dB. This is the corner frequency. As the frequency decreases below this value, the magnitude increases. The slope of the asymptotic dB plot is −6 dB/octave or −20 dB/decade up to the corner frequency. In the gain expression of Eq. (3.8), this term appears in the denominator. We will then be interested in the reciprocal of this factor. Figure 3.17 shows the asymptotic frequency response for the reciprocal of this term. Again,

Figure 3.17
Approximate frequency response for $(1 - jf_1/f)$.

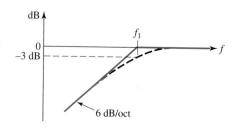

a more accurate curve could be generated using the complex magnitude equation, but the simple sketch is often accurate enough to be useful in many applications. The accurate expression for gain of this factor in dB is

$$A_{dB}\left(\frac{1}{(1-jf_1/f)}\right) = -20\log\left(1 + f_1^2/f^2\right)^{1/2}$$

We might also note here that when this factor appears in the denominator of an expression, it leads to low-frequency falloff of the expression. A term such as

$$\frac{K}{\left(1 - j\frac{f_1}{f}\right)}$$

can be manipulated to lead to

$$\frac{jfK/f_1}{1 + j\frac{f}{f_1}}$$

Both of these factors have been discussed before and, taken together, also cause falloff of an expression as frequency is lowered below f_1.

3.3.5 THE COMPOSITE CURVE

Figure 3.18 plots the complete curve for a gain expressed by several of the factors previously discussed:

$$A(f) = \frac{-25\left(j\frac{f}{10}\right)}{\left(1 + j\frac{f}{10}\right)\left(1 + j\frac{f}{10^4}\right)}$$

The asymptotic response of each of these factors is plotted individually; then all plots are summed, taking numerator factors as positive and denominator factors as negative to obtain the overall response. The lower corner frequency of the curve is 10 Hz and the upper corner frequency is 10,000 Hz. These points are the frequencies at which the gain has dropped 3 dB below the midband gain. The value of midband gain is 28 dB and the bandwidth is

$$\Delta f = 10,000 - 10 = 9990\,\text{Hz}$$

The denominator factor $(1 + j\frac{f}{10^4})$ causes the high-frequency 3-dB point to be at $f = 10^4$ Hz and leads to the asymptotic falloff of 6 dB/octave above this frequency. The two terms $jf/10$ in the numerator and $(1 + j\frac{f}{10})$ in the denominator lead to a lower corner frequency of 10 Hz and an asymptotic slope of 6 dB/octave from very low frequencies up

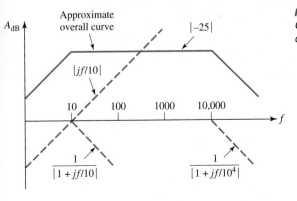

A_{dB}

Approximate overall curve

$|{-25}|$

$|jf/10|$

10 100 1000 10,000

f

$\dfrac{1}{|1 + jf/10|}$

$\dfrac{1}{|1 + jf/10^4|}$

Figure 3.18
Composite frequency response of several factors.

to this point. Again, these two terms are equivalent to

$$\frac{1}{\left(1 - j\frac{10}{f}\right)}$$

The following section will demonstrate how to derive the equation for amplifier gain from the circuit configuration and parameters. Once the equation is obtained and manipulated to the proper form, the methods of the preceding paragraphs can be applied.

EXAMPLE 3.2

An amplifier has a frequency response that can be expressed as

$$A(\omega) = \frac{-200(2 + j\omega/2 \times 10^7)}{(1 - j35/\omega)(3 + j\omega/1.5 \times 10^6)}$$

(a) What is the value of midband gain?

(b) Express the midband gain in dB.

(c) Plot the approximate magnitude response in dB as a function of frequency in Hz (Bode plot). Use a log scale for frequency.

SOLUTION The expression for gain is first factored to allow each frequency factor to have a real part that equals unity. This results in

$$A(\omega) = \frac{-200 \times 2(1 + j\omega/4 \times 10^7)}{(1 - j35/\omega) \times 3(1 + j\omega/4.5 \times 10^6)} = \frac{-133.3(1 + j\omega/4 \times 10^7)}{(1 - j35/\omega)(1 + j\omega/4.5 \times 10^6)}$$

(a) The midband gain is now easily identified as $A_{MB} = -133.3$.

(b) In dB, the midband gain is

$$A_{dB} = 20 \log_{10} |A_{MB}| = 20 \log_{10} 133.3 = 42.5 \, \text{dB}$$

(c) Since the frequency is to be expressed in Hz rather than in rad/s, the variable ω is converted to f in the expression for gain. This leads to

$$A(f) = \frac{-133.3(1 + jf/6.37 \times 10^6)}{(1 - j5.57/f)(1 + jf/7.16 \times 10^5)}$$

The resulting response is shown in Fig. 3.19.

The high-frequency gain is found by allowing f to approach infinity in the expression for gain. This gives

$$A_{high} = \lim A(f)|_{f \to \infty} = \frac{-133.3 \times 7.16 \times 10^5}{6.37 \times 10^6} = -15$$

This value converts to 23.5 dB.

Figure 3.19
Frequency response of gain in Example 3.2.

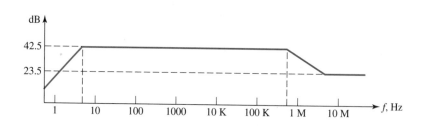

PRACTICAL Considerations

A slope of 6 dB/oct implies a direct variation of gain with frequency. In a region of +6 dB/oct slope, an increase of frequency by a factor of 10 results in an increase of gain by this same factor of 10.

A slope of −6 dB/oct implies an inverse variation of gain with frequency. In a region of −6 dB/oct slope, an increase in frequency by a factor of 10 results in a decrease in gain by this same factor of 10.

3.4 Low-Frequency Response of an Amplifier

IMPORTANT Concepts

1. The lower 3-dB frequency of an amplifier depends on the values of the source and input resistances and the value of the coupling capacitor. This frequency point can be set by the designer with the proper selection of coupling capacitor.
2. Multiple coupling capacitors create multiple critical frequencies. The overall lower 3-dB frequency can be calculated by an iterative method for this situation.

The frequency-dependent expression for gain could be derived from the model of Fig. 3.10, which includes both coupling and parasitic capacitances. This type of model might be appropriate for computer analysis of the circuit, but becomes somewhat cumbersome for manual analysis. It is more convenient for manual analysis to analyze the midband model of Fig. 3.11 to find the midband gain, analyze the low-frequency model of Fig. 3.12 to find the lower corner frequency, and then analyze the high-frequency model of Fig. 3.13 to find the upper corner frequency.

In this section, we want to find the lower corner frequency; thus, the model of Fig. 3.12 pertains. Let us first consider the effect of the coupling capacitor nearest the source while ignoring the effect of the capacitor nearest the load. In fact, we will temporarily assume that the load capacitor is removed. Figure 3.20 shows this configuration.

The overall voltage gain from the source to the amplifier output can be written as

$$A_o = \frac{v_{\text{out}}}{v_{\text{in}}} = \frac{v_{cd}}{v_{\text{in}}} = \frac{v_{ab}}{v_{\text{in}}} \times \frac{v_{cd}}{v_{ab}} = \frac{v_{ab}}{v_{\text{in}}} \times A$$

The value of v_{ab}/v_{in} is found from simple voltage division to be

$$\frac{v_{ab}}{v_{\text{in}}} = \frac{R_{\text{in}}}{R_{\text{in}} + R_g + 1/j\omega C_{c1}}$$

PRACTICE Problems

3.10 Plot the expression in dB for

$$A = \frac{260}{\left(1 - j\frac{100}{f}\right)\left(1 + j\frac{f}{10^6}\right)}$$

Use a log scale for frequency.

3.11 Repeat Problem 3.10 if the factor

$$\left(1 + j\frac{f}{2 \times 10^6}\right)$$

is added to the denominator.

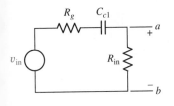
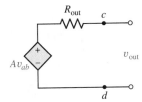

Figure 3.20
A circuit with a single coupling capacitor.

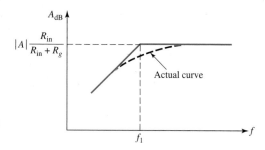

Figure 3.21
Low-frequency response of the stage of Fig. 3.20.

The gain from source to amplifier output can then be written as

$$A_o = \frac{R_{in}}{R_{in} + R_g + 1/j\omega C_{c1}} A \tag{3.11}$$

As ω becomes smaller, the denominator of the above equation becomes larger. Physically this is due to the fact that the impedance of the capacitor increases, dropping more of the applied voltage across C_{c1}. The overall gain decreases at lower values of ω, approaching a value of zero when ω approaches zero. As ω increases, the gain increases to a point. At higher values of frequency, the term $1/j\omega C_{c1}$ becomes very small compared to $R_{in} + R_g$, and the gain approaches the midband value of

$$A_{MB} = \frac{R_{in}}{R_{in} + R_g} A \tag{3.12}$$

for all higher values of ω.

For lower frequencies, Eq. (3.11) can be written as

$$A_o = \frac{R_{in}}{R_{in} + R_g} \times \frac{1}{1 - j\frac{f_1}{f}} \times A \tag{3.13}$$

where

$$f_1 = \frac{1}{2\pi C_{c1}(R_{in} + R_g)} \tag{3.14}$$

The frequency f_1 is the lower 3-dB point or lower corner frequency of the amplifier. A plot of this low-frequency response is shown in Fig. 3.21.

EXAMPLE 3.3

Select R_1 in Fig. 3.22 to result in a lower 3-dB frequency of 50 Hz.

SOLUTION In the circuit of Fig. 3.22, there is a single coupling capacitor that would cause a lower 3-dB frequency of

$$f_1 = \frac{1}{2\pi C_c(R_1 + 1000)} = 50 \, \text{Hz}$$

Since C_c is given as 1 μF, this equation can be solved for R_1 to give $R_1 = 2183 \, \Omega$.

Figure 3.22
Amplifier for Example 3.3.

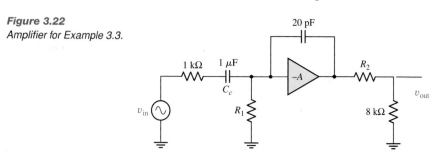

In a more formal manner, we could note that the input loop results in an attenuation factor of

$$\frac{R_1}{R_1 + 1000 + \frac{1}{j\omega C_c}} = \frac{R_1}{R_1 + 1000} \frac{1}{\left(1 - j\frac{1}{\omega C_c(R_1 + 1000)}\right)}$$

As we have seen earlier, this leads to the 3-dB frequency calculated above.

Before proceeding to the high-frequency response of an amplifier we will return to the situation depicted in Fig. 3.12, wherein two coupling capacitors are used. Each coupling capacitor will cause a lower corner frequency. The expression for overall gain can be expressed as

$$A_o = \frac{R_{\text{in}}}{R_{\text{in}} + R_g} A \frac{R_L}{R_L + R_{\text{out}}} \frac{1}{1 - jf_1/f} \frac{1}{1 - jf_1'/f} \tag{3.15}$$

where

$$f_1 = \frac{1}{2\pi C_{c1}(R_{\text{in}} + R_g)}$$

and

$$f_1' = \frac{1}{2\pi C_{c2}(R_L + R_{\text{out}})}$$

Equation (3.15) can also be written as

$$A_o = A_{MB} \frac{1}{1 - jf_1/f} \frac{1}{1 - jf_1'/f} \tag{3.16}$$

Both of the frequency-dependent factors will affect the overall lower 3-dB frequency of the amplifier. This frequency can be found by equating the magnitude of the denominator to $\sqrt{2}$ and solving for the resulting frequency. When both sides of this equation are squared, it can be written as

$$\left[1 + \left(\frac{f_1}{f_{\text{low}}}\right)^2\right]\left[1 + \left(\frac{f_1'}{f_{\text{low}}}\right)^2\right] = 2 \tag{3.17}$$

where f_{low} is the overall lower 3-dB frequency. The factors on the left of the equation represent the square of the magnitude; thus the right side of the equation becomes 2 rather than $\sqrt{2}$.

Generally we will know or can choose the values of f_1 and f_1' and will want to find f_{low}, the lower 3-dB frequency. This can be done by iteratively estimating the value of f_{low} and checking to see whether Eq. (3.17) is satisfied. After two or three iterations, the correct value of f_{low} can be found.

EXAMPLE **3.4**

An amplifier has a midband voltage gain of 22 V/V, an input impedance of 10 kΩ, and an output impedance of 325 Ω. A source with an output resistance of 600 Ω is coupled to the amplifier through a 10-μF capacitor. The amplifier output is coupled through a 40-μF capacitor to a 1-kΩ load. Calculate

(a) the midband voltage gain of the circuit
(b) the lower corner frequency caused by the 10-μF capacitor
(c) the lower corner frequency caused by the 40-μF capacitor

(d) the overall lower corner frequency of the voltage gain of the circuit

(e) Then write the expression for voltage gain as a function of frequency

SOLUTION

(a) At midband frequencies the capacitors are considered to be short circuits. Figure 3.11 and Eq. (2.3) can be applied to this situation, giving

$$A_{MB} = \frac{R_{in}}{R_{in} + R_g} A \frac{R_L}{R_L + R_{out}} = \frac{10}{10 + 0.6} \times 22 \times \frac{1}{1 + 0.325} = 15.7 \, \text{V/V}$$

(b) The lower corner frequency due to the 10-μF capacitor is found as

$$f_1 = \frac{1}{2\pi C_{c1}(R_{in} + R_g)} = \frac{1}{2\pi \times 10^{-5} \times 10,600} = 1.50 \, \text{Hz}$$

(c) The second lower corner frequency is

$$f_1' = \frac{1}{2\pi C_{c2}(R_{out} + R_L)} = \frac{1}{2\pi \times 4 \times 10^{-5} \times 1325} = 3.00 \, \text{Hz}$$

(d) The overall lower corner frequency is found from Eq. (3.17),

$$\left[1 + \left(\frac{f_1}{f_{low}}\right)^2\right]\left[1 + \left(\frac{f_1'}{f_{low}}\right)^2\right] = \left[1 + \left(\frac{1.5}{f_{low}}\right)^2\right]\left[1 + \left(\frac{3.0}{f_{low}}\right)^2\right] = 2$$

By iteration, f_{low} is found to be 3.57 Hz.

(e) Using Eq. (3.16), the voltage gain is written as

$$A_o = 15.7 \times \frac{1}{1 - j1.5/f} \times \frac{1}{1 - j3.0/f}$$

PRACTICE Problems

3.12 Calculate the lower 3-dB frequency of the voltage gain for the circuit of Fig. 3.20 if $R_g = 1 \, \text{k}\Omega$, $R_{in} = 20 \, \text{k}\Omega$, and $C_{c1} = 10 \, \mu\text{F}$. If $A_{MB} = 100 \, \text{V/V}$, write an expression for gain as a function of frequency. *Ans:* $f_1 = 0.76 \, \text{Hz}$, $A_o = 100/(1 - j0.76/f)$.
3.13 If $f_1 = 50 \, \text{Hz}$ and $f_1' = 80 \, \text{Hz}$ in Eq. (3.16), find the lower 3-dB frequency of the gain. *Ans:* $f_{low} = 102 \, \text{Hz}$.

PRACTICAL Considerations

In discrete circuit design, the input coupling capacitor is used to keep any DC voltage at the amplifier input from reaching the input signal generator. An output coupling capacitor may be used to keep any DC voltage at the amplifier output from reaching the load. If the load is a speaker, for example, distortion can be produced in the audio output of the speaker by a DC current flowing through the speaker. The coupling capacitor solves this problem. A common configuration of coupling capacitors was shown previously in Fig. 3.5.

3.5 High-Frequency Response of an Amplifier

IMPORTANT Concepts

1. The upper 3-dB frequency is determined by the shunt capacitance and the equivalent resistance seen by the capacitance.

2. If the circuit has two upper critical frequencies that are relatively close, the overall upper 3-dB frequency is calculated by iterative methods. If one of the critical frequencies is greater than the other by a factor of 5, the lower critical frequency approximates the overall upper 3-dB point.
3. The Miller effect must be considered when a capacitor bridges the input and output terminals of an amplifying element.

From the high-frequency model of Fig. 3.13 we note that the shunt capacitors reduce the voltage across the nodes to which they are connected as the frequency becomes higher. The impedances of these elements become smaller, and they can no longer be considered to be open circuits as they can at low and intermediate frequencies.

3.5.1 A SINGLE-POLE CIRCUIT

We will first consider a simplified model of Fig. 3.13 that includes only a single capacitor. This circuit is shown in Fig. 3.23. This simple model can often be used for the BJT or the MOSFET stage. The overall gain of this stage can be written as

$$A_o = \frac{Z_{ab}}{Z_{ab} + R_g} A \frac{R_L}{R_L + R_{out}} \tag{3.18}$$

where Z_{ab} is the impedance between terminals a and b and is

$$Z_{ab} = \frac{R_{in}}{1 + j\omega C_{in} R_{in}}$$

Manipulating Eq. (3.18) allows it to be written in the more conventional form of

$$A_o = \frac{R_{in}}{R_{in} + R_g} A \frac{R_L}{R_L + R_{out}} \frac{1}{1 + j\omega C_{in} \frac{R_{in} R_g}{R_{in} + R_g}} \tag{3.19}$$

This equation can be identified with the high-frequency form discussed in Section 3.3. In simplified form, Eq. (3.19) is expressed as

$$A_o = A_{MB} \frac{1}{1 + jf/f_2} \tag{3.20}$$

The single frequency factor in the denominator is a pole factor. The type of response represented by Eq. (3.20) is then a single-pole response.

The midband gain is the same as calculated using the midband model, while the upper 3-dB frequency is

$$f_2 = \frac{1}{2\pi C_{in} \frac{R_{in} R_g}{R_{in} + R_g}} \tag{3.21}$$

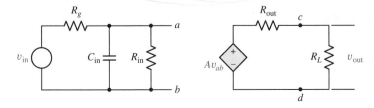

Figure 3.23
*A single time constant
high-frequency amplifier model.*

It is significant to note that both R_g and R_{in} affect the upper corner frequency of the stage. Equation (3.21) shows that the parallel combination of these two resistors is used in calculating f_2. This resistance also equals the Thévenin equivalent resistance seen by the capacitor.

A second interesting point to note is the relationship between upper corner frequency and the time constant of the amplifier when a step function of voltage is applied. This relationship was mentioned in Chapter 2. Assuming that a small step of amplitude ΔV is applied to the input, the output voltage will be an exponential given by

$$v_{out} = \Delta V\, A_{MB} \left(1 - e^{-t/\tau}\right) \tag{3.22}$$

where

$$\tau = C_{in} \frac{R_{in} R_g}{R_{in} + R_g} \tag{3.23}$$

We note that this time constant is equal to the reciprocal of the upper corner frequency expressed in rad/sec (Eq. (3.21)). This is a general result in a single time constant stage with high-frequency falloff.

3.5.2 A DOUBLE-POLE CIRCUIT

When a second capacitor appears in parallel with the load resistor, the circuit has two high-frequency poles. Figure 3.13 applies to this situation. The voltage gain in this case becomes

$$A_o = \frac{A_{MB}}{(1 + jf/f_2)(1 + jf/f_2')} \tag{3.24}$$

There are now two critical frequencies or two poles in the response. The overall amplifier upper 3-dB point is found by equating the magnitude of the denominator to $\sqrt{2}$ and solving for the resulting frequency, which results in the equation

$$\left[1 + \left(\frac{f_{high}}{f_2}\right)^2\right]\left[1 + \left(\frac{f_{high}}{f_2'}\right)^2\right] = 2 \tag{3.25}$$

Table 3.2 shows some examples of amplifiers with two shunt capacitances causing two individual corner frequencies along with the resulting upper 3-dB frequency of the amplifier.

If the two poles are widely separated, the lower frequency pole is called the *dominant pole*. In such a case, the magnitude of the gain begins decreasing as the frequency increases above the dominant pole frequency. The magnitude of the gain may drop below unity before the second pole frequency is reached.

In calculating the overall upper 3-dB frequency, we note that, if the upper pole frequency is five times larger than the lower pole frequency, the overall upper 3-dB frequency can be approximated by the lower pole frequency. This approximation leads to an error less than 0.2 dB. For example, if f_2 is 120 kHz and f_2' is 600 kHz, f_{high} can be approximated as $f_2 = 120$ kHz.

It can also be shown that if the higher pole frequency is five or more times the lower pole frequency, the new overall 3-dB point will be within 5% of the lower pole frequency (see Problem 3.38).

PRACTICE Problem

3.14 If $R_g = 1$ kΩ, $R_{in} = 3$ kΩ, $R_{out} = 4$ kΩ, $R_L = 6$ kΩ, $C_{in} = 300$ pF, and $A = -30$, calculate the upper 3-dB frequency of the circuit in Fig. 3.23. Express the overall gain as a function of frequency. *Ans:* $f_{high} = 707$ kHz, $A_o = -13.5/(1 + jf/707\,\text{kHz})$.

Table 3.2 Upper 3-dB Frequency for Two-Pole Circuit

f_2 kHz	f_2' kHz	f_{high} kHz
40	30	22
680	680	435
680	1360	569

EXAMPLE 3.5

The amplifier from Example 3.4 has a midband voltage gain of 22 V/V, an input impedance of 10 kΩ, a parasitic input capacitance of 20 pF, and an output impedance of 325 Ω. A source with an output resistance of 600 Ω is coupled to the amplifier through a 10-μF capacitor. The amplifier output is coupled through a 40-μF capacitor to a 1-kΩ load. Find the upper corner frequency of the voltage gain of the circuit. Write an expression for voltage gain of the circuit as a function of frequency, including both high- and low-frequency effects.

SOLUTION Equation (3.21) is used to calculate the upper corner frequency due to C_{in}. Since there is only one capacitance that causes high-frequency falloff, this frequency also equals the overall upper corner frequency of the circuit. We write

$$f_{high} = f_2 = \frac{1}{2\pi C_{in} \frac{R_{in} R_g}{R_{in} + R_g}} = \frac{1}{2\pi \times 20 \times 10^{-12} \times (10 \| 0.6)} = 14.06\,\text{MHz}$$

Using the midband voltage gain and lower corner frequencies calculated in Example 3.4, the voltage gain is then expressed as

$$A_o = 15.7 \times \frac{1}{1 - j1.5/f} \times \frac{1}{1 - j3.0/f} \times \frac{1}{1 + jf/14.06 \times 10^6}$$

3.5.3 THE CAPACITIVE MILLER EFFECT

The Miller effect was introduced in Chapter 2. In Chapter 3 we are interested in the capacitive Miller effect on the high-frequency response of an amplifier. The circuit of Fig. 3.24 demonstrates the reflection of a bridging capacitor to the input side of the circuit. The bridging capacitor has a value of 5 pF, but it is reflected to the input side as a 405-pF capacitor. This result comes from the multiplication of the bridging capacitance by a factor of $(1 + A)$, where $A = 80$ in this circuit; that is, $C_{in} = 5 \times 81 = 405$ pF.

From Eq. (3.21), this value of capacitance will cause an upper 3-dB frequency in the voltage gain of the circuit at

$$f_2 = \frac{1}{2\pi R_{eq} C_{in}}$$

where C_{in} is the reflected capacitance and R_{eq} is the parallel combination of the 1-kΩ and 5.5-kΩ resistors. The result is an upper corner frequency of 464 kHz.

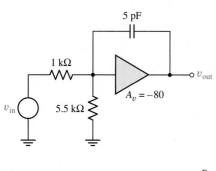

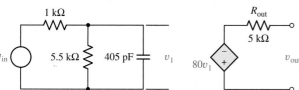

Figure 3.24
Using the Miller effect.

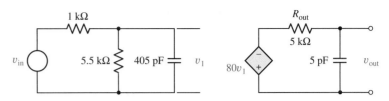

Figure 3.25
A more accurate equivalent circuit for a bridging capacitor.

The Miller effect is particularly important in high-frequency amplifier design. Both BJTs and MOSFETs exhibit a capacitance that bridges the input and output terminals in some important circuit configurations. The Miller multiplication of this value must be considered in the design of such stages.

Although the capacitance reflected to the amplifier input accounts for the proper input current required to drive the capacitance, we have neglected the effect on the output current. As derived in Chapter 2, a value equal to the bridging capacitor should be reflected to the output side of the circuit. Thus, when the bridging capacitor is reflected to both input and output sides of the amplifier, a more accurate circuit is that of Fig. 3.25. The gain element is assumed to have an output impedance of 5 kΩ.

The equivalent circuit of Fig. 3.25 will have two high-frequency poles rather than one. The first equals the pole frequency of the amplifier input circuit of Fig. 3.24—that is, 464 kHz. A second pole frequency results from the output circuit. This value is given by

$$f_2' = \frac{1}{2\pi\, R_{\text{out}}\, C} = 6.37\,\text{MHz}$$

In many cases, as in this one, this output pole frequency will be much larger than the input pole frequency and can be neglected. Only if it becomes less than five times the input pole frequency will it be considered. Consequently, the reflection of the bridging capacitor to the output side of the circuit is not often done if R_{out} is small. For IC amplifier stages, the output resistance can be quite large and the output pole frequency may dominate the high-frequency response. We consider this matter in Chapters 9 and 10.

PRACTICAL Considerations

An amplifier gain of

$$A_o = \frac{A_{MB}}{\left(1 + j\frac{f}{f_2}\right)\left(1 + j\frac{f}{5f_2}\right)}$$

has one pole frequency at f_2 and another at $5 f_2$. At the lower pole frequency, the first factor contributes 3.02 dB to the falloff of gain. At this same frequency, the higher pole contributes

$$20 \log \sqrt{1 + 0.04} = 0.17\,\text{dB}$$

to the falloff of gain. Neglecting this second term leads to an error in voltage gain at the lower pole frequency of less than 6%.

The accurate 3-dB frequency when including both factors will be about 0.96 f_2. Ignoring the larger pole factor results in a 4% error in upper 3-dB frequency.

EXAMPLE 3.6

Given that the amplifying block in Fig. 3.26 represents an amplifier with infinite input and zero output resistance, select the values R_1, R_2, and A to create an overall midband voltage gain of −50, an upper 3-dB frequency of 100 kHz, and a lower 3-dB frequency of 50 Hz.

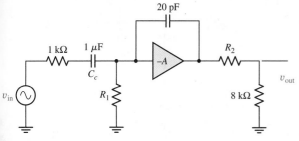

Figure 3.26
Amplifier circuit for Example 3.6.

SOLUTION The equation for midband gain is

$$A_{MB} = -50 = \frac{R_1}{R_1 + 1}(-A)\frac{8}{8 + R_2}$$

This equation accounts for the loading or attenuation of R_1 in the input loop and the attenuation due to R_2 in the output loop, but there are three unknowns in the equation.

The equation for lower corner frequency is

$$f_1 = \frac{1}{2\pi C_c(R_1 + 1000)} = 50\,\text{Hz}$$

and was used in an earlier example. This equation for lower corner frequency contains only one unknown, R_1, and was used earlier to calculate

$$R_1 = 2183\,\Omega$$

The equation for upper corner frequency can only be written after the Miller effect is taken into account. In this circuit, the capacitor links the input of the amplifying element to the output loop. This capacitor can be reflected to the input side, in parallel with R_1, with a value of

$$(1 + A) \times 20\,\text{pF}$$

The upper corner frequency can now be expressed as

$$f_2 = \frac{1}{2\pi\,(1 + A)\,20 \times 10^{-12} \times 1000R_1/(1000 + R_1)} = 100\,\text{kHz}$$

Using the value found earlier for R_1 in the equation for f_2 allows the solution of A. The necessary voltage gain is found to be -115. With this value of voltage gain, the Miller capacitance seen between the input terminal of the amplifier and ground is

$$C = 116 \times 20 = 2320\,\text{pF}$$

We note that the corner frequency is determined by the Thévenin resistance seen by the Miller capacitance. This value is calculated as the parallel combination of R_1 and the 1-kΩ resistor.

In calculating the lower 3-dB frequency, the coupling capacitor sees R_1 in series with the 1-kΩ resistor; this is the Thévenin equivalent resistance seen at the terminals of the coupling capacitor.

Finally, the equation for overall midband gain—that is,

$$A_{MB} = -50 = \frac{R_1}{R_1 + 1}(-A)\frac{8}{8 + R_2}$$

is used to find that

$$R_2 = 4.62\,\text{k}\Omega$$

We point out that the circuit in this example has no high-frequency falloff in the output side of the amplifier. Although the 20-pF capacitor should be reflected to appear between the output terminal and ground, the Thévenin resistance seen by this capacitor is zero because the amplifier output resistance is zero. If the amplifier had a finite rather than a zero output impedance, a second pole would have resulted from the output circuit.

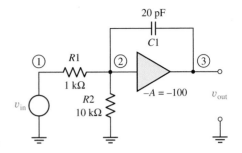

Figure 3.27
A Miller effect amplifying circuit.

Before we leave this section, we will consider two Spice simulations that deal with the Miller effect and the dominant pole concept. The first simulation demonstrates the Miller effect as well as the fact that a one-pole circuit has a time constant that equals the reciprocal of the 3-dB frequency expressed in rad/s; that is,

$$\tau = \frac{1}{\omega_{high}}$$

Figure 3.27 shows the circuit diagram for a simulation using schematic capture. The circuit of Fig. 3.28 shows the circuit after reflecting the effects of the bridging capacitor to the input.

The netlists used in the simulation of these circuits are indicated next.

```
MILLER.CIR                         Explanation
*MILLER EFFECT CIRCUIT
R1 1 2 1K
R2 2 0 10K
C1 2 3 20P
VIN 1 0 AC 0.05                    Comment out for trans. resp.
VIN 1 0 PWL(0 0 0.1U 0.05 10M      Comment out for freq. resp.
0.05)
EAMP 3 0 VALUE=-100*V(2)           Amplifier gain=-100

*EQUIVALENT CIRCUIT
R3 1 4 1K
R4 4 0 10K
C2 4 0 2020P
.AC DEC 10 1000 1MEG               Comment out for trans. resp.
.TRAN 1U 0.1M                      Comment out for freq. resp.
                                   1 μs steps—stops at 1 ms

.PROBE
.END
```

Figure 3.28
*Equivalent circuit with
capacitance reflected to input.*

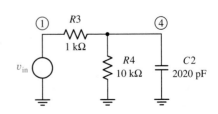

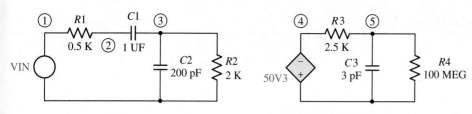

Figure 3.29
Circuit with two high-frequency poles.

Results: The Miller circuit has a midband gain of -90.9 V/V. The upper 3-dB frequency of the circuit is 86.36 kHz. For the equivalent circuit that reflects the bridging capacitance to the input of the amplifier, the 3-dB frequency is also 86.36 kHz. This circuit reflects the 20-pF capacitor to the input terminals (nodes 4, 0 in the simulation) with a value of 2020 pF. Using the step function input, the Miller circuit time constant was found to be 1.881 μs. The equivalent circuit also had a time constant of 1.881 μs. Note that $1/\tau = 531.6$ krad/s $=$ 84.61 kHz. This value closely approximates the 3-dB frequency from the simulation, which is 86.36 kHz.

The next simulation is done for a circuit having two high-frequency poles, but one is negligible because it is far above the other pole frequency. The simulation is first done for the circuit of Fig. 3.29 and leads to the actual upper 3-dB frequency of the amplifier.

The simulation is repeated after eliminating the 200-pF capacitor that causes the dominant pole; this will lead to a 3-dB frequency caused by the nondominant pole. It is then repeated after adding the 200-pF capacitor to the circuit and removing the 3-pF capacitor. This shows the dominant pole or the 3-dB frequency that results when the 3-pF capacitor is neglected.

```
DOM.CIR                    Explanation
*DOMINANT POLE CIRCUIT
R1 1 2 0.5K
R2 3 0 2K
R3 4 5 2.5K
C1 2 3 1U
C2 3 0 200P                This element is removed
                           for 2nd sim.
C3 5 0 3P                  This element is removed
                           for last sim.
R4 5 0 100MEG              This element replaces C3
                           when C3 removed
VIN 1 0 AC 0.01 0.01       V AC input
EAMP 4 0 VALUE=-50*V(3)    Amplifier gain=-50
.AC DEC 10 1000 100MEG
.PROBE
.END
```

Results: The midband gain is -39.9 V/V. The low-frequency point is found to be 63.5 Hz after changing the starting and ending points of the frequency sweep statement. The high-frequency 3-dB point is 1.995 MHz. When C3 is removed, the 3-dB frequency is 1.997 MHz. When C2 is removed, the 3-dB frequency is 21.23 MHz, which is over 10 times the frequency caused by C2 and can be neglected. Thus, the frequency caused by C2 is the dominant frequency.

3.6 Multistage Amplifiers

In many applications a single stage will not produce sufficient amplification to meet the specified gain. In such cases, several stages of amplification may be used. A typical configuration of a multistage amplifier appears in Fig. 3.30 for midband frequencies. The input of a given stage is connected to the output of the previous stage and is referred to as *cascading* stages. If more than two stages are required, the output of the second stage drives the input of a third stage. Some high-gain amplifiers might include eight to ten stages. Each stage will experience attenuation similar to a single stage coupling a generator resistance to a load resistance. The expression for overall midband gain is

$$A_{MBo} = A_1 A_2 A_3 \frac{R_{in1}}{R_{in1} + R_g} \frac{R_{in2}}{R_{in2} + R_{out1}} \frac{R_{in3}}{R_{in3} + R_{out2}} \frac{R_L}{R_L + R_{out3}} \quad (3.26)$$

where A_1, A_2, and A_3 are the unloaded midband gains of the individual stages—that is, the gains from input to output of the amplifier stages when no loads are applied to their outputs.

If coupling capacitors are used to couple all stages, there will be several critical corner frequencies. For example, one of these frequencies for the coupling capacitor between stage 2 and stage 3 will be

$$f_{13} = \frac{1}{2\pi C_{c3}(R_{out2} + R_{in3})}$$

where R_{out2} is the output resistance of the second stage and R_{in3} is the input impedance to the third stage. Once all critical frequencies are calculated, an equation similar to Eq. (3.17) can be written to solve for the lower 3-dB frequency.

The same procedure can be used to find the upper 3-dB frequency after all the higher corner frequencies are found. In this case, an equation similar to Eq. (3.25) would be solved to find the upper 3-dB point.

Figure 3.30
A three-stage amplifier.

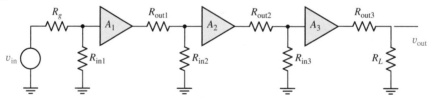

We now return to the Demonstration Problem introduced at the beginning of this chapter. The schematic for this amplifier is shown here for convenience. The calculation of the midband voltage gain can be done by considering the coupling capacitors as short circuits and the bridging

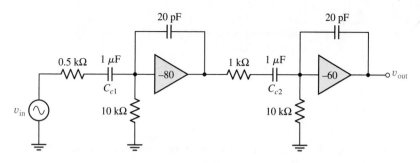

An amplifier.

capacitors as open circuits. The equivalent circuit at midband frequencies is indicated in Fig. 3.31. From this circuit, the overall midband gain is found to be

$$A_{MBo} = \frac{10}{10+0.5}(-80)\frac{10}{10+1}(-60) = 4156 \text{ V/V}$$

We do not know the frequency range over which this voltage gain calculation is valid until we calculate the lower and upper 3-dB points of the circuit. The lower critical frequencies are found from the low-frequency equivalent circuit of Fig. 3.32. The bridging capacitors appear as open circuits at low frequencies and are not shown in the equivalent circuit of Fig. 3.32. There are two lower corner frequencies in this circuit: one caused by C_{c1} and one caused by C_{c2}. These two frequencies are calculated as

$$f_1 = \frac{1}{2\pi \times 10,500 \times 10^{-6}} = 15.2 \text{ Hz}$$

and

$$f_1' = \frac{1}{2\pi \times 11,000 \times 10^{-6}} = 14.5 \text{ Hz}$$

It should be emphasized that, in finding the corner frequencies due to the coupling capacitors, the resistance used is the series resistance of the respective loops. This is the Thévenin equivalent resistance seen at the capacitor terminals. Equation (3.17) can now be used to find the overall lower corner frequency caused by these two coupling capacitors. This gives a value of

$$f_{low} = 23.1 \text{ Hz}$$

The high-frequency equivalent of Fig. 3.33 can be used to find the two individual high corner frequencies. In Fig. 3.33, the two bridging capacitors are reflected to the respective amplifier stage inputs. The first stage reflected capacitance is found from Eq. (2.25) to be

$$C_1 = (1 + 80)20 = 1620 \text{ pF}$$

The second stage reflected capacitance is

$$C_2 = (1 + 60)20 = 1220 \text{ pF}$$

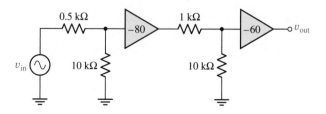

Figure 3.31
Midband equivalent circuit.

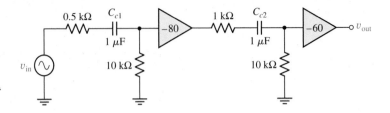

Figure 3.32
Low-frequency equivalent
circuit for amplifier.

The resistances seen by these capacitors are the parallel combinations of the two loop resistors. The capacitor C_1 sees a Thévenin equivalent resistance of

$$R_1 = 10 \| 0.5 = 476\,\Omega$$

Capacitor C_2 sees a resistance of

$$R_2 = 10 \| 1 = 909\,\Omega$$

The two individual stage upper corner frequencies are then

$$f_2 = \frac{1}{2\pi \times 476 \times 1.62 \times 10^{-9}} = 206.4\,\text{kHz}$$

and

$$f_2' = \frac{1}{2\pi \times 909 \times 1.22 \times 10^{-9}} = 143.5\,\text{kHz}$$

Using Eq. (3.25), the overall upper 3-dB frequency is found to be

$$f_{\text{high}} = 108.2\,\text{kHz}$$

The 3-dB bandwidth is then

$$\Delta f = f_{\text{high}} - f_{\text{low}} = 108,200 - 23 \approx 108,200\,\text{Hz}$$

The expression for overall voltage gain as a function of frequency can be written as

$$A_o = \frac{4156}{\left(1 - j\frac{14.5}{f}\right)\left(1 - j\frac{15.2}{f}\right)\left(1 + j\frac{f}{143,500}\right)\left(1 + j\frac{f}{206,400}\right)}$$

The mathematical expression for variation of voltage gain is seen to depend on the midband gain and the individual stage corner frequencies.

A good understanding of the principles involved in analyzing this Demonstration Problem will serve as preparation for subsequent chapters.

Figure 3.33
The high-frequency
equivalent circuit for
the amplifier.

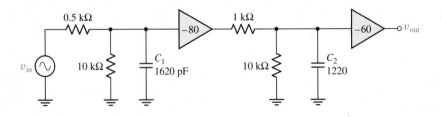

SUMMARY

➤ From this chapter we learn that for voltage amplification, the parameters R_{in}, R_{out}, and A (the unloaded voltage gain) for the amplifying element are very important parameters. As we discuss amplifying elements, we will generally want to evaluate these parameters for the element used.

➤ The coupling capacitors determine the lower 3-dB frequency, and parasitic capacitances determine the upper 3-dB frequency.

➤ It is important to be able to write the overall amplifier gain in terms of frequency-dependent factors because doing so allows the evaluation of the upper and lower 3-dB frequencies.

➤ The Miller effect is used to find the high-frequency limitations in a circuit that has a capacitor bridging input and output nodes of an inverting amplifier.

PROBLEMS

SECTION 3.1 THE GENERAL MODEL OF AN AMPLIFYING ELEMENT

3.1 An input signal of $v_{in} = 0.02 \sin \omega t$ V is applied to an amplifier stage. The output current, $I_{OUT} = (1 + 0.1 \sin \omega t)$ mA, flows through a 5-kΩ resistor to produce the output voltage.

(a) What is the voltage gain of the amplifier?

(b) What is the value of dc voltage at the output?

3.2 An amplifying element is configured as shown. If $v_{in} = 0.02 \sin \omega t$ V, resulting in an output current of $I_{OUT} = (1 + 0.1 \sin \omega t)$ mA,

(a) Calculate the incremental output voltage.

(b) Calculate the voltage gain.

(c) What is the dc voltage at the output?

3.3 A coupling capacitor and load resistor are added to the amplifier of Problem 3.2 as indicated. What is the midband voltage gain of the amplifier?

Figure P3.3

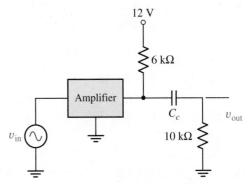

Problems 3.4 through 3.11 refer to Fig. 3.10. Unless otherwise stated, assume the following element values are used: $R_g = 200\,\Omega$, $R_{in} = 12\,k\Omega$, $R_{out} = 3\,k\Omega$, $R_L = 10\,k\Omega$, $C_{c1} = 5\mu F$, $C_{c2} = 1\mu F$, $C_{in} = 200\,pF$, and $C_{out} = 40\,pF$. The value of A is 160 V/V.

Figure P3.2

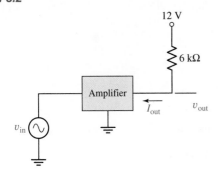

SECTION 3.2.2 THE PRACTICAL AMPLIFYING ELEMENT

3.4 Calculate the magnitudes of the impedances of C_{c1} and C_{c2} at a frequency of $f = 1$ kHz in Fig. 3.10. Compare these magnitudes to the pertinent resistances to show that the voltage drops across the coupling capacitors are negligible at this frequency. Do these elements affect the midband voltage gain? Use the values given above.

3.5 Calculate the magnitudes of the impedances of C_{in} and C_{out} at a frequency of $f = 1$ kHz in Fig. 3.10. Compare these magnitudes to the pertinent resistances to show that the parasitic capacitances do not affect the midband voltage gain. Use the values given above.

SECTION 3.2.3 THE MIDBAND MODEL

3.6 Calculate the midband voltage gain of the amplifier of Fig. 3.10. Use the values given above.

3.7 Calculate the midband voltage gain of the amplifier of Fig. 3.10 if R_L is changed to 1 kΩ. Use the values given above.

3.8 Calculate the midband voltage gain of the amplifier of Fig. 3.10 if R_g is changed to 10 kΩ. Use the values given above.

3.9 For efficient voltage amplification, loading should be minimized. How should R_g relate to R_{in} and how should R_{out} relate to R_L to minimize loading? Explain why it may not be possible to satisfy the conditions to minimize loading in a practical amplifier design.

SECTION 3.2.4 THE LOW-FREQUENCY MODEL

3.10 Write the frequency-dependent expression for voltage gain for the amplifier of Fig. 3.10 at lower frequencies. Use the values given above.

SECTION 3.2.5 THE HIGH-FREQUENCY MODEL

3.11 Write the frequency-dependent expression for voltage gain for the amplifier of Fig. 3.10 at higher frequencies. Use the values given above.

SECTION 3.3 FREQUENCY-DEPENDENT FACTORS

3.12 Plot the asymptotic frequency magnitude response for the gain

$$A = \frac{100}{1 + j\frac{40}{f}}$$

Use a log frequency scale extending from 1 Hz to 1000 Hz. What is the value of the 3-dB frequency?

3.13 Repeat Problem 3.12 for the gain

$$A = \frac{100}{4 + j\frac{220}{f}}$$

3.14 Plot the asymptotic frequency magnitude response for the gain

$$A = \frac{560}{1 + j\frac{f}{20,000}}$$

What is the value of the 3-dB frequency?

3.15 Repeat Problem 3.14 for a gain of

$$A = \frac{1000}{12 + j0.000010f}$$

3.16 Repeat Problem 3.14 for a gain of

$$A = 30\frac{1 + j\frac{f}{10}}{1 + j\frac{f}{134}}$$

3.17 Repeat Problem 3.14 for a gain of

$$A = 42\frac{3.2 + j\frac{f}{15}}{3.8 + j0.00024f}$$

3.18 Plot the asymptotic frequency magnitude response for the gain

$$A = \frac{36}{\left(1 + j\frac{40}{f}\right)\left(1 + j\frac{f}{42,000}\right)}$$

What are the values of the 3-dB frequencies?

3.19 Plot the asymptotic frequency magnitude response for the gain

$$A = 12\frac{5 + j\frac{f}{10}}{\left(8 + j\frac{f}{45}\right)\left(2 + j\frac{f}{50,000}\right)}$$

D 3.20 Write the expression for voltage as a function of frequency for the response plotted in the figure.

Figure P3.20

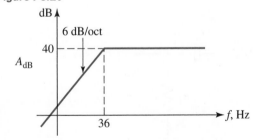

☆ D 3.21 Write the expression for voltage as a function of frequency for the response plotted in the figure. Calculate the lower 3-dB frequency.

Figure P3.21

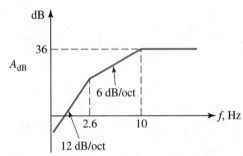

D 3.22 Write the expression for voltage as a function of frequency for the response plotted in the figure.

Figure P3.22

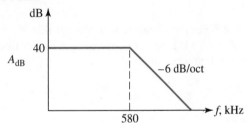

☆ D 3.23 Write the expression for voltage as a function of frequency for the response plotted in the figure. Calculate the upper 3-dB frequency.

Figure P3.23

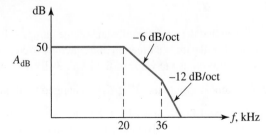

D 3.24 Write the expression for voltage as a function of frequency for the response plotted in the figure.

Figure P3.24

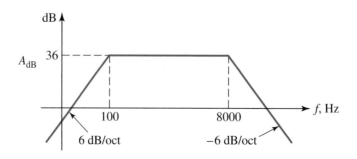

SECTION 3.4 LOW-FREQUENCY RESPONSE OF AN AMPLIFIER

3.25 In the circuit of Fig. 3.20, assume that $R_g = 1$ kΩ, $R_{in} = 10$ kΩ, and the midband voltage gain is -12 V/V. If C_{c1} is chosen to be 0.5 μF, calculate the lower 3-dB frequency of the stage. Write the expression for voltage gain as a function of frequency.

D 3.26 Replace C_{c1} in Problem 3.25 with a value that results in a 3-dB frequency of 6 Hz.

☆ **D 3.27** For the circuit shown, assume that $R_g = 1$ kΩ, $R_{in} = 10$ kΩ, $R_{out} = 10$ kΩ, and $R_L = 10$ kΩ. Select C_{c1} to cause a corner frequency of 50 Hz. Select C_{c2} to cause a corner frequency of 20 Hz.

Figure P3.27

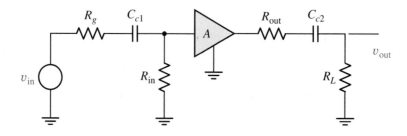

Given that the overall midband voltage gain is -10 V/V, write the analytic expression for the overall gain as a function of frequency. Sketch an asymptotic frequency magnitude response. What is the value of the lower 3-dB frequency?

D 3.28 Repeat Problem 3.27 if R_L is changed to 40 kΩ and $A_{MB} = -16$ V/V.

D 3.29 Repeat Problem 3.27 for the case where both corner frequencies are selected to be 50 Hz. How many dB below the midband value will the gain be at 50 Hz?

☆ **3.30** Given that an n stage amplifier contains n coupling capacitors, all of which are chosen to have individual corner frequencies of f_1, prove that the overall lower corner frequency is

$$f_{low} = \frac{f_1}{\sqrt{2^{1/n} - 1}}$$

How many dB will the gain at f_1 be down from the midband value?

D 3.31 Choose C_{c1} in Fig. 3.20 to cause a lower 3-dB frequency that falls between 80 and 100 Hz. Assume $R_{in} = 10$ kΩ and $R_g = 1$ kΩ.

D 3.32 Repeat Problem 3.31 if the lower 3-dB frequency is to fall in the range 6 to 8 Hz.

D 3.33 Choose C_{c1} and C_{c2} in Fig. 3.12 to cause a lower 3-dB frequency that falls between 80 and 86 Hz. Assume that $R_{in} = 10$ kΩ, $R_g = 2$ kΩ, $R_{out} = 3$ kΩ, and $R_L = 8$ kΩ.

D 3.34 Repeat Problem 3.33 if the lower 3-dB frequency is to fall in the range 20 to 24 Hz.

☆ **D 3.35** In Problem 3.33, the capacitor C_{c2} is selected to be 1 μF. Select C_{c1} to lead to a lower 3-dB frequency that falls in the range of 30 to 36 Hz.

SECTION 3.5 HIGH-FREQUENCY RESPONSE OF AN AMPLIFIER

D 3.36 In the circuit of Fig. 3.23, $R_g = 200$Ω and $R_{in} = 6$ kΩ. If the upper corner frequency of the circuit with C_{in} present is 2 MHz, add enough capacitance across R_{in} to lower this frequency to 1.2 MHz.

☆ **D 3.37** In the circuit of Fig. 3.13, $R_{in} = 10$ kΩ, $R_g = 2$ kΩ, $R_{out} = 3$ kΩ, and $R_L = 8$ kΩ. If $C_{in} = 100$μF, select C_{out} to lead to an upper corner frequency of 500 kHz.

3.38 In the circuit of Fig. 3.13, the input capacitance causes a pole frequency of f_2 while C_{out} causes a pole frequency of $f_2' = Xf_2$. Calculate the value of X that makes $f_{high} = 0.95 f_2$.

☆ **3.39** Given that an n stage amplifier has n individual corner frequencies of f_2. Prove that the overall upper corner frequency is

$$f_{high} = f_2\sqrt{2^{1/n} - 1}$$

SECTION 3.6 MULTISTAGE AMPLIFIERS

D 3.40 An amplifying stage has an input impedance of 6 kΩ and an output impedance of 1 kΩ. If the source impedance is 2 kΩ and the load impedance is 4 kΩ, calculate the necessary gain of each stage for an amplifier with an overall gain of 260. Assume the amplifier uses three identical stages.

D 3.41 Repeat Problem 3.40 if the required overall gain is 420.

D 3.42 Each stage of the three-stage amplifier of Problem 3.40 must be coupled to adjacent resistances by

coupling capacitors. Select a single value for all four of these capacitors to cause the overall lower 3-dB frequency to fall between 30 and 40 Hz.

D 3.43 Each stage of the three-stage amplifier of Problem 3.40 must be coupled to adjacent resistances by coupling capacitors. Set each individual critical frequency to 20 Hz by proper selection of the four coupling capacitors. What will be the overall lower 3-dB frequency of the amplifier?

DEMONSTRATION PROBLEM

3.44 If C_{c1} in the Demonstration Problem amplifier is changed to 10 μF, calculate the new overall lower corner frequency.

☆ **D 3.45** Change the voltage gain on the first stage of the Demonstration Problem amplifier to create equal individual stage upper corner frequencies. To what value must this gain be changed? What is the new value of overall midband gain? What is the new value of the overall upper corner frequency?

3.46 If the 1-kΩ resistance between the two stages in the Demonstration Problem amplifier is replaced by a short circuit, calculate the new value of midband voltage gain. Find the overall upper corner frequency.

3.47 Explain why it is unnecessary to reflect the bridging capacitors to the output sides of the stages in Problem 3.46.

Modeling and the Operational Amplifier

4

In Chapter 3 we examined in detail an incremental model for an amplifying element. This model consisted of input and output resistances, an ideal gain element, and parasitic capacitors. The amplifying element could be a MOSFET, a BJT, or an op amp. The op amp might be made up of 20 MOS or bipolar devices. An understanding of the complex circuitry within the op amp is not necessary to use this amplifying circuit in the construction of an amplifier. We need only to understand key performance parameters of the op amp to apply the models developed in Chapter 3.

Modeling is one of the most significant concepts in the study of electronic circuits. The use of modeling is imperative in designing circuits and systems. The ability to apply modeling techniques is a necessity for the design engineer. Because the op amp closely approximates an ideal, linear amplifying element, we will introduce it in this chapter to expand on the concept of modeling. A later chapter will consider the actual design of the op amp IC chip. The semiconductor diode will be considered in Chapter 5 in order to demonstrate modeling of a nonlinear device.

4.1 Modeling

4.2 The Operational Amplifier

4.3 Op Amp Circuit Examples

4.4 Designing for Maximum Bandwidth

DEMONSTRATION PROBLEM

At the end of this chapter we will discuss the analysis of the circuit shown. For this circuit, both op amp gains are expressed as

$$A = \frac{300,000}{1 + j\frac{f}{10}}$$

We need to evaluate the overall midband voltage gain, the overall lower corner frequency, and the overall upper corner frequency of the amplifier. We then want to express the voltage gain as a function of frequency.

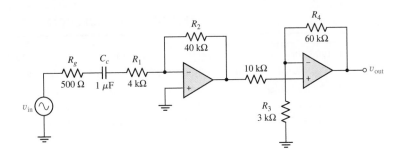

Demonstration circuit.

In an attempt to maximize the overall upper corner frequency of this amplifier with respect to R_4, we need to calculate the value of R_4 that makes the individual corner frequency of the second stage equal that of the first stage. We must then calculate the new overall upper corner frequency and midband voltage gain.

In order to analyze this problem, we must understand

1. The factors that determine the midband gain of each op amp stage
2. The factors that determine input and output impedances
3. The factors that limit the high-frequency performance of each stage

We must also understand how to deal with individual op amp stages before we can analyze the Demonstration Problem. Calculation of midband voltage gain and upper corner frequency for single op amp stages is necessary in order to determine the performance of a multistage amplifier. Thus, we will first center our discussion on single op amp stages.

A typical problem in single-stage amplifier design using an op amp is the following:

Select the values of R_1, R_F, and C_c in the circuit of Fig. 4.1 to result in a midband voltage gain of -10, an upper 3-dB frequency that exceeds 300 kHz, and a lower corner frequency that is smaller than 100 Hz. Specify the minimum gain-bandwidth product, GBW, to satisfy this design.

A familiarization with the operation of the op amp is necessary in order to complete this design problem. The basic approach in modeling is to replace the op amp with equivalent circuits that are easy to analyze, allowing solution of the problem. It is the goal of this chapter to develop an appropriate equivalent circuit or model for the op amp.

Figure 4.1
A typical op amp design problem.

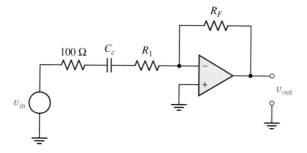

4.1 Modeling

IMPORTANT Concepts

1. Common elements used in modeling linear circuits are resistors, capacitors, inductors, independent sources, and dependent sources.
2. Manual analysis of electronic circuits containing nonlinear devices can be accomplished by using elements that are linear over the region of operation to model the approximate characteristics of the nonlinear devices.

4.1.1 WHAT IS MODELING?

Electronic devices are governed by equations that are typically nonlinear and very complex. Even the simple semiconductor diode has a highly nonlinear voltage-current (V-I) relationship expressed by

$$I = I_S \left(e^{q\,V/kT} - 1 \right) \tag{4.1}$$

where q is the charge of an electron, k is Boltzmann's constant, and T is the absolute temperature. This type of equation contrasts with the linear V-I relationship of a resistor, $I = V/R$, or the constant V-I relationship of a power supply, $V = E_{ps}$ for all currents.

In general, it is very difficult to analyze components that obey nonlinear equations, much less develop design methods for these components. On the other hand, good design methods already exist for networks made up of passive components such as resistors, inductors, and capacitors, plus voltage or current sources. These methods are based on Ohm's law and Kirchhoff's laws.

The basic idea in modeling an electronic device is to replace the device in the circuit with linear elements that approximate the V-I characteristic of the device. Once all nonlinear elements are replaced by linear elements, the design methods that apply to linear networks can be used. On occasion more than one model is applied for a given device depending on the operating conditions, the accuracy required, or the frequency of interest. We will discuss these possibilities in more detail as we introduce each device model in succeeding chapters.

For use in this textbook we can define a model in the following way.

A model is a collection of simple components or elements used to represent a more complex electronic device.

The "simple" components must be simple enough to allow conventional methods of analysis to be efficiently applied.

4.1.2 THE NEED FOR MODELS

In electronic circuits, there are two major reasons for using models. As mentioned previously, the first is that replacing a nonlinear device with appropriate linear elements leads to simpler equations. Another reason is that circuit operation is easier to visualize or understand when modeled in terms of known elements. In order to extend the performance of a circuit, a complete understanding is necessary. Modeling aids the designer in gaining an understanding of the circuit operation.

Nonlinear problems are immensely more difficult to solve than are linear problems. Of course, the computer can solve nonlinear problems and can thus be utilized to advantage in electronic circuit design. However, the solution of nonlinear circuits is often very time-consuming even with a computer, and most popular circuit analysis programs apply linear models to such circuits when possible.

4.1.3 MODELING CONSIDERATIONS

Levels of Modeling From a circuit designer's point of view, the simplest model is one that represents an electronic device such as a diode or a transistor. In this instance, a single device is replaced by a few elements that can be easily analyzed.

The next higher level is modeling of an entire circuit such as an amplifier. An operational amplifier or op amp might include an interconnection of 20 transistors and 20 resistors. In some applications, this complex circuit can be modeled by one or two simple components. As long as the important performance properties of the op amp are represented by the model, the simple model can be used for analysis and design.

An even higher level of modeling is that of replacing an electronic system with an appropriate model. The microprocessor chip includes many complex circuits such as registers, adders, gates, and buffers. Although not all functions of the microprocessor can be modeled by components, certain characteristics of this system are representable by models. Each input or output pin can be modeled by some electrical circuit consisting of simple components as far as the pin's effect on external circuits is concerned.

This textbook will concentrate on models used for electronic devices such as diodes and transistors and models used for op amps or other circuits. System models will not be considered.

Important Parameters in Modeling

The model chosen to represent an electronic device should have the same V-I characteristics at the terminals of interest as the device has, over the range of frequency of interest. Above all, the model should be as simple as possible. In many cases, a low-frequency model will be different from a high-frequency model of the same device simply to make the model less complex when used in low-frequency applications.

Although an equivalent circuit approximates the V-I characteristics of a nonlinear device, there is generally little physical correspondence between the model and the device. Furthermore, the elements chosen for the model need not be implemented; they must only be easy to characterize. For example, a current-dependent voltage source may be chosen to model some aspect of the device. Although this element may simplify the analysis, it is difficult to construct an ideal source of this type. Fortunately, this device is only used in analyzing the circuit.

There are two important types of equivalent circuit that are utilized in network design. Large-signal networks that operate over a nonlinear portion of the device characteristics are modeled by piecewise-linear equivalents. Other circuits that operate over a limited region of the nonlinear characteristics may be represented by one linear segment of the piecewise-linear approximation. Such a circuit is called a linear or small-signal equivalent circuit and is much easier to analyze than the large-signal model. Linear amplifiers make up a significant class of networks to which the small-signal model applies.

When manual analysis or design is necessary, many simplifying assumptions are made in developing an equivalent circuit. If the use of approximations leads to errors smaller than 10%, the results are generally acceptable.

With the availability of high-performance personal computers, circuit simulation by Spice-based programs has become very popular. Although more accurate models are generally used in simulation programs, inclusion of too many elements to achieve high accuracy can lead to excessive computer time. Often some compromise between accuracy and simplicity is made for computer models, though not to the extent made for manual analysis models.

A number of networks are designed to operate at frequencies far below the maximum operating frequencies of the devices used. In this instance a midband model for the device is appropriate. As noted in Chapter 3, the midband equivalent circuit generally contains no frequency-dependent elements such as capacitors or inductors. In higher frequency applications, capacitors or inductors are added to represent frequency effects of the device.

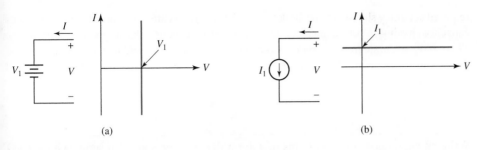

Figure 4.2
*Ideal dc sources: (a) voltage
source, (b) current source.*

4.1.4 ELEMENTS USED IN MODELING

Passive Elements Resistors, capacitors, and inductors are used to model various electronic devices. For manual analysis, constant values of resistors, capacitors, and inductors are used. Nonlinear circuits can be modeled with these linear elements using piecewise-linear methods. Computer analysis may allow passive elements having values that depend on the voltages across the elements.

Independent Sources The ideal voltage source and ideal current source are popular elements in equivalent circuits. These elements along with their V-I characteristics are indicated in Fig. 4.2. Although independent ac sources are also possible, most models require the dc sources of Fig. 4.2.

The characteristic of the voltage source implies that there is no series resistance associated with the source. The terminal voltage remains constant at V_1 regardless of the value of the current through the source. It is important to note that the voltage across the source is determined by the value of the source, but the current through the source depends on the external circuit connected to the terminals.

The current source determines the current flowing from the source terminals, but the voltage across the terminals depends on the external circuit connected to the terminals. This behavior suggests that the current source offers an infinite impedance since voltage across this source does not modify the terminal current.

Dependent Sources A dependent source generates either a voltage or a current that is directly proportional to a voltage across two nodes or to a branch current. These choices lead to the four possible types of dependent sources depicted in Fig. 4.3. The voltage-dependent voltage source generates a voltage that is proportional to the voltage from node a to node b. Nodes a and b are two nodes contained in the overall circuit. The current-dependent voltage source generates a voltage proportional to the current through branch 2 of the overall circuit. The voltage-dependent current source generates a current proportional to the voltage across two nodes of the circuit. The current-dependent current source produces a current proportional to some branch current. Although all of these dependent sources can be physically constructed, the circuitry involved is quite complex. It should be emphasized that they need not be physically constructed to use as elements in an equivalent circuit or model.

The following sections and the following chapter will consider the development of equivalent circuits based on the elements previously discussed. There are many possible models for each device. In choosing a model, it is important that the simplest model consistent with

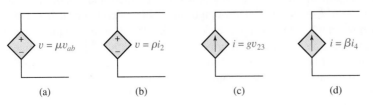

Figure 4.3
*Dependent sources: (a) voltage-
dependent voltage source,
(b) current-dependent voltage
source, (c) voltage-dependent
current source, (d) current-
dependent current source.*

required accuracy should always be selected. When circuits are constructed with resistors or capacitors having 10% tolerance, it is unnecessary to use models that approach 1% accuracy. These more accurate models lead to more complex calculations and may greatly extend the design time. Simpler models, leading to errors that are less than component tolerances, will often result in minimal design time.

EXAMPLE **4.1**

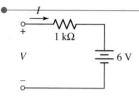

Figure 4.4
A series circuit.

Write an expression for V in terms of I for a 6-V voltage source in series with a 1-kΩ resistor as indicated in Fig. 4.4.

SOLUTION It is conventional to define current flow into the upper terminal as positive and a voltage polarity as shown in the figure. Since the two elements are in series, equal current will flow through both. Using Ohm's law, the voltage across the terminals is written as $V = 6 + I$ where I is expressed in mA.

EXAMPLE **4.2**

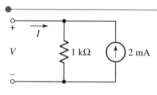

Figure 4.5
A parallel circuit.

Write an expression for V in terms of I for the 2-mA current source in parallel with the 1-kΩ resistor of Fig. 4.5.

SOLUTION Since both elements are in parallel, the terminal voltage will appear across both. Using Kirchhoff's current law at the node above the resistor indicates that the current through the resistor will be

$$I_R = (I + 2) \text{ mA}$$

The voltage across the resistor equals the terminal voltage and can be written from Ohm's law as

$$V = (2 + I) \text{ V}$$

4.2 **The Operational Amplifier**

PRACTICE **Problem**

4.1 The circuit of Fig. 4.4 is placed in series with the input terminal of the circuit in Fig. 4.5. Write an expression for input voltage, V, as a function of input current, I.
Ans: $V = 8 + 2I$.

IMPORTANT **Concepts**

1. The op amp is a very important element in low- to moderate-frequency amplifier applications. A simple model can be used to represent this complex device.
2. The model can be applied in calculating input impedance, output impedance, and voltage gain of the popular inverting and noninverting amplifier stages.

Amplification of electrical signals is a very common and important function in the electronics field. As noted earlier, the output signal of a microphone is in the millivolt range and may only supply a microamp of current. A loudspeaker may require several volts and several amps to drive it to its full volume. If the microphone signal is to drive the speaker, an amplifier must be inserted between the components to magnify or amplify the signal. The signal generated by speaking into the mouthpiece of a telephone must be amplified before it can be transmitted over long distances. A radio or television set receives a very weak signal that must be amplified before the transmitted information can be extracted.

One of the functions that can be performed by the operational amplifier or op amp is that of amplification. This useful device performs other functions and is perhaps the most popular single IC chip in the field of electronics today. Although limited to low- and medium-frequency operation, this device is used for amplifiers, comparators, Schmitt triggers, digital-to-analog converters, analog-to-digital converters, and many other analog and digital circuits.

The op amp is used in many applications that require only a simple model for analysis and design. Because of the simplicity of this model, the op amp serves as a vehicle to further demonstrate the use of models in the electronics area. The nonideal effects that must be considered for more critical design will be treated in a later chapter. We will consider the use of the op amp for amplification in the following sections.

All electronic amplifiers receive energy from one or more dc sources and use this energy to generate an amplified version of the input signal. Often this input signal is an ac signal, but it can also be a dc or slowly varying signal. The op amp requires at least one dc power supply to function properly. The more common configuration of an op amp uses both a positive and a negative dc power supply. In the following discussion of the op amp, it will be assumed that dc power supplies are properly connected to the device.

PRACTICAL Considerations

The output voltage swing of an op amp is limited by the dc power supplies used. A typical op amp may use +12 V and −12 V supplies. The output voltage swing is then limited to a range of values from approximately −11 V to approximately +11 V.

Because the op amp can exhibit a very high voltage gain, it is important to minimize dc power supply voltage fluctuations. These fluctuations might arise as the supply current value changes or if some type of noise is present. Generally, a small capacitor, perhaps 0.01 μF to 0.1 μF, is placed between ground and the IC pins that connect to each dc power supply voltage. This tends to hold the supply voltages applied to the chip at a constant voltage. The accompanying figure shows the usual configuration for the op amp and dc power supplies, which is referred to as *decoupling* the power supplies.

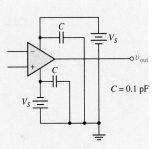

Decoupling capacitors.

4.2.1 THE BASIC OP AMP

A near-ideal op amp has a very high gain, a high input impedance, and a low output impedance. The symbol for this device generally does not include the dc supply terminals and is shown in Fig. 4.6 along with its model or equivalent circuit. The "+" sign marks the noninverting input terminal and the "−" sign identifies the inverting terminal.

The op amp always has a differential input stage. A differential stage has two inputs, and the amplifier output is proportional to the difference in voltage applied to the inputs. The output voltage can be expressed as

$$v_{\text{out}} = A(v_2 - v_1) \tag{4.2}$$

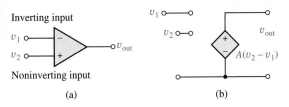

(a)

Figure 4.6
(a) Symbol for the op amp.
(b) Idealized equivalent circuit.

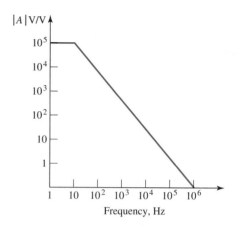

Figure 4.7
Voltage gain as a function of frequency.

where v_2 is the voltage applied to the noninverting input and v_1 is the voltage applied to the inverting input. The gain A might be 10,000 V/V to 1,000,000 V/V at dc or very low frequencies, but generally begins decreasing above a corner frequency that might be as low as 5 Hz. Figure 4.7 shows a typical variation of gain with frequency.

The model shown in Fig. 4.6(b) indicates that the op amp has zero output impedance and infinite input impedance. Although these conditions do not accurately describe the actual op amp, the simple model shown can be used in a surprisingly large number of practical applications.

The op amp gain can be written as

$$A = \frac{A_{MBoa}}{1 + jf/f_{2oa}}$$

where A_{MBoa} is the midband or low-frequency gain of the op amp and f_{2oa} is the upper 3-dB frequency.

Gain-Bandwidth Product The *gain-bandwidth product* or *GBW* of an amplifier is defined as the product of the magnitude of midband gain and the 3-dB bandwidth of the amplifier. The units of *GBW* can be either Hz or rad/s, depending on the units used to express bandwidth. For an op amp with the frequency response of Fig. 4.7, the midband gain is 10^5 and the bandwidth is 10 Hz, resulting in $GBW_{oa} = 10^6$ Hz.

There is a *GBW* for the op amp itself, a *GBW* for the noninverting stage, and a *GBW* for the inverting stage. These figures may or may not be equal for each stage as shown in the following sections.

4.2.2 THE NONINVERTING AMPLIFIER

A popular configuration for a noninverting amplifier is shown in Fig. 4.8. The resistors R_F and R_2 constitute a feedback network that determines the voltage gain of the amplifier. The feedback configuration is used to lower the very high gain of the stage to a usable value. As we shall see later, the reduction of gain is accompanied by an increased bandwidth. In order to calculate the voltage gain of the noninverting stage, the model of Fig. 4.6(b) is used as shown in Fig. 4.9.

The output voltage at low frequencies can be written as

$$v_{out} = A_{MBoa}(v_2 - v_1)$$

The input voltage equals v_2 and the voltage v_1 can be found as

$$v_1 = \frac{R_2}{R_2 + R_F} \times v_{out}$$

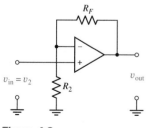

Figure 4.8
A noninverting amplifier.

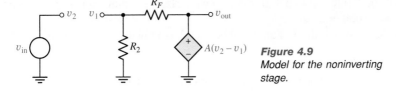

Figure 4.9
Model for the noninverting stage.

Allowing the resistive voltage divider factor to be represented by

$$F = \frac{R_2}{R_2 + R_F} \tag{4.3}$$

leads to an output voltage expression of

$$v_{out} = A_{MBoa}(v_{in} - F v_{out})$$

The quantity F is also called the feedback factor because it expresses the ratio of voltage fed back from the output to the input. Solving for the overall low-frequency midband voltage gain of the noninverting amplifier, A_{MBni}, gives

$$A_{MBni} = \frac{v_{out}}{v_{in}} = \frac{A_{MBoa}}{1 + A_{MBoa}F} \tag{4.4}$$

In the typical case the product $A_{MBoa}F$ will be much larger than unity, and Eq. (4.4) can be reduced to

$$A_{MBni} = \frac{1}{F} = \frac{R_2 + R_F}{R_2} = 1 + \frac{R_F}{R_2} \tag{4.5}$$

This equation is very useful in design work to set specified gain with minimal calculation.

The amplifier gain A_{MBni} is generally set to be in the range of $1 - 100$ V/V whereas A_{MBoa} may have a value of 200,000 V/V. One might question the use of R_2 and R_F to reduce the gain of the stage. There are two reasons for doing so. If the gain is not reduced, the exceptionally high value of gain would cause the output signal to clip as the voltage reaches the limit of the active region. With a gain of 2×10^5 V/V, an input signal of only 0.16 mV would lead to a 32-V output signal, and clipping would occur in several op amps for this value of input. Although there are occasional applications that require amplification of such a small signal, the frequency limitation also virtually eliminates the usefulness of the op amp as an amplifier when there is no feedback. Without feedback, the gain A is constant only up to frequencies of 5 or 10 Hz as shown in Fig. 4.7. Such a small frequency range is rarely acceptable for a practical amplifier; thus feedback is used to extend the useful frequency range while reducing the gain of the amplifier.

Noninverting Stage Frequency Dependence

We will now derive the equation for frequency dependence of the noninverting stage amplifier after first stating the result.

When an op amp with low-frequency gain A_{MBoa} is used in the noninverting configuration, the midband voltage gain is reduced from A_{MBoa} to A_{MBni}. The upper 3-dB frequency of the noninverting stage will be greater than that of the op amp by the same factor of gain reduction.

Thus, if the low-frequency gain of the op amp is $A_{MBoa} = 200,000$ V/V and the resistors are selected to result in a noninverting stage low-frequency gain of $A_{MBni} = 40$ V/V, the gain is reduced by a factor of 5000. The upper 3-dB frequency will then increase by this

PRACTICE Problem

4.2 The noninverting op amp of Fig. 4.8 uses a value of $R_2 = 2$ kΩ. The midband voltage gain of the op amp is $A_{MBoa} = 180,000$ V/V. If it is desired to improve the upper corner frequency from 10 Hz for the op amp to 12 kHz for the noninverting stage, what value of R_F is required? What is the midband voltage gain of the stage?
Ans: 249 kΩ, 150 V/V.

factor of 5000. If the op amp 3-dB frequency is 5 Hz, the noninverting stage 3-dB frequency will be $5 \times 5000 = 25$ kHz.

For the following discussion it will be useful to emphasize the definitions:

A—The gain of the op amp as a function of frequency

A_{MBoa}—The midband gain of the op amp (no feedback)

f_{2oa}—The 3-dB bandwidth of the op amp (no feedback)

GBW_{oa}—The gain-bandwidth product of the op amp (no feedback)

A_{ni}—The gain of the noninverting stage as a function of frequency

A_{MBni}—The midband gain of the noninverting amplifier with feedback

f_{2ni}—The 3-dB bandwidth of the noninverting amplifier with feedback

GBW_{ni}—The gain-bandwidth product of the noninverting amplifier with feedback.

To determine the frequency-dependent gain of the noninverting stage, we first assume zero output impedance for the op amp and high input impedance at the inverting terminal. The voltage fed back from output to the inverting terminal is then

$$v_1 = \frac{R_2}{R_2 + R_F} \times v_{\text{out}} = F v_{\text{out}}$$

where F is the feedback factor or the transfer function of the feedback network. We can now derive the overall voltage gain by noting that

$$v_{\text{out}} = A(v_2 - v_1) = A(v_{\text{in}} - F v_{\text{out}})$$

and then solving for $v_{\text{out}}/v_{\text{in}}$, which gives

$$A_{ni} = \frac{v_{\text{out}}}{v_{\text{in}}} = \frac{A}{1 + AF} \tag{4.6}$$

This is the well-known formula for closed-loop gain of a feedback amplifier.

The frequency-dependent expression for A is

$$A = \frac{A_{MBoa}}{1 + jf/f_{2oa}} \tag{4.7}$$

If we now substitute Eq. (4.7) into Eq. (4.6), we find that the gain of the noninverting stage is

$$A_{ni} = \frac{A_{MBoa}}{1 + A_{MBoa}F} \frac{1}{1 + jf/(1 + A_{MBoa}F)f_{2oa}} \tag{4.8}$$

The overall midband gain of the amplifier is the same as in the low-frequency derivation of gain; that is,

$$A_{MBni} = \frac{A_{MBoa}}{1 + A_{MBoa}F} \tag{4.4}$$

whereas the upper 3-dB frequency is

$$f_{2ni} = f_{2oa}(1 + A_{MBoa}F) \tag{4.9}$$

Equation (4.7) shows that the op amp itself has a low-frequency gain of A_{MBoa} and a bandwidth of f_{2oa}. For the 741, a very popular older op amp, $A_{MBoa} = 2 \times 10^5$ and

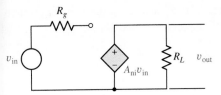

Figure 4.10
Noninverting stage model.

$f_{2oa} = 5$ Hz. If feedback resistors with values $R_F = 99$ kΩ and $R_2 = 1$ kΩ are used, the feedback factor is $F = 0.01$. The overall midband voltage gain given by Eq. (4.4) is

$$A_{MBni} = \frac{2 \times 10^5}{1 + 2 \times 10^5 \times 10^{-2}} = 100 \text{ V/V}$$

and the corner frequency is

$$f_{2ni} = 5(1 + 2 \times 10^5 \times 10^{-2}) = 10{,}005 \text{ Hz}$$

The gain after feedback has been reduced by a factor of 2000 from 200,000 to 100, while the bandwidth is increased by this same factor from 5 Hz to approximately 10,000 Hz.

The GBW is an important figure of merit for an amplifying stage. Using the noninverting configuration results in a nearly constant value of GBW_{ni}. That is, gain can be decreased by controlling the resistors R_F and R_2, but the bandwidth increases by the same factor of gain decrease. This is demonstrated from Eqs. (4.4) and (4.9). The gain-bandwidth product of the noninverting stage is

$$GBW_{ni} = A_{MBni} \; f_{2ni} = \frac{A_{MBoa}}{1 + A_{MBoa}F} \; f_{2oa}(1 + A_{MBoa}F)$$

$$= A_{MBoa} \; f_{2oa} = GBW_{oa} \qquad (4.10)$$

Note that the op amp with feedback has the same GBW as the op amp with no feedback. Gain and bandwidth are varied with F, but the GBW does not change with feedback factor. The input impedance of the op amp in the noninverting configuration is very high, ranging from tens of megohms for bipolar op amps to hundreds of megohms for FET input stage op amps.

A model for the noninverting stage is shown in Fig. 4.10. The input impedance is approximated as an open circuit while the output impedance is taken as zero. The gain A_{ni} is found from Eq. (4.8).

EXAMPLE 4.3

Calculate the overall midband voltage gain of the circuit in Fig. 4.11.

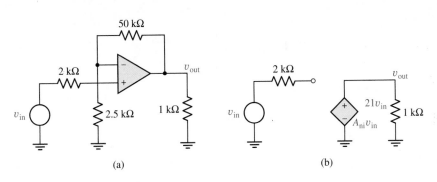

Figure 4.11
(a) An amplifier using the noninverting stage.
(b) Equivalent circuit.

(a)

(b)

SOLUTION The noninverting stage model of Fig. 4.10 is used to produce the equivalent circuit of Fig. 4.11(b). The midband voltage gain, A_{MBni}, is calculated as

$$A_{MBni} = 1 + \frac{50}{2.5} = 21$$

Since the input impedance is infinite and the output impedance is zero, no loading occurs and the overall gain equals the noninverting stage gain.

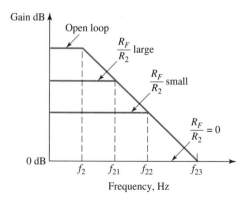

The Buffer Stage There is an important configuration derived from the noninverting amplifier. For the circuit of Fig. 4.8, when R_F approaches zero (a short circuit) and R_2 approaches infinity (an open circuit), the feedback factor becomes

$$F = \frac{R_2}{R_2 + R_F} \rightarrow 1$$

From Eq. (4.4), with $F = 1$, the low-frequency gain becomes

$$A_{MBni} = \frac{A_{MBoa}}{1 + A_{MBoa}} \rightarrow 1$$

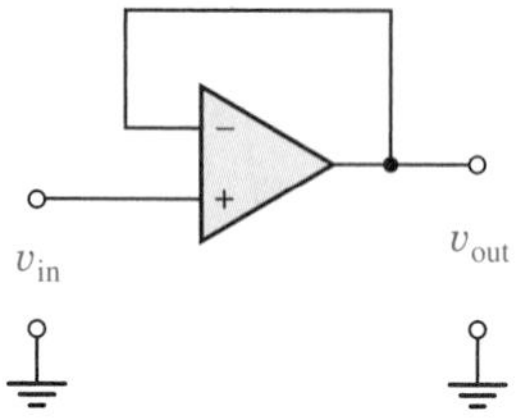

Figure 4.12 shows this unity gain amplifier or buffer stage. Although the voltage gain is unity, the current gain is quite high. This implies that the input impedance is high and the output impedance low.

Figure 4.12
A buffer stage.

PRACTICE **Problems**

4.3 An op amp with $GBW_{oa} = 4 \times 10^6$ Hz is used in the configuration of Fig. 4.8 to achieve a gain of 12 V/V. If $R_2 = 4$ kΩ, select the value of R_F. Calculate the bandwidth of the stage. *Ans:* 44 kΩ, 333 kHz.

4.4 If $R_F = 86$ kΩ and $R_2 = 3.3$ kΩ in the amplifier of Fig. 4.8, calculate the midband voltage gain and the bandwidth, assuming $GBW_{oa} = 4 \times 10^6$ Hz. *Ans:* 27.1 V/V, 148 kHz.

PRACTICAL **Considerations**

Microprocessor and microcontroller chips are often used to control an electronic device. Microcontrollers now control such appliances as dishwashers, microwave ovens, and clothes dryers. Since most of these processors are fabricated with small CMOS devices, the output current ratings are very low, often in the microamp range. The semiconductor switches used in appliances may require tens or hundreds of mA to activate. The op amp buffer stage can be used to interface between the processor and the switches. This stage is quite popular in low-frequency buffer stages requiring mA of current. When higher current devices must be driven, the op amp may not directly satisfy the current requirement, but can be followed by other high-current amplifiers.

For the noninverting stage, the adjustment of the ratio of R_F to R_2 affects both gain and bandwidth as shown in Fig. 4.13. We will show later that the noninverting stage has higher input and lower output impedances than the op amp with no feedback.

Figure 4.13
Effect of R_F/R_2 on gain and bandwidth.

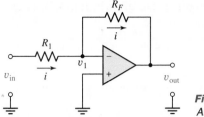

Figure 4.14
An inverting amplifier.

4.2.3 THE INVERTING AMPLIFIER

If phase inversion is needed or if a smaller value of input impedance is important, the stage of Fig. 4.14 can be used. Because the input impedance across the two input terminals, called the differential input impedance, is generally very high, little current flows into the inverting terminal. Thus, all input current flowing through R_1 can be assumed to travel toward the output terminal through R_F.

This assumption is further strengthened by the Miller effect on R_F. This feedback resistance bridges the inverting node and the output node that has a voltage dependent on the inverting node voltage. The current through R_F can be calculated from

$$i = \frac{v_{R_F}}{R_F} = \frac{v_1 - v_{\text{out}}}{R_F} = \frac{v_1 - (-Av_1)}{R_F} = \frac{v_1}{R_F}(1 + A)$$

The reflected impedance of R_F to the input side is given by

$$R_R = \frac{v_1}{i} = \frac{v_1}{v_1(1 + A)/R_F} = \frac{R_F}{1 + A}$$

If this reflected impedance is placed between the inverting terminal and ground, the same current would be drawn that flows through R_F in the actual circuit. Thus, all input current travels through R_F.

For typical values of R_F and A_{MBoa}, the reflected impedance has very low values. For example, if $R_F = 20$ kΩ and $A_{MBoa} = 200{,}000$, the value of R_R is

$$R_R = \frac{20{,}000 \ \Omega}{200{,}001} \approx 0.1 \ \Omega$$

Because this reflected impedance is so small, the inverting terminal is said to be a virtual ground in this configuration. It will be emphasized that the near short circuit behavior at this point is not due to current that enters the inverting terminal. The current actually flows through R_F.

The overall gain of the inverting stage can be calculated by equating the currents through R_1 and R_F, which leads to

$$\frac{v_{\text{in}} - v_1}{R_1} = \frac{v_1 - v_{\text{out}}}{R_F}$$

Noting that $v_1 = -v_{\text{out}}/A$ allows this equation to be expressed as

$$v_{\text{in}} + \frac{v_{\text{out}}}{A} = -\frac{R_1}{R_F}\left(\frac{v_{\text{out}}}{A} + v_{\text{out}}\right)$$

Solving for overall gain gives

$$A_i = \frac{v_{\text{out}}}{v_{\text{in}}} = \frac{-AR_F}{R_1(1 + A) + R_F}$$

If we now use the expression for op amp gain given by Eq. (4.7), we can write the gain as

$$A_i = \frac{-A_{MBoa}R_F}{R_1(1 + A_{MBoa}) + R_F + j(R_1 + R_F)f/f_{2oa}}$$

Since A_{MBoa} is a very large number, $(A_{MBoa} + 1)R_1$ is much larger than R_F, and $A_{MBoa} + 1$ can be approximated by A_{MBoa}. These observations allow the gain to be written

$$A_i = \frac{-R_F}{R_1} \frac{1}{1 + j\frac{f}{f_{2oa}(1 + A_{MBoa}R_1/[R_1 + R_F])}} \tag{4.11}$$

The closed-loop, low-frequency gain is given by

$$A_{MBi} = \frac{-R_F}{R_1} \tag{4.12}$$

and the closed-loop bandwidth is

$$f_{2i} = f_{2oa}(1 + A_{MBoa}R_1/[R_1 + R_F]) \tag{4.13}$$

For closed-loop gains considerably greater than unity, which ensures that $R_F \gg R_1$, gain and bandwidth can again be directly exchanged.

We should note that, if a signal generator resistance, R_g, is present, this value should be added to the series resistor R_1 in both Eqs. (4.12) and (4.13). This value will lower the midband gain and raise the bandwidth if R_g is significant compared to R_1.

For an inverting stage, if $R_F/R_1 = 1000$, $A_{MBoa} = 2 \times 10^5$, and $f_{2oa} = 5$ Hz, the closed-loop midband gain becomes

$$A_{MBi} = \frac{-R_F}{R_1} = -1000 \text{ V/V}$$

The gain decreases from the open-loop value by a factor of 200; hence, the bandwidth increases by this same factor to 1000 Hz.

If R_F is not much larger than R_1, the *GBW* of the inverting stage, designated GBW_i, can be calculated from

$$GBW_i = |A_{MBi}| f_{2i} \approx A_{MBoa} f_{2oa} \frac{R_F}{R_F + R_1} = GBW_{oa} \frac{R_F}{R_F + R_1} \tag{4.14}$$

For high-bandwidth, low-gain applications, the gain-bandwidth product of the inverting stage drops from the op amp *GBW*. In such instances, the noninverting stage should be used to maximize the amplifier voltage gain.

A comparison of the *GBW* as a function of midband voltage gain between the inverting and noninverting stages is shown in Fig. 4.15. At a voltage gain of unity, the noninverting stage will exhibit twice the bandwidth as that of the inverting stage.

Figure 4.15
GBW vs. voltage gain for inverting and noninverting stages.

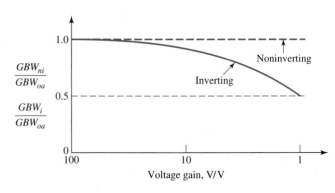

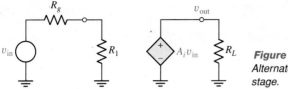

Figure 4.16
Alternate model for the inverting
stage.

Both the inverting and noninverting stages allow us to easily check the appropriateness of an op amp for a required amplifier design. If an amplifier with a gain of 100 and a bandwidth of 20 kHz is needed, an amplifying element with *GBW* equal to or greater than 2×10^6 Hz must be used. The 741 with a *GBW* of 10^6 would not be appropriate for this amplifier unless more than one stage were used.

An alternate model for the inverting stage amplifier is shown in Fig. 4.16. The midband value of gain, A_{MBi}, in this model is given by

$$A_{MBi} = \frac{-R_F}{R_g + R_1}$$

The bandwidth is given by Eq. (4.13) with the sum of R_1 and R_g replacing R_1; that is,

$$f_{2i} = f_{2oa} \left(1 + \frac{A_{MBoa}[R_g + R_1]}{[R_g + R_1 + R_F]} \right) \tag{4.15}$$

The noninverting stage is generally used in amplifier applications because its input impedance is high and its bandwidth exceeds that of the inverting stage at lower gains. The inverting stage is popular where a low input impedance is important such as in summer or difference circuits or integrators.

EXAMPLE 4.4

Calculate the overall midband voltage gain of the circuit of Fig. 4.17.

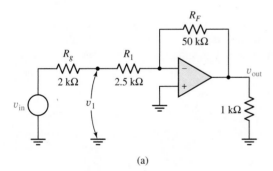

Figure 4.17
An inverting stage and model.

(a)

(b)

SOLUTION The output voltage in this case is given by

$$v_{out} = -\frac{R_F}{R_1} v_1 = -\frac{50}{2.5} v_1 = -20v_1$$

PRACTICE Problems

4.5 An amplifier uses the configuration of Fig. 4.14. If $R_F = 10$ kΩ and $R_1 = 1$ kΩ, calculate the midband voltage gain and the bandwidth of the inverting stage. Assume that the GBW of the op amp is 4×10^6 Hz.
Ans: -10 V/V, 364 kHz.

4.6 Repeat Practice Problem 4.5 with a source resistance of 1 kΩ placed in series with R_1.
Ans: -5 V/V, 667 kHz.

The voltage v_1 is

$$v_1 = \frac{2.5}{2 + 2.5} \, v_{in} = 0.56 v_{in}$$

Substituting this value into the previous equation results in

$$v_{out} = -20 v_1 = -20 \times 0.56 v_{in} = -11.1 v_{in}$$

Dividing both sides of this equation by v_{in} allows the overall gain to be written as

$$A_{MBi} = \frac{v_{out}}{v_{in}} = -11.1 \text{ V/V}$$

A second method of calculating this result is to combine the 2-kΩ resistor with the 2.5-kΩ resistor in the input loop to result in an equivalent resistance of 4.5 kΩ. This resistance can now be taken as R_1. The overall gain is then the negative of the ratio of R_F to 4.5 kΩ or

$$A_{MBi} = -\frac{50}{4.5} = -11.1 \text{ V/V}$$

4.2.4 INPUT AND OUTPUT RESISTANCES

We have noted the significant effects of loading on overall amplifier gain in Chapters 2 and 3. It is important to know the input and output impedances of the two op amp stages to calculate the loading effect.

The output impedance of an inverting or noninverting amplifier is essentially zero; consequently, loading of the output is negligible when a load resistance is used. We will derive this later in connection with feedback amplifiers in Chapter 12; however, a derivation will be given here to prove this important result. Figure 4.18 shows the equivalent circuit of an op amp with resistive feedback. Both noninverting and inverting inputs of the amplifier are connected to ground. This is done to allow a test voltage to be applied at the output to determine the current that results. The ratio of the voltage to the current equals the output impedance. Independent input sources are shorted for this calculation of the Thévenin equivalent resistance at the output.

The output impedance for the op amp itself is shown as R_{oa}. Typically, this value might be 100 to 200 Ω for modern op amps. The current resulting from a test voltage of v_t applied to the output is $i_1 + i_2$. The current i_1 is easy to calculate by Ohm's law, resulting in

$$i_1 = \frac{v_t}{R_F + R_1}$$

The impedance of this path is simply $R_F + R_1$.

The current i_2 is found by noting that when v_t is applied, an attenuated voltage reaches the inverting input of the op amp. This voltage is

$$v_1 = \frac{R_1}{R_1 + R_F} \, v_t$$

Figure 4.18
Op amp model used to calculate output impedance.

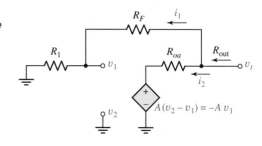

This voltage will be amplified and inverted, giving a voltage at the left side of R_{oa} of $-Av_1$. The current i_2 can now be calculated as

$$i_2 = \frac{v_t - (-A)\frac{R_1}{R_1 + R_F}v_t}{R_{oa}} = \frac{v_t\left(1 + A\frac{R_1}{R_1 + R_F}\right)}{R_{oa}}$$

The impedance due to this component of current is

$$R = \frac{v_t}{i_2} = \frac{R_{oa}}{1 + A\frac{R_1}{R_1 + R_F}}$$

An examination of this component reveals that R is a very small resistance. The value of

$$1 + A\frac{R_1}{R_1 + R_F}$$

is very large for general purpose op amps and R_{oa} may be 100 Ω. The resulting value of R is hundredths or tenths of an Ohm. While this resistance is paralleled by $R_F + R_1$ to obtain a value for R_{out}, the output impedance can be accurately approximated by

$$R_{\text{out}} = \frac{R_{oa}}{1 + A\frac{R_1}{R_1 + R_F}} \qquad (4.16)$$

When a noninverting or inverting op amp is used, this value is taken as zero.

Although the frequency dependence of A was not considered in the preceding derivation, it is easy to show that the very small output impedance becomes even smaller at higher frequencies.

The input resistance is another matter. For the noninverting stage, as mentioned previously, the input resistance may range from tens of megohms for bipolar op amps to hundreds or thousands of megohms for FET op amps. Feedback increases this impedance even more in the noninverting stage, so for normal applications, infinite input resistance is assumed.

The inverting stage is considerably different from the noninverting stage relative to input resistance. The input resistance to the inverting stage is approximately equal to R_1. The impedance from the inverting terminal to ground is determined by the reflected impedance of R_F. In Chapter 2 we saw that this resistance that links the input to the output will cause an equivalent reflected resistance from the negative input terminal to ground of

$$R_R = \frac{R_F}{1 + A}$$

The total input impedance of the inverting stage is then given by

$$R_{\text{in}} = R_1 + \frac{R_F}{A + 1} \approx R_1 \qquad (4.17)$$

PRACTICE Problems

4.7 A source of resistance 1 kΩ is coupled to a load of 1 kΩ. If a noninverting stage such as that of Fig. 4.8 is used with $R_F = 20$ kΩ and $R_2 = 1$ kΩ, what is the voltage gain from source to load? *Ans:* 21 V/V.

4.8 Repeat Practice Problem 4.7 for an inverting stage such as that of Fig. 4.14 with $R_F = 20$ kΩ and $R_1 = 1$ kΩ. *Ans:* -10 V/V.

4.3 Op Amp Circuit Examples

This section includes three examples and some Spice simulations to solidify the concepts developed in the preceding discussions. The first example will demonstrate the use of the op amp models in a simple design problem.

EXAMPLE 4.5

An amplifier is to be designed to couple a microphone to a resistive load. The microphone generates a peak output of 50 mV for a typical voice input level and has a 10-kΩ output impedance. The output voltage across the 2-kΩ load is to have a peak value of 10 V. The bandwidth of the voltage gain should be at least 40 kHz. If the *GBW* of the op amp used is 3×10^6 Hz, calculate the overall bandwidth of the final design.

SOLUTION From the specifications, the overall midband voltage gain must be

$$A_{MBo} = \frac{10}{0.05} = 200 \text{ V/V}$$

If a single noninverting stage were used to construct an amplifier with a midband gain of 200, the upper corner frequency would be

$$f_{2ni} = \frac{GBW_{ni}}{A_{MBni}} = \frac{GBW_{oa}}{A_{MBni}} = \frac{3 \times 10^6}{200} = 15 \text{ kHz}$$

This value of bandwidth does not satisfy the specification; thus more than one stage must be used.

One possible solution might consist of two stages, the first having a gain of 10 and the second a gain of 20. Since the noninverting stages present negligible loading to source and load resistances in the kΩ range, the overall voltage gain would then be 200. Figure 4.19 shows this amplifier.

In order to set the proper gains, the following two equations must be satisfied:

$$1 + \frac{R_{F1}}{R_{21}} = 10$$

$$1 + \frac{R_{F2}}{R_{22}} = 20$$

There are four unknowns and only two equations here, a situation common in design problems. This situation allows us to select two of the resistances and then calculate the remaining two. If R_{21} and R_{22} are chosen (somewhat arbitrarily) to be 2 kΩ, then R_{F1} and R_{F2} can be found from the equations to be $R_{F1} = 18$ kΩ and $R_{F2} = 38$ kΩ.

The remaining calculation is that of the upper 3-dB frequency, which can be done by first finding the 3-dB frequency of each stage by dividing the respective gains into GBW_{ni} or GBW_{oa}. The bandwidth of stage 1 is

$$f_{2ni1} = \frac{3 \times 10^6}{10} = 300 \text{ kHz}$$

Figure 4.19
A two-stage amplifier.

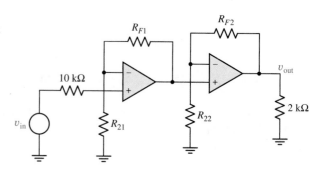

The second stage bandwidth is

$$f_{2ni2} = \frac{3 \times 10^6}{20} = 150 \text{ kHz}$$

The overall gain as a function of frequency can be written as

$$A_o(\omega) = \frac{10}{1 + jf/3 \times 10^5} \frac{20}{1 + jf/1.5 \times 10^5}$$

From Chapter 3 we know that the magnitude of the denominator must equal $\sqrt{2}$ at the upper 3-dB point, f_{2o}. The result is the equation

$$2 = \left(1 + \frac{f_{2o}^2}{9 \times 10^{10}}\right) \left(1 + \frac{f_{2o}^2}{2.25 \times 10^{10}}\right)$$

Using trial-and-error methods, the overall 3-dB frequency of the amplifier is found to be $f_{2o} = 126$ kHz.

PRACTICAL Considerations

An observation drawn from Example 4.5 will illustrate a basic difference between analysis and design. In analysis of an electronic circuit, there is a single correct answer. For this design problem, an infinite number of possible resistor values would satisfy the specifications. There are typically fewer constraining equations in design problems than values to be selected; thus all values cannot be calculated from a set of simultaneous equations. For this situation some values must first be selected before the remaining values can be calculated. The intelligent selection of these starting values is an important part of effective circuit design.

It has been shown (D. J. Comer, *Modern Electronic Circuit Design,* Addison-Wesley, Reading, MA, 1977) that in order to optimize overall circuit bandwidth, the individual stage bandwidths should be equal. For identical op amps, the bandwidths will be equal if the individual voltage gains are equal.

When upper 3-dB frequencies are equal, the formula for bandwidth shrinkage can be applied (see Problem 3.39). This equation is

$$f_{2o} = f_2 \sqrt{2^{1/n} - 1} \tag{4.18}$$

where n is the number of identical stages, f_{2o} is the overall upper 3-dB frequency, and f_2 is the 3-dB frequency of each individual stage. The derivation of this equation is done in a later section of this chapter. An example will demonstrate the use of this equation as well as the improvement in overall bandwidth that results from using identical individual stage bandwidths.

EXAMPLE 4.6

Using the op amp stages of Example 4.5, realize the overall voltage gain of 200 V/V with equal gains and bandwidths of both stages. Calculate the overall upper 3-dB frequency of this two-stage amplifier.

SOLUTION For identical stages, the gain of each stage must be set to

$$A_{MB1} = A_{MB2} = \sqrt{200} = 14.14 \text{ V/V}$$

For the noninverting stages of Fig. 4.19, these gains will be satisfied if R_{F1} and R_{F2} are chosen to be 26.3 $k\Omega$ and R_{21} and R_{22} are chosen to be 2 $k\Omega$. The bandwidth of each stage will then be

$$f_{2ni1} = f_{2ni2} = f_{2ni} = \frac{GBW_{ni}}{A_{MBni}} = \frac{GBW_{oa}}{A_{MBni}} = \frac{3 \times 10^6}{14.14} = 212 \text{ kHz}$$

Using the bandwidth shrinkage equation we find that the overall bandwidth of the amplifier is

$$f_{2o} = f_{2ni} \sqrt{2^{1/2} - 1} = 0.64 f_{2ni} = 0.64 \times 212 = 136 \text{ kHz}$$

This value should be compared to that from Example 4.5, which is $f_{2o} = 126$ kHz. When bandwidth maximization is important, identical bandwidths are imperative.

EXAMPLE 4.7

The amplifier of Example 4.5 is to be ac coupled to the source and the load with coupling capacitors. Choose reasonable values to set the lower corner frequency between 90 and 110 Hz.

SOLUTION Figure 4.20 shows an appropriate configuration for the circuit. The 500-$k\Omega$ resistor is used at the input of the first amplifier to establish a dc reference for this stage. The input impedance of a noninverting stage can be so high that a charge can build up at this point if not drained to ground. The value of this resistor is high enough to present negligible loading to the source resistance.

The critical frequencies caused by C_{c1} and C_{c2} are found to be

$$f_1 = \frac{1}{2\pi C_{c1}(10^4 + 5 \times 10^5)}$$

and

$$f_1' = \frac{1}{2\pi C_{c2}(2 \times 10^3)}$$

If we choose $C_{c1} = C_{c2}$, the frequency f_1 will be much lower than f_1'. Thus, f_1' will determine the overall lower 3-dB frequency. Setting this value to 100 Hz results in $C_{c2} = C_{c1} = 0.8 \ \mu F$.

Figure 4.20
Amplifier with coupling capacitors.

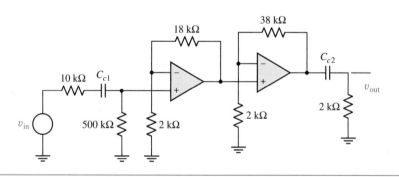

We will now consider a Spice example of both amplifier configurations. In Spice, it is possible to express gains in terms of frequency-dependent equations. Using the Laplace method of analysis, the variable s equals $j\omega$ for sinusoidal analysis. Spice allows a frequency-dependent gain to be expressed under its Laplace option. This will be demonstrated using

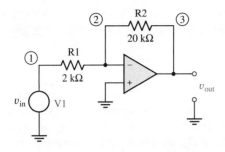

Figure 4.21
Schematic used for simulation.

an op amp that is equivalent to a 741 op amp. This device has a dc gain of 200,000 and a 3-dB bandwidth of 5 Hz (31.4 rad/s). This gain can be expressed as

$$A = \frac{200,000}{1 + j\omega/31.4} = \frac{200,000}{1 + s/31.4}$$

We will find the midband gains and bandwidths of the inverting op amp and the noninverting op amp circuits shown. Figure 4.21 is the schematic used for Windows© pSpice© simulation, and the Spice netlist is shown here.

```
OABW.CIR
*INVERTING OP AMP STAGE
R1 1 2 2K
R2 2 3 20K
EAMP 3 0 LAPLACE V(2,0)=-2E5/(1+S/31.4) Amplifier gain
V1 1 0 AC 0.1
.AC DEC 10 1000 10MEG
.PROBE
.END
```

Figure 4.22 shows the noninverting schematic. The Spice netlist follows.

```
NI.CIR
*NONINVERTING OP AMP STAGE
R1 0 2 2K
R2 2 3 18K
EAMP 3 0 LAPLACE V(1,0)=2E5/(1+S/31.4) Amplifier gain
V1 1 0 AC 0.1
.AC DEC 10 1000 10MEG
.PROBE
.END
```

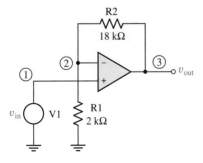

Figure 4.22
Noninverting stage used for simulation.

The results are $A_{MBi} = -10$ V/V, $A_{MBni} = 10$ V/V, $f_{2i} = 90.3$ kHz, and $f_{2ni} = 99.5$ kHz. Do these results make sense? Why is the bandwidth of the inverting stage less than that of the noninverting stage? Compare Eqs. (4.10) and (4.14).

A Typical Single-Stage Design Problem
We postulated a typical single-stage design problem at the beginning of this chapter. The specifications given in connection with Fig. 4.1 can be used to emphasize the essence of electronic circuit design. In order to analyze or design a circuit, all complex devices are replaced by appropriate models. After this is done, the circuit will consist only of elements that can be analyzed by conventional methods such as Ohm's or Kirchhoff's laws.

The model of Fig. 4.16 for the inverting amplifier is appropriate for this design problem. The original circuit can be redrawn using this model, as shown in Fig. 4.23. In this circuit, the midband gain is given by

$$A_{MBi} = -\frac{R_F}{R_g + R_1}$$

The lower corner frequency is related to other circuit parameters by

$$f_1 = \frac{1}{2\pi C_c(R_g + R_1)}$$

and the upper 3-dB frequency is found from Eq. (4.13) to be

$$f_{2i} = f_{2oa}\left(1 + \frac{A_{MBoa}[R_g + R_1]}{R_g + R_1 + R_F}\right) \tag{4.19}$$

From these equations it can be seen that both midband gain and lower corner frequency of the amplifier can be set without a knowledge of the op amp gain or bandwidth. If R_1 is chosen to be 2 kΩ, then R_F is found from the gain equation to be 21 kΩ. The coupling capacitor is found from the equation for lower corner frequency as

$$C_c = \frac{1}{2\pi f_1(R_g + R_1)} = 0.76 \ \mu F$$

A 1-μF capacitor would lead to a corner frequency below 100 Hz and can be used in this design.

Using the selected values in Eq. (4.15) for upper 3-dB frequency gives

$$f_{2i} = f_{2oa}[1 + 0.0909 A_{MBoa}] = f_{2oa} + 0.0909 A_{MBoa} f_{2oa}$$

Assuming that f_{2oa}, the op amp bandwidth, is small compared to $0.909 A_{MBoa} f_{2oa}$, this equation can be approximated as

$$f_{2i} = 0.0909 A_{MBoa} f_{2oa} = 0.0909 G B W_{oa} \geq 300 \text{ kHz}$$

Solving for the op amp GBW gives

$$GBW_{oa} \geq 3.3 \text{ MHz}$$

Figure 4.23
Equivalent circuit for the single-stage circuit.

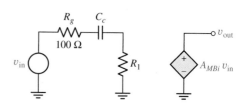

The demonstration problem circuit extends the previous discussion of single-stage amplifiers to a two-stage circuit. The schematic of the Demonstration Problem circuit is repeated here. The circuit of Fig. 4.24 is the midband equivalent circuit for this amplifier.

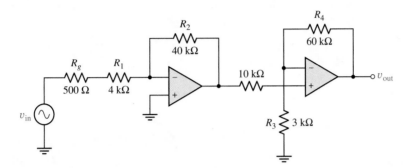

Demonstration problem amplifier.

We note that the 10-kΩ resistance between the two stages will have no effect on the gain since the input impedance to the noninverting stage is very high. The overall midband voltage gain is

$$A_{MBo} = -\frac{R_2}{R_g + R_1}\left(1 + \frac{R_4}{R_3}\right) = -\frac{40}{4.5}\left(1 + \frac{60}{3}\right) = -187 \text{ V/V}$$

Since the single coupling capacitor appears in the first stage, only this stage will exhibit a lower corner frequency. This frequency will then equal the overall lower corner frequency of the amplifier. The coupling capacitor sees a total resistance in the input loop of the input resistance to the first stage, R_1, and the signal generator resistance, R_g.

$$f_{\text{low}} = \frac{1}{2\pi C_c(R_g + R_1)} = \frac{1}{2\pi \times 10^{-6} \times 4500} = 35.4 \text{ Hz}$$

Each op amp stage will exhibit a separate upper corner frequency. In general, each frequency can be found by dividing the GBW of the stage by the midband gain of the stage. The first stage is an inverting stage; hence Eq. (4.14) is used to find the GBW. This results in

$$GBW_i = GBW_{oa}\frac{R_2}{R_2 + R_1 + R_g}$$

The op amp GBW is found from the gain expression

$$A_{oa} = \frac{300,000}{1 + j\frac{f}{10}}$$

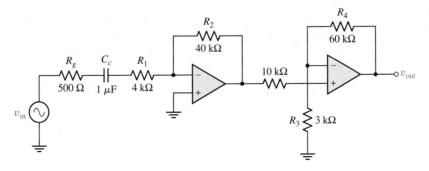

Figure 4.24
Midband equivalent circuit.

The product of midband gain (300,000) and bandwidth (10) is 3 MHz. The GBW of the inverting stage is then

$$GBW_i = 3 \; \frac{40}{40 + 4 + 0.5} = 2.7 \text{ MHz}$$

The bandwidth of this stage is calculated by dividing the absolute value of gain of the stage into the GBW, to result in

$$f_{2i} = \frac{GBW_i}{R_2} \; (R_g + R_1) = \frac{2.7}{40} \; 4.5 = 304 \text{ kHz}$$

The bandwidth of the noninverting stage equals the op amp GBW figure, giving a bandwidth of

$$f_{2ni} = \frac{GBW_{ni}}{1 + \frac{R_4}{R_3}} = \frac{3}{1 + \frac{60}{3}} = 143 \text{ kHz}$$

The overall upper corner frequency is found from Eq. (3.17) to be

$$f_{2o} = 122 \text{ kHz}$$

The voltage gain as a function of frequency is expressed as

$$A_o(f) = \frac{-187}{\left(1 - j\frac{35.4}{f}\right) \left(1 + j\frac{f}{304,000}\right) \left(1 + j\frac{f}{143,000}\right)}$$

4.4 Designing for Maximum Bandwidth

Because the op amp has a somewhat limited bandwidth, a useful result for maximizing the bandwidth of a multistage amplifier will be derived in this section. This derivation applies to any device having a constant GBW with gain changes such as the noninverting op amp stage.

A circuit consisting of n identical stages is called an *iterative* stage amplifier. Iterative stages are often chosen to satisfy an amplifier design because equal bandwidths for each stage lead to the maximum overall bandwidth for a specified overall voltage gain. In wideband amplifiers, the lower 3-dB frequency is either zero or very small compared to the upper corner frequency. Consequently, the bandwidth is

$$BW = f_2 - f_1 \approx f_2$$

We will use the term *upper corner frequency* interchangeably with *bandwidth* for wideband stages, since the two parameters are almost equal. In practice the two major specifications for an amplifier are the overall voltage gain and the overall upper corner frequency.

4.4.1 BANDWIDTH SHRINKAGE IN ITERATIVE STAGES

We will consider the amplifier configuration of Fig. 4.25. In this n-stage amplifier, if each stage is down from its midband gain by 3-dB at a frequency of f_2, the overall amplifier will be down by $3n$ dB at f_2. For a five-stage circuit, the amplifier will be down from its midband gain by 15 dB at f_2. The frequency at which the overall midband voltage gain is down by

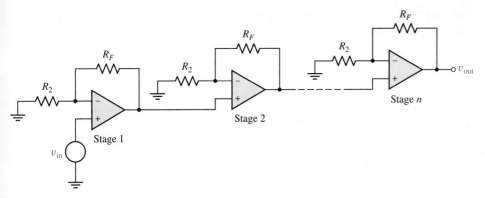

Figure 4.25
An iterative stage amplifier.

3 dB is called f_{2o}. This overall upper 3-dB point is always smaller than f_2. The decrease in f_{2o} as more identical stages are added is referred to as *bandwidth shrinkage*. We will now derive the relationship between overall upper corner frequency, f_{2o}, and individual stage upper corner frequency, f_2.

The individual stage gains in the amplifier of Fig. 4.25 can be written as

$$A_1 = A_2 = A_3 = \cdots = A_n = \frac{A_{MB}}{1 + j\frac{f}{f_2}} \tag{4.20}$$

The overall voltage gain equals this expression raised to the nth power or

$$A_o = \left[\frac{A_{MB}}{1 + j\frac{f}{f_2}}\right]^n = \frac{A_{MB}^n}{\left(1 + j\frac{f}{f_2}\right)^n} = \frac{A_{MBo}}{\left(1 + j\frac{f}{f_2}\right)^n} \tag{4.21}$$

The magnitude of the overall gain is a function of frequency and can be expressed in dB as

$$A_{dB} = 20 \ \log \ |A_o| = n \ \log \ \left|\frac{A_{MB}}{1 + j\frac{f}{f_2}}\right| \tag{4.22}$$

At low and midband frequencies, Eq. (4.22) becomes

$$A_o = A_{MBo} = A_{MB}^n \tag{4.23}$$

At higher frequencies the denominator increases, dropping the overall gain. The upper 3-dB frequency is the value of f that causes the magnitude of the complex denominator of Eq. (4.22) to equal $\sqrt{2}$. When this occurs, the gain magnitude is

$$|A_o| = \frac{A_{MBo}}{\sqrt{2}}$$

and is 3 dB below the overall midband gain, A_{MBo}. We can equate the magnitude of the denominator to $\sqrt{2}$ and solve for the frequency that satisfies this relationship. This frequency will be f_{2o}.

As we do this we note that the magnitude of a complex number raised to the nth power equals the magnitude of the number raised to the nth power. The result is

$$|(1 + jf_{2o}/f_2)^n| = |1 + jf_{2o}/f_2|^n = \sqrt{2}$$

Squaring both sides of this equation gives

$$|1 + jf_{2o}/f_2|^{2n} = 2$$

The magnitude of the complex term is

$$|1 + jf_{2o}/f_2| = \sqrt{1 + \frac{f_{2o}^2}{f_2^2}}$$

Consequently we can write

$$\left(\sqrt{1 + \frac{f_{2o}^2}{f_2^2}}\right)^{2n} = \left(1 + \frac{f_{2o}^2}{f_2^2}\right)^n = 2$$

and

$$f_{2o} = f_2 \sqrt{2^{1/n} - 1} \tag{4.18}$$

This is the well-known bandwidth shrinkage equation.

To understand the magnitude of this effect, if two stages are used in the amplifier ($n = 2$) the overall bandwidth is 0.64 times the individual stage bandwidth. Three stages shrink the bandwidth to 51% of the individual value. Equation (4.18) can be approximated for $n \geq 3$ by the simpler expression

$$f_{2o} = \frac{f_2}{1.2\sqrt{n}} \tag{4.24}$$

4.4.2 OPTIMIZATION OF OVERALL BANDWIDTH

An interesting problem encountered in iterative stage design is the following:

> *Given a specified overall midband voltage gain and overall bandwidth for an amplifier, what number of stages should be used to achieve this design?*

If a high value of individual stage gain is used, only a few stages are needed to realize the gain, thereby minimizing the bandwidth shrinkage. However, since $f_2 = GBW/A_{MB}$, a high vale of A_{MB} leads to a low value of f_2. If A_{MB} is decreased to increase f_2, more stages are needed and the bandwidth shrinkage effect becomes more pronounced.

One would expect that there exists some value of gain or some number of stages that will optimize the overall bandwidth while realizing the specified value of overall gain. Such an optimizing value does exist and will be derived next.

We begin the derivation with Eq. (4.24):

$$f_{2o} = \frac{f_2}{1.2\sqrt{n}} \tag{4.24}$$

Since the gain-bandwidth product of the stage is

$$GBW = A_{MB} \times f_2$$

we can rewrite Eq. (4.24) as

$$f_{2o} = \frac{GBW}{1.2n^{1/2} A_{MB}} = \frac{GBW}{1.2n^{1/2} A_{MBo}^{1/n}} \qquad (4.25)$$

Noting that GBW and A_{MBo} are constants, we could differentiate f_{2o} with respect to n and set this derivative equal to zero to find the value of n that gives the maximum value of f_{2o}. The differentiation is simplified if we let $A_{MBo} = e^k$ where $k = \ln A_{MBo}$. This allows us to write the overall bandwidth as

$$f_{2o} = \frac{GBW}{1.2n^{1/2} e^{k/n}} \qquad (4.26)$$

Differentiating gives

$$\frac{df_{2o}}{dn} = \frac{GBW}{1.2n^{3/2} e^{k/n}} \left[\frac{k}{n} - \frac{1}{2} \right] \qquad (4.27)$$

This derivative is zero only if n takes on a value of

$$n = 2k = 2 \ln A_{MBo} \qquad (4.28)$$

With this value of n, the individual stage gain becomes

$$A_{MB} = A_{MBo}^{1/n} = e^{k/n} = e^{1/2} = 1.65 \text{ V/V} \qquad (4.29)$$

A quick check proves that this value of individual stage gain leads to a maximum rather than a minimum value of bandwidth.

In a given problem of amplifier design we assume an individual stage gain of 1.65, then calculate the required number of stages. The number of stages can be found from solving

$$1.65^n = A_{MBo}$$

giving

$$n = \frac{\ln A_{MBo}}{\ln 1.65} = 2 \ln A_{MBo} \qquad (4.30)$$

The individual stage bandwidth is found to be $f_2 = GBW/1.65$. The bandwidth shrinkage equation is then applied to find the maximum overall bandwidth. An example will demonstrate this method.

EXAMPLE 4.8

An amplifier requires a gain of 1000 V/V. If the op amps used have a GBW of 6 MHz, calculate the maximum overall bandwidth that can be achieved and the number of stages that must be used.

SOLUTION A single stage gain of 1.65 V/V is required to lead to the maximum overall bandwidth. An overall gain of 1000 leads to a value of

$$n = 2 \ln 1000 = 13.8$$

Since n must be an integer, 14 stages will be used. The individual stage bandwidth is

$$f_2 = \frac{GBW}{1.65} = \frac{6 \times 10^6}{1.65} = 3.636 \text{ MHz}$$

Applying the bandwidth shrinkage equation gives an overall bandwidth of

$$f_{2o} = \frac{f_2}{1.2n^{1/2}} = \frac{3.635 \times 10^6}{1.2 \times 14^{1/2}} = 810 \text{ kHz}$$

Let us suppose that the overall bandwidth resulting in Example 4.8 is not sufficient to satisfy the amplifier bandwidth. Using these op amps, this design cannot be realized. The only way to increase the overall bandwidth above the calculated maximum is to use op amps with higher *GBW* values, if available.

On the other hand, if the bandwidth resulting from the design is higher than needed for the amplifier, the gain per stage can be increased, lowering the number of required stages. Often, the number of stages can be reduced considerably by trial-and-error methods to achieve the correct amplifier bandwidth. Thus, the major purpose of this theory is to verify that a given op amp can satisfy a design specification and suggest that a smaller number of stages than given by Eq (4.30) can be used.

SUMMARY

➤ Modeling of a nonlinear device uses linear electronic elements to represent the device. This model may be valid only over a limited region of operation of the nonlinear device.

➤ The op amp is a complex circuit consisting of many internal transistors. Fortunately, the op amp can be modeled by a rather simple equivalent circuit that is valid over a large range of output voltages.

➤ The gain-bandwidth product of a noninverting stage is constant with gain. For an inverting stage, the gain-bandwidth product falls at lower values of gain.

PROBLEMS

SECTION 4.1.4 ELEMENTS USED IN MODELING

4.1 Draw the V-I characteristics for the series combination of a 6-V voltage source and a 100-Ω resistance. Plot for currents ranging from −100 to +100 mA.

4.2 Draw the V-I characteristics for the parallel combination of a 6-V voltage source and a 100-Ω resistance. Plot for currents ranging from −100 to +100 mA. What effect does a parallel resistance have on the terminal voltage of the source?

4.3 Draw the V-I characteristics for the parallel combination of a 50-mA current source and a 100-Ω resistance. Plot for voltages ranging from −6 V to +6 V.

4.4 Draw the V-I characteristics for the series combination of a 50-mA current source and a 100-Ω resistance. Plot for voltages ranging from −6 V to +6 V. What effect does a series resistance have on the terminal current of the current source?

D 4.5 Synthesize (implement) a circuit to realize the characteristics shown in the figure.

4.8 If $v_{in} = 0.5 \cos \omega t$, calculate v_{out} for each circuit in the figure.

Figure P4.5

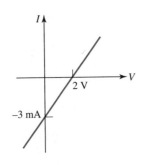

Figure P4.8

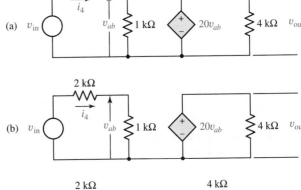

(a) v_{in} ... i_4 ... v_{ab} ... 1 kΩ ... 20v_{ab} ... 4 kΩ ... v_{out}

(b) v_{in} ... 2 kΩ ... i_4 ... v_{ab} ... 1 kΩ ... 20v_{ab} ... 4 kΩ ... v_{out}

(c) v_{in} ... 2 kΩ ... i_4 ... v_{ab} ... 1 kΩ ... 20v_{ab} ... 4 kΩ ... v_{out}

D 4.6 Synthesize (implement) a circuit to realize the characteristics shown in the figure.

Figure P4.6

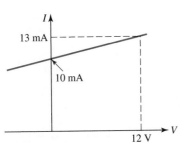

4.9 If the voltage-dependent voltage source of Problem 4.8 is replaced by a current-dependent current source, dependent on i_4 with $\beta = 150$ (see Fig. 4.3(d)), calculate v_{out} for each circuit.

☆ **4.7** Plot I_{out} as a function of V_{out} for the circuit shown as V_{out} varies from 0 V to 10 V in 2-V increments. Plot this characteristic for $I_b = 10\ \mu A,\ 20\ \mu A$, and $30\ \mu A$.

Figure P4.7

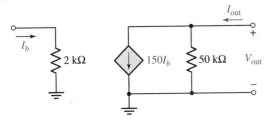

SECTION 4.2 THE OPERATIONAL AMPLIFIER

4.10 Calculate the output voltage for the circuit shown.

Figure P4.10

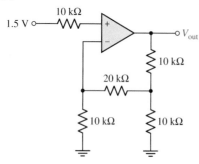

4.11 Write the expression for gain A of an op amp that has a dc gain of 400,000 V/V and a bandwidth of 10 Hz. Calculate the magnitude of the gain at a frequency of 5 kHz.

4.12 Assuming $A_{MBoa} = 200{,}000$ V/V, find R_{in}, R_{out}, and A_{MBi} for the circuit shown. If the op amp GBW is 4×10^6 Hz, what is the upper 3-dB frequency of the amplifier? (*Hint: R_{out} can be approximated.*)

Figure P4.12

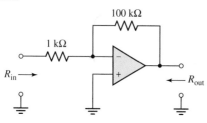

4.13 Assuming $A_{MBoa} = 200{,}000$ V/V, find R_{in}, R_{out}, and A_{MBi} for the circuit shown. If the op amp GBW is 4×10^6 Hz, what is the upper 3-dB frequency of the amplifier? (*Hint: R_{out} can be approximated.*)

Figure P4.13

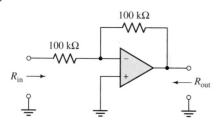

4.14 Assuming $A_{MBoa} = 200{,}000$ V/V, find R_{in}, R_{out}, and A_{MBni} for the circuit shown. If the op amp GBW is 4×10^6 Hz, what is the upper 3-dB frequency of the amplifier? (*Hint: R_{in} and R_{out} can be approximated.*)

Figure P4.14

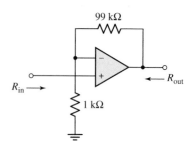

4.15 Either of the unity-gain stages shown can be used to couple the 10-kΩ resistance to the 1-kΩ load. Calculate the resulting midband value of voltage gain, v_{out}/v_{in}, for both stages.

Figure P4.15

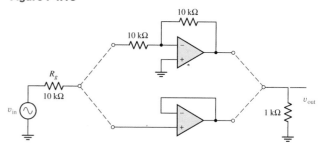

D 4.16 Using an op amp with $GBW_{oa} = 4 \times 10^6$ Hz, design a noninverting amplifier with a midband voltage gain of 75 V/V. What is the bandwidth of the amplifier?

D 4.17 Using an op amp with $GBW_{oa} = 4 \times 10^6$ Hz, design an inverting amplifier with a midband voltage gain of -10 V/V. What is the bandwidth of the amplifier?

4.18 For the given circuit:

(a) Calculate the midband value of voltage gain of the amplifier,

$$A_{MBi} = \frac{v_{out}}{v_{in}}$$

(b) Does the source resistance, R_g, affect the midband gain?

(c) Calculate the upper corner frequency of the overall voltage gain if the op amp GBW is 3.2 MHz.

(d) Does the source resistance affect the upper corner frequency of the amplifier?

Figure P4.18

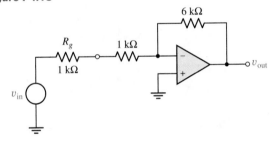

4.19 Repeat Problem 4.18 for the circuit shown.

Figure P4.19

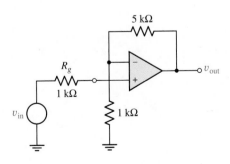

☆ **4.20** For the amplifier of the figure:

(a) Calculate $A_{MB} = v_{out}/v_{in}$.

(b) If the GBW of the op amp is 4 MHz, what is the upper corner frequency of the voltage gain?

(c) Write an expression for the voltage gain as a function of frequency.

Figure P4.20

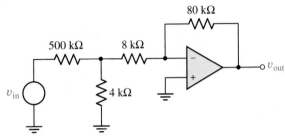

4.21 The upper corner frequency of the voltage gain for the circuit shown is found to be 150 kHz. What is the GBW of the op amp?

Figure P4.21

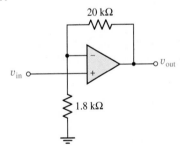

4.22 The upper corner frequency of the voltage gain for the circuit shown is found to be 150 kHz. What is the GBW of the op amp?

Figure P4.22

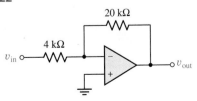

D 4.23 Design an inverting stage to amplify a 0.2-V peak amplitude signal to a 3-V peak amplitude signal. The source has a 4-kΩ impedance and the load is 2 kΩ. Assuming that the input source is referenced to ground and can be dc coupled to the op amp stage, use only one resistor with the op amp to complete this design. Calculate the bandwidth of your amplifier if the op amp *GBW* is 6×10^6 Hz.

D 4.24 Repeat Problem 4.23 using a noninverting stage and two resistors.

4.25 Calculate the midband voltage gain of the noninverting circuit shown and the upper 3-dB frequency, assuming that the op amp *GBW* is 2×10^6 Hz.

★ 4.27 The op amps are identical with $GBW_{oa} = 3.6$ MHz.

(a) Calculate the overall midband voltage gain of the amplifier.

(b) Calculate the upper 3-dB frequency of the amplifier voltage gain.

(c) Write an expression for voltage gain of the amplifier as a function of frequency.

(d) Modify the resistors R_1 and R_2 to keep the midband voltage gain constant while maximizing the upper corner frequency.

(e) Calculate this maximum upper corner frequency.

Figure P4.25

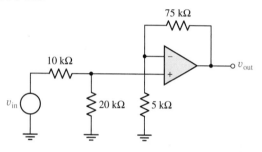

Figure P4.27

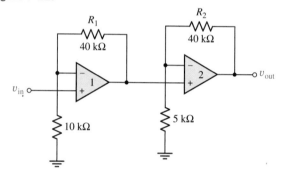

4.26 Calculate the midband voltage gain of the inverting circuit shown and the upper 3-dB frequency, assuming that the op amp *GBW* is 2×10^6 Hz.

★ 4.28 Repeat Problem 4.27 if $GBW_{oa1} = 3.6$ MHz and $GBW_{oa2} = 2.2$ MHz.

Figure P4.26

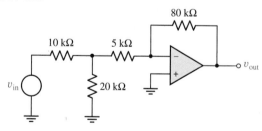

DEMONSTRATION PROBLEM AND DESIGN PROBLEMS

☆ **D 4.29** Change R_4 in the Demonstration Problem circuit to achieve equal upper corner frequencies for both stages. What is the value of the overall bandwidth? What is the overall midband voltage gain?

☆ **D 4.30** Change the two feedback resistors in the Demonstration Problem circuit to achieve an overall midband gain of -140 V/V while leading to the maximum overall bandwidth obtainable. What is the value of this bandwidth?

D 4.31 Identical op amp stages with $GBW_{oa} = 10^7$ Hz are available to construct an amplifier with overall midband gain of 600 V/V and an overall bandwidth of 800 kHz or more.

(a) What is the minimum number of stages required?

(b) What is the midband voltage gain of each stage?

D 4.32 Repeat Problem 4.31 if the required midband gain is to be 1000 V/V.

D 4.33 An amplifier is to be designed with an overall midband gain of 800 V/V and an overall bandwidth of 300 kHz. If six identical op amp stages are used to implement this design, what is the minimum GBW of the op amp?

D 4.34 Repeat Problem 4.33 if four stages are used.

5 | The Semiconductor Diode and Nonlinear Modeling

Most material discussed in the first four chapters of this textbook has been directed toward amplifier circuits. In this chapter we depart from that theme by discussing a device that is rarely used in amplifiers—the diode. We will treat the diode in some detail over the next few pages for several important reasons:

1. The base-emitter junction of the bipolar junction transistor behaves as a forward-biased diode in amplifying applications. Thus a knowledge of diode operation will aid the understanding of the bipolar junction transistor.

2. The behavior of the diode when reverse biased is the key to the fabrication of integrated circuits. This behavior is utilized to isolate from each other the many separate circuits created on a semiconductor chip.

3. The diode is a nonlinear device, and the diode model is somewhat different from that of linear devices such as constant value resistors or capacitors. The important concept of piecewise-linear modeling will be applied to the diode in most applications. The concept of small-signal and large-signal models will be introduced.

4. The diode is used in many important nonamplifier applications. A few of these will be considered later in this chapter.

DEMONSTRATION PROBLEM

In this chapter we will discuss principles that will allow us to solve the following problem.

Circuit for Demonstration Problem.

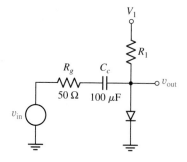

In the circuit shown, V_1 is a variable dc control voltage. The input voltage is an ac signal of frequency 5 kHz and a peak value of 20 mV. Calculate V_A and R_1 so that

$$\text{when } V_1 = V_A \text{ V}, \quad v_{\text{out}} = 12 \text{ mV}$$

$$\text{when } V_1 = (V_A + 2) \text{ V}, \quad v_{\text{out}} = 8 \text{ mV}$$

Also calculate the lower corner frequency of the circuit when $V_1 = V_A$.

This problem requires a knowledge of the small-signal or incremental model of the diode as well as the large-signal behavior of this device. We will develop these topics in this chapter.

Prior to discussing the operation of the diode, it is necessary to consider the conduction process that takes place in the semiconductor material from which the diode is constructed.

5.1 Semiconductor Material and Doping

IMPORTANT Concepts

1. A pure semiconductor material is a rather poor conductor of current, with a conductivity falling between those of a good insulator and a good metal conductor.
2. In order to make a semiconductor useful in electronics, impurity atoms are added to increase the conductivity of the material. This process is referred to as doping the semiconductor.
3. If a voltage is applied to a doped semiconductor material, current is carried by the mechanism of drift, similar to conduction in a metal.
4. If free carriers are injected into a semiconductor material to create a nonuniform density of these carriers, current can be carried by diffusion.

A semiconductor is a material possessing an electrical conductivity that falls between that of an insulator and that of a conductor. A good conductor might have a conductivity of 6×10^5 S/cm, whereas the conductivity of an insulator may be 6×10^{-18} S/cm. The most popular semiconductor material, silicon, has a conductivity at room temperature that is around 6×10^{-6} S/cm. The conductivity of silicon increases rather rapidly with temperature.

The difference in conductivity relates not to the total number of electrons in each material but to the concentration of free electrons that are not bound to the lattice atoms. These electrons are forced to move when an electric field is applied, resulting in charge flow or electric current. Copper has a very high concentration of free electrons, whereas insulators have a very low concentration. The insulator contains many electrons bound to the immobile lattice structure that makes up the material.

A microscopic examination of the silicon lattice shows that each atom is surrounded by four neighboring atoms in a diamond-type structure. Each individual lattice atom has the maximum number of electrons allowed to completely fill each inner orbit. The outermost orbit of a silicon atom is only half full, having only four electrons but calling for eight electrons to be completely full. These outer-orbit electrons are called valence electrons.

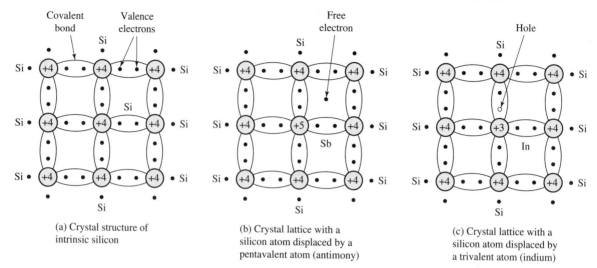

(a) Crystal structure of intrinsic silicon

(b) Crystal lattice with a silicon atom displaced by a pentavalent atom (antimony)

(c) Crystal lattice with a silicon atom displaced by a trivalent atom (indium)

Figure 5.1
Schematic representation of silicon structure.

Figure 5.1(a) shows a two-dimensional schematic that illustrates the atomic arrangement for a pure silicon crystal. In this structure, each atom shares one of its outer-orbit electrons with four surrounding atoms. The net effect is that each atom appears to have the required eight electrons in the outer orbit to satisfy the valence requirements of the atom.

Such an arrangement of shared electrons in the outer orbits of different atoms is called *covalent* bonding. All of the electrons in the inner orbits are tightly bound to the nucleus of the atom, whereas those in the outer orbit are only weakly bound to the lattice structure due to the covalent bond.

Generation and Recombination of Carriers At temperatures above absolute zero, each electron possesses some finite thermal energy. The thermal energy of an electron can cause the covalent bond to break, allowing an electron to move away from its parent atom. At room temperature, there is sufficient electron energy to produce several broken bonds. Each broken bond creates a free electron and a vacancy, or hole, in the outer orbit of the lattice atom. Free electrons are constantly being *generated* due to thermal energy. If a free electron moves near a hole, *recombination* can occur to again put eight electrons in the outer orbit. Recombination also takes place continually. At a given temperature, the rate of recombination equals the rate of generation of electrons and the average number of free electrons remains constant. This average concentration of free electrons, which equals the average concentration of holes, is called the *intrinsic carrier concentration* and is designated n_i. For silicon, near room temperature, $n_i = 1.5 \times 10^{10}/\text{cm}^3$.

Mobility of Electrons and Holes When an electric field is impressed across an intrinsic semiconductor material, current is carried by the free electrons and by the holes. Lattice collisions limit the velocity of the electrons for the given applied field. A measure of the ease with which the particles can move through the lattice is called the *mobility* of the particle. The mobility is given by

$$\mu = |v/E| \tag{5.1}$$

where v is the drift velocity of the carriers and E is the applied electric field. At room temperature, the mobility of free electrons in intrinsic silicon is about 1300 cm²/volt-sec.

The process of hole conduction is more complex than that of free electron conduction. A hole can be filled by a valence electron from a neighboring atom, creating a new hole at the site of the atom from which the electron came. All valence electrons move over a relatively large range at room temperature, and a given hole can move randomly as electrons move from one atom to another. Application of an electric field influences the direction of movement of any valence electrons that break their bonds and move to fill holes. The holes move in the opposite direction to that of the electrons. This type of electron movement is slower and more involved than the movement of free electrons. The mobility of holes is considerably less than the mobility of free electrons. For silicon at room temperature a typical value of hole mobility is about 480 cm^2/volt-sec.

The conductivity of a semiconductor material can be expressed as

$$\sigma = q(n\mu_n + p\mu_p) \tag{5.2}$$

where p and n are hole and electron densities, respectively, and q is the charge of a single electron (1.6×10^{-19} coulomb). For an intrinsic material, $n = p = n_i$, and Eq. (5.2) can be written as

$$\sigma_i = qn_i(\mu_n + \mu_p) \tag{5.3}$$

With an intrinsic concentration of $n_i = 1.5 \times 10^{10}$, the conductivity of silicon at room temperature is calculated to be 4.27×10^{-6} S/cm.

5.1.1 INCREASING THE DENSITY OF FREE ELECTRONS OR HOLES (DOPING)

The conductivity of silicon can be greatly increased by the addition of the proper impurity. The controlled addition of an impurity to a semiconductor material is called *doping*. The number of free electrons can be increased by doping with donor atoms, and the number of holes can be increased by doping with acceptor atoms. If excess electrons are created, the material is called *n*-type; if excess holes are created, the material becomes *p*-type.

Donor Atoms: A donor atom is one such as antimony that has five valence electrons in its outer orbit. When added to silicon, the antimony atoms fit into the lattice without modifying the lattice structure. Since only four valence electrons per atom are required for covalent bonding, the fifth electron of the doping atom is only loosely bound to the parent atom as shown in Fig. 5.1(b). At room temperature, the extra electron has enough thermal energy to break its association with the parent atom and become a free electron. If the density of donor impurities, N_D, is high compared to n_i, the total concentration of free electrons is approximately equal to N_D. Values of N_D may range from 10^{13} to 10^{20} per cm^3.

A well-known relationship between the carrier densities in a semiconductor is

$$pn = n_i^2 \tag{5.4}$$

For a donor-doped material $n = N_D$, leading to a density of holes of

$$p = \frac{n_i^2}{N_D} \tag{5.5}$$

Since $N_D \gg n_i$, the density of holes is much smaller than n_i. Using Eq. (5.2), the conductivity of an *n*-type material is

$$\sigma_n = q\mu_n N_D \tag{5.6}$$

The *majority current carrier* in *n*-type silicon is the electron, and the *minority carrier* is the hole.

Acceptor Atoms: Acceptor atoms such as indium have only three valence electrons. When added to the silicon material, the indium atom has an outer orbit with four electrons shared by the four silicon neighbor atoms plus the three electrons of the indium atom. Since eight electrons are required in the outer orbit to satisfy bonding requirements, a hole is created at the site of each added indium atom as shown in Fig. 5.1(c). The density of holes is then given by

$$p = N_A \tag{5.7}$$

where N_A is the density of acceptor atoms.

The conductivity of the p-type semiconductor is

$$\sigma_p = q\mu_p N_A \tag{5.8}$$

since the number of free electrons, given by

$$n = \frac{n_i^2}{N_A} \tag{5.9}$$

is much smaller than N_A.

When an electric field is applied to a p-type semiconductor, current is carried primarily by holes. Holes are then majority carriers, whereas free electrons are minority carriers.

EXAMPLE 5.1

If $n_i = 1.5 \times 10^{10}/\text{cm}^3$ for a silicon material doped with $N_A = 2 \times 10^{14}/\text{cm}^3$, determine the concentrations of free electrons and holes. Calculate the conductivity of the material due only to free electrons. Calculate the total conductivity of the material. Assume that $\mu_n = 1300 \text{ cm}^2/\text{volt-sec}$ and $\mu_p = 480 \text{ cm}^2/\text{volt-sec}$.

SOLUTION Since N_A is considerably larger than n_i, the concentration of free holes is

$$p = N_A = 2 \times 10^{14}/\text{cm}^3$$

Using Eq. (5.4), the concentration of free electrons is

$$n = \frac{n_i^2}{p} = \frac{(1.5 \times 10^{10})^2}{2 \times 10^{14}} = 1.125 \times 10^6/\text{cm}^3$$

We note that the concentration of free electrons is eight orders of magnitude less than the concentration of free holes.

The conductivity due to free electrons is

$$\sigma_n = qn\mu_n = 1.6 \times 10^{-19} \times 1.125 \times 10^6 \times 1300 = 2.34 \times 10^{-10} \text{ S/cm}$$

This component of conductivity is even less than the intrinsic conductivity of silicon and contributes negligibly to the total conductivity, which is

$$\sigma = \sigma_p = qp\mu_p = 1.6 \times 10^{-19} \times 2 \times 10^{14} \times 480 = 1.54 \times 10^{-2} \text{ S/cm}$$

5.1.2 SPACE-CHARGE NEUTRALITY

All atoms in the semiconductor have a net charge of zero, if no electrons have broken free. When an electron breaks free and takes a net charge of $-q$ with it, the atom now has a charge of $+q$. When an electron is trapped in a hole, the atom takes on a charge of $-q$ and the newly created hole has a charge of $+q$. The total charge in any region of a semiconductor material must equal zero, provided there is no significant voltage difference across the region. Any movement of charge in the semiconductor material, whether by thermal motion or an external field, will be accompanied by an electric field, tending to pull balancing charges into the region to maintain total charge neutrality. We note that the charge carried by electrons or holes is mobile charge, and that resulting from the lattice atoms is immobile charge.

5.1.3 CURRENT FLOW IN A SEMICONDUCTOR

Drift Current: Conduction in a semiconductor material may occur by either of two mechanisms. The first is referred to as current by *drift* and has been discussed previously. When an electric field is applied to the material, the free holes and free electrons move under the influence of the field. Drift current is governed by the ohmic relationship

$$J = \sigma E \tag{5.10}$$

where J is the current density, σ is the conductivity, and E is the applied electric field. We have seen that the conductivity is determined by the free carrier concentration and the mobility. The conductivity and the current density can be controlled by the addition of impurity atoms.

Diffusion Current: The second mechanism leading to current flow in a semiconductor material is that of *diffusion*. Diffusion of carriers occurs in a semiconductor when there is a difference in the concentration of carriers in adjacent regions of the material. Since all carriers exhibit a random thermal motion causing collision with other carriers, they naturally diffuse away from regions of higher concentration because of more frequent collisions in these regions. This movement of carriers is similar to that of gas molecules with an initial concentration that is nonuniform over some region. The molecules move from the region of high concentration until a uniform distribution of molecules exists.

It has been shown that the current carried by diffusion at some point x in a semiconductor material is related to the concentration gradient of free carriers at that point. The equation for diffusion current carried by free electrons is

$$I_n = q D_n \frac{dn}{dx} A \tag{5.11}$$

where A is the cross-sectional area of the material, D_n is the diffusion constant for electrons, and dn/dx is the slope of the free electron concentration at point x.

For free holes, the diffusion current is given by

$$I_p = -q D_p \frac{dp}{dx} A \tag{5.12}$$

where D_p is the concentration gradient of free holes at point x.

The diffusion constants are related to the carrier mobilities. Typical values for silicon at room temperature are

$$D_n = 34 \text{ cm}^2/\text{sec}$$

$$D_p = 13 \text{ cm}^2/\text{sec}$$

Drift current is the component of importance in MOSFET operations. Although both components are important in the diode and the BJT, their operation is limited by diffusion current flow.

5.1.4 RECOMBINATION OF CHARGE CARRIERS

For an intrinsic semiconductor material, the concentrations of free electrons and holes are both equal to n_i and, although recombination and generation of free carriers take place continuously, these concentrations remain statistically constant. For a material doped with donor impurities the situation is similar, except now $n \gg p$. The recombination and generation rates are again equal, and both n and p remain constant at a given temperature.

In the operation of semiconductor devices, a given region will have excess carriers injected into it, upsetting the equilibrium conditions. When this occurs, the recombination rate increases and the excess carriers ultimately recombine over a period of time, returning the concentration of these carriers to the equilibrium value.

5.1.5 DIFFUSION CURRENT IN THE PRESENCE OF RECOMBINATION

Let us consider a p-type material into which electrons are injected at one end and absorbed at the other end as shown in Fig. 5.2. We will assume that no electric field is applied to the material.

Since recombination takes place throughout the region, we expect the density of electrons to decrease from the injection density, $n_p(0)$, to the equilibrium value of n_{po} as the electrons move away from the injection end. The value of n_{po} is equal to n_i^2/p_p, which is very small for the p-type material. We will take this value to be zero.

It has been shown that the recombination causes an exponential falloff of electron concentration from the left edge to the right edge of the material. If $x = 0$ at the point of injection, the electron density is given by

$$n_p(x) = n_p(0)e^{-x/L_n} \tag{5.13}$$

where L_n is the diffusion length for the electrons. We note that the concentration gradient decreases in magnitude as x increases. This gradient can be found by differentiating Eq. (5.13) to give

$$\frac{dn_p}{dx} = \frac{-n_p(0)}{L_n} e^{-x/L_n} \tag{5.14}$$

The diffusion current is proportional to the negative gradient and from Eq. (5.11) is found to be

$$I_n = -D_n Aq \frac{n_p(0)}{L_n} e^{-x/L_n} \tag{5.15}$$

The total current throughout the material must be constant, but Eq. (5.15) shows that the diffusion current falls with x. To understand the means by which total current flow remains constant, we must return to the principle of space charge neutrality. If there is no electric field impressed across the material, the total space charge at any point x must be zero. The injection of electrons at the left edge of the block must be accompanied by a flow

Figure 5.2
Electron distribution in the presence of recombination.

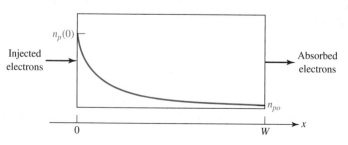

of holes into the right edge of the block from the external circuit. This assumes that holes are prevented from entering with the injected electrons at the left edge, an assumption that is true for the *pn* junction. The holes that enter assume the same concentration profile as the injected electrons at every point along the *x*-axis. These holes then move with the electrons from left to right and flow out of the material with the electrons. The holes that flow into the block and recombine with electrons result in a net recombination current. The holes that diffuse to the right side of the material and out to the external circuit contribute no net charge movement and no current flow. The constant total current at any point *x* is the sum of the electron diffusion current and the hole recombination current. These two components vary with *x*, but the sum is constant.

The electron diffusion current has a negative sign, indicating a conventional current flow from right to left even though the negatively charged electrons move from left to right. The hole recombination current flows under the influence of an induced electric field that occurs as the electrons are injected, disturbing neutrality. Thus, the recombination current is a drift current.

To find the constant total current, we evaluate the diffusion current at the point of injection, since no recombination takes place until the electrons begin diffusing. This gives

$$I_T = I_n|_{x=0} = \frac{-D_n A q n_p(0)}{L_n} \tag{5.16}$$

The total current flow is then proportional to the number of electrons injected at $x = 0$. As diffusion current decreases farther into the material, the total current remains constant because the induced hole drift current increases.

Diffusion of holes injected into an *n*-type material takes place by the same mechanism just described. In this case, the total current is given by

$$I_T = \frac{D_p A q p_n(0)}{L_p} \tag{5.17}$$

With this background material, we are prepared to discuss the current flow through a *pn* junction.

EXAMPLE 5.2

For the two concentrations of free electrons of Fig. 5.3, calculate the diffusion current at $x = 0$, $x = 0.001$ cm, and $x = 0.01$ cm. Assume that the cross-sectional area is $A = 10^{-4}$ cm^2.

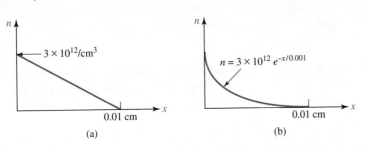

Figure 5.3
Free electron concentrations.

SOLUTION For the distribution of Fig. 5.3(a), the slope of the free electron concentration is constant with *x* at a value of

$$\frac{dn}{dx} = \frac{\Delta n}{\Delta x} = \frac{-3 \times 10^{12}}{0.01} = -3 \times 10^{14}/\text{cm}^3$$

PRACTICE Problems

5.1 Calculate the conductivity of a p-type material doped with $N_A = 10^{18}/cm^3$. *Ans:* 75.9 S/cm.

5.2 An n-type material has a conductivity of 100 S/cm. Find the concentration of donor atoms in this material. *Ans:* $4.8 \times 10^{17}/cm^3$.

From Eq. (5.11), the electron diffusion current is

$$I_n = qD_n\frac{dn}{dx}A = 1.6 \times 10^{-19} \times 34 \times (-3 \times 10^{14}) \times 10^{-4} = -1.63 \times 10^{-7}A$$

The negative sign again indicates that conventional current flows from right to left when electrons diffuse from left to right. This current is constant from $x = 0$ to $x = 0.01$ cm since the slope is constant over this range.

For the distribution of Fig. 5.3(b), the slope varies and is found by differentiation to be

$$\frac{dn}{dx} = 3 \times 10^{12} \, (-10^3) \, e^{-x/0.001} = -3 \times 10^{15} \, e^{-x/0.001}$$

The current is again calculated from Eq. (5.11) and is

$$I_n = qD_n\frac{dn}{dx}A = 1.6 \times 10^{-19} \times 34 \times \left(-3 \times 10^{15} \, e^{-x/0.001}\right) \times 10^{-4} = -1.63 \times 10^{-6} \, e^{-x/0.001}$$

Equation (5.15) could also have been used for this calculation.

At $x = 0$, this current is -1.63×10^{-6} A. At $x = 0.001$ cm, the current is

$$-1.63 \times 10^{-6} \, e^{-1} = -0.6 \times 10^{-6} \, A$$

At $x = 0.01$ cm, the current is

$$-1.63 \times 10^{-6} \, e^{-10} = -7.4 \times 10^{-11} \, A$$

The diffusion current drops as the magnitude of the slope of free electron distribution drops.

5.2 The *pn* Junction

IMPORTANT Concepts

1. A *pn* junction is called a diode. With no electrical connections to the diode, free electrons and holes flow across the physical junction. This current flows due to the built-in junction voltage and due to diffusion. The net current flow is zero since no external current flows.

2. The net current flow through a diode in the forward direction varies exponentially with the voltage applied to the junction.

3. There are two sources of capacitance across the *pn* junction. These are the depletion-layer and diffusion capacitances.

The two types of rectifying junction of major import in electronics are the semiconductor *pn* junction and the metal-semiconductor junction. Since the rectifying properties of both junctions are somewhat similar to those of a vacuum tube diode, the name diode has been applied to these rectifiers.

The metal-semiconductor junction has great historical importance because it led to the development of the point-contact transistor one or two years prior to the development of the junction transistor. This device is no longer used. A metal-semiconductor junction that is presently important is the well-known Schottky diode. This device is applied in the TTL digital logic family to decrease switching times. The *pn* junction is currently the most widely used, especially when serving as the basis for the bipolar junction transistor.

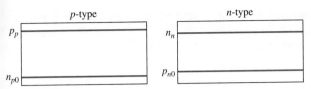

Figure 5.4
p- and n-type materials showing majority and minority carrier concentrations.

5.2.1 JUNCTION FORMATION

A semiconductor diode is made up of two differently doped regions abutting each other. One of these regions is doped with donor atoms to create an *n*-type region having free electrons. The other region is doped with acceptor atoms to create a *p*-type region containing free holes. The border of the two regions is the physical junction of the diode; however, the electrical junction includes narrow regions on either side of the border referred to as depletion regions.

In the early days of semiconductor fabrication, the junction was formed by bringing together two separate doped materials and fusing them together at high temperatures. Modern fabrication methods for smaller diodes generally apply vapor diffusion techniques to create *pn* junctions within a single mass of semiconductor material. This process leads to impurity profiles within the two regions that are not constant. To simplify the discussion of diode operation and highlight the more important mechanisms, we will assume that the junction is created by bringing together two separate, uniformly doped regions. Practical departures from the behavior of this idealized junction will be treated later in this chapter.

The separate *p*- and *n*-type materials are shown in Fig. 5.4. The majority carrier concentrations can be controlled during doping and will be given by

$$p_p = N_A \quad \text{and} \quad n_n = N_D$$

where N_A is the concentration of acceptor atoms and N_D is the concentration of donor atoms.

The minority carrier concentrations can be calculated as

$$p_{n0} = \frac{n_i^2}{n_n} = \frac{n_i^2}{N_D} \quad \text{and} \quad n_{p0} = \frac{n_i^2}{p_p} = \frac{n_i^2}{N_A}$$

The concentrations of minority carriers are many orders of magnitude smaller than the concentrations of majority carriers created by doping. For example, p_p might be $10^{18}/cm^3$ whereas n_{p0} would be $2.25 \times 10^2/cm^3$. In the *n*-type region, n_n could be $10^{16}/cm^3$ and p_{n0} would be $2.25 \times 10^4/cm^3$.

Note that there is a very sharp gradient of free carriers at the end points of each region. The concentration of p_p, for example, appears as in Fig. 5.5. There is a very strong diffusive tendency for these holes, but they cannot leave the *p*-type region, because they are bound by the extremes of the material. The same reasoning applies to the *n*-type material also; that is, the electrons have a strong diffusive tendency due to the high concentration gradient, but are constrained from leaving the material by the extremes of the material.

If the *p*- and *n*-type materials are joined together, a *pn* junction is formed. The holes that are near the edge of the *p*-type material can now diffuse. They will tend to move into

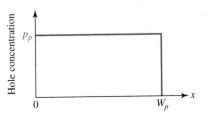

Figure 5.5
Majority-carrier distribution in p-type material.

the n-type material. The electrons near the edge of the n-type material will flow into the p-type region. As these carriers move, the concentration gradient decreases. If there were no electrostatic forces acting on the particles—that is, if they behaved like gas molecules—the diffusion would proceed until no gradient existed. The carriers would then be uniformly distributed throughout both regions. Fortunately, there is an electrostatic effect, which limits this diffusion current and ultimately allows external control of the junction current flow.

5.2.2 THE DEPLETION REGION

It is important to recognize that prior to joining, free holes exist in one region and free electrons in the other, but both regions are space-charge neutral. The negative charge of each free electron is offset by the positive charge of an immobile parent atom in the n-type region. The positive charge of each free hole is offset by the negative charge of an immobile parent atom in the p-type region. After joining the regions, the diffusion of free carriers causes a charge to build up near the junction. When a hole leaves the p region, it uncovers a parent acceptor atom, which now has a negative charge. When an electron leaves the n region, it uncovers a parent or donor atom, which now has a positive charge. The doping atoms are bound to the lattice and hence are immobile. The resulting junction appears in Fig. 5.6.

There is a region near the junction where few free carriers are present due to the fact that they have diffused across the junction. This region is called the depletion region or transition region. The voltage appearing across the depletion region is called the barrier voltage. Its presence is due to the net charge in the depletion region. The barrier voltage opposes the flow of carriers due to diffusion and eventually limits the depletion of further area. The steps involved in this limiting process can be listed:

1. A hole-diffusion current and an electron-diffusion current exist across the depletion region of the diode due to the concentration gradients of the carriers.

2. The removal of free carriers by diffusion current upsets space-charge neutrality; hence a voltage is developed across this area.

3. This barrier voltage causes drift currents for both electrons and holes, which cancel the diffusion components exactly.

4. The cancellation of the drift and diffusion currents is a "built-in" feature of the junction, as can be demonstrated by the following assumption.

Assume that the diffusion components are larger than the drift components of current. This means that more free carriers will leave their associated impurity atoms, thereby widening the depletion region. The barrier voltage will then increase, causing the drift current to increase. It is obvious that this process will result in an equilibrium situation wherein the width of the depletion region is constant in the absence of an applied voltage.

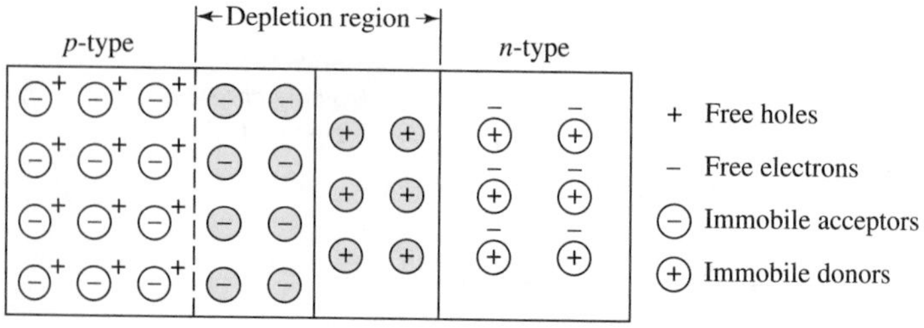

Figure 5.6
Schematic representation of carriers and doping atoms in a pn junction.

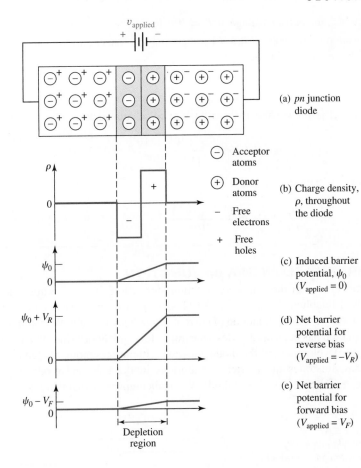

(a) *pn* junction diode

(b) Charge density, ρ, throughout the diode

⊖ Acceptor atoms

⊕ Donor atoms

− Free electrons

+ Free holes

(c) Induced barrier potential, ψ_0 ($V_{applied} = 0$)

(d) Net barrier potential for reverse bias ($V_{applied} = -V_R$)

(e) Net barrier potential for forward bias ($V_{applied} = V_F$)

Depletion region

Figure 5.7
(a) pn junction. (b) Charge density throughout diode. (c) Induced barrier voltage with no applied voltage. (d) Net barrier potential for reverse bias. (e) Net barrier potential for forward bias.

Under equilibrium conditions, the barrier voltage is given by

$$\psi_o = \frac{kT}{q} \ln \frac{n_n p_p}{n_i^2} = \frac{kT}{q} \ln \frac{N_D N_A}{n_i^2} = V_T \ln \frac{N_D N_A}{n_i^2} \qquad (5.18)$$

where k is Boltzmann's constant, T is the absolute temperature, and q is the electronic charge. This quantity kT/q is generally designated V_T, and at $T = 300°K$ the value of V_T is approximately 0.026 V. The value of the barrier voltage for a silicon diode might be 0.8–0.9 V.

The depletion region is two to three orders of magnitude narrower than the p- or n-type regions of a diode. In Fig. 5.7 the extent of this region is greatly exaggerated to demonstrate the carrier concentrations in the depletion region. Typically the depletion region width is less than a small fraction of a micron (10^{-6} m).

If the applied voltage is V, the net barrier voltage is $\psi = \psi_o - V$. For a reverse voltage of $V = -V_R$, the net voltage difference across the junction is $\psi_o + V_R$. The application of a forward voltage, $V = V_F$, that does not exceed ψ_o results in a voltage difference across the junction of $\psi_o - V_F$. Figure 5.7 demonstrates the voltage across the depletion region for no applied voltage, a reverse voltage, and a forward voltage.

EXAMPLE 5.3

A diode can be approximated as an abrupt junction with $N_A = 3 \times 10^{17}/\text{cm}^3$ and $N_D = 10^{15}/\text{cm}^3$.

(a) Calculate ψ_o, the barrier voltage at $T = 300°$K.

(b) If it is desired to make $\psi_o = 0.8$ V by changing the concentration of acceptor atoms, what is the required value of N_A?

SOLUTION

(a) Equation (5.18) allows the barrier voltage to be calculated as

$$\psi_o = 0.026 \ \ln \frac{3 \times 10^{17} \times 10^{15}}{(1.5 \times 10^{10})^2} = 0.726 \text{ V}$$

(b) Solving Eq. (5.18) for N_A gives

$$N_A = \frac{n_i^2}{N_D} e^{\psi_o/V_T} = \frac{(1.5 \times 10^{10})^2}{10^{15}} \ e^{0.8/0.026} = 5.19 \times 10^{18}/\text{cm}^3$$

5.2.3 CURRENT FLOW IN A *pn* JUNCTION

Within a short time after the discovery of "transistor action" by John Bardeen and Walter Brattain of Bell Telephone Labs in the late 1940s, the project supervisor, William Shockley, developed a theory for the conduction of current across a *pn* junction. With slight modifications, this equation has been used to describe transistors and diodes since that time. We will summarize the development of the diode equation without presenting all details involved.

Figure 5.8 illustrates the carrier distributions in the unbiased diode, the forward-biased diode, and the reverse-biased diode. We again emphasize that there is a tremendous

Figure 5.8
Majority and minority concentrations for (a) an unbiased diode, (b) a forward-biased diode, (c) a reverse-biased diode.

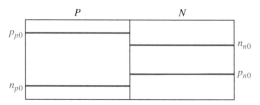

(a) Carrier concentration for zero bias

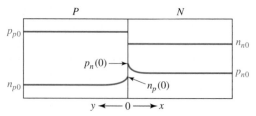

(b) Carrier concentration for forward bias

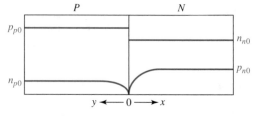

(c) Carrier concentration for reverse bias

difference in concentrations of majority carriers compared to minority carriers in each region. The depletion region is not shown in this simplified representation of the *pn* junction.

When the diode is unbiased, the net current flow is zero. When the external voltage is applied to the diode, the voltage is absorbed across the junction. The *p* and *n* regions have negligible drops. The barrier voltage is lowered by a forward bias, and this leads to less drift current for both holes and electrons. The diffusion components now exceed the drift components, and more holes move from the *p* region to the *n* region and more electrons move from the *n* region to the *p* region.

The concentration of holes injected into the left edge of the *n* region depends on the minority hole concentration and on the applied voltage. The same can be said of the electrons injected into the right edge of the *p* region. The equations governing these injected quantities are

$$p_n(0) = p_{n0}\, e^{V/V_T} \tag{5.19}$$

and

$$n_p(0) = n_{p0}\, e^{V/V_T} \tag{5.20}$$

where V is the applied voltage and is positive for forward bias and negative for reverse bias.

For the forward-biased diode, the densities of both injected holes and electrons increase exponentially over the unbiased case. Since V_T is so small, an applied voltage of 0.6 V can result in a huge increase of injected carriers over the equilibrium values as shown in Fig. 5.8(b). If a reverse voltage is applied, the densities of holes and electrons at the edges of the two regions are driven to zero as shown in Fig. 5.8(c).

As the holes and electrons diffuse away from the junction, recombination and resulting drift currents take place. However, the total current flow can be evaluated by summing the hole diffusion current at the junction edge and the electron diffusion current at the junction edge. Using the diffusion equations for holes and electrons, Eqs. (5.11) and (5.12), the current is expressed as

$$I = I_n + I_p = q\,D_n A\left[\frac{dn}{dx}\right]_{x=0} - q\,D_p A\left[\frac{dp}{dx}\right]_{x=0} \tag{5.21}$$

As discussed in an earlier section, when excess carriers are injected into a region, these carriers recombine as they diffuse away from the point of injection. The carrier concentration exhibits an exponential decay with distance from the point of injection down to the equilibrium value. The equation for the hole distribution in the *n* region is then given by

$$p_n(x) = (p_n(0) - p_{n0})\,e^{-x/L_p} + p_{n0} = p_{n0}\left(e^{V/V_T} - 1\right)e^{-x/L_p} + p_{n0} \tag{5.22}$$

Differentiating this expression with respect to x and evaluating this gradient at $x = 0$ gives

$$\left[\frac{dp_n}{dx}\right]_{x=0} = -\frac{p_{n0}}{L_p}\left(e^{V/V_T} - 1\right) \tag{5.23}$$

Substituting this expression into the diffusion equation for holes results in a hole diffusion current at the point of injection of

$$I_p = \frac{q D_p A p_{n0}}{L_p} \left(e^{V/V_T} - 1\right) \tag{5.24}$$

Similar consideration for the electron diffusion current at the point of injection gives

$$I_n = \frac{q D_n A n_{p0}}{L_n} \left(e^{V/V_T} - 1\right) \tag{5.25}$$

The total current is the sum of the hole and electron components and can be written as

$$I = Aq \left[\frac{n_{p0} D_n}{L_n} + \frac{p_{n0} D_p}{L_p}\right] \left(e^{V/V_T} - 1\right) \tag{5.26}$$

At a given temperature, the constant terms can be identified with a single constant called the saturation current and expressed as

$$I_S = Aq \left[\frac{n_{p0} D_n}{L_n} + \frac{p_{n0} D_p}{L_p}\right] = Aq n_i^2 \left[\frac{D_n}{N_A L_n} + \frac{D_p}{N_D L_p}\right] \tag{5.27}$$

The diode equation, sometimes called the Shockley diode equation, becomes

$$I = I_S\left(e^{qV/kT} - 1\right) \tag{5.28}$$

The parameter I_S is a function of doping and other constants of the material used. For a typical small silicon diode, it may be equal to 10^{-14} A at room temperature. I_S is a very strong function of temperature. For a silicon diode, I_S doubles with each increase in temperature of 5–10°C. An increase of temperature by 35–40°C could lead to an increase in I_S by a factor of as much as 2^7 or 128.

Near room temperature, the quantity kT/q is about 0.025 V. The diode is generally at a slightly higher temperature than ambient temperature due to its power dissipation. The quantity kT/q is often taken as 0.026 V for this reason.

Equation (5.28) predicts a diode current of zero for an applied voltage of zero since the expression in parentheses gives

$$e^0 - 1 = 0$$

When V is negative as it is for a reverse bias and much larger in magnitude than 0.026 V, the exponential term in Eq. (5.28) approaches zero. For example, if $V = -0.2$ with $I_S = 10^{-14}$, the current is calculated as

$$I = I_S\left(e^{-0.2/0.026} - 1\right) = 10^{-14} \times (0.00046 - 1) = -0.99954 \times 10^{-14} \approx -I_S$$

The diode current approaches a value of $-I_S$ as V becomes more negative and is very near this value even for small values of reverse-bias voltages. For all reverse voltages over a few tenths of a volt, the diode current can be accurately approximated by $-I_S$. A typical value of reverse current for a low-power silicon diode is 10^{-14} A. When used in most applications, this small current can be taken to be zero. Thus, when it is stated that no diode current flows for the reverse-biased condition, the assumption that I_S is negligible is invoked.

As a forward voltage is applied to the junction, the exponential term in Eq. (5.28) increases very rapidly with V. This term becomes much larger than unity for quite small values of V. The equation for current in this situation can accurately be expressed as

$$I = I_S e^{qV/kT} \tag{5.29}$$

Even though the exponential term rapidly increases with V, the I_S term is very small, resulting in a very small forward current for small values of V. In fact, for most typical applications, we can assume that I is zero for values of V up to a value of several tenths of a volt. Often this value is taken to be 0.6–0.7 V for silicon diodes. Once I increases to an appreciable value, perhaps in the range of tens of μA, the curve for I breaks upward quite abruptly with increasing V. For voltages above 0.7–0.8 V, the curve has a very steep slope.

We can calculate the value of V to cause I in Eq. (5.29) to reach 100 μA for $I_S = 10^{-14}$, which gives

$$V = \frac{kT}{q} \times \ln \frac{I}{I_S} = 0.026 \times \ln 10^{10} = 0.6 \text{ V}$$

As the voltage increases from 0.6 V to 0.7 V, the current for this diode increases from 100 μA to 4.93 mA. The small increase of 0.1 V leads to a current increase by a factor of 49.3. Whereas an ideal diode shows a vertical line segment in the forward-biased region, the theoretical diode has a finite, but very steep, increasing slope with V.

EXAMPLE 5.4

In the circuit of Fig. 5.9, the diode has a value of $I_S = 10^{-12}$ mA at room temperature (300°K).

(a) Approximate the current I, assuming the voltage drop across the diode is 0.7 V.

(b) Calculate the accurate value of I.

(c) If I_S doubles for every 6°C increase in temperature, repeat part (b) if the temperature increases by 40°C.

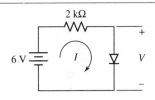

Figure 5.9
Diode circuit.

SOLUTION

(a) The resistor will have an approximate voltage of $6 - 0.7 = 5.3$ V. Ohm's law then gives a current of

$$I = \frac{5.3}{2} = 2.65 \text{ mA}$$

In many situations, this is the preferred method of calculation because it is easy and fairly accurate.

(b) This part of the example demonstrates the difficulty in solving a nonlinear equation. The current through the resistor must equal the diode current, allowing us to write

$$I = \frac{6 - V}{2} \text{ (resistor current)}$$

$$I = I_S e^{V/V_T} \text{ (diode current)}$$

and

$$\frac{6 - V}{2} = 10^{-12} e^{V/0.026}$$

This is a nonlinear equation in V and must be solved by iterative methods. In this instance the solution to three-place accuracy is

$$V = 0.744 \text{ V}$$

Using this voltage in the equation for resistor current leads to

$$I = \frac{6 - V}{2} = \frac{6 - 0.744}{2} = 2.63 \text{ mA}$$

Note that this value differs from the approximate value of part (a) by 0.02 mA, a difference of less than 1%.

(c) When the temperature changes, both I_S and V_T will change. Since $V_T = kT/q$ varies directly with T, the new value is

$$V_T(340) = V_T(300) \times \frac{340}{300} = 0.0295 \text{ V}$$

The value of I_S doubles for each 6°C increase, thus the new value of I_S is

$$I_S(340) = I_S(300) \times 2^{40/6} = 1.016 \times 10^{-10} \text{ mA}$$

The equation for I is now

$$I = \frac{6 - V}{2} = 1.016 \times 10^{-10} \times e^{V/0.0295}$$

The iterative solution to this equation gives

$$V = 0.640 \text{ V} \qquad \text{and} \qquad I = 2.68 \text{ mA}$$

5.2.4 DEPARTURES FROM IDEAL BEHAVIOR

It is found that Eq. (5.28) approximates the experimental V-I curve of a diode. Figure 5.10 shows the ideal V-I characteristics of the diode plotted from this equation along with a dashed curve that represents a more practical diode.

The four major reasons why the actual diode characteristics do not correspond exactly to the ideal are listed here.

1. Ohmic resistance and contact resistance in series with the diode cause the V-I characteristics to become linear at high forward currents.

2. Avalanche or Zener breakdown takes place at high reverse voltages, causing an abrupt increase in current. This characteristic can be controlled and used to advantage in the

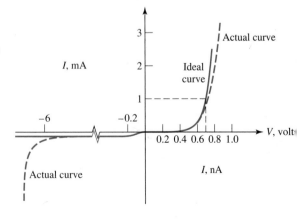

Figure 5.10
V-I characteristics of a silicon diode. Note the change in current scale between positive and negative currents.

breakdown diode to create constant voltage drops in applications such as dc power supplies.

3. Surface contaminants cause an ohmic layer to form across the junction. As reverse voltage is increased, the reverse current increases slightly, instead of remaining constant at I_S.

4. Recombination of current carriers in the depletion region takes place due to "traps" brought about by irregularities of the lattice. This effect can be accounted for by introducing a number N in the exponent of Eq. (5.28) to give

$$I = I_S\left(e^{qV/NkT} - 1\right) \tag{5.30}$$

The number N ranges from 1 to 2. For a silicon diode, it is near 1, whereas it may approach 2 for the *pn* junction of a transistor.

The dashed curves of Fig. 5.10 demonstrate these effects.

5.2.5 DIODE CAPACITANCE

The basic equation for capacitance is

$$C = \frac{dQ}{dV} \tag{5.31}$$

where dQ is the change in charge resulting from a change in voltage dV. There are two sources of capacitance in the semiconductor diode. One is associated with the change in charge of the depletion region when the voltage across the junction is changed. This is called depletion-layer capacitance. The other capacitance results from the charge required to change the minority carrier distributions when junction voltage is changed. Since the minority carrier distributions determine the diffusion current that flows through the diode, this capacitance is called the diffusion capacitance.

Depletion-Layer Capacitance: The depletion-layer capacitance can be calculated by an equation identical to that for the parallel plate capacitance; that is,

$$C = \frac{\varepsilon A}{W}$$

In this equation, ε is the dielectric constant, A is the area of the plates, and W is the separation of the plates. For the *pn* junction, A and ε are constant, but the width of the depletion region is a function of applied voltage. This equation is therefore not very useful for calculating depletion-layer capacitance.

It has been shown that the depletion-layer capacitance can be expressed as

$$C_T = \frac{K_1}{(\psi_o - V)^m} \tag{5.32}$$

where K_1 is a constant called the zero-bias capacitance, ψ_o is the zero-bias barrier voltage, and V is the applied voltage. This voltage is taken to be positive for a forward applied voltage and negative for a reverse applied voltage. The constant m is 0.5 for an abrupt junction and is nearer 0.33 for graded junctions. An abrupt junction corresponds to the "textbook" junction that assumes uniform doping of the two regions. A graded junction is one that is formed in a modern integrated circuit process where the doping densities are not uniform.

PRACTICE Problems

5.3 The two regions of a silicon diode are doped with concentrations of $N_D = 10^{16}/cm^3$ and $N_A = 10^{18}/cm^3$. Calculate the minority carrier concentrations in each region at $T = 300°K$. *Ans:* $p_n = 2.25 \times 10^4$, $n_p = 2.25 \times 10^2$.

5.4 Calculate the barrier voltage across the depletion region of a silicon diode at $T = 300°K$ given that $N_A = 10^{16}/cm^3$ and $N_A = 10^{18}/cm^3$. *Ans:* 0.817 V.

5.5 If I_S for a silicon diode is 1×10^{-14} A, calculate the applied junction voltage at room temperature if the diode current is 1.4 mA. *Ans:* 0.663 V.

For a low-power silicon diode, a typical capacitance for a 6-V reverse bias is 5 pF. Since this capacitance is a function of the cross-sectional area, larger diodes used for higher current applications will exhibit larger depletion-layer capacitances for comparable applied voltages.

Diffusion Capacitance: The minority carrier distributions in both regions of a diode are functions of the applied voltage. The majority carrier concentrations are constant with voltage. Consequently, the charge due to minority carrier distributions gives rise to a capacitance. This capacitance is referred to as the diffusion capacitance. It is straightforward to show that this capacitance is a function of the current through the diode and can be written as

$$C_D = K_2 I \tag{5.33}$$

where K_2 is a constant and I is the current through the diode. For a forward current of 1 mA, the diffusion capacitance of a low-power diode might be 100 pF. It is, of course, a function of cross-sectional area.

In comparing the depletion-layer and diffusion capacitances, note that the former is more significant for reverse biases whereas the latter is more significant for forward biases. The diffusion capacitance is very small for the low currents resulting in the reverse-biased diode, and the depletion-layer capacitance will dominate. In the forward-biased diode, the larger currents lead to larger diffusion capacitances, and this component dominates the forward-biased case.

5.3 Nonlinear Modeling

IMPORTANT Concepts

1. Nonlinear problems are difficult to solve mathematically or on the computer.
2. The diode is a nonlinear device.
3. Piecewise-linear models can simplify the solution of nonlinear circuit problems such as those involving the diode.

5.3.1 THE PURPOSE OF MODELING

In the previous chapters of this text, all circuits considered had linear V-I relationships at a given frequency. The resistor has linear V-I characteristics at all frequencies, whereas the capacitor and inductor have linear characteristics at a given frequency. Amplifiers involving the op amp have linear relationships between input and output voltages and can be modeled in terms of other linear elements. The diode introduced in this chapter is the first element to be considered that has nonlinear V-I characteristics. Equation (5.28) indicates this highly nonlinear relationship between current and voltage.

Nonlinear problems are much more difficult to solve than linear ones. These problems could be impossible to solve manually and could require huge amounts of time if solved on a computer. Let us consider the circuits of Fig. 5.11 to demonstrate the difficulty in solving nonlinear problems.

If the problem consists of solving for I and V_{out}, only Ohm's law need be applied to the first circuit. The current I is

$$I = \frac{V_1}{R_1 + R_2} = \frac{6}{(200 + 300)} = 0.012 \text{ A}$$

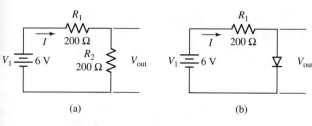

Figure 5.11
(a) A linear circuit. (b) A nonlinear circuit.

The output voltage is then

$$V_{\text{out}} = IR_2 = 0.012 \times 300 = 3.6 \text{ V}$$

For the second circuit, the diode used has a V-I relationship near room temperature of

$$I = 10^{-10}\left(e^{V/0.026} - 1\right)$$

Although this is a valid mathematical relationship, there is no closed form solution. As in Example 5.4, there are two equations that must be solved simultaneously, namely

$$V_1 = IR_1 + V_{\text{out}}$$

and

$$I = 10^{-10}\left(e^{V_{\text{out}}/0.026} - 1\right)$$

It will be left to the reader to show that the quickest method of solving this problem is a trial-and-error iterative method that leads to $I = 0.02747 \approx 0.027$ A and $V_{\text{out}} = 0.505215 \approx 0.5$ V.

Once the difficulty in solving nonlinear equations is appreciated, it becomes obvious why modeling is important. If we can approximate the nonlinear relationship with a model that has a linear relationship, we can apply analysis methods with which we are familiar. The thrust of nonlinear modeling is directed toward this end.

One possible model for the forward-biased diode of Fig. 5.11(b) is a simple 0.6-V voltage source. When this model replaces the diode, the circuit appears as shown in Fig. 5.12 and is very easy to analyze. From this circuit the current is calculated to be

$$I = \frac{V_1 - 0.6}{200} = 0.027 \text{ A}$$

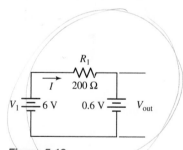

Figure 5.12
The circuit of Fig. 5.11 with a simplified diode model.

and the output voltage is 0.6 V. These values compare well to the results calculated from the exact equations, but are much easier to obtain.

The model not only simplifies the solution of I or V_{out}, it also allows the designer to understand how the circuit behaves. For example, if the input source V_1 is increased, it is easy to see that I will increase directly with V_1 while V_{out} remains constant. This would not be immediately obvious from the actual circuit of Fig. 5.11(b). Thus, modeling often increases the conceptual understanding of circuit operation.

PRACTICAL Considerations

An alternate and more traditional graphical method to analyze a circuit containing a nonlinear element is that of using a load line. This method is especially useful as a conceptual aid. The series circuit shown here can be split into a nonlinear element and the remaining external circuit.

PRACTICE Problems

5.6 Use a load line to find the voltage across the diode in Fig. 5.11(b) if the diode characteristics are expressed by

$$I = 0.0025 \, V^2$$

where I is expressed in amps and V is in volts. What is the value of diode current? *Ans:* $V_{\text{out}} = 2.6$ V; $I = 16.9$ mA.

5.7 Repeat the previous problem using iterative (trial-and-error) methods.

5.8 Repeat Practice Problem 5.6 if the diode characteristics are expressed by

$$I = 10^{-11} \left(e^{V/0.026} - 1 \right)$$

where I is in amps and V is in volts. Is the -1 term significant? Why? *Ans:* $V_{\text{out}} = 0.54$ V; $I = 27$ mA.

5.9 Repeat Practice Problem 5.8 using iterative methods.

A load line approach.

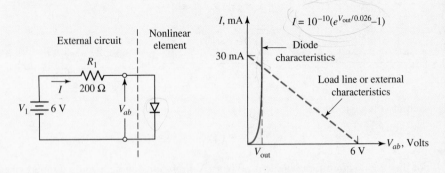

We note that the current through the external circuit equals that through the diode. The voltage V_{ab} appears simultaneously across the diode and across the external circuit. Consequently, the characteristics of each part of the circuit can be plotted on the same graph of V versus I. The plot of the external circuit V-I relationship is called the load line. The load line equation is

$$V_{ab} = V_1 - R_1 I$$

and can be easily plotted by connecting a straight line between the extreme points $V_{ab} = V_1$ when $I = 0$ and $I = V_1/R_1$ when $V_{ab} = 0$.

The intersection of the load line with the nonlinear characteristics of the diode is the only point that satisfies the characteristics of both sections of the circuit. This point gives the output voltage and the current through the circuit.

5.3.2 ELEMENTS USED IN NONLINEAR MODELING

Many of the same elements used in modeling linear circuits are also used in nonlinear models. Passive components such as resistors, capacitors, and inductors are important elements in both types of modeling. These components generally have constant values for manual models, but may have nonlinear values for models to be used in computer analysis programs. Independent and dependent sources are also used in nonlinear modeling.

One element used in nonlinear modeling that is never used in linear models is the ideal diode. The ideal diode is a voltage-controlled switch that allows current in the forward direction with no forward voltage drop. In the reverse direction, no current flows regardless of the reverse voltage applied. The characteristics of this device are shown in Fig. 5.13 (solid line). The dashed line shows the characteristics of an actual diode.

In effect, the ideal diode appears as an open switch or open circuit when negative voltage is applied. For forward current, this device appears as a closed switch or short circuit. The ideal diode is an important element in the modeling of piecewise-linear circuits.

Figure 5.13
V-I characteristics of an ideal diode.

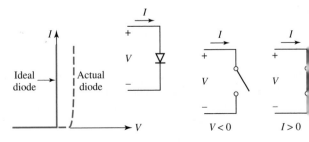

EXAMPLE 5.5

Develop a model that results in the V-I characteristics of Fig. 5.14.

SOLUTION The fact that no current flows for negative voltages suggests that the ideal diode should be included in this model. For positive voltages, the ideal diode would offer zero resistance rather than the finite value indicated by the V-I characteristics. A series circuit consisting of an ideal diode and a resistor will exhibit the characteristics of Fig. 5.14. This circuit is shown in Fig. 5.15.

The slope of the V-I characteristics in the forward voltage direction is

$$\text{slope} = \frac{\Delta I}{\Delta V} = \frac{4}{1} = 4 \text{ mA/V}$$

which is equivalent to a resistance of

$$R = \frac{\Delta V}{\Delta I} = \frac{1}{4} \text{ k}\Omega = 250 \ \Omega$$

Choosing a value of $R = 250 \ \Omega$ in the circuit of Fig. 5.15 completes the model.

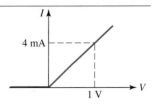

Figure 5.14
V-I characteristics.

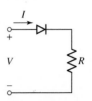

Figure 5.15
A model to generate the curve of Fig. 5.14.

5.4 The Diode Equivalent Circuit

IMPORTANT Concepts

1. Several models can be used for the diode. The particular model selected depends on the frequency and magnitude of the applied signal, the region of operation, and the method of analysis to be used.

2. The large-signal diode model may contain as few as two piecewise-linear regions. Higher accuracy can be achieved by adding more piecewise-linear regions, but analysis becomes correspondingly more complex and such accuracy is rarely necessary.

3. The small-signal dynamic diode resistance can represent the forward-biased, low-frequency characteristics of the diode over a small region of operation.

5.4.1 PRELIMINARY INFORMATION

The following points represent an overview of the diode.

1. The diode is a two-terminal semiconductor (silicon) device that conducts current freely in one direction but blocks current flow in the opposite direction. The diode is often used in rectifier circuits that convert an ac current to a dc current.

2. The name diode is a holdover from vacuum tube days. A vacuum tube diode consisted of two electrodes, an anode and a cathode, hence the name "diode." The terminal to which a positive voltage must connect for forward current flow is called the anode, and the remaining terminal is called the cathode in the semiconductor device also.

3. The diode is seldom used in a small-signal mode, but a thorough understanding of operation in this mode is important since it forms the basis of operation for a bipolar junction transistor.

There are many possible models for a diode depending on accuracy, frequency, or other requirements pertaining to the circuit at hand. This section begins with the most basic

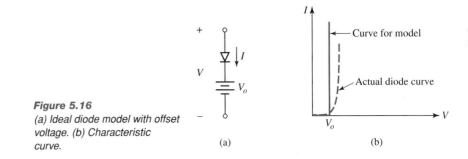

Figure 5.16
(a) Ideal diode model with offset voltage. (b) Characteristic curve.

(a)   (b)

models and progresses to more complex models. The choice of the equivalent circuit to be used in a given problem rests with the designer.

5.4.2 LOW-FREQUENCY, LARGE-SIGNAL DIODE MODELS

At low to moderate frequencies the equivalent circuit of the diode must reflect only the essential features of the diode characteristic curve as given by Eq. (5.28). This equation is plotted in Fig. 5.10 as the solid line. The symbol for the diode is also shown with anode and cathode terminals identified. A very simple model that can be proposed to approximate the actual V-I curve of the diode is the ideal diode or voltage-controlled switch. This switch is open if the applied anode to cathode voltage is negative and becomes a short circuit if the applied voltage tends to be positive. This model and its characteristics are shown in Fig. 5.13. The solid curve is the V-I characteristics of the ideal diode. The current of this model rarely departs from the actual diode current by more than 10 nA in the reverse direction, and the voltage rarely departs from the actual diode voltage by more than a few tenths of a volt in the forward direction.

The ideal diode model is useful in many applications. Obviously, if the forward resistance of a diode is much less than other series resistances in a circuit and the back resistance is much greater than the series circuit resistance, the ideal diode model is appropriate. This assumes also that the small forward-bias voltage required to conduct appreciable diode current is negligible. If the forward offset voltage must be included in the analysis, the slightly more complex circuit of Fig. 5.16 can be used.

This model was applied in Fig. 5.12. Since current was positive in that circuit, the ideal diode was represented by a short circuit. The model then consisted of a 0.6-V voltage source. We note that both diode models discussed thus far are piecewise-linear models with characteristics made up of two line segments, one with zero slope and one with infinite slope.

In addition to the practical utility of the ideal diode model, this equivalent circuit also serves as a component of more complex models. For example, if an offset voltage and a finite slope for forward current is required, the diode model of Fig. 5.17 can be applied. The slope of the curve is determined by the choice of R_f.

More accuracy and additional complexity of the model can be achieved by combining parallel branches similar to that of Fig. 5.18. A two-break-point model is shown in this

Figure 5.17
(a) Ideal diode model with offset voltage and finite forward slope. (b) Characteristic curve.

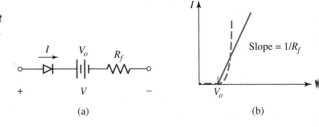

(a)   (b)

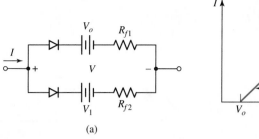 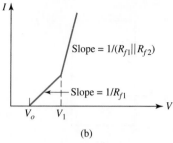

(a)   (b)

Figure 5.18
(a) Two-break-point model.
(b) Characteristic curve.

figure. The slope for voltages above V_1 is equal to the reciprocal of the parallel equivalent resistance, $R_{f1}R_{f2}/(R_{f1} + R_{f2})$.

With the addition of new parallel branches, any degree of accuracy in representing the actual diode characteristics can be achieved. There are, however, few applications that require this type of accuracy. It is worth noting that almost any practical curve can be approximated in the manner outlined above. This fact is used to advantage in analog computers where a precise V-I curve is required. Twenty or more parallel paths may actually be constructed to generate the desired V-I curve.

The circuits previously discussed exhibit infinite back resistance. In some instances, a lower actual back resistance may lead to errors. To correct this situation, the model of Fig. 5.19 can be applied. The back resistance of the diode is equal to R_r in this model.

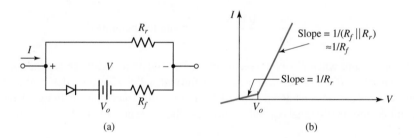

(a)   (b)

Figure 5.19
(a) Diode model with finite back resistance. (b) Characteristic curve.

5.4.3 LOW-FREQUENCY, SMALL-SIGNAL DIODE MODEL

When the diode is used in an application where the forward dc or bias current is considerably greater than the magnitude of the ac current, small-signal operation is in effect. Although the V-I curve for the diode is quite nonlinear, any region of this curve small enough to be considered linear can be represented by a resistance. For example, let us assume that a diode with the characteristics represented by Eq. (5.28) is used in the circuit of Fig. 5.20(a). The characteristic curve is shown in part (b) of the figure.

We want to find the ac current flowing into the diode. For this type of circuit it is customary to define a small-signal resistance that relates the ac current to the ac voltage. If the ac voltage is small compared to the dc voltage across the diode, the V-I curve will have a constant slope throughout the cycle. In this case, the current and voltage are related by the slope of the V-I curve as

$$\frac{\Delta I}{\Delta V} = \text{slope} = \frac{dI}{dV} \tag{5.34}$$

Since $\Delta I/\Delta V$ has units of conductance, Eq. (5.34) can be inverted to give units of resistance. We can then write

$$r_d = \frac{dV}{dI} = \frac{1}{dI/dV} \tag{5.35}$$

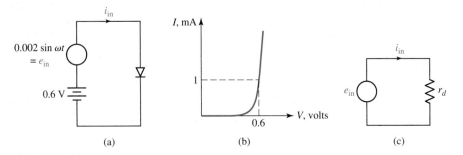

Figure 5.20
Small-signal diode circuit:
(a) actual circuit, (b) diode V-I
characteristics, (c) diode model.

where r_d is defined as the small-signal resistance and relates small changes of current to small changes in voltage. We can easily find r_d by approximating the diode equation. In the forward direction

$$I = I_S\left(e^{qV/kT} - 1\right) \approx I_S e^{qV/kT} \qquad (5.36)$$

therefore

$$\frac{dI}{dV} = \left(\frac{q}{kT}\right) I_S e^{qV/kT} = \frac{q}{kT} I$$

The resistance r_d equals the reciprocal of dI/dV; thus we can write

$$r_d = \frac{kT}{qI} = \frac{0.026}{I} \text{ (near room temperature)} \qquad (5.37)$$

where r_d is in ohms when I is expressed in amps. This dynamic resistance depends on the dc bias current of the diode. Once the bias current through a diode has been determined, the dc voltage sources can be considered to be short circuits for succeeding small-signal analysis. For example, the small-signal equivalent circuit of the diode network of Fig. 5.20(a) appears in Fig. 5.20(c).

For the circuit of Fig. 5.20(a), the dc source applies $0.6V$ to the diode, resulting in a dc bias current of 1 mA according to Fig. 5.20(b). The dynamic resistance is then

$$r_d = \frac{0.026}{1 \times 10^{-3}} = 26\ \Omega$$

The input ac current can be found from

$$i_{\text{in}} = \frac{e_{\text{in}}}{r_d} = \frac{0.002\ \sin\ \omega t}{26} = 0.077\ \sin\ \omega t\ \text{mA}$$

We note that this value of ac current can be changed, even when e_{in} remains constant, if the dc bias voltage or current is changed. For example, if the 0.6-V signal is increased slightly, causing the dc current to increase to 1.8 mA, the dynamic resistance becomes

$$r_d = \frac{0.026}{1.8 \times 10^{-3}} = 14.4\ \Omega$$

The ac current will now be $0.139\ \sin\ \omega t$ mA.

The dynamic resistance is sometimes called the incremental resistance or the resistance to incremental signals. An incremental signal implies that the increments of change of voltage across the diode and current through the diode are small compared to the dc values.

In some small-signal applications of the diode, notably in communication circuits, the device is used as a voltage variable resistance. The dc current through the diode determines the resistance to incremental current flow. The resistance can then be adjusted by controlling the dc current level.

PRACTICAL Considerations

We must be careful to use the concept of dynamic resistance only when the ac current is small compared to the dc bias current. If there is a question whether this condition holds prior to the analysis, we simply compare results after we calculate the two values. In the preceding two situations, we had a dc bias current of 1 mA with an ac peak current of 0.080 mA followed by a change of bias current to 1.8 mA with an ac peak current of 0.144 mA. The small-signal equivalent circuit can be safely applied to situations wherein the peak value of ac current is less than 10% of the dc bias circuit.

Since most electronic circuits deal with current magnitudes in the mA range, it is convenient to express Eq. (5.37) as follows:

$$r_d = \frac{26}{I} \quad I \text{ in mA}$$

From a practical standpoint, the dc bias voltage would not be established by an expensive dc voltage source of 0.6–0.7 V as shown in Fig. 5.20(a). It is difficult to accurately determine a dc current by applying a dc voltage since the diode current varies so markedly with applied voltage. In most situations, it is less costly and more accurate to establish the diode bias current with a resistor connected between an existing power supply voltage and the diode as shown in the accompanying figure.

The dc bias current value is found by estimating the forward bias diode voltage to be 0.6 V. The dc current can then be calculated from the equation

$$I_{dc} = \frac{12 - 0.6}{10} = 1.14 \text{ mA}$$

We note that if the estimate of diode voltage is slightly incorrect, little error results. For example, if the diode voltage were 0.7 V instead of 0.6 V, the current value would be 1.13 mA, a very small difference.

The capacitor blocks any dc current from flowing through the ac source but, if chosen correctly, exhibits negligible impedance to the flow of ac current.

Diode biasing.

It should be recognized that the large-signal piecewise-linear models can be used for small-signal analysis if desired. There are times when this should be done, although there are limitations to this procedure also. Piecewise-linear models such as those shown in Figs. 5.17, 5.18, and 5.19 are more involved than the small-signal model of Fig. 5.20. The analysis is complicated by the use of such models.

EXAMPLE 5.6

Select the resistor R in Fig. 5.21 to result in an output voltage of 1.2-V peak value when the sinusoidal input signal has a peak value of 0.05 V and a frequency of 5 kHz.

SOLUTION We must first determine whether the diode is operating in the large-signal mode or the small-signal mode. The dc current flowing through the 10-kΩ resistance must also

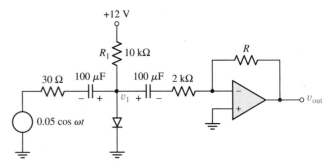

Figure 5.21
Circuit of Example 5.6.

flow through the diode since the capacitors block dc current flow. The diode will obviously be forward biased; thus we will assume a dc drop of 0.6 V across this device. The voltage drop across the 10-kΩ resistance is now easily found to be the difference of the voltage on each side or

$$V_R = 12 - 0.6 = 11.4 \text{ V}$$

The dc diode current equals the resistor current and is

$$I_{dc} = \frac{11.4}{10} = 1.14 \text{ mA}$$

The small-signal resistance of the diode is given by

$$r_d = \frac{26}{1.14} = 22.8 \ \Omega$$

With the very small ac input voltage, the small-signal model of the diode is appropriate, and this device can be replaced with a 22.8-Ω resistor to model its behavior.

We note that the capacitors can be approximated as short circuits. The dynamic resistance of the diode will have a 10-kΩ resistor (R_1) and a 2-kΩ resistor in parallel. The latter value is the input impedance to the op amp. Since both resistors are much larger than 22.8 Ω we can neglect this loading and use voltage division to calculate v_1. This value is

$$v_1 = \frac{22.8}{30 + 22.8} \ 0.05 = 21.6 \text{ mV}$$

The op amp must have a gain that amplifies this peak signal into one with a 1.2-V peak value. The required amplification is

$$A = \frac{1.2}{0.0216} = 55.6 \text{ V/V}$$

Since the magnitude of the gain of an inverting stage equals the ratio of R to the 2-kΩ input resistor, R is found to be

$$R = 55.6 \times 2 = 111 \text{ k}\Omega$$

5.4.4 THE HIGH-FREQUENCY DIODE MODEL

The high-frequency model of the diode must include the depletion-layer capacitance and the diffusion capacitance. Both of these capacitors appear across the junction, and both are nonlinearly related to voltage across the junction.

One version of the high-frequency diode model is shown in Fig. 5.22. The ohmic resistance in series with the junction is represented by R_S, and R_D is the junction resistance. The depletion region capacitance, C_T, arises from the change in charge in this region resulting from a voltage change as explained earlier. This capacitance varies with voltage as shown in Eq. (5.32)

$$C_T = \frac{K_1}{(\psi_o - V)^m} \tag{5.38}$$

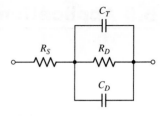

Figure 5.22
A high-frequency diode model.

The junction voltage, V, has a positive value for forward voltage and a negative value for reverse voltage. The quantity m is called the grading coefficient and varies from 0.5 for an abrupt junction to 0.3 for a linearly graded junction.

The diffusion capacitance, C_D, is due to the change in minority carrier distribution in the two regions as a result of change in applied voltage. This capacitance is a function of diode current and is given by

$$C_D = K_2 I \tag{5.33}$$

where K_2 is constant for a given diode and I is the steady-state diode current.

As mentioned previously, when the diode is reverse biased the current is very small, leading to a negligible diffusion capacitance. For this condition, the depletion-region capacitance dominates the reactive behavior. When the diode is forward biased, the diffusion component becomes more significant.

For small-signal operation, C_D and C_T can be considered constant at the values determined by I and V. The circuit can then be analyzed using conventional circuit analysis methods.

The more difficult situation occurs when V or I vary significantly, causing C_T and C_D to vary greatly. In such situations an average capacitance can be calculated from

$$C_{av} = \frac{\Delta Q}{\Delta V} = \frac{Q(V_1) - Q(V_2)}{V_1 - V_2} \tag{5.39}$$

This approach is difficult to apply to manual calculations, but is appropriate for computer calculations using a model such as that of the following section.

5.4.5 THE SPICE DIODE MODEL

The diode model used in most versions of Spice is shown in Fig. 5.23 [A. Vladimirescu, *The Spice Book,* Wiley, New York, 1994, Chapter 3]. Although the total voltage across the diode, including the ohmic drop across R_S, is V_D, the junction voltage V_J is used in the diode equation; That is,

$$I_D = I_S \left(e^{q V_J / NkT} - 1\right) \tag{5.40}$$

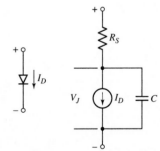

Figure 5.23
The Spice diode model.

The capacitance is the sum of the diffusion component and the depletion region component. Both of these elements are nonlinear, with the diffusion capacitance dominating in the forward direction and the depletion region dominating in the reverse direction. Spice allows a specification of a zero-bias depletion region capacitance, a barrier voltage, and a grading coefficient or exponent m. For large-signal swings, the capacitance is averaged using an equation similar to Eq. (5.39).

5.5 Applications of the Diode

IMPORTANT Concepts

---

1. An important use of the diode is in rectifier circuits. These circuits change an ac waveform that contains no dc value to a rectified waveform that contains a dc value.
2. Clipping and clamping circuits can be constructed using diodes.

5.5.1 RECTIFICATION

One of the major uses of the diode is in rectifier circuits such as that shown in Fig. 5.24. If v_{in} is a sinusoid with a 100-V peak value, the output will consist only of the positive portion of the sinusoid. As the input voltage goes positive, the diode conducts with a small or negligible drop. Essentially all of the positive-going portion of the sinusoid is passed to the resistor. Since the total voltage drop across the forward-biased diode is expected to be small compared to the total applied voltage, we can use a simple diode model. The ideal diode model is applicable in this circuit. Had we assumed a 0.6-V drop across the diode when forward biased, the peak value of output would be 99.4 V rather than the approximate value of 100 V shown in Fig. 5.24(b). Obviously, little accuracy is lost if the forward diode drop is neglected. As v_{in} goes negative, the diode behaves as an open switch, conducting negligible current. With no current flow through R, there is no voltage drop across the resistor, and v_{out} equals zero for all negative values of v_{in}. In this situation, all applied voltage drops across the reverse-biased diode.

The half-wave rectifier circuit is used in dc power supplies to change an input ac waveform with no dc or average value to a rectified waveform that contains a dc component. The dc or average value of a rectified sinusoid with a peak value of V_m is given by

$$V_{dc} = \frac{V_m}{\pi} \quad \text{(half-wave rectification)} \qquad (5.41)$$

It is possible to change a sinusoidal waveform to an output signal containing twice the dc voltage given by Eq. (5.41) with the full-wave rectifiers of Fig. 5.25. The voltages appearing across the secondary from point a to ground and point b to ground are 180° out of phase, as shown in Fig. 5.26.

The positive portion of v_{ag} passes through diode D1 to the load resistor when v_{ag} is positive. The voltage v_{bg} is negative during this time and is blocked from reaching the load by diode D2. As v_{ag} goes negative and v_{bg} goes positive, diode D1 blocks v_{ag}, but v_{bg} reaches the load through diode D2. The output signal is shown in Fig. 5.26(b). This full-wave rectified signal contains two times the dc component contained in the half-wave

Figure 5.24
A half-wave rectifier: (a) rectifier circuit, (b) output waveform.

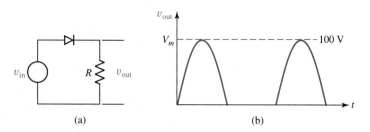

(a) (b)

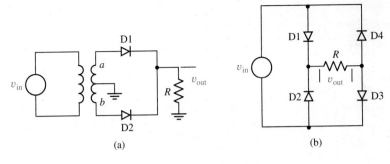

(a) (b)

Figure 5.25
*Full-wave rectifiers:
(a) center-tapped transformer
input, (b) bridge rectifier.*

rectified signal, or

$$V_{dc} = \frac{2V_m}{\pi} \quad \text{(full-wave rectification)} \qquad (5.42)$$

The bridge rectifier eliminates the center-tapped transformer at the expense of requiring one ground for the load and another for the ac input voltage reference. When v_{in} is positive, diodes D1 and D3 are forward biased and v_{in} drops across R. Diodes D2 and D4 are reverse biased during this time. As v_{in} goes negative, diodes D2 and D4 become forward biased while D1 and D3 are reverse biased. During each half-cycle of the input signal, current flows through R in the same direction, leading to a full-wave rectified v_{out}.

5.5.2 CLIPPING AND CLAMPING CIRCUITS

Two other applications of the diode are clipping and clamping. The circuit of Fig. 5.27 is a clipping circuit that converts an ac signal to the output shown. As long as the input signal falls in the range between −4.6 V and +4.6 V, the diodes are reverse biased. Negligible current flows through R, leading to an output voltage that equals the input voltage. When v_{in} exceeds 4.6 V, diode D1 is forward biased with a drop of approximately 0.6 V across it. The output is forced to remain at 4.6 V even as the input becomes more positive. When v_{in} is less than −4.6 V, diode D2 becomes forward biased, forcing the output to remain at −4.6 V. The resulting waveform exhibits sharply clipped maximum and minimum values. The clipping-voltage levels can be controlled by the voltage sources. The ideal diode model with offset voltage is used in this analysis. This model is more appropriate in this instance because the voltage sources are not considerably larger than the forward diode drop.

Figure 5.26
*(a) Secondary voltages.
(b) Rectified signal.*

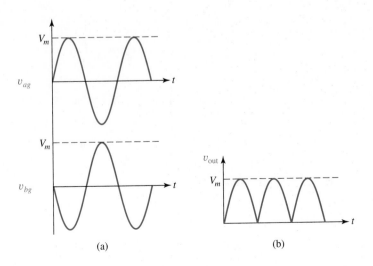

(a) (b)

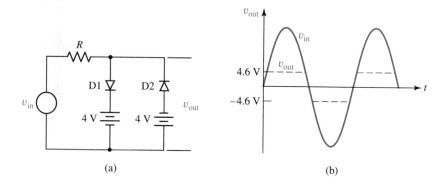

Figure 5.27
(a) A clipping circuit. (b) Input and output waveforms.

The circuit of Fig. 5.28 is a clamping circuit that changes the dc level of an ac waveform. Again, the ideal diode model with offset voltage can be used to obtain the output waveform. As v_{in} goes negative the output attempts to follow v_{in}. The diode becomes forward biased and forces v_{out} to remain at -0.6 V. The capacitor charges to approximately $-V_m$ on the source side. As v_{in} goes positive from its most negative value, the output also moves positive. The diode now becomes reverse biased and allows v_{out} to follow the input to a peak value of approximately $2V_m$.

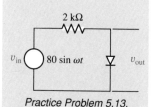
5.5.3 DIODE ISOLATION IN INTEGRATED CIRCUITS

One of the more important applications of the diode is in the fabrication of the integrated circuit. A silicon integrated circuit includes diodes, transistors, resistors, and capacitors in or on the silicon chip. The various regions of the circuit must not be connected electrically, except as prescribed by the configuration, even though all regions share the same chip volume. Junction isolation is the prevalent method of isolating these regions.

Figure 5.29 indicates a portion of an integrated circuit. The transistor shown is an *npn* device fabricated on a *p* substrate. The p^+ regions are more heavily doped than the *p* region to minimize the extension of the depletion region into the *p* region, which allows individual devices to be placed closer together. The substrate is connected to the most negative power supply voltage used. This *p*-type substrate then forms a reverse-biased junction with the *n*-type regions of other devices. Only high-impedance paths through reverse-biased junctions connect individual devices within the substrate. These reverse-biased diodes electrically isolate the various regions within the chip.

Figure 5.28
(a) A clamping circuit. (b) Input and output waveforms.

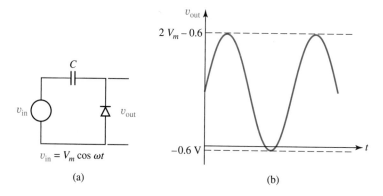

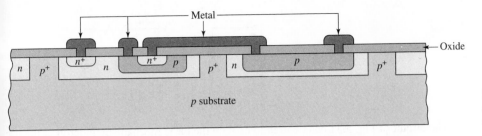

Figure 5.29
A portion of an integrated circuit.

5.5.4 THE BREAKDOWN OR ZENER DIODE

The breakdown diode was originally called the Zener diode, named after the man who developed the early theory of this device. The mechanism that defines its behavior was referred to as Zener breakdown. Later, a second mechanism, avalanche breakdown, was found to play the most significant part in the device in a particular voltage range. Zener breakdown applies for diodes with a breakdown in one range, whereas avalanche breakdown occurs for diodes in another voltage range. The term breakdown diode was suggested for this device, but it had become known as the Zener diode and that name is still popularly applied. In this section we will use the name Zener for the breakdown diode. The Zener is most often used in power supplies to *regulate* a voltage or maintain it at some constant value.

The V-I characteristics of the Zener diode are shown in Fig. 5.30. The forward region is identical to that of the simple diode. When an increasing negative or reverse voltage is applied to the Zener diode, a point is reached, designated V_Z, where the diode breaks down. At voltages less negative than V_Z, essentially zero current flows. At voltages slightly more negative than V_Z, the diode conducts a great deal of reverse current. The breakdown voltage can be controlled in the manufacturing process and can be made to fall anywhere in the range from about 2.6 V to 200 V. At reverse voltages greater than V_Z, the curve exhibits a rather constant but steep slope. The reciprocal of this slope is equal to the dynamic resistance of the Zener diode. This value is given by

$$r_Z = \frac{\Delta V}{\Delta I} \tag{5.43}$$

where ΔV and ΔI are taken from that portion of the curve with voltage values more negative than V_Z. For small diodes, r_Z may be in the range of tens of ohms, whereas r_Z for larger diodes may be a few ohms or less.

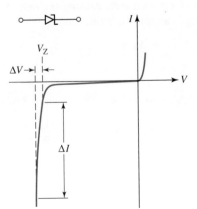

Figure 5.30
Zener diode characteristics.

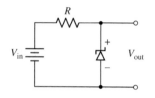

Figure 5.31
Zener regulator.

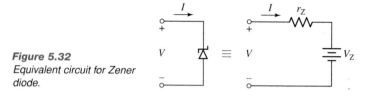

Figure 5.32
Equivalent circuit for Zener
diode.

The circuit of Fig. 5.31 can be used to explain the basic mechanism underlying the use of the Zener diode as a regulator. If V_{in} is smaller than the breakdown voltage, the diode will behave as a normal, reverse-biased diode. Very little current flows, and the drop across the resistor, R, is negligible. The entire input voltage appears across the diode. The device does not regulate in this mode of operation since the output voltage varies directly with input voltage. As V_{in} increases to higher values than V_Z, the diode breaks down and conducts current, maintaining an almost constant voltage of V_Z across the diode. The current flowing through R can be found from Ohm's law as

$$I = \frac{V_{in} - V_Z}{R}$$

Increasing V_{in} does not change the output voltage significantly, but the current increase will lead to a slight voltage change. From Fig. 5.30 we can see that the current increase causes a diode voltage increase that is dependent on the slope of the curve or on the dynamic resistance. A zero resistance or infinite slope would indicate perfect regulation as input voltage varies. This discussion suggests that the equivalent circuit shown in Fig. 5.32 is applicable to the Zener diode in the regulating mode. Example 5.7 demonstrates the effectiveness of the Zener diode in regulating output voltage as input voltage varies.

EXAMPLE 5.7

In the circuit of Fig. 5.31 the resistance $R = 1$ kΩ, $V_Z = 10$ V at 1 mA, and $r_Z = 30$ Ω. Given that V_{in} changes from 11 V to 20 V, calculate the Zener current change and the output voltage change.

SOLUTION The circuits of Fig. 5.33 can be helpful in working this example. In the first circuit, the output voltage is equal to V_Z. The current is

$$I_Z = \frac{V_{in} - V_{out}}{1} = \frac{11 - 10}{1} = 1 \text{ mA}$$

When V_{in} increases to 20 V, the Zener current increases, causing the output voltage to be slightly higher than V_Z. The output voltage can now be written as

$$V_{out} = V_Z + \Delta I r_Z \tag{5.44}$$

Figure 5.33
Circuits for Example 5.7:
(a) 11-V input, (b) 20-V input.

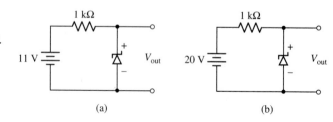

(a)   (b)

where ΔI is the increase in current resulting from the increase in V_{in}. The new current is then

$$I_Z = \frac{V_{in} - (V_Z + \Delta I r_Z)}{R} = \frac{V_{in} - (V_Z + (I_Z - 0.001)r_Z)}{R}$$

From this equation, I_Z is found to be 9.76 mA.

From a practical standpoint, the Zener current can be approximated by assuming that $V_{out} \approx V_Z$ for all values of input voltage. This value of current can then be used in Eq. (5.44) to find the output voltage. Thus,

$$I_Z = \frac{V_{in} - V_Z}{R} = \frac{20 - 10}{1} = 10 \text{ mA}$$

The change in Zener current is $\Delta I = 10 - 1 = 9$ mA, giving an output voltage of

$$V_{out} = 10 + 0.009 \times 30 = 10.27 \text{ V}$$

when $V_{in} = 20$ V. The input voltage has almost doubled while the output voltage changes only

$$\frac{\Delta V}{V} \times 100 = \frac{0.27}{10} \times 100 = 2.7\%$$

We have seen that the Zener diode can regulate an output voltage, but we have yet to examine the limitations on this device. The maximum current that can be allowed to flow through the Zener diode is limited by the allowable device dissipation. The power dissipated by the Zener diode in its regulating region is

$$P_Z = V_Z I_Z \qquad (5.45)$$

The maximum allowable power, P_{Zmax}, specified by the manufacturer limits the maximum current to

$$I_{Zmax} = \frac{P_{Zmax}}{V_Z} \qquad (5.46)$$

The minimum current that can flow and still cause the Zener diode to be in the regulating region is usually taken to be one-tenth of the maximum current or

$$I_{Zmin} = \frac{1}{10} I_{Zmax} \qquad (5.47)$$

The characteristic curve of the Zener diode is shown again in Fig. 5.34 with these key points noted.

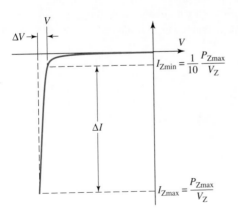

Figure 5.34
Zener diode characteristics
showing key values.

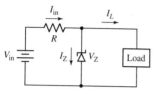

Figure 5.35
Zener diode regulating circuit.

To demonstrate the effect of the current limitations on circuit performance, we will consider the circuit of Fig. 5.35. If the Zener diode is in its regulating region and V_{in} remains constant, the input current will remain constant. This input current divides into two components, one through the Zener diode and one to the load. The load current may vary from 0 to a maximum of I_{Lmax}. Since I_{in} is constant, the Zener current must decrease as load current increases to keep the sum of these currents equal to I_{in}. The two extremes of current through the Zener diode can be found by considering the extremes of load current. When $I_L = I_{Lmax}$, minimum current will flow through the Zener diode. This value must equal or exceed I_{Zmin}; thus

$$I_{in} = I_{Lmax} + I_{Zmin}$$

At the other extreme, when $I_L = 0$, maximum Zener current flows and

$$I_{in} = I_{Zmax}$$

Combining these equations to find I_{Lmax} results in

$$I_{Lmax} = I_{Zmax} - I_{Zmin} = 0.9 I_{Zmax} \qquad (5.48)$$

The maximum current that can be supplied to the load is nine-tenths of the maximum diode current. If this current limitation is too severe for a given application, the current can be amplified by a BJT to lead to a much greater current range.

DISCUSSION OF THE DEMONSTRATION PROBLEM

The circuit for the demonstration problem is repeated here. We now understand that the voltage V_1 and the resistor R_1 will determine the dc current through the diode. This dc current will set the incremental resistance of the diode. Along with the 50-Ω resistor, this incremental resistance forms a voltage divider for the input signal, producing an output voltage of

$$v_{out} = v_{in} \frac{r_d}{r_d + 50}$$

where r_d is the incremental diode resistance. A change in the voltage V_1 will change the incremental resistance of the diode and modify the value of the output voltage, assuming the value of R_1 is sufficiently large compared to the diode resistance that it can be neglected.

Circuit for Demonstration Problem.

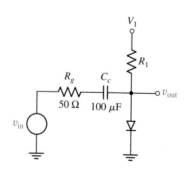

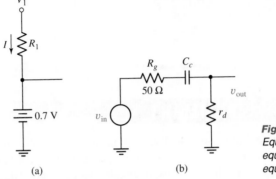

Figure 5.36
Equivalent circuits; (a) dc
equivalent, (b) incremental
equivalent.

Figure 5.36 indicates both the dc and incremental equivalent circuits for this problem. From the dc equivalent circuit, we can write the dc diode current as

$$I = \frac{V_1 - 0.7}{R_1}$$

When $V_1 = V_A$, the diode current is

$$I_1 = \frac{V_A - 0.7}{R_1}$$

When $V_1 = V_A + 2$, the diode current is

$$I_2 = \frac{V_A + 2 - 0.7}{R_1} = \frac{V_A + 1.3}{R_1}$$

We can determine the values required for I_1 and I_2 from a consideration of the necessary incremental diode resistances. Using the incremental equivalent circuit we see that, to cause an ac output of 12 mV, the equation

$$12 = 20 \frac{r_{d1}}{r_{d1} + 50}$$

must be satisfied. Likewise, for an output of 8 mV, the equation

$$8 = 20 \frac{r_{d2}}{r_{d2} + 50}$$

must be satisfied. In these equations, r_{d1} and r_{d2} are the incremental diode resistances corresponding to dc diode currents of I_1 and I_2, respectively.

From these latter two equations, the diode resistances can be found to be $r_{d1} = 75\Omega$ and $r_{d2} = 33.3\Omega$. The corresponding dc diode currents are found from

$$r_d = \frac{26}{I}$$

This leads to $I_1 = 0.347$ mA and $I_2 = 0.780$ mA. Simultaneous solution for V_A and R_1 based on the earlier equations for dc diode current results in $V_A = 2.30$ V and $R_1 = 4.62$ kΩ.

The lower corner frequency can be calculated from

$$f_{low} = \frac{1}{2\pi C_c(r_{d1} + R_g)} = 12.7 \text{ Hz}$$

SUMMARY

➤ The conductivity of pure silicon is rather poor, but it can be improved by introducing impurity atoms into the silicon material. The conductivity can be controlled by the concentration of impurity atoms added.

➤ The movement of charges such as free electrons results in electrical current. A second type of movement takes place as an electron leaves a vacancy and moves to another point where a vacancy previously existed. This type of current is said to be carried by holes.

➤ A voltage drop across a semiconductor material containing free electrons or holes leads to a drift current flow. Free charge carriers can also lead to current in regions that have no voltage drop. This current is said to be carried by diffusion and results if the carrier distribution is nonuniform.

➤ When a p region and an adjacent n region are created, a pn junction or diode is formed. The current flow resulting from an applied source is very asymmetrical relative to the polarity of the voltage. A forward voltage bias can create huge currents, whereas a reverse bias leads to essentially no current.

➤ The relationship between forward diode current and applied voltage is highly nonlinear. Current varies exponentially with applied voltage. For forward-biased diodes, a small-signal dynamic resistance can be defined as $r_d = \Delta V / \Delta I$. In this equation, the ΔV and ΔI must be small enough to be considered linear.

➤ At higher frequencies, two sources of capacitance in the pn junction must be considered. These capacitors are called depletion region capacitance and diffusion capacitance. Both are nonlinear functions of applied voltage.

➤ There are several possible models for the diode using elements that are easier to analyze. The purpose of these models is to avoid the necessity of solving nonlinear equations.

➤ Large-signal applications of the diode include rectification and clipping or limiting. The most important small-signal application occurs in the base-emitter junction of a transistor, which will be considered in a later chapter.

PROBLEMS

SECTION 5.1.1 INCREASING THE DENSITY OF FREE ELECTRONS OR HOLES (DOPING)

5.1 If $n_i = 1.5 \times 10^{10}/\text{cm}^3$ at room temperature for silicon and a silicon wafer is doped with $N_A = 2 \times 10^{13}/\text{cm}^3$, determine the concentrations of free electrons and free holes in the wafer.

5.2 If $n_i = 1.5 \times 10^{10}/\text{cm}^3$ at room temperature for silicon and a silicon wafer is doped with $N_D = 3 \times 10^{16}/\text{cm}^3$, determine the concentrations of free electrons and free holes in the wafer.

5.3 Calculate the conductivity of the wafer of Problem 5.1, assuming that $\mu_p = 480 \text{ cm}^2/\text{volt-sec}$.

5.4 Calculate the conductivity of the wafer of Problem 5.2, assuming that $\mu_n = 1300 \text{ cm}^2/\text{volt-sec}$.

5.5 A bar of n-type silicon of 1 cm^2 cross-sectional area and 10 cm long has 1 V applied across its length. A current of 100 mA results. Calculate the concentration of free electrons in this material.

5.6 A wafer of silicon has 10^{16} arsenic atoms (donor) per cubic centimeter and 7.4×10^{15} boron atoms (acceptor) per cubic centimeter. Calculate the hole density in this material and compare it to intrinsic silicon at room temperature.

5.7 It is desired to create a conductivity of $\sigma = 10 \text{ S/cm}$ in a silicon bar. Calculate the concentration of impurity atoms necessary to satisfy this requirement for n-type material.

SECTION 5.1.3 CURRENT FLOW IN A SEMICONDUCTOR

5.8 The concentration of holes at the right edge of the silicon bar of unit area shown in the figure is maintained at zero. Given that the hole-diffusion current flowing through the bar is 1 mA, find the concentration of holes at the left edge of the bar.

Figure P5.8

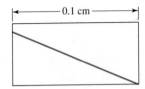

5.9 If the concentration of holes at the left edge of the bar of Problem 5.8 is $2 \times 10^{14}/\text{cm}^3$ and drops with constant slope to $4 \times 10^{13}/\text{cm}^3$ at the right edge, calculate the hole-diffusion current flow.

SECTION 5.1.5 DIFFUSION CURRENT IN THE PRESENCE OF RECOMBINATION

5.10 If 10 μA of electron-diffusion current flows at $x = 0$ in the bar of unit area shown, calculate the concentration of electrons that must be maintained at the left edge of the bar. Assume that the equilibrium concentration of electrons is negligible and $L = 0.01$ cm.

Figure P5.10

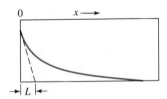

5.11 Electrons are injected into a bar of doped silicon semiconductor material with $p_p = 10^{15}/\text{cm}^3$. The bar is 1 cm square in area and infinitely long. The electron concentration at the point of injection ($x = 0$) is $10^{14}/\text{cm}^3$. Calculate and plot drift and diffusion currents between $x = 0$ and $x = 0.1$ cm. Assume that $L_n = 0.01$ cm.

SECTION 5.2.1 JUNCTION FORMATION

5.12 The two regions of a silicon diode are doped with concentrations of $N_D = 10^{15}/\text{cm}^3$ and $N_A = 10^{18}/\text{cm}^3$. Calculate the minority carrier concentrations in each region at room temperature.

SECTION 5.2.2 THE DEPLETION REGION

5.13 Calculate the barrier voltage across the depletion region of a silicon diode at $T = 300°K$ given that $N_D = 10^{15}/cm^3$ and $N_A = 10^{18}/cm^3$.

SECTION 5.2.3 CURRENT FLOW IN A *pn* JUNCTION

☆ **5.14** Given that $I_S = 2 \times 10^{-14}$ A at room temperature, $T = 300°K$, and doubles with each $7°K$ increase in T.

(a) Calculate I_S at $T = 350°K$.

(b) Find the temperature at which $I_S = 8 \times 10^{-14}$ A.

5.15 For the diode of Problem 5.14 at room temperature, calculate the diode current if a voltage of 0.65 V is applied in the forward direction.

5.16 For the diode of Problem 5.14 at room temperature, calculate the diode voltage if a current of 2 mA flows in the forward direction.

5.17 A diode has the characteristics given by Eq. (5.30) where N is an empirical constant to be determined. At a dc current of 0.5 mA, the slope of the V-I curve is found to be 14.7 mA/V. Assuming room temperature operation,

(a) Determine the value for N.

(b) When the diode bias is 0.64 V, it conducts 100 μA of current. Determine the value of I_S.

5.18 The depletion-layer capacitance for an abrupt junction diode is measured to be 5.2 pF at a reverse bias of 6 V. If the barrier voltage is 0.9 V, calculate the expected capacitance for a reverse bias of 12 V.

5.19 Repeat Problem 5.18 for a graded junction with $m = 0.33$.

SECTION 5.3.1 THE PURPOSE OF MODELING

5.20 If R_1 of Fig. 5.11(b) is changed to 2 kΩ, find V_{out} and I by iterative methods to at least two-place accuracy.

5.21 Repeat Problem 5.20 using a load line.

5.22 If the diode of Fig. 5.11(b) is replaced by a device that satisfies the equation $I = 2V^3$ mA, find V_{out} and I. Find the voltage across the 200-Ω resistor.

SECTION 5.3.2 ELEMENTS USED IN NONLINEAR MODELING

5.23 Sketch the output voltage of the circuit shown in the figure. Use the ideal diode model.

Figure P5.23

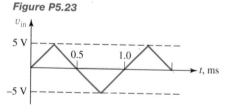
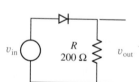

SECTION 5.4.2 LOW-FREQUENCY, LARGE-SIGNAL DIODE MODELS

5.24 Sketch the output voltage of the circuit of Problem 5.23 using the diode model of Fig. 5.16(a) with $V_o = 0.6$ V.

5.25 Sketch the output voltage of the circuit of Problem 5.23 using the diode model of Fig. 5.17(a) with $V_o = 0.6$ V and $R_f = 20\Omega$.

5.26 In the circuit of Problem 5.23 the 200-Ω resistor is replaced by a 10-kΩ resistance. The input voltage is increased from a 5-V amplitude to a 50-V amplitude. How does the accuracy of the ideal diode model compare to the more complex models in finding v_{out} in this problem? Explain.

5.27 Using the diode model of Fig. 5.19(a) with $V_o = 0.5$ V, $R_r = 2$ MΩ, and $R_f = 30$ Ω, sketch the output waveform for the circuit shown. Now sketch the output waveform using the ideal diode model. Compare results.

Figure P5.27

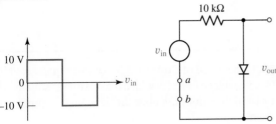

5.28 Reverse the positions of the diode and resistor in Problem 5.27 and sketch the output voltage across the resistor using the ideal diode model.

SECTION 5.4.3 LOW-FREQUENCY, SMALL-SIGNAL DIODE MODEL

5.29 Ignore diode capacitance and assume that the coupling capacitor in the circuit shown appears as a short circuit to the input pulse. If V_o for the diode is 0.5 V, sketch the incremental output voltage as a function of time.

Figure P5.29

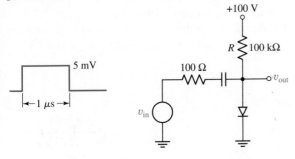

5.30 Repeat Problem 5.29 if R is changed to 50 kΩ. Explain why the amplitude of the output voltage changes when R is changed.

5.31 Repeat Problem 5.29 if the 100-V source is changed to 50 V. Explain why the amplitude of the output voltage changes when the dc voltage is changed.

5.32 In the circuit of Fig. 5.20(a) the dc voltage is increased to a value that results in 1.4 mA of dc current. If v_{in} remains constant at $0.002 \sin \omega t$, find the ac component of diode current.

5.33 If the forward-biased diode drop is 0.5 V (dc) in the circuit shown, plot the ac output voltage as a function of time.

Figure P5.33

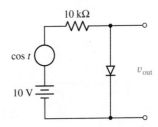

SECTION 5.4.4 THE HIGH-FREQUENCY DIODE MODEL

5.34 If the small-signal depletion-layer capacitance for an integrated circuit diode is 8 pF at $V = -8$ V with $\psi_o = 1$ V, what will the capacitance be when (a) $V = -3$ V; (b) $V = -16$ V?

5.35 Repeat Problem 5.34 for a forward-bias voltage of $V = 0.5$ V.

5.36 If the diffusion capacitance of a forward-biased diode is $C_D = 120$ pF when the diode current is 2 mA, calculate C_D when the current is (a) 1 mA; (b) 3 mA.

☆ **5.37** The depletion-layer capacitance of Problem 5.34 has a reverse-bias voltage across it that swings from 3 V to 12 V. What is the average value of C_T?

SECTION 5.5.1 RECTIFICATION

5.38 In the half-wave rectifier shown, the forward-biased diode can be represented as a 0.6-V voltage source. Calculate the dc voltage component of the output.

Figure P5.38

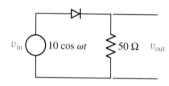

v_{in} 10 cos ωt 50 Ω v_{out}

5.39 If the diode of Problem 5.38 can be represented by a 7-Ω resistance in the forward direction, calculate the dc voltage component of the output.

5.40 If the diode of Problem 5.38 can be represented by a 6-Ω resistance in series with a 0.5-V voltage source in the forward direction, calculate the dc voltage component of the output.

5.41 Calculate the dc voltage of a full-wave rectifier if the forward resistance of each diode is taken as 10 Ω and the load resistance is 100 Ω. The secondary voltage referenced to the center tap is 220 V (rms).

5.42 The input signal to the bridge rectifier of Fig. 5.25(b) is 100 cos ωt. If the load resistance is 50 Ω and the forward resistance of the diode is 5 Ω, calculate the dc output voltage. Assume the forward diode voltage is 0.7 V.

SECTION 5.5.2 CLIPPING AND CLAMPING CIRCUITS

5.43 Plot the output voltage of the clipping circuit shown as a function of time. Assume the forward diode voltage is 0.7 V and $\omega = 15,000$ rad/s.

5.44 Plot the approximate output voltage of the clamping circuit shown for two cycles of input signal. What will the dc voltage of the output become after several input cycles? Assume the forward diode voltage is 0.7 V.

Figure P5.43

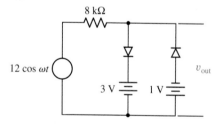

8 kΩ

12 cos ωt v_{out}

3 V   1 V

Figure P5.44

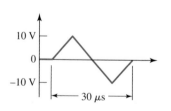

10 V

0

−10 V

$\leftarrow$ 30 μs $\rightarrow$

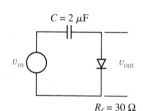

$C = 2\ \mu F$

v_{in} v_{out}

$R_f = 30$ Ω

SECTION 5.5.4 THE BREAKDOWN OR ZENER DIODE

5.45 A Zener diode regulator circuit, such as that shown in Fig. 5.35, is used to supply 10 V to a load having a variable current requirement from 0 to 20 mA. Given that the input dc voltage is 15 V, calculate the required value of R and the maximum power dissipated by the Zener diode.

5.46 Repeat Problem 5.45 assuming that the load current must vary from 0 to 1 mA.

5.47 Repeat Problem 5.45 assuming that the load current must vary from 0 to 100 mA.

5.48 Repeat Problem 5.45 if $V_{in} = 18$ V. If the input signal changes to 17.5 V and r_Z is 25 Ω, calculate the output voltage change.

6 — The MOSFET

The junction field-effect transistor was introduced commercially a few years after the introduction of the bipolar-junction transistor (BJT). This device operates in the *depletion mode;* that is, the gate voltage controls the channel resistance by removing or depleting free charge carriers from the channel. The junction field-effect transistor has a very high input impedance, but it has a rather limited voltage gain. Thus the dominance of the BJT was never threatened by this device.

In the mid-1960s, the metal-oxide-semiconductor field-effect transistor or MOSFET was introduced. Originally, a metal gate was used in this device, but the metal gate later evolved into a silicon gate. In spite of this change, the name for the device remained the same. Throughout the 1970s, this device was used primarily in digital circuits, even though its operating speed was much lower than that of the BJT. Two advantages of the MOSFET were apparent very early; the low power dissipation and the small size of the MOS device were very attractive for digital circuits.

As device size decreased with improved fabrication technology, operation speed improved and the MOSFET began to dominate the computer area. In addition, it became popular for use in analog circuits. The year 2000 found the MOSFET becoming dominant in the analog circuit area as well as the digital. This chapter will introduce the MOSFET and discuss some simple discrete implementations of analog amplifiers.

DEMONSTRATION PROBLEM

The circuit in the figure represents a two-stage discrete MOSFET amplifier. The MOSFETs are identical with $\mu C_{ox} W/2L = 1.6$ mA/V^2, $V_T = 0.8$ V, and $\lambda = 0.018$ V^{-1}. We are to select

Amplifier for Demonstration Problem.

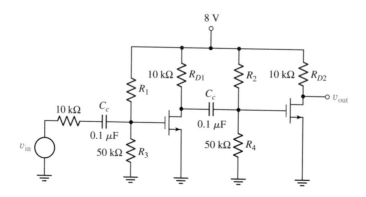

R_1 and R_2 to result in an overall midband voltage gain in the range of 100–120 V/V within a 10% tolerance. The output stage should be biased to allow a peak voltage swing of 1 V. After completing this portion of the design, calculate the overall lower corner frequency.

In order to complete this problem we must understand

1. How to bias the MOSFET stage
2. How incremental parameters such as transconductance are determined by the bias current
3. How to use the incremental model of the MOSFET

6.1 The Field-Effect Transistor

IMPORTANT Concepts

1. The MOSFET is an important amplifying element on which to base amplifier design.
2. The MOSFET is extremely important in the construction of digital computer circuits.
3. Four distinct regions make up the physical MOSFET: the gate, source, drain, and substrate.

In Chapter 4 we discussed op amp circuits as applied to amplifier design. These integrated circuit (IC) chips may use 10 to 20 amplifying elements to create the op amp. In addition, amplifying elements may be used to create amplifiers that are not based on op amp circuits. The two popular amplifying elements of today are the bipolar-junction transistor (BJT) and the metal-oxide-semiconductor field-effect transistor (MOSFET).

A second important use of the MOSFET is in the construction of digital computer circuits. Two features of this device make it significant in this application. The first is small size and the second is low power dissipation. Each individual device is so small that millions of MOSFETs can be fabricated on a single computer chip. This number exceeds by one or two orders of magnitude the number of BJTs that can be put on a comparable chip. The dc source power dissipated by a large number of MOSFETs is also much less than that dissipated by the same number of BJTs.

Because of these advantages, the MOSFET is primarily useful in the fabrication of IC chips. Rarely are discrete MOSFET circuits used, except in the higher power area that uses large devices. This chapter will discuss the basic operation of the MOSFET and then consider a few discrete circuit configurations. Although this type of design is uncommon, it allows the discussion of principles that are important to the IC design of later chapters.

There are several different field-effect devices. The first commercially important field-effect device was the junction field-effect transistor or JFET that gained acceptance in the early 1960s. This device provided a high input impedance in the tens of megohms and a voltage gain of 8 to 16 in typical applications. The high-frequency performance was rather poor compared to that of the BJT. The JFET has a relationship between current and voltage that makes this device a good mixer of two signals in communication applications. The output current varies as the square of the input voltage in one region of operation, referred to as a square-law relationship. The JFET was never as popular as the BJT, but became useful in the fabrication of high-input-impedance IC op amps in the 1960s and 1970s. It is still used in some configurations of the op amp.

Figure 6.1
(a) Four-terminal symbol for the nMOSFET. (b) Three-terminal symbol for the nMOSFET. (c) Four-terminal symbol for the pMOSFET. (d) Three-terminal symbol for the pMOSFET.

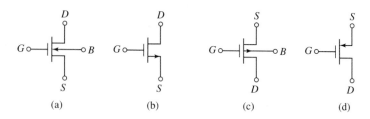

(a)   (b)   (c)   (d)

The metal-oxide-semiconductor field-effect transistor or MOSFET became the most popular field-effect device in the 1980s and 1990s. Originally, this device was formed by depositing an oxide layer on silicon followed by the deposition of a metal layer on the oxide. The popular processes of today replace the metal layer with a layer of polysilicon. Even though the acronym MOSFET is no longer completely accurate, it is still applied to the silicon-oxide-silicon devices of today.

There are two different types of FET device behavior that can be synthesized in modern devices: enhancement mode operation and depletion mode operation. In enhancement mode, as the control voltage across the gate and source terminals increases, the output current flow increases. In depletion mode, an increase of the control voltage causes a decrease in output current flow. Different variations of these modes can be incorporated into a circuit, but the most significant mode at the present time is the enhancement mode, which is the mode used in the very important complementary MOSFET or CMOS circuit. The remainder of this chapter will consider only the enhancement mode MOSFET device.

The output current of the MOSFET is controlled by an input voltage. No input current flows into the input or control terminal, making it a very high input impedance device. The MOSFET consists of four separate regions, although in many applications two of the regions are tied together electrically.

There are several symbols for the MOSFET device. The ones to be used in this textbook are the three-terminal symbols shown in Figs. 6.1(b) and (d). In these symbols, the terminal with the arrowhead is the source. If the arrow points away from the device, it signifies an nMOS device. An arrow pointing toward the device signifies a pMOS device. The gate terminal is the single line coming to one side of the device. The drain terminal is opposite to the source terminal, but has no arrow associated with it. The remaining terminal is the bulk or substrate terminal. Often, the bulk terminal is connected to the source terminal although this is not always possible. The symbols in Figs. 6.1(a) and (c) are used in some textbooks, but we will not use these symbols in succeeding material.

6.2 Qualitative Description of MOSFET Operation

IMPORTANT Concepts

1. The MOSFET is fabricated by creating drain and source regions in the substrate by doping. A thin layer of oxide followed by gate material is deposited on the substrate.

2. The voltage applied across the gate and source terminals determines the current flow between source and drain.

3. There are three regions of operation in which the MOSFET can operate: the cutoff region, the triode region, and the active region.

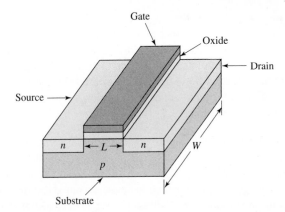

Figure 6.2
*Physical layout of an nMOS
transistor.*

The MOSFET is called a field-effect device because an electric field between the gate and the substrate, established by the applied gate voltage, controls the current flow between the source and drain terminals. Figure 6.2 indicates the approximate physical layout of an nMOS transistor. The p substrate is only lightly doped with acceptor atoms, whereas the source and drain n-type regions are doped with a higher concentration of donor atoms. The oxide layer forms an insulator that creates an extremely high resistance between the gate electrode and the other regions.

PRACTICAL Considerations

A key feature of the MOSFET is the channel length, L. This parameter is used to characterize the device. For example, when a MOSFET is referred to as a 0.5-μ device, this means that the channel is 0.5 μ in length.

For microprocessors that must include millions of devices on a single chip, the channel length is small, as is the channel width, W. Small gate areas that minimize the product $W \times L$ also minimize parasitic capacitances and power dissipation. In order to minimize the space required or the *real estate* for each device on a microprocessor chip, the dimension of W is normally of the same order of magnitude as L.

For amplifier and other analog circuit design using the MOSFET, the value of W may be many times that of L. In these circuits, the number of devices on a chip is rather small compared to the number of devices in a microprocessor. Real estate is not a major consideration for most analog circuits; consequently, the value of W can be used to optimize circuit performance. Large ratios of W to L lead to higher voltage gains in many amplifier configurations. Thus, it is not uncommon to see channel widths that are 1000 times larger than channel lengths in MOSFET analog circuits.

The three regions of operation of the MOSFET are the *cutoff region,* the *triode region,* and the *active region*. Digital circuits often make excursions into all three regions whereas analog circuits, such as amplifiers, typically only use the active region. The MOSFET is in the cutoff region when there is no current flow between source and drain terminals. It is in the triode region when the current flow is a strong function of both the gate-to-source voltage and the drain-to-source voltage. When in the active region, the drain current is primarily a function of the gate-to-source voltage. A weak dependence on drain-to-source voltage may be exhibited in the active region.

Other names for the triode region are the linear region and the ohmic region. The active region is sometimes referred to as the saturation region or the pinchoff region.

Cutoff Region: The cutoff region occurs when the magnitude of the gate-to-source voltage is less than the threshold voltage. This voltage may be around 1 V, but can be controlled in the manufacturing process. In cutoff, no current flows between any pair of terminals. This region is used in switching circuits to emulate a switch that is turned off. If a voltage is applied between source and drain, no current flows because the source to substrate forms a *pn* junction while the substrate to drain forms a second *pn* junction. Negligible current flows through back-to-back series connected diodes.

Inversion of Channel: It would be possible for significant current to flow from drain to source if the substrate between these two regions could be changed to *n*-type. This would remove the *pn* junctions and create an *n*-type region that would allow current to flow under the influence of a drain-to-source voltage. This *n*-type region between source and drain can be created by a voltage applied from gate to substrate. As the gate voltage of an nMOS device goes sufficiently positive, it creates a thin region in the substrate, generating a concentration of electrons that is larger than that of the holes. The gate voltage is said to invert the region from *p*-type to *n*-type. This inversion begins when V_{GS} reaches V_T, the threshold voltage, and increases as V_{GS} increases above this value. The difference between these two voltages is called the effective voltage; that is,

$$V_{\text{eff}} = V_{GS} - V_T \tag{6.1}$$

Inversion begins when V_{eff} equals zero.

With no applied V_{GS}, the number of free holes in the substrate is considerably greater than the number of free electrons; consequently, a very high electric field is required for inversion or to generate more electrons than holes. The magnitude of electric field intensity is proportional to the applied voltage divided by the distance over which the voltage drops. The electric field intensity becomes greater with higher applied voltages and with smaller distances over which the voltage is applied. For the MOSFET, a high intensity is achieved by creating a small distance from gate to substrate. The oxide layer between these regions is made extremely thin. Thus, the electric field intensity can be very high even for relatively low values of V_{GS}, in the range of tenths of a volt to a few volts.

When the gate voltage is positive but smaller than V_T, very few electrons are induced in the channel and little current can flow. Although this condition leads to operation in cutoff, this region is sometimes referred to as the subthreshold region, and only picoamps or nanoamps of drain current can flow for typical values of applied drain voltage.

When the gate voltage is increased to the point where the channel has become an *n*-type channel, electrons can flow from source to drain through the channel, if a positive voltage is applied from drain to source. The result is a conventional current flow from drain to source. As gate voltage is increased to exceed V_T by several tenths of a volt, larger concentrations of electrons are induced in the channel, and conduction from drain to source can become appreciable. Applied gate voltages above the threshold value result in a more highly conductive channel.

Triode Region: When the voltages at both source end and drain end of the channel exceed the threshold voltage, the MOSFET is in the triode region. The current flowing from drain to source is a function of the applied gate voltage and also the applied drain voltage.

Active Region: When the voltage at the source end of the channel exceeds the threshold voltage but the voltage at the drain end is less than the threshold voltage, the MOSFET is in the active region. The current flowing from drain to source is now primarily a function of the applied gate voltage. The drain voltage has a small, often negligible, effect on current.

These regions will be discussed in more detail in the next section.

6.3 Mathematical Description of MOSFET Characteristics

IMPORTANT Concepts

1. A mathematical derivation of the drain current as a function of gate-to-source and drain-to-source voltages can be done for the triode region.
2. A simple extension of the triode region equation can be made for the active region.

6.3.1 THE TRIODE REGION

Figure 6.3 shows an idealized cross-section of an nMOSFET for the case where the source terminal is at ground potential, V_{GS} exceeds the threshold voltage V_T, and the drain-to-source voltage is zero or very small. The bulk or substrate voltage is also assumed to be at ground potential. These conditions lead to operation in the triode region.

The voltage from gate to source, from gate to bulk, and from gate to drain will be approximately the same for this device when $V_{DS} \approx 0$ V. When V_{GS} exceeds V_T, electrons begin to accumulate near the surface of the silicon, just below the oxide layer. The induced channel or the electron distribution is fairly uniform for this situation. The conductance of the channel depends on the gate voltage, becoming higher as V_{GS} is increased.

As the drain-to-source voltage, V_{DS}, increases to cause channel current, the gate-to-drain voltage decreases. This voltage is given by $V_{GD} = V_{GS} - V_{DS}$. The concentration of electrons induced at the drain end of the channel is less than that at the source end as V_{DS} becomes more positive. The induced charge density is no longer constant over the length of the channel, as shown in Fig. 6.4(a).

Also shown in Fig. 6.4(b) is a differential unit of the electron concentration in the channel. The thickness is dx, the height is y_1, and the depth is W.

We want to find the drain current as a function of V_{GS} and V_{DS}. Before we begin this derivation of drain current as a function of applied voltages, we will review some pertinent relationships. In Chapter 5, the mobility of free electrons within a lattice was defined as

$$\mu = \frac{-v}{E} \tag{6.2}$$

where v is the average velocity of the carriers and E is the electric field intensity. We will be concerned with the movement of electrons between source and drain; thus, the electric field

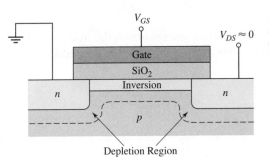

Figure 6.3
nMOSFET with conductive channel.

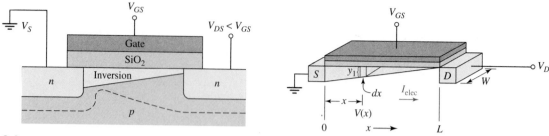

Figure 6.4
nMOSFET with larger drain
voltage.

intensity relates to the voltage drop along the length of the channel, $V(x)$. Mathematically, this quantity can be expressed as

$$E = -\frac{dV(x)}{dx}$$ (6.3)

Combining these two equations results in

$$\mu_n = \frac{v}{\frac{dV(x)}{dx}}$$ (6.4)

A second useful relationship is that of current to charge contained in the differential thickness of Fig. 6.4. If the total charge in this differential element is dQ_T and moves under the influence of the electric field, the channel current or drain current can be written as

$$I_D = \frac{dQ_T}{dt}$$ (6.5)

The quantity dt is the time taken for the differential charge to move across the differential element. This distance is dx and, since the average velocity of the electrons is v, we can write the time as

$$dt = \frac{dx}{v}$$ (6.6)

The drain current can then be written as

$$I_D = \frac{v\,dQ_T}{dx}$$ (6.7)

The charge can be related to the effective voltage across the oxide layer between the gate and channel by the capacitance. This relationship can be expressed as

$$C = \frac{dQ_T}{V_{GS} - V_T - V(x)} = \frac{dQ_T}{V_{\text{eff}} - V(x)}$$ (6.8)

Since no electronic charge is induced in the channel until the voltage from gate to channel exceeds V_T, this parameter must appear in Eq. (6.8).

With these relationships we begin the derivation of current by substituting for v, found by Eq. (6.4), into Eq. (6.7), which gives

$$I_D = \left(\mu_n \frac{dV}{dx} \right) \frac{dQ_T}{dx} \tag{6.9}$$

We substitute into this equation the value of Q_T found from Eq. (6.8). The result is

$$I_D = \frac{\mu C[V_{GS} - V_T - V(x)]}{dx} \frac{dV(x)}{dx} \tag{6.10}$$

We now write the capacitance of the differential element, C, as a capacitance per unit area, C_{ox}, multiplied by the area of the differential element looking from the gate; that is,

$$C = C_{ox} W \, dx \tag{6.11}$$

Equation (6.10) can be written as

$$I_D = \mu W C_{ox}[V_{GS} - V_T - V(x)] \frac{dV(x)}{dx} \tag{6.12}$$

Transposing dx to the opposite side of the equation leads to an expression that can be integrated. This equation is

$$\int_0^L I_D \, dx = \mu W C_{ox} \int_0^{V_{DS}} [V_{GS} - V_T - V(x)] \, dV(x) \tag{6.13}$$

Performing these integrations, substituting the proper limits, and solving for I_D leads to a final equation of

$$I_D = \mu \frac{W}{L} C_{ox} \left[(V_{GS} - V_T)V_{DS} - \frac{V_{DS}^2}{2} \right] \tag{6.14}$$

This equation expresses the V-I characteristics of the nMOSFET in the triode region. It is based on the existence of a finite charge density throughout the channel, which implies that the gate-to-channel voltage must exceed V_T over the channel length. Since the minimum gate-to-channel voltage will exist between gate and drain, this equation is valid up to values of V_{DS} that satisfy

$$V_{GS} - V_{DS} = V_T \tag{6.15}$$

The maximum value of V_{DS} in the triode region is called the pinchoff value and is given by

$$V_{DSP} = V_{GS} - V_T \tag{6.16}$$

For values of V_{DS} less than V_{DSP} and V_{GS} greater than V_T, the nMOSFET is in the triode region; that is,

$$V_{GS} > V_T \tag{6.17}$$

and

$$V_{DS} \leq V_{GS} - V_T = V_{DSP} \tag{6.18}$$

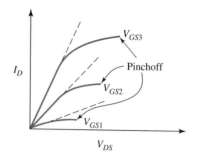

Figure 6.5
Triode region operation for the nMOS device.

This region is also referred to as the linear region, but the variation of I_D with V_{DS} is linear only for small values of V_{DS}. From Eq. (6.14), when V_{DS} is small enough that the term $(V_{GS} - V_T)V_{DS}$ is larger than $V_{DS}^2/2$, the equation approximates a linear variation. As V_{DS} increases, the term V_{DS}^2 becomes larger and decreases the value that I_D would have if a linear variation continued. Figure 6.5 shows the V-I characteristics of an nMOS device in the triode region. These curves are based on Eq. (6.14) with a different value of V_{GS} used to generate each curve. The dashed lines are included to demonstrate the departure of the curves from a linear variation as the pinchoff voltage is approached. As V_{GS} is increased to generate another curve, the value of V_{DSP} also increases.

The triode region is used in applications requiring a voltage-controlled resistance. The resistance is a function of the applied gate-to-source voltage as seen in Fig. 6.5 or from Eq. (6.14) for small values of V_{DS}. The triode region is rarely used in typical amplifier stages.

When values of V_{DS} are low enough to limit operation to the linear portion of the curves, the resistance can be calculated from an approximation of Eq. (6.14). If the second-order term in V_{DS} is negligible, Eq. (6.14) is approximated as

$$I_D = \mu \frac{W}{L} C_{ox} [(V_{GS} - V_T)V_{DS}] \tag{6.19}$$

For this equation to be relatively accurate, the first-order V_{DS} term in Eq. (6.14) should be about 10 times the second-order term or

$$(V_{GS} - V_T)V_{DS} \geq 10 \left[\frac{V_{DS}^2}{2} \right] \tag{6.20}$$

This leads to the relationship that

$$V_{GS} - V_T \geq 5V_{DS} \tag{6.21}$$

The equivalent conductance of the linear portion of the curve in the triode region can be found by differentiating I_D in Eq. (6.19) with respect to V_{DS}. The result is then inverted to give resistance.

Carrying out this differentiation leads to

$$\frac{\partial I_D}{\partial V_{DS}} = \mu \frac{W}{L} C_{ox}(V_{GS} - V_T) \tag{6.22}$$

Inverting this value to find resistance results in

$$R_{lin} = \frac{\partial V_{DS}}{\partial I_D} = \frac{1}{\mu \frac{W}{L} C_{ox}(V_{GS} - V_T)} \tag{6.23}$$

EXAMPLE 6.1

For the nMOS device of Fig. 6.6, $V_T = 1$ V and $\mu C_{ox} W/L = 0.1$ mA/V^2. This circuit is designed as a variable-resistance circuit.

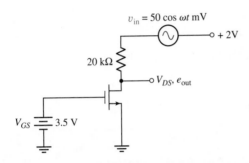

Figure 6.6
Circuit for Example 6.1.

(a) Calculate the dc value of V_{DS}.

(b) Calculate the dc value of I_D.

(c) Calculate the ac output voltage.

(d) Calculate a new value of V_{GS} to lead to a peak ac output voltage of 6 mV.

SOLUTION If we were certain that this circuit operated in the linear part of the triode region, we could use Eq. (6.19) to express the drain current. This is a linear equation as opposed to the more accurate Eq. (6.14). We will use the latter equation to find V_{DS}.

(a) The drain current is

$$I_D = 0.1 \left[(3.5 - 1)V_{DS} - \frac{V_{DS}^2}{2} \right] = 0.1 \left[2.5\, V_{DS} - 0.5\, V_{DS}^2 \right] \text{ mA}$$

The external circuit forces the dc drain current to be

$$I_D = \frac{2 - V_{DS}}{20} \text{ mA}$$

Equating these two values of drain current and solving for V_{DS} from the resulting equation leads to $V_{DS} = 0.354$ V. From Eq. (6.21), we see that the value of $V_{GS} - V_T$, which is 2.5 V, is more than five times the value of V_{DS}; thus, the simpler equation, Eq. (6.19), could have been used.

(b) The drain current can be found from the voltage drop across the external resistor, giving

$$I_D = \frac{2 - 0.354}{20} = 82 \ \mu A$$

(c) The ac output voltage is calculated by finding the resistance from drain to source of the MOSFET and recognizing that this resistance forms a voltage divider with the 20-kΩ external resistor. The resistance from drain to source is found from Eq. (6.23) as

$$R_{lin} = \frac{1}{0.1 \times 2.5} = 4 \text{ k}\Omega$$

The value of e_{out} is then

$$e_{out} = \frac{4}{20 + 4}\, 50 \cos \omega t = 8.33 \cos \omega t \text{ mV}$$

(d) To find the value of V_{GS} to cause a peak output of 6 mV, the necessary MOSFET resistance is first calculated. This is done with the voltage division equation

$$6 = \frac{R_{\text{lin}}}{R_{\text{lin}} + 20} \, 50$$

Solving for R_{lin} gives

$$R_{\text{lin}} = 2.73 \text{ k}\Omega$$

Equation (6.23) is now applied to find the value of V_{GS} that leads to this value of R_{lin}. The result is

$$V_{GS} = \frac{1}{\mu C_{\text{ox}} W/L \, R_{\text{lin}}} + V_T = \frac{1}{0.1 \times 2.73} + 1 = 4.66 \text{ V}$$

PRACTICAL Considerations

PRACTICE Problems

6.1 If $V_{GS} - V_T = 1.8$ V in Eq. (6.14), calculate the value of V_{DS} for which the second-order term is 10% of the linear term. *Ans:* 0.36 V.
6.2 Calculate the percentage difference between the actual current given by Eq. (6.14) and the current calculated by neglecting the second-order term at $V_{DS} = V_{DSP} = 1.8$ V. *Ans:* 100%.

One application of the MOSFET in the triode region is that of a voltage-controlled resistance. The strategic placement of this device in a larger circuit can result in an amplifier with a voltage gain that can be controlled by a dc voltage applied to the gate of the MOSFET.

The signal generators used for laboratory work contain an oscillator that generates the output signal. An amplitude control sets the desired magnitude of this output signal. As temperature changes, the amplitude of this output signal may change. One well-known manufacturer uses a MOSFET to minimize the amplitude change. The output signal is continuously converted to a dc signal proportional to the magnitude of the output signal. This dc voltage serves as the control voltage to drive the gate of the MOSFET. If the output signal magnitude drops in value, the decreased value of control voltage is used to change the resistance of the MOSFET. The MOSFET is connected such that the change in resistance between drain and source will increase the gain of the oscillator and restore the signal magnitude to the desired value.

The idea of volume control in an oscillator or an amplifier is important to several areas of electronics including communication circuits. The MOSFET, operating in the triode region, is often used for this purpose.

6.3.2 THE ACTIVE REGION

Amplification of a signal requires that the amplified signal be linearly proportional to the input signal. This situation can be approximated when the MOSFET operates in the active region. For a constant value of applied V_{GS}, if V_{DS} increases above the pinchoff value, the gate-to-drain voltage decreases to the point that the drain end of the channel has a zero electron density. The resistance of an incremental length of the channel at the drain end at pinchoff is very high compared to the resistance of an equal length at the source end of the channel. As V_{DS} is further increased above V_{DSP}, most of this increased voltage drops across the pinched-off drain end of the channel. This condition is shown in Fig. 6.7.

Conditions along the channel toward the source change little as V_{DS} increases above pinchoff; thus, the drain current hardly changes with V_{DS}. One approximation often used for the MOSFET is that the drain current remains constant at the pinchoff value for any V_{DS} above V_{DSP}. This region is sometimes referred to as the saturation region, but in this book we will refer to it as the active region. Figure 6.8 shows the V-I characteristics for an nMOS device in both the triode and the active regions.

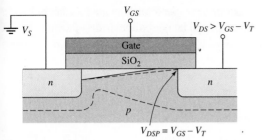

Figure 6.7
Induced electron density at or above pinchoff.

The dashed lines show a slight increase in drain current as V_{DS} increases above V_{DSP}. These lines are a better approximation to the actual characteristics of the MOSFET than are the solid lines.

The channel resistance increases dramatically as the electron density at the drain end approaches zero. This situation is referred to as *drain pinchoff* or simply as *pinchoff* and occurs when the voltage from gate to drain equals the threshold voltage as expressed by Eq. (6.16):

$$V_{DSP} = V_{GS} - V_T \quad \text{at pinchoff} \tag{6.24}$$

The equation for drain current for an nMOSFET in the region where V_{DS} is less than the pinchoff value is given by Eq. (6.14). When pinchoff is reached as shown in Fig. 6.5, Eq. (6.14) can be used with the drain-to-source voltage of Eq. (6.24) to find the current at pinchoff. This gives

$$I_{DP} = \mu \frac{W}{2L} C_{ox}(V_{GS} - V_T)^2 \tag{6.25}$$

Note that the pinchoff current is a function of the ratio of channel width to channel length or W/L. This gives the designer some control over the V-I characteristics of the MOSFET.

Equation (6.25) is useful in a great deal of design work, even though it is not a precise equation. It is particularly useful in manual analysis of MOS devices with longer channel lengths—for example, over 1 μ in length.

Because the actual curves in the active region have a slight upward slope, Eq. (6.25) is sometimes modified to reflect this finite slope by means of a multiplicative factor. The resulting equation for active region current is

$$I_D = I_{DP} \left[1 + \lambda(V_{DS} - V_{DSP})\right] = \mu \frac{W}{2L} C_{ox}[V_{GS} - V_T]^2 \left[1 + \lambda(V_{DS} - V_{DSP})\right] \tag{6.26}$$

where λ is the channel length modulation factor. This equation is only valid for values of V_{DS} above V_{DSP}. The appearance of λ in the equation for drain current leads to a slight

PRACTICE **Problems**

6.3 If $\mu \frac{W}{2L} C_{ox} = 300 \, \mu$ A/V^2 and $V_T = 0.9$ V, calculate the pinchoff value of drain current when $V_{GS} = 1.4$ V.
Ans: 75 μA.

6.4 Calculate the drain current for the situation of Practice Problem 6.3, assuming that $V_{DS} = 5$ V and $\lambda = 0.015 V^{-1}$.
Ans: 80.1 μA.

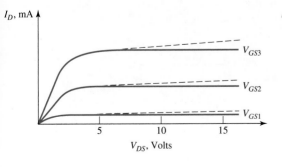

Figure 6.8
V-I characteristics in the active region.

upward slope of current with V_{DS}. The upward slope is due to the change in channel length as the voltage V_{DS} is increased. When the drain voltage increases above the value that causes pinchoff, the pinched-off end of the channel extends slightly further from the drain toward the source end of the channel. The channel length shortens accordingly and more current flows through this shorter channel for a given value of V_{GS}. This is called the channel length modulation effect and is indicated in Fig. 6.7 by the solid line representing electron distribution in the channel.

6.4 The MOSFET as an Amplifying Element

IMPORTANT Concepts

1. Although small, incremental signals are to be amplified, a dc bias is applied to the MOSFET amplifying stage to establish the no-signal or quiescent output voltage near the middle of the active region.

2. An incremental equivalent circuit can be used to model the MOSFET in the common-source amplifier. Voltage gain, input impedance, and output impedance can be calculated from this equivalent circuit.

3. The quiescent or dc values of drain current and gate-to-source voltage help determine the parameter values in the model.

4. The MOSFET must remain in the active region over all output voltages for the model to be valid. It is important to check the output voltage to ensure that neither the triode region nor cutoff is reached.

6.4.1 RELATION OF THE GENERAL AMPLIFIER MODEL TO THE MOSFET

The general amplifier model of Fig. 3.4 is redrawn in Fig. 6.9. This model will be compared to the common-source amplifier stage of Fig. 6.10. Resistors R_1 and R_2 are used to create a no-signal or quiescent dc value of gate-to-source voltage, V_{GSQ}, that will set the quiescent drain-to-source voltage, V_{DSQ}, in the active region. The active region extends from cutoff to the edge of pinchoff or the triode region. Cutoff occurs when $I_D = 0$, leading to no voltage drop across R_D and an output voltage level of $V_{DS} = V_{DD}$. The edge of pinchoff is reached when I_D is increased to the point that V_{DS} reaches its pinchoff value as calculated by Eq. (6.24). Thus, the maximum level of output voltage is $V_{DS\max} = V_{DD}$, which might be several volts, and the minimum level is $V_{DS\min} = V_{DSP} = V_{GS} - V_T$, which might fall between a few tenths of a volt to a few volts.

Figure 6.9
The general amplifier model.

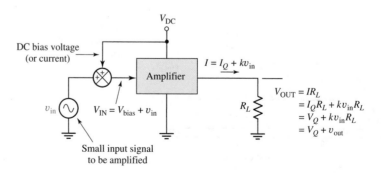

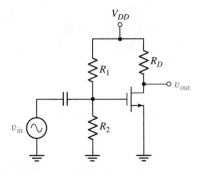

Figure 6.10
A common-source amplifying stage.

The load line construction of Fig. 6.11 demonstrates these values. The load line represents the characteristics of the dc power supply voltage and the resistance R_D. It obeys the relationship

$$V_{DS} = V_{DD} - I_D R_D \qquad (6.27)$$

The external circuit constrains the output current and voltage of the MOSFET to vary along the load line. A voltage of V_{GSQ} applied to the gate leads to a particular quiescent drain current, found from Eq. (6.26). This value of I_{DQ} must be on the load line and results in the value of V_{DSQ} shown in Fig. 6.11.

If the input voltage decreases, decreasing I_D to cutoff, the output voltage swings to V_{DD} along the load line. If the input voltage increases, the output voltage swings along the load line to V_{DSmin}, the pinchoff value. The calculation of this value will be deferred to the next section. We generally set the quiescent voltage to be somewhere between V_{DSmin} and V_{DSmax}.

Once the quiescent or dc level of V_{DS} is set, we can apply the incremental signal to be amplified. This voltage adds to the quiescent gate voltage to give

$$V_{GS} = V_{GSQ} + v_{in} \qquad (6.28)$$

If v_{in} is a sinusoid, the gate voltage becomes

$$V_{GS} = V_{GSQ} + B \sin \omega t \qquad (6.29)$$

The value of B must be much smaller than V_{GSQ} to be considered a small, incremental signal.

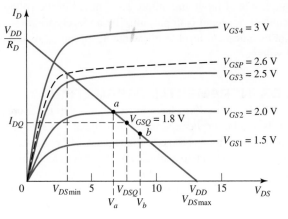

Figure 6.11
Load line for the stage of Fig. 6.10.

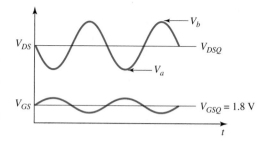

Figure 6.12
Input and output voltages of MOSFET stage.

For the device represented by Fig. 6.11, the total voltage applied to the gate might be

$$V_{GS} = 1.8 + 0.2 \sin \omega t$$

As the signal voltage swings to its peak value of 0.2 V, the total value of V_{GS} becomes equal to 2.0 V. This moves the output to point a on the load line, decreasing the output voltage from V_{DSQ} to V_a. As the ac voltage swings to its most negative value of -0.2 V, V_{GS} swings to 1.6 V and the output swings to point b with an output voltage of V_b. Figure 6.12 plots the input and output voltages as a function of time.

Although the V-I characteristics of the MOSFET are nonlinear, if the output voltage is limited in magnitude, the ac swing around V_{DSQ} is fairly symmetric. Thus the negative peak output value equals the positive peak value or

$$V_{DSQ} - V_a = V_b - V_{DSQ}$$

As the input ac signal magnitude is increased, the output magnitude also increases as well as the distortion of this output voltage. If the output is driven to cutoff or pinchoff, the distortion becomes very abrupt. Even before these extreme points are reached, unacceptable distortion may be present. When this is the case, the output can no longer be considered an incremental signal. However, the output may reach several volts in magnitude before distortion becomes unacceptable.

In Fig. 6.12, the peak output voltage may be 1.4 V, leading to a voltage gain of

$$A_v = \frac{V_{p-\text{out}}}{V_{p-\text{in}}} = \frac{-1.4}{0.2} = -7 \text{ V/V}$$

The negative sign occurs because an increasing value of V_{GS} results in a decreasing value of V_{DS}.

We note that the dc voltage levels of input and output are ignored in the voltage gain calculation. Only the incremental or ac signal is important. The dc levels are present only to establish a proper bias point or quiescent level to allow the ac output to swing in both directions from the quiescent voltage.

6.4.2 SMALL-SIGNAL OR INCREMENTAL PARAMETERS

When the quiescent dc values of V_{GSQ} and V_{DSQ} have been established, the incremental input signal is applied. This leads to a larger incremental output signal that swings around a dc quiescent level. The incremental output voltage is the signal of importance. The dc voltage is of no interest. An important approach to analysis and design of electronic circuits is that of the incremental equivalent circuit. This method ignores the dc voltages while basing all calculations on the incremental parameters of the circuit. These incremental parameters relate incremental changes of current to incremental changes of voltage. Analysis of voltage gain, input impedance, and output impedance can be based on these incremental quantities. Two important parameters are the *incremental drain-to-source resistance* and the *transconductance*.

Incremental Drain-to-Source Resistance: Returning to Fig. 6.8, we see that there is a slight upward slope in the drain current as V_{DS} is increased, as indicated by the dashed lines. For a specific value of V_{GS}, this slope is constant over a large range of V_{DS}. Over this range, we can define an incremental value of drain-to-source resistance, r_{ds}. This value can be calculated from the characteristic curves, if available, or from Eq. (6.26). We define r_{ds} mathematically as

$$r_{ds} = \left. \frac{\partial V_{DS}}{\partial I_D} \right]_{V_{GS}=K}$$

The value of V_{GS} is held constant as we take this derivative. Since we have I_D expressed as a function of V_{DS} in Eq. (6.26), it is easier to find the derivative of current with respect to voltage, and then take the reciprocal of this value to get r_{ds}. This results in

$$r_{ds} = \left. \frac{1}{\frac{\partial I_D}{\partial V_{DS}}} \right]_{V_{GS}=K} = \frac{1}{\lambda \mu \frac{W}{2L} C_{ox} [V_{GS} - V_T]^2} = \frac{1}{\lambda I_{DP}} \tag{6.30}$$

The drain-to-source resistance will decrease as V_{GS} increases, leading to higher values of I_{DP}.

Transconductance: Another important parameter of the MOSFET is the transconductance, which is designated g_m and relates the change in V_{GS} to the resulting change in I_D while holding V_{DS} constant. This parameter is defined mathematically as

$$g_m = \left. \frac{\partial I_D}{\partial V_{GS}} \right]_{V_{DS}=K} = \mu \frac{W}{L} C_{ox} [V_{GS} - V_T] \left[1 + \lambda (V_{DS} - V_{DSP}) \right]$$

$$= \frac{2I_D}{V_{GS} - V_T} = \frac{2I_D}{V_{\text{eff}}} \tag{6.31}$$

Often λ can be taken as zero to give an approximation for transconductance of

$$g_m = \mu \frac{W}{L} C_{ox} (V_{GS} - V_T) = \mu \frac{W}{L} C_{ox} V_{\text{eff}} \tag{6.32}$$

Neglecting λ allows this equation to be written in another useful form. Equation (6.25) expresses the pinchoff current in terms of V_{eff}, but with $\lambda = 0$, the drain current equals the pinchoff current at all drain-to-source voltages. Solving for V_{eff} leads to

$$V_{\text{eff}} = \sqrt{\frac{I_D}{\mu \frac{W}{2L} C_{ox}}} \tag{6.33}$$

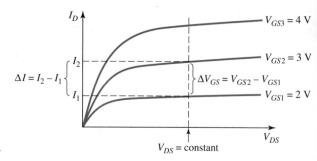

Figure 6.13
Graphical determination of g_m.

Substituting this value into Eq. (6.32) allows g_m to be expressed in terms of drain current as

$$g_m = \sqrt{2\mu C_{ox}(W/L)I_D} \tag{6.34}$$

Since the active region is generally used for amplifying applications, the parameters r_{ds} and g_m are important in determining amplifier behavior.

The pMOSFET operates much like the nMOSFET. For this device, the threshold voltage is negative and a negative gate-to-source voltage must be applied to lead to channel conduction. When in the triode or active regions, drain current flows from the drain terminal rather than into the drain terminal as for the nMOS device.

Whereas g_m can be calculated from Eqs. (6.31), (6.32), or (6.34) and r_{ds} can be found from Eq. (6.30), Figs. 6.13 and 6.14 show graphical methods to determine these parameters from the characteristic curves. In Fig. 6.13 a constant drain-to-source voltage is indicated by the vertical dashed line. The intersection of this line with V_{GS1} and V_{GS2} defines the changes in drain current. The value of g_m is

$$g_m = \left. \frac{\Delta I_D}{\Delta V_{GS}} \right]_{V_{DS=K}} = \frac{I_2 - I_1}{V_{GS2} - V_{GS1}} \tag{6.35}$$

Figure 6.14 shows a single curve for a constant value of V_{GS}. The drain-to-source resistance is

$$r_{ds} = \left. \frac{\Delta V_{DS}}{\Delta I_D} \right]_{V_{GS=K}} = \frac{V_{DS2} - V_{DS1}}{I_2 - I_1} \tag{6.36}$$

Of course, the curves must be given in order to do these graphical evaluations.

Figure 6.14
Graphical determination of r_{ds}.

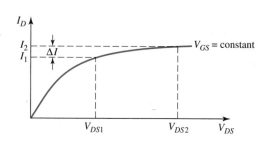

EXAMPLE 6.2

If $V_{DS} = 4$ V, $V_{GS2} = 3$ V, $V_{GS1} = 2$ V, $I_2 = 200$ μA, and $I_1 = 100$ μA in Fig. 6.13, evaluate g_m. Find the approximate value of $\mu C_{ox} W/L$ for this device at $I_D = 150$ μA.

SOLUTION Equation (6.35) is used with the given values to find that

$$g_m = \frac{200 - 100}{3 - 2} = 100 \ \mu A$$

If we assume this value holds at $I_D = 150$ μA, a somewhat shaky assumption, we can substitute values into Eq. (6.34) to result in

$$g_m = 100 \ \mu A/V = \sqrt{2\mu C_{ox}(W/L)I_D} = 17.3\sqrt{\mu C_{ox} W/L}$$

Solving for $\mu C_{ox} W/L$ leads to

$$\frac{\mu C_{ox} W}{L} = \left(\frac{100}{17.3}\right)^2 = 33.3 \ \mu A/V^2$$

This method used rather large increments of V_{GS} and would be expected to be less accurate than the equations for g_m.

PRACTICE Problems

6.5 If $\mu W C_{ox}/2L = 50 \ \mu$ A/V^2, $V_T = 0.8$ V, $V_{GS} = 2$ V, and $\lambda = 0.02$/V, calculate r_{ds}.
6.6 For the MOSFET of Practice Problem 6.5, evaluate g_m.

6.4.3 THE MIDBAND INCREMENTAL MOSFET MODEL

In this section we will consider the model for the MOSFET at low and midband frequencies. In a later section we will consider the modifications to the model necessary at high frequencies.

A common-source stage, resistive load amplifier is shown in Fig. 6.15. As explained earlier, the MOSFET must be biased into the active region before an input signal is applied. If the output drain-to-source voltage can swing from a minimum of 2 V (at the edge of the triode region) up to 10 V (cutoff region), the quiescent drain-to-source voltage might be set near 6 V. This would allow the voltage to swing 4 V in either direction from the quiescent value before serious distortion takes place. The desired value of V_{DSQ} is established by setting the proper value of V_{GSQ}.

The resistors R_1 and R_2 are used to establish the necessary gate-to-source quiescent bias voltage, V_{GSQ}. These resistors can be very large to avoid loading since no gate current is required. Because the gate of the nMOSFET draws no current, the bias voltage is established by voltage division of the two bias resistors.

The standard way to analyze this stage is to use the small-signal equivalent circuit of the MOSFET rather than the graphical method. This model is indicated in Fig. 6.16. The

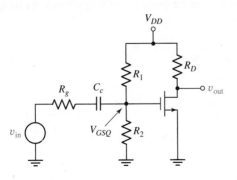

Figure 6.15
An nMOSFET amplifier.

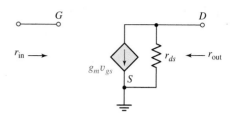

Figure 6.16
The equivalent circuit or model of the common-source MOSFET stage.

impedance between the gate and source terminals is represented by an open circuit, reflecting the fact that there is no current flow across the oxide layer between gate and channel. The effect of gate-to-source voltage on the drain current is modeled by the dependent current generator of value $g_m v_{gs}$. The resistance r_{ds} accounts for the incremental resistance between drain and source. This model is an incremental model that applies to the active region when active region values of g_m and r_{ds} are used.

This circuit is the basic incremental model for the MOSFET device in the active region. It is used every time a MOSFET stage is manually analyzed. While the values of g_m and r_{ds} are determined by the quiescent drain current, no dc quantities appear in this model.

The equivalent circuit of the complete amplifier stage is completed by replacing the MOSFET by its model at terminals G, S, and D. All dc power supply voltages are considered to be incremental signal grounds. The resulting circuit is shown in Fig. 6.17.

The input impedance to the stage is determined by the bias resistors since the gate-to-source impedance of the MOSFET is so high. This input impedance is

$$R_{in} = \frac{R_1 R_2}{R_1 + R_2} \tag{6.37}$$

This value can be made quite high if necessary. Theoretically, the total impedance is this quantity in parallel with r_{in}, the gate-to-source impedance, which is approximated as an open circuit.

The output impedance of the MOSFET looking into the drain terminal is

$$r_{out} = r_{ds} \tag{6.38}$$

The total output impedance of the amplifier is r_{ds} in parallel with the external drain resistance R_D; that is,

$$R_{out} = \frac{r_{ds} R_D}{r_{ds} + R_D} \tag{6.39}$$

The voltage gain from gate to drain can be found by writing the output voltage in terms of the current, which leads to

$$v_{out} = -g_m v_{gs} \frac{r_{ds} R_D}{r_{ds} + R_D}$$

Figure 6.17
The equivalent circuit for the amplifier stage.

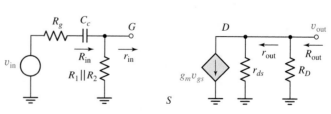

The gate-to-drain voltage gain is then

$$A_{gd} = \frac{v_{out}}{v_{gs}} = -g_m \times \frac{r_{ds} R_D}{r_{ds} + R_D} \tag{6.40}$$

Ignoring bias considerations, this voltage gain increases as R_D is increased. The maximum gain would then occur for an infinite value of R_D. This maximum gain is sometimes designated A_μ and is expressed as

$$A_\mu = g_m r_{ds} \tag{6.41}$$

The voltage amplification of any practical MOSFET stage cannot exceed this value of A_μ.

The overall voltage gain must account for any loading on the signal generator resistance by the bias resistors, leading to an overall midband voltage gain of

$$A_{MB} = \frac{v_{out}}{v_{in}} = \frac{-R_B}{R_B + R_g} \times g_m \times \frac{r_{ds} R_D}{r_{ds} + R_D} \tag{6.42}$$

where R_B is equal to the parallel combination of the bias resistors.

The equivalent circuit of Fig. 6.17 should be compared to the general low-frequency amplifier model of Fig. 3.12. The earlier model shows a dependent voltage source and series resistance whereas the equivalent circuit of Fig. 6.17 shows a dependent current source in parallel with r_{ds} and R_D. If a Thévenin equivalent of the current source and resistance is taken, the circuit appears as shown in Fig. 6.18. The dependent voltage source has a value of

$$g_m R_{out} v_{gs}$$

with the polarity shown in the figure. The output impedance is

$$R_{out} = \frac{r_{ds} R_D}{r_{ds} + R_D}$$

Normally, the dependent current source version of the equivalent circuit is used in analysis, but it is easier to use the dependent voltage source circuit in a few special cases.

The analysis of a MOSFET amplifier stage will be demonstrated by assuming the following values for the stage of Fig. 6.15:

$$R_1 = 50 \text{ k}\Omega, \quad R_2 = 15 \text{ k}\Omega, \quad R_D = 4 \text{ k}\Omega, \quad R_g = 4 \text{ k}\Omega$$

$$g_m = 2.7 \text{ mA/V}, \quad r_{ds} = 50 \text{ k}\Omega, \quad C_c = 0.1 \ \mu\text{F}, \quad V_{DD} = 10 \text{ V}$$

It is desired to find the midband voltage gain and the lower corner frequency of the circuit.

The midband voltage gain is found from Eq. (6.42) after calculating R_B, the parallel combination of R_1 and R_2. This gives

$$R_B = \frac{50 \times 15}{50 + 15} = 11.5 \text{ k}\Omega$$

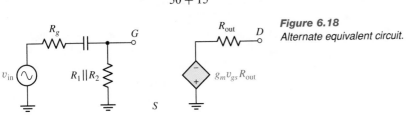

Figure 6.18
Alternate equivalent circuit.

The gain is calculated from Eq. (6.42) as

$$A_{MB} = \frac{-11.5}{11.5 + 4} \times 2.7 \times \frac{50 \times 4}{50 + 4} = -7.42 \text{ V/V}$$

From Chapter 3, the lower corner frequency is calculated from

$$f_{\text{low}} = \frac{1}{2\pi C_c [R_g + R_{\text{in}}]} = \frac{1}{2\pi \times 10^{-7} \times 15.5 \times 10^3} = 103 \text{ Hz}$$

Although this analysis is fairly straightforward, an actual design problem is considerably more complex. This is primarily due to the fact that g_m and r_{ds} are both functions of the quiescent operating point. The choice of quiescent drain current is influenced by R_D, which may not be known at the start of the design. If R_D is known, the design can proceed as indicated in Example 6.3.

EXAMPLE 6.3

Assume that the circuit of Fig. 6.15 has a power supply of $V_{DD} = 8$ V, $R_D = 4$ kΩ, and $R_g = 10$ kΩ. Also assume that the drain current is related to the gate-to-source voltage as in Eq. (6.25); that is,

$$I_D = I_{DP} = \mu \frac{W}{2L} C_{\text{ox}} (V_{GS} - V_T)^2$$

If $V_T = 0.8$ V and $\mu W C_{\text{ox}}/2L = 0.4$ mA/V^2,

(a) Calculate the required value of V_{GSQ} to establish V_{DSQ} at 5 V.
(b) Calculate suitable values of R_1 and R_2 to establish this bias point. Assume that $R_B = R_1 \parallel R_2$ is equal to 100 kΩ.
(c) Calculate the small-signal midband voltage gain for this bias point.
(d) Calculate the largest peak value of input voltage that can be applied without causing the output to leave the active region.

SOLUTION (a) From the load line equation, $V_{DS} = V_{DD} - R_D I_D$. Solving this equation with $V_{DSQ} = 5$ V results in $I_{DQ} = 0.75$ mA. We now substitute this value of quiescent drain current into Eq. (6.25) and solve for V_{GSQ}, which results in a value of $V_{GSQ} = 2.17$ V.

(b) Since the source terminal of this stage is at 0 V, setting the gate voltage to 2.17 V will establish the proper value of V_{GSQ}. This can be accomplished by picking resistor values to satisfy

$$V_{GSQ} = V_{DD} \frac{R_2}{R_1 + R_2}$$

However, in order to result in $R_B = 100$ kΩ, we must also satisfy the equation

$$R_B = \frac{R_1 R_2}{R_1 + R_2} = 100 \text{ k}\Omega$$

Solving these two equations simultaneously yields values of $R_1 = 369$ kΩ and $R_2 = 137$ kΩ.

(c) In order to evaluate the voltage gain, the equivalent circuit of Fig. 6.16 is used. Note that Eq. (6.25) implies a zero value of λ and, therefore, an infinite value for r_{ds} from Eq. (6.30). We need to find a value for g_m, and this can be done with the aid of Eq. (6.32):

$$g_m = 2 \times 0.4 \times (V_{GSQ} - V_T) = 0.8 \times (2.17 - 0.8) = 1.096 \text{ mA/V}$$

Using Eq. (6.42) we now calculate the voltage gain as

$$A_{MB} = \frac{-100}{100 + 10} \times 1.096 \times 4 = -3.99 \text{ V/V}$$

(d) In order to remain in the active region as the incremental output voltage varies, the stage must avoid cutoff as current decreases and the triode region as current increases. With the given bias point of $V_{DSQ} = 5$ V, a peak voltage of 3 V can be developed in the cutoff direction. This would lead to a cutoff value of $V_{DS} = V_{DD} = 8$ V. We next find how far toward the triode region this output can swing. To do this, we need to find the point where V_{DS}, as calculated from the load line equation, equals a value of V_{DSP}, the pinchoff value. At pinchoff, we have noted that $V_{DSP} = V_{GSP} - V_T$. We can use the load line equation to write

$$V_{DSP} = 8 - 4I_{DP}$$

Equation (6.25) must also be satisfied; that is,

$$I_{DP} = 0.4(V_{GSP} - 0.8)^2 = 0.4(V_{DSP})^2$$

Solving these equations simultaneously for V_{DSP} leads to a value of 1.94 V for this voltage. The largest peak output voltage swing in the negative direction is the quiescent value less this value or $5 - 1.94 = 3.06$ V.

The output swing before serious distortion is almost the same in both directions, with the 3-V peak value toward cutoff taken as the limiting value. To relate this to the largest possible peak input voltage, we simply divide by the voltage gain, giving

$$|v_{in}| = \frac{3}{3.99} = 0.75 \text{ V}$$

Although an applied voltage of this magnitude or larger would result in serious distortion, even smaller values will result in modest distortion due to the nonlinear relationship between I_D and V_{GS}. Distortion decreases as the applied signal magnitude decreases.

The next example demonstrates the interaction of the dc bias point and the small-signal parameters, g_m and r_{ds}.

EXAMPLE 6.4

The transconductance and drain-to-source resistance for the MOSFET of Fig. 6.19 were measured at a drain current of $I_D = 100 \ \mu A$. The measured values were $g_m = 2$ mA/V and $r_{ds} = 180$ kΩ. The threshold voltage is 0.8 V.

(a) Choose R_1 to set $V_{DSQ} = 6$ V.
(b) Calculate the midband voltage gain for the bias of part (a).

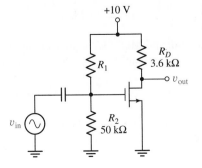

Figure 6.19
Amplifier for Example 6.4.

SOLUTION (a) To set V_{DSQ} to 6 V, a drain current of

$$I_{DQ} = \frac{V_{DD} - V_{DQ}}{R_D} = \frac{10 - 6}{3.6} = 1.11 \text{ mA}$$

is required. The value of V_{GSQ} to establish this drain current could be approximated from Eq. (6.25) except that a value for $\mu C_{ox} W/L$ has not been given. This value can be calculated from Eq. (6.34)

$$g_m = \sqrt{2\mu C_{ox}(W/L)I_D}$$

Since $g_m = 2$ mA/V for a current of $I_D = 0.1$ mA, we solve Eq. (6.34) for $\mu C_{ox} W/L$ to get

$$\mu C_{ox} W/L = \frac{g_m^2}{2I_D} = \frac{2^2}{2 \times 0.1} = 20 \text{ mA/V}^2$$

Using this number in Eq. (6.25), we can solve for the voltage to result in $I_{DQ} = 1.11$ mA, which leads to

$$V_{GS} = V_T + \sqrt{\frac{2LI_D}{\mu C_{ox} W}} = 0.8 + \sqrt{\frac{2 \times 1.11}{20}} = 1.13 \text{ V}$$

The resistor R_1 is chosen to obtain this value of quiescent gate voltage by the equation

$$V_{GSQ} = \frac{R_2}{R_1 + R_2} V_{DD} = \frac{500}{R_1 + 50}$$

The result is $R_1 = 391$ kΩ.

Since the drain current of the actual circuit is higher than the value at which g_m and r_{ds} were measured, we must calculate new values of these parameters to use in the equivalent circuit. From Eq. (6.34) we can write

$$\frac{g_{m2}}{g_{m1}} = \frac{\sqrt{2\mu C_{ox}(W/L)I_{D2}}}{\sqrt{2\mu C_{ox}(W/L)I_{D1}}} = \sqrt{\frac{I_{D2}}{I_{D1}}}$$

where g_{m2} is the transconductance at I_{D2} and g_{m1} is the transconductance at I_{D1}. If g_{m1} at I_{D1} is known, then g_{m2} at I_{D2} can be found from

$$g_{m2} = \sqrt{\frac{I_{D2}}{I_{D1}}} g_{m1} = \sqrt{\frac{1.11}{0.1}} \times 2 = 6.66 \text{ mA/V}$$

The new value of r_{ds} can be approximated from Eq. (6.30), assuming that $I_D \approx I_{DP}$. Using this equation, we can write

$$\frac{r_{ds2}}{r_{ds1}} = \frac{\frac{1}{\lambda I_{D2}}}{\frac{1}{\lambda I_{D1}}} = \frac{I_{D1}}{I_{D2}}$$

The value of r_{ds} at $I_{DQ} = 1.11$ mA is

$$r_{ds2} = \frac{I_{D1}}{I_{D2}} r_{ds1} = \frac{0.1}{1.11} \times 180 = 16.2 \text{ k}\Omega$$

(b) After evaluating g_m and r_{ds}, the midband MOSFET equivalent circuit of Fig. 6.20 can be used to evaluate the voltage gain. The midband voltage gain is found to be

$$A_{MB} = -g_m \frac{r_{ds} R_D}{r_{ds} + R_D} = -6.66 \times \frac{16.2 \times 3.6}{16.2 + 3.6} = -19.6 \text{ V/V}$$

This example demonstrates the fact that g_m and r_{ds} are strong functions of the quiescent drain current. Once these values are determined for the bias current used, the equivalent

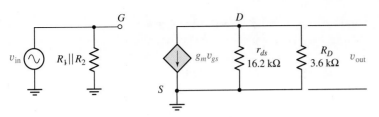

Figure 6.20
Equivalent circuit for Example 6.4.

circuit can be used to find incremental parameters such as the voltage gain. It should be emphasized that no dc quantities appear in the equivalent circuit.

PRACTICAL Considerations

A large majority of signals to be amplified are incremental, ac signals, rather than dc signals. If amplifying elements such as the MOSFET or BJT could amplify these ac signals without requiring dc voltages or currents, amplifier design would be much simpler. Unfortunately, these devices require dc voltages and currents to move the operating point to the active region. Thus, dc considerations are necessary as well as incremental considerations.

In addition to setting the correct quiescent point, the dc bias also determines values of g_m and r_{ds}. Although ac considerations are the driving force in an amplifier design, dc considerations are also very important.

PRACTICAL Considerations

In some amplifier stages, the dc power supply voltage is determined by other considerations to be a specified value. For example, a 12-V source may be required by other circuits in the overall system. If the voltage gain of a stage is to be maximized, one might attempt to use a high value of drain current to maximize the transconductance since

$$g_m = \sqrt{2\mu C_{ox}(W/L)I_D}$$

However, with a fixed power supply voltage, the active region size is largely determined. With a 12-V supply, a quiescent voltage of 8 V might be selected. The dc voltage drop across R_D is then fixed at 4 V, which implies that there is a tradeoff between achieving a high value of I_D or g_m and achieving a high value of R_D.

The gain equation, assuming that r_{ds} is much larger than R_D, is given by

$$A_{MB} = -g_m R_D$$

If the fixed dc voltage drop across R_D is V_1, then $R_D = V_1/I_D$ and we can write this expression as

$$A_{MB} = -\sqrt{2\mu C_{ox}(W/L)I_D}\ \frac{V_1}{I_D} = -V_1\sqrt{\frac{2\mu C_{ox}}{L}\frac{W}{I_D}}$$

For the constraints mentioned, we see that the midband voltage gain decreases as I_D increases. Although g_m increases proportionally to $\sqrt{I_D}$, the load resistance decreases as $1/I_D$. Thus, we conclude that larger drain resistances, accompanied

PRACTICE Problems

6.7 Rework Example 6.3 with $V_{DD} = 12$ V. Explain why the voltage gain is higher. *Ans:* (a) 2.89 V, (b) $R_1 = 415$ kΩ, $R_2 = 132$ kΩ, (c) −6.08 V/V, (d) 0.42 V.

6.8 Rework Example 6.3 with $V_{DD} = 12$ V and $R_D = 6$ kΩ. *Ans:* (a) 2.51 V, (b) $R_1 = 478$ kΩ, $R_2 = 126$ kΩ, (c) −7.5 V/V, (d) 0.40 V.

by lower drain currents to achieve the correct bias, lead to higher voltage gains. Lower drain currents also decrease the dissipation of the power supply.

As in many engineering situations, the results of a parameter change to optimize one variable may not be beneficial in all respects. We will later see that larger drain resistors can decrease the upper corner frequency of the stage.

6.5 Other Amplifier Configurations

IMPORTANT Concepts

1. The source follower has a voltage gain less than unity and is used as a buffer stage. It exhibits a high current gain.
2. The common-gate configuration has a low input impedance. It is used occasionally as an amplifier.

The MOSFET amplifier stage discussed earlier in this chapter is called the common-source stage since the source terminal is connected to common or circuit ground. Two other configurations of MOSFET stages are the common-drain or source follower and the common-gate. While these stages are not as popular as the common-source stage, both are used in specific applications.

Both the source-follower and the common-gate stage allow a signal voltage to appear between source and ground. For discrete devices, the substrate and the source are tied together inside the package. For MOSFET circuits fabricated on a chip, the substrates of all n-devices are tied to ground, while the source will have a signal voltage for the source follower and the common-gate stages. This leads to a modification of the model due to a change in threshold voltage with source-to-substrate voltage.

6.5.1 THE SOURCE FOLLOWER

Figure 6.21 demonstrates a simple source-follower circuit. The basic characteristic of this configuration is that it has the same high current gain as the common-source circuit, but exhibits a voltage gain that is always less than unity.

Without going into further detail relative to the dc design of the circuit, we note that V_{GSQ} is no longer equal to the gate voltage V_{GQ}. The dc drop across the source resistance must be considered to find V_{SQ} and then calculate $V_{GSQ} = V_{GQ} - V_{SQ}$.

A significant difference between this circuit and the common-source circuit is the appearance of an incremental voltage at the source terminal. In the common-source stage, the

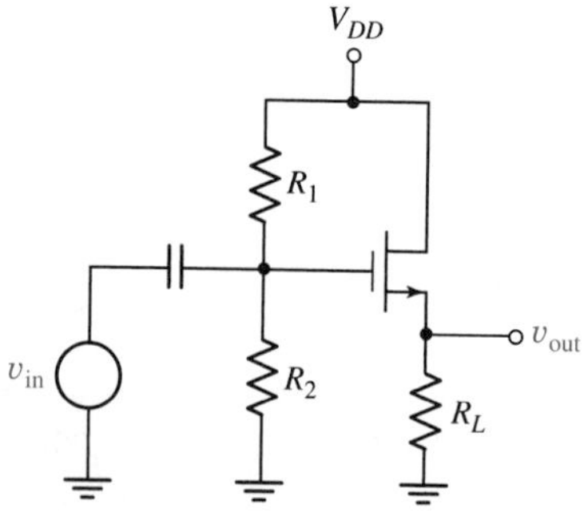

Figure 6.21
A source follower.

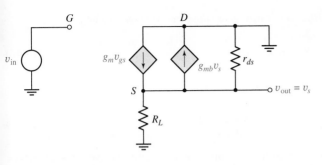

Figure 6.22
The equivalent circuit of a source-follower stage for zero substrate voltage.

source voltage is zero since this terminal is tied to ground. The source follower develops the output voltage across the source resistance, and the source voltage equals the output voltage.

If the source is tied to substrate, as in discrete devices, there is little consequence of the source voltage being nonzero. On the other hand, IC circuits generally tie the substrate to ground. The source terminal in a source-follower circuit has a signal voltage appearing on this terminal. As the source voltage changes, the voltage between source and bulk also changes. This influences the channel conductivity. This effect is called the *body effect* or the *back-gate effect*. It is represented in the equivalent circuit as a dependent current generator that is proportional to the source voltage, as indicated in Fig. 6.22.

For a discrete device with source and substrate tied together, the body effect current generator would be zero. The remaining discussion of this section will assume that the substrate is tied to ground as in the case of an IC nMOS device.

In the equivalent circuit of Fig. 6.22, the current generator of value $g_{mb}v_s$ could be replaced by a conductance of this same value. When a voltage is applied to the gate of the source follower, the voltage appearing at the source is found to be

$$v_{\text{out}} = \frac{g_m \, v_{gs}}{G} = \frac{g_m(v_{\text{in}} - v_{\text{out}})}{G} \tag{6.43}$$

where $G = g_{ds} + g_{mb} + G_L$. The parameter g_{ds} is simply the reciprocal of r_{ds}, and G_L is the reciprocal of R_L. Transposing all terms involving v_{out} to one side of the equation and solving for the voltage gain from gate to source gives

$$A_{GS} = \frac{g_m}{g_m + G} = \frac{g_m}{g_m + g_{mb} + g_{ds} + G_L} \tag{6.44}$$

In some applications the load resistor R_L can be chosen to be large enough so that G_L is negligible. Typically, the value of g_{ds} is also small compared to g_m. Neglecting these conductances allows Eq. (6.44) to be written as

$$A_{GS} = \frac{g_m}{g_m + g_{mb}} \tag{6.45}$$

for

$$g_{ds} + G_L \leq g_m + g_{mb}$$

The value of g_{mb} is generally 10% to 15% of g_m, leading to gains of 0.75 to 0.9 for the source follower. Although the voltage gain is rather small, this stage is often used as an output buffer stage because of the high current gain.

The output impedance of the source-follower stage can be shown to be

$$R_{\text{out}} = \frac{1}{g_m} \parallel \frac{1}{g_{mb}} \parallel r_{ds} \parallel R_L \qquad (6.46)$$

The output impedance looking back into the source terminal is

$$r_{\text{out}} = \frac{1}{g_m} \parallel \frac{1}{g_{mb}} \parallel r_{ds} \qquad (6.47)$$

This low impedance makes the source follower an excellent stage to drive low-impedance loads including high-capacitance loads.

EXAMPLE 6.5

For the MOSFET in the circuit of Fig. 6.23, the following values apply:

$$K = \frac{\mu C_{\text{ox}} W}{2L} = 1.6 \text{ mA/V}^2$$

$$V_T = 1 \text{ V}$$

Assume that $\lambda = 0$, $g_{mb} = 0.15\, g_m$, and neglect the body effect on dc current. Also assume that the substrate is tied to ground.

Figure 6.23
Source follower for Example 6.5.

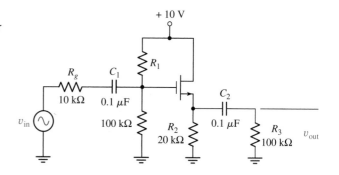

(a) Select R_1 to set V_{SQ} at 4 V.
(b) Calculate the midband voltage gain.
(c) Calculate the lower 3-dB frequency of the voltage gain.

SOLUTION (a) Setting $V_{SQ} = 4$ V requires a drain or source current of

$$I_D = \frac{V_{SQ}}{R_2} = \frac{4}{20} = 200 \ \mu\text{A}$$

Neglecting the body effect on dc current allows the gate-to-source voltage to be calculated from

$$I_D = K[V_{GS} - V_T]^2$$

Substituting the given value of V_T and the required value of I_D into this equation results in $V_{GS} = 1.35$ V. Since the source voltage is 4 V, the gate voltage must be 5.35 V. The value of R_1 is found from the voltage division equation

$$V_G = \frac{100}{100 + R_1} \times 10 = 5.35 \text{ V}$$

Solving for R_1 results in $R_1 = 87 \text{ k}\Omega$.

(b) The midband voltage gain from gate to source can be found from Eq. (6.44), noting that if $\lambda = 0$, the resistance r_{ds} is infinite and $g_{ds} = 0$. At midband frequencies the conductance between the source terminal and ground, G_L, will be determined by the parallel combination of the 20-kΩ resistor and the 100-kΩ resistor. Thus,

$$G_L = \frac{1}{20} + \frac{1}{100} = 0.06 \text{ mA/V}$$

We must now find g_m for this device. Using Eq. (6.34), we can write

$$g_m = \sqrt{2\mu C_{ox}(W/L)I_D} = \sqrt{4KI_D} = \sqrt{4 \times 1.6 \times 0.2} = 1.13 \text{ mA/V}^2$$

leading to a value of $g_{mb} = 0.15\, g_m = 0.17$ mA/V. These transconductances and conductances can be substituted into the gain equation to give a gate-to-source gain of

$$A_{GS} = \frac{1.13}{1.13 + 0.17 + 0.06} = 0.83 \text{ V/V}$$

The overall voltage gain will be further reduced by the loading on the source resistance by the input impedance. Input resistance is calculated as the parallel combination of the bias resistors and is $R_{in} = 46.5$ kΩ. The overall midband voltage gain is

$$A_{MB} = \frac{R_{in}}{R_g + R_{in}} A_{GS} = \frac{46.5}{46.5 + 10} \times 0.83 = 0.68 \text{ V/V}$$

(c) There will be two separate lower 3-dB frequencies: one caused by the input circuit and one caused by the output circuit. The input 3-dB frequency is

$$f_1 = f_{in-low} = \frac{1}{2\pi C_1(R_{in} + R_g)} = \frac{1}{2\pi \times 10^{-7} \times 56.5 \times 10^3} = 28 \text{ Hz}$$

The output resistance looking into the source terminal of the MOSFET is required to calculate the output 3-dB frequency. From Eq. (6.46) this resistance is

$$R_{out} = \frac{1}{g_m} \parallel \frac{1}{g_{mb}} \parallel R_2 = 0.88 \parallel 5.9 \parallel 20 = 0.74 \text{ k}\Omega$$

The output 3-dB frequency is

$$f_1' = f_{out-low} = \frac{1}{2\pi C_2(R_{out} + R_3)} = \frac{1}{2\pi \times 10^{-7} \times (740 + 10^5)} = 16 \text{ Hz}$$

The overall lower corner frequency is found by iterating Eq. (3.17), which gives

$$f_{low} = 35 \text{ Hz}$$

6.5.2 THE COMMON-GATE AMPLIFIER

The common-gate stage is not used often by itself, but is used in conjunction with a common-source stage in the cascode connection, to be discussed in a later chapter. The common-gate stage has a current gain of unity and a positive voltage gain that is similar in magnitude to that of the common-source stage. The upper corner frequency of the common-gate stage can also be higher than that of the common-source stage. Figure 6.24 shows the basic common-gate stage.

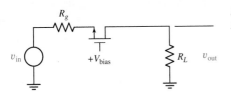

Figure 6.24
A common-gate stage.

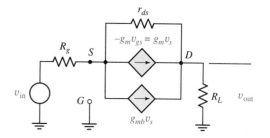

Figure 6.25
An equivalent circuit of the common-gate stage.

The voltage gain can be calculated from the equivalent circuit of Fig. 6.25. In this situation, the incremental gate-to-source voltage is equal to $-v_s$, allowing the dependent source direction to be reversed with a strength of $g_m v_s$. The drain current through the load resistance can be found by applying Kirchhoff's current law to the output node. This current equals the sum of the currents generated by the dependent sources minus the current through the conductance g_{ds},

$$i_d = (g_m + g_{mb})v_s - (v_{out} - v_s)g_{ds}$$

We also note that the source terminal voltage can be written as

$$v_s = v_{in} - R_g i_s = v_{in} - R_g i_d$$

Using this information allows the midband voltage gain to be expressed as

$$A_{MB} = \frac{g_m + g_{mb} + g_{ds}}{G_L + g_{ds} + (g_m + g_{mb} + g_{ds})G_L/G_g} \tag{6.48}$$

In most cases, $g_{ds} \ll (g_m + g_{mb})$, and the gain can be approximated as

$$A_{MB} = \frac{(g_m + g_{mb})R_L}{1 + (g_m + g_{mb})R_g} \tag{6.49}$$

This voltage gain for the common-gate stage is comparable to that of the common-source stage. The current gain from source to drain is unity for the common-gate stage, whereas the current gain from gate to drain is essentially infinite for the common-source stage.

6.6 Biasing of Discrete MOSFET Stages

IMPORTANT Concepts

1. The voltage divider designed to establish the necessary gate-to-source voltage is the most common biasing method for discrete MOSFET circuits.
2. A voltage source in the gate circuit can also be used for biasing. This method is more common in IC design, where the source is a simple voltage reference circuit.

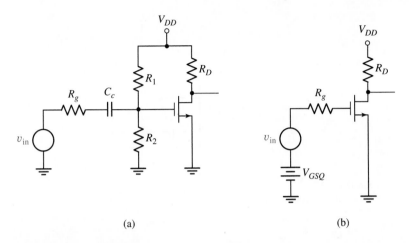

Figure 6.26
(a) A stage with a voltage divider for bias. (b) Series voltage source for bias.

6.6.1 RESISTIVE BIASING

The voltage divider of Fig. 6.15 is a common method of biasing a discrete MOSFET amplifier stage. The primary basis in the selection of R_1 and R_2 is the establishment of the desired value of V_{GSQ}. Secondary considerations include the minimization of loading on the source resistance, the partial determination of lower corner frequency, or the minimization of noise in the circuit.

Another configuration used in IC design is a simple voltage reference circuit placed in series with the gate circuit. Both methods are shown in Fig. 6.26. The voltage divider establishes the gate-to-source voltage for the circuit of Fig. 6.26(a). This scheme has been considered in connection with the circuit of Fig. 6.15. Since gate current is negligible, no loading on this voltage divider occurs, and establishing the correct value of V_{GSQ} is quite straightforward.

In IC design, the use of resistors is minimized. Voltage reference circuits to generate V_{GS} in Fig. 6.26(b), based on MOSFET circuits, have been developed over the years but will not be discussed here.

PRACTICE Problem

6.9 If $\mu W C_{ox}/2L = 500\ \mu A/V^2$, $V_T = 0.8$ V, $V_{DD} = 10$ V, and $R_2 = 20$ kΩ, calculate the value of R_1 in Fig. 6.26(a) to lead to $V_{DSQ} = 6$ V. Assume that $R_D = 4$ kΩ. *Ans:* 70.5 kΩ

6.7 High-Frequency Model of the MOSFET

IMPORTANT Concepts

1. Parasitic capacitances exist between each pair of terminals and between each terminal and substrate.

2. Most parasitic capacitances are nonlinear with bias voltage. Simulation programs are generally used to analyze the high-frequency response of MOSFET amplifiers.

The high-frequency model of the MOSFET is shown in Fig. 6.27. This model shows capacitance between every possible set of terminals. Making the model even more complex is the fact that many of these capacitors are functions of the voltages across these terminals.

There are basically two kinds of capacitance involved in this model. The first is the parallel plate capacitance, which arises from the separation of two conductive materials by

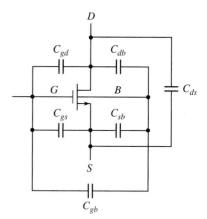

Figure 6.27
The high-frequency MOSFET model.

a dielectric that is relatively independent of voltage applied. The second is the capacitance associated with the charge included in a depletion region, which is highly voltage dependent. Both of these capacitances are functions of dimensions.

The origins of the capacitances between each set of terminals are considered in the next few paragraphs.

C_{gs}: In the active or triode regions, this capacitance is made up mainly of parallel plate capacitance. The largest component of this capacitance is that between the gate and the inverted channel. Since the charge that flows through the channel is supplied from the source terminal, this capacitance essentially appears between gate and source. It has been shown that this component of C_{gs} is given by

$$C_{gsch} = \frac{2}{3}WLC_{ox} \tag{6.50}$$

where W is the channel width, L is the channel length, and C_{ox} is the capacitance per unit area due to the oxide layer between gate and substrate.

A second, generally smaller component of C_{gs} is due to the overlap of the gate electrode and the source area. As the source and drain are formed by diffusion or implantation of impurity atoms, these impurities diffuse under the gate by a distance L_{ov}. This overlap of the gate and source is much smaller than the channel length L except for very high-frequency, short-channel devices. This capacitance is given by

$$C_{gsov} = WL_{ov}C_{ox} \tag{6.51}$$

C_{gd}: The capacitance between the gate and drain terminals is due primarily to overlap capacitance and is generally equal to the gate-to-source overlap capacitance given by Eq. (6.51). Although the gate-to-drain overlap capacitance is typically small, it can be multiplied by the Miller effect in the common-source configuration to limit the high-frequency amplification.

C_{db}: The drain-to-body capacitance consists of depletion layer capacitance. There is a reverse-biased junction between drain and substrate. This capacitance also includes the depletion layer capacitance between drain and the more heavily doped regions between adjacent MOS devices. This capacitance depends on the voltage from drain to substrate as well as on the areas involved.

C_{sb}: The source-to-body capacitance is similar to the drain-to-body capacitance except that it includes the depletion layer capacitance from channel to body. The source-to-body component becomes unimportant when both source and body are connected to ground. The channel-to-body component is actually distributed over the length of the channel.

Because of the complexity of the capacitances, it is difficult to manually analyze the high-frequency performance of the MOSFET. Generally, a Spice-based simulation program is used to evaluate and modify MOSFET stages. We will consider this topic in Chapters 8 and 9.

It is possible to derive approximate frequency performance in some cases, as demonstrated by the following example.

EXAMPLE 6.6

For the MOSFET in the circuit of Fig. 6.28, $g_m = 3$ mA/V and $r_{ds} = 63$ kΩ. Assuming that $C_{gs} = 1$ pF and $C_{gd} = 0.1$ pF and these are the dominant capacitors, calculate

(a) The midband voltage gain
(b) The upper corner frequency for the amplifier

SOLUTION Two equivalent circuits for this stage are shown in Fig. 6.29.
(a) The midband gain can be calculated simply as

$$A_{MB} = -g_m(r_{ds} \parallel R_D) = -3 \times 7.1 = -21.3 \text{ V/V}$$

(b) There will be two upper corner frequencies for this circuit, one caused by the input loop and one caused by the output loop. The Miller effect capacitance reflected to the input loop is

$$C_M = (1 + |A_{gd}|)C_{gd} = 22.3 \times 0.1 = 2.23 \text{ pF}$$

The upper corner frequency due to the input loop is

$$f_{\text{in-high}} = \frac{1}{2\pi R_g(C_{gs} + C_M)} = \frac{1}{2\pi \times 10,000 \times 3.23 \times 10^{-12}} = 4.93 \text{ MHz}$$

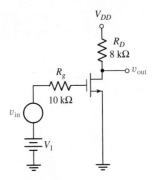

Figure 6.28
Amplifier stage for Example 6.6.

Figure 6.29
(a) An equivalent circuit.
(b) After using the Miller effect.

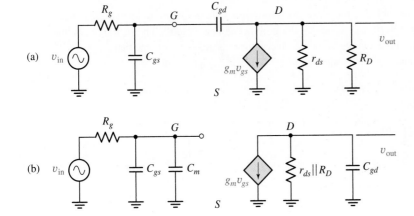

The upper corner frequency due to the output loop is approximately

$$f_{out-high} = \frac{1}{2\pi C_{gd}(r_{ds} \parallel R_D)} = \frac{1}{2\pi \times 10^{-13} \times 7,100} = 224 \text{ MHz}$$

Since this latter frequency is so much greater than the input corner frequency, the overall upper corner frequency is

$$f_{high} = f_{high-in} = 4.93 \text{ MHz}$$

DISCUSSION OF THE DEMONSTRATION PROBLEM

The amplifier of the Demonstration Problem is repeated here for convenience. A midband incremental equivalent circuit for the amplifier is shown in Fig. 6.30.

Amplifier for Demonstration Problem.

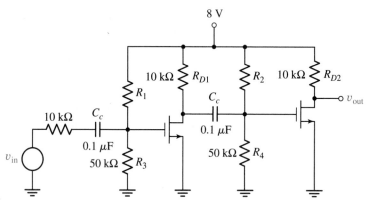

The midband voltage gain of this amplifier can only be calculated if R_1 and R_2 are known. These values would establish the dc drain currents of both devices, and the currents would establish the values of g_m and r_{ds} for both devices. We emphasize the dependence of voltage gain, an incremental quantity, on the dc design of the devices.

As in most design problems, there are many possible values for the unknown elements that will satisfy the specifications. Generally, the problem can be simplified by making certain acceptable assumptions. For example, in this problem it is reasonable to assume that R_1 and R_2 can be chosen to be equal. This will establish equal values of g_m and r_{ds} for both devices.

Figure 6.30
Midband incremental equivalent circuit.

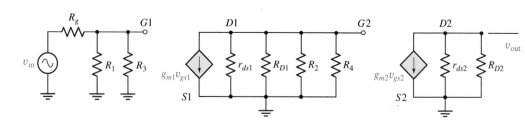

We now make some assumptions to allow an approximation of the necessary drain currents. Since R_1 is typically much larger than R_3, we will assume that $R_B = R_1 \| R_3 = R_2 \| R_4 \approx 40\,\text{k}\Omega$. We will further assume that r_{ds} is much larger than R_D. These assumptions allow the gain to be written as

$$A_{MBo} = \frac{R_B}{R_B + R_g}\, g_m (R_B \| R_D)\, g_m R_D$$

Substituting the actual or assumed values into this equation, the necessary value of g_m to set an overall gain of 100 V/V is

$$g_m = 1.25\ \text{mA/V}$$

Equation (6.34) allows us to calculate the necessary drain current to set g_m to this value, giving

$$I_D = g_m^2\, \frac{L}{2\mu C_{\text{ox}} W} = 0.244\ \text{mA}$$

We will select $I_D = 0.3$ mA to lead to a slightly higher value of g_m. This higher value will compensate for attenuation due to r_{ds} that was not considered in our approximations.

We will now determine an accurate value of overall gain when both drain currents are set to 0.3 mA. We will first find the required value of V_{GS}. From this value, R_1 and R_2 can be found. The values of g_m and r_{ds} can be found from the drain current; then the overall voltage gain can be accurately calculated.

Theoretically, the required value of V_{GS} to set $I_D = 0.3$ mA can be found from Eq. (6.26); that is,

$$I_D = \mu \frac{W}{2L} C_{\text{ox}} [V_{GS} - V_T]^2\, [1 + \lambda(V_{DS} - V_{DSP})] \tag{6.26}$$

where

$$V_{DSP} = V_{GS} - V_T \tag{6.16}$$

Equation (6.26) is difficult to solve directly. A simpler method of finding V_{GS} is to neglect the channel-length modulation factor λ to find an approximate value for V_{GS}. Once this is done, a few iterations using Eq. (6.26) will provide an accurate value for V_{GS}.

With the given parameter values we solve

$$I_D = \mu \frac{W}{2L} C_{\text{ox}} [V_{GS} - V_T]^2 = 1.6[V_{GS} - 0.8]^2 = 0.3\ \text{mA}$$

This yields an approximate value of $V_{GS} = 1.233$ V. We now lower this value slightly, as our first iteration, and substitute it into Eq. (6.26). A few iterations result in $V_{GS} = 1.216$ V.

The resistors R_1 and R_2 can be found from the voltage divider equation

$$V_{GS} = 1.216 = \frac{50}{50 + R_1} \times 8$$

This gives $R_1 = R_2 = 279\,\text{k}\Omega$ and $R_B = 42\,\text{k}\Omega$.

Equation (6.34) is used to find that the transconductance is

$$g_m = 1.39 \text{ mA/V}$$

After finding the drain pinchoff current to be 0.277 mA for this bias, the value of r_{ds} from Eq. (6.30) is found to be

$$r_{ds} = \frac{1}{0.018 \times 0.277} = 201 \text{ k}\Omega$$

The more accurate equation for midband voltage gain is

$$A_{MBo} = \frac{R_B}{R_B + R_g} g_m (R_B \| R_D \| r_{ds}) g_m (R_D \| r_{ds})$$

Substituting all known values into this equation gives

$$A_{MBo} = 116 \text{ V/V}$$

which satisfies the given specification on voltage gain.

The output stage has a quiescent voltage of $V_{DSQ} = 5$ V. This follows from setting the drain current to 0.3 mA and dropping 3 V across the 10-kΩ drain resistance. This voltage can swing to 8 V before hitting cutoff. The method of Example 6.3 is used to find that the voltage can swing downward to about 0.68 V before reaching the triode region. Thus, a peak value of 1 V on the output can appear while remaining safely within the active region.

The lower corner frequency is caused by the two coupling capacitors. The first stage has a corner frequency of

$$f_1 = f_{\text{low1}} = \frac{1}{2\pi C_c (R_g + R_B)} = 30.4 \text{ Hz}$$

The corner frequency of the second stage is

$$f_1' = f_{\text{low2}} = \frac{1}{2\pi C_c ([r_{ds} \| R_D] + R_B)} = 30.6 \text{ Hz}$$

These frequencies are essentially equal. Using Eq. (3.17), the overall lower corner frequency is found to be

$$f_{\text{low}} = 48 \text{ Hz}$$

SUMMARY

➤ The MOSFET is an important amplifying element in electronic circuit design.

➤ The voltage applied between gate and source of the MOSFET is the major factor in determining the current flow between the drain and source terminals.

➤ The three operating regions of the MOSFET are the cutoff region, the triode region, and the active region. For amplifier applications, the MOSFET generally operates in the active region.

➤ A dc bias is required to set the output voltage to a proper voltage within the active region. The drain current influences the small-signal values of transconductance and drain-to-source resistance.

➤ The incremental or small-signal equivalent circuit of the MOSFET is used to find the voltage gain, the input impedance, and the output impedance of the amplifier circuit.

➤ Although the common-source amplifier stage is used more than the others, the source follower (common-drain) and the common-gate configurations are used in some amplifiers.

➤ High-frequency performance of the MOSFET is limited by the parasitic capacitances from terminal to terminal and from terminal to substrate.

PROBLEMS

SECTION 6.3.1 THE TRIODE REGION

6.1 If $\mu W C_{ox}/L = 0.4$ mA/V^2 and $V_T = 0.8$ V in Eq. (6.14), calculate the drain current for $V_{GS} = 2$ V and $V_{DS} = 0.5$ V. Is Eq. (6.19) valid for this value of V_{DS}? Explain.

6.2 For the parameters of Problem 6.1 but allowing V_{DS} to vary:

(a) Calculate the pinchoff drain current.

(b) What is the minimum value of V_{DS} that leads to pinchoff?

6.3 For the conditions of Problem 6.1, except $V_{DS} = 0.1$ V, calculate the small-signal resistance between drain and source.

SECTION 6.3.2 THE ACTIVE REGION

☆ **6.4** For the MOSFET circuit shown, $\mu W C_{ox}/2L = 80 \, \mu$A/V^2, $V_T = 0.9$ V, and $\lambda = 0$.

(a) What value must V_{GS} have to bring the device from the active region to the edge of the cutoff region?

(b) What value must V_{GS} have to bring the device from the active region to the edge of the triode region?

Figure P6.4

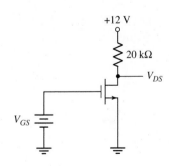

6.5 In the circuit of Problem 6.4, identify the region in which the MOSFET exists for the following conditions:

(a) $V_{GS} = 2.84$ V, $V_{DS} = 6$ V

(b) $V_{GS} = 3.27$ V, $V_{DS} = 3$ V

(c) $V_{GS} = 3.46$ V, $V_{DS} = 2$ V

6.6 The threshold voltage for a MOSFET is 0.86 V. In which region does the device exist for

(a) $V_{GS} = 2$ V, $V_{DS} = 1$ V

(b) $V_{GS} = 1.8$ V, $V_{DS} = 1.8$ V

(c) $V_{GS} = 1.5$ V, $V_{DS} = 4$ V

6.7 A MOSFET with $V_T = 1$ V has the following measured data taken:

V_{GS} V	V_{DS}, V	I_D, μA
2	1	80
2	8	91

Calculate λ for this device.

SECTION 6.4.2 SMALL-SIGNAL OR INCREMENTAL PARAMETERS

6.8 Using Eq. (6.25) with $V_{GS} = 2.4$ V, calculate the drain current for $V_{DS} = 4$ V and again at $V_{DS} = 8$ V. Now use Eq. (6.26) to make these same calculations. Explain any differences. Use values of $\mu W C_{ox}/2L = 0.5$ mA/V^2, $\lambda = 0.025$/V, and $V_T = 0.8$ V.

6.9 With $V_{GS} = 2.4$ V, calculate r_{ds} at $V_{DS} = 4$ V and again at $V_{DS} = 8$ V using $r_{ds} = 1/\lambda I_{DP}$, and compare to the approximate values at these voltages given by $r_{ds} = 1/\lambda I_D$. Use values of $\mu W C_{ox}/2L = 0.5$ mA/V^2, $\lambda = 0.025$/V, and $V_T = 0.8$ V.

6.10 If $g_m = 3.2$ mA/V at $I_D = 120$ μA, calculate g_m when $I_D = 820$ μA.

6.11 If $r_{ds} = 80$ kΩ at $I_D = 500$ μA, estimate r_{ds} when $I_D = 100$ μA.

6.12 If $\mu W C_{ox}/2L = 0.4$ mA/V^2, $\lambda = 0.025$/V, and $V_T = 0.8$ V, calculate r_{ds} at $V_{DS} = 4$ V for V_{GS} values of 2 V and 4 V.

☆ **6.13** For the conditions of Problem 6.12, calculate the two values of g_m. Compare these values to those found by assuming $\lambda = 0$.

SECTION 6.4.3 THE MIDBAND INCREMENTAL MOSFET MODEL

6.14 A MOSFET is used in a common-source configuration with a drain resistance of $R_D = 8$ kΩ. If the drain-to-source resistance is $r_{ds} = 70$ kΩ, what must the value of g_m be to result in a midband voltage gain of -18 V/V? If the quiescent drain current is 1 mA, what is the value of $\mu C_{ox} W/2L$?

6.15 If the quiescent drain current in Problem 6.14 is decreased to 0.1 mA, what are the new values of g_m and r_{ds}? What is the new value of midband voltage gain?

☆ **6.16** A device with $\mu W C_{ox}/2L = 0.4$ mA/V^2, $\lambda = 0.025$/V, and $V_T = 0.8$ V is used in a common-source amplifier stage with $V_{GS} = 2.5$ V. If the power supply is 8 V and $R_D = 4$ kΩ, calculate the voltage gain from gate to drain. Compare this result to that found by assuming $\lambda = 0$.

☆ **6.17** Repeat Problem 6.16 if the power supply is changed to 12 V and the load resistance is changed to 6 kΩ.

☆ **6.18** For the amplifier of Problem 6.16, calculate the minimum and maximum possible output voltages before serious distortion is encountered (triode region and cutoff).

☆ **6.19** For the amplifier of Problem 6.17, calculate the minimum and maximum possible output voltages before serious distortion is encountered (triode region and cutoff).

☆ **D 6.20** For the amplifier shown, assume that $\mu W C_{ox}/2L = 0.2$ mA/V^2, $\lambda = 0$/V, and $V_T = 1$ V. The quiescent drain voltage is to be in the range of 4 V to 9 V, and the quiescent drain current should be in the range of 0.1 mA to 4 mA. Select R_D to achieve the maximum possible midband voltage gain magnitude. Calculate A_{MB}.

Figure P6.20

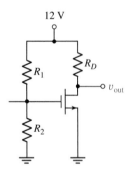

☆ **6.21** For the amplifier shown, $g_m = 2$ mA/V and $r_{ds} = 60$ kΩ at the quiescent bias point. Find the midband voltage gain of the amplifier.

Figure P6.21

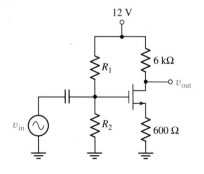

☆ **6.22** Assume that $g_m = 3 \text{ mA/V}$, $r_{ds} = 40 \text{ k}\Omega$, and $V_T = 0.8$ V for the MOSFET of the circuit.

(a) Calculate the quiescent drain current, I_{DQ}, neglecting the effects of λ.

(b) Calculate the quiescent drain-to-source voltage, V_{DSQ}.

(c) Draw the incremental equivalent circuit.

(d) Calculate the midband incremental voltage gain, $A_{MB} = v_{\text{out}}/v_{\text{in}}$.

Figure P6.22

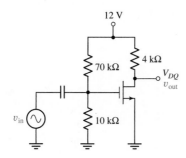

☆ **D 6.23** For the MOSFET shown, $V_T = 1$ V, $\lambda = 0$, and $\mu C_{\text{ox}} W/2L = 0.1 \text{ mA/V}^2$. Select R_D and R_1 to result in a midband voltage gain of -4 V/V and $V_{DSQ} = 7$ V.

Figure P6.23

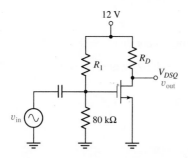

☆ **6.24** For the amplifier stage shown, $\mu W C_{\text{ox}}/2L = 75 \text{ } \mu\text{A/V}^2$, $V_T = 1$ V, and $\lambda = 0$.

(a) Calculate V_{DQ} and V_{SQ}.

(b) Calculate the ac output voltage at midband frequencies.

(c) Find the lower corner frequency.

Figure P6.24

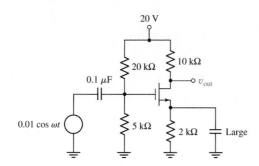

☆ **6.25** Repeat Problem 6.24 if $\lambda = 0.02/\text{V}$.

D 6.26 A MOSFET has values of $\mu C_{ox} W / 2L = 0.75 \text{mA/V}^2$, $V_T = 1$ V, and $\lambda = 0$. Design an amplifier to have a midband voltage gain of -10 V/V. Assume that a resistance of 5 kΩ is placed between source terminal and ground and is bypassed by a large capacitor at midband frequencies. The power supply voltage is 12 V.

☆ **6.27** For the MOSFET in the circuit, $\mu W C_{ox} / 2L = 2$ mA/V^2, $\lambda = 0.02/\text{V}$, and $V_T = 0.8$ V. Calculate the midband voltage gain of the stage.

Figure P6.27

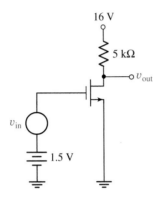

6 **6.28** Repeat Problem 6.27 if all parameters remain the same except that $\lambda = 0$.

☆ **6.29** For the amplifier shown, $\mu W C_{ox} / 2L = 3$ mA/V^2, $\lambda = 0.02/\text{V}$, and $V_T = 1.0$ V.

(a) Calculate V_{DSQ} when $V_{GSQ} = 1.5$ V.
(b) Calculate the midband voltage gain for the stage.

Figure P6.29

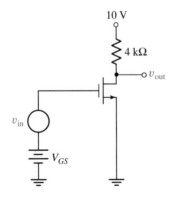

6.30 Repeat Problem 6.29 if all parameters remain the same except that $\lambda = 0$.

☆ **D 6.31** For the MOSFET in the circuit, $\mu W C_{ox}/2L = 2.1$ mA/V^2, $\lambda = 0.02$/V, and $V_T = 0.94$ V. Select R_1 to result in a midband gain of $A_{MB} = -25$ V/V $\pm 5\%$.

Figure P6.31

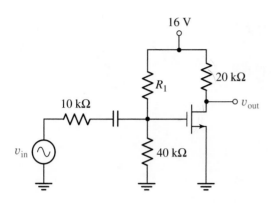

SECTION 6.5 OTHER AMPLIFIER CONFIGURATIONS

6.32 Derive Eq. (6.47).

6.33 Design a common-gate stage having a midband voltage gain of 10 V/V $\pm 10\%$ using a load resistance of 5 kΩ and a power supply of 10 V. The input signal is 75 mV peak. The device used has $\mu C_{ox}W/2L = 3$ mA/V^2, $g_{mb} = 0.15\ g_m$ mA/V, $V_T = 1$ V, and $\lambda = 0.03$/V.

☆ **6.34** Derive an expression for R_{out} in the circuit shown, assuming that g_m, g_{mb}, and r_{ds} are known. Find R_{out} if $g_m = 2$ mA/V, $g_{mb} = 0.3$ mA/V, $r_{ds} = 60$ kΩ, and $R_S = 1$ kΩ.

Figure P6.34

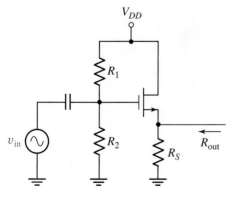

☆ **6.35** The circuit shown is a cascode amplifier. Both devices have values of $\mu W C_{ox}/2L = 0.5$ mA/V^2, $V_T = 0.8$ V, and $\lambda = 0$. Calculate the voltage gain for the circuit. Ignore the body effect.

Figure P6.35

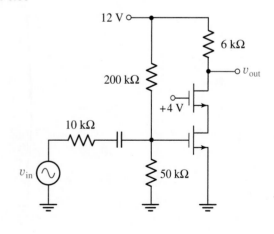

SECTION 6.7 HIGH-FREQUENCY MODEL OF THE MOSFET ———————

6.36 Assume that $g_m = 3$ mA/V, $r_{ds} = 100$ kΩ, $C_{gs} = 1$ pF, $C_{gd} = 0.5$ pF, and $V_{DSQ} = 6$ V for the MOSFET of the circuit. The input voltage is a positive-going 20-mV step function.

Figure P6.36

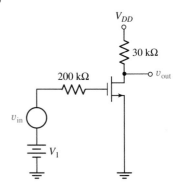

(a) Calculate the time constant, τ, and sketch the output voltage with time showing values of voltage levels.

(b) If the circuit is now driven with a small sinusoidal input signal, what is the upper corner frequency of the circuit, f_{high}?

(c) How do τ and f_{high} relate to each other?

THE BJT | 7

Chapter 3 introduced the concept of modeling and developed simple models for the voltage amplifier. Op amp models were introduced in Chapter 4 and diode models in Chapter 5. The MOSFET amplifying element was discussed in Chapter 6. Op amps provide amplification and can be used without a knowledge of the internal components that comprise this device. On the other hand, some electrical engineers design integrated circuits and must understand the operation of the components and devices used in the op amp.

An important amplifying element of the op amp is the transistor. Some op amps are based on the bipolar junction transistor (BJT), some are based on the field-effect transistor (JFET or MOSFET), and some combine both of these devices.

Although op amps are used in many amplification applications, these devices exhibit relatively low frequency and power limits. Consequently, when high frequency or high power amplification is needed, discrete amplifier circuits must be designed or a high-frequency integrated circuit amplifier must be used. These circuits use transistors along with resistors, capacitors, and other components to create high-frequency amplifiers. Circuit models must be known and understood for all devices included in these amplifiers. This chapter will develop models for the BJT.

The integrated circuit MOSFET amplifier has become more popular than the IC BJT amplifier. However, for discrete amplifiers, the BJT is used more than the MOSFET. Heterojunction BJT stages are integrated for use as high-frequency amplifiers in the tens of GHz range, but MOSFETs are rapidly approaching the heterojunction BJT in frequency performance.

The underlying physical principles governing the behavior of many electronic devices are fairly complex, although the electrical behavior of a given device is generally rather straightforward. For example, solid-state physics is used to derive the relationship of current flow through a diode to voltage applied to the diode. The concepts of drift and diffusion currents and a knowledge of Maxwell–Boltzmann statistics are used to find this voltage–current relationship. Once we have found an equation for the current as a function of voltage, the device can be used in a circuit without referring back to the basic concepts. Only the current–voltage equation is necessary to make further calculations of circuit quantities. When we have obtained the V-I characteristics of a device, both static and dynamic, we can produce a valid model to be used.

DEMONSTRATION PROBLEM

The BJT amplifier of the figure is to have a midband voltage gain in the range of -20 to -24 V/V and a lower corner frequency below or equal to 40 Hz. The maximum peak value of the output signal should be 1 V. Choose all element values to satisfy these specifications.

Amplifier for demonstration problem.

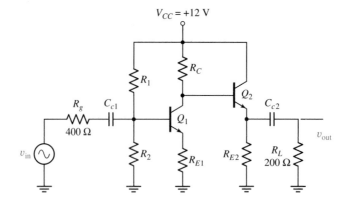

In order to complete this problem we must understand

1. How to select and then establish proper bias voltage levels for the BJT
2. How to determine midband voltage gain for both BJT stages
3. How to calculate input and output impedances to be used for frequency response calculations

7.1 Bipolar Junction Transistor Operation

IMPORTANT Concepts

1. The BJT is in its active region when the base-emitter junction is forward biased and the base-collector junction is reverse biased.
2. The current through the base-emitter junction is governed by the diode equation. The magnitude of the current through the base-collector junction is approximately equal to that through the base-emitter junction.
3. The common-base configuration exhibits good voltage gain but unity current gain. The common-emitter configuration exhibits both good voltage gain and current gain.

7.1.1 BJT BEHAVIOR

The BJT is a three-region device consisting of either *npn* or *pnp* regions. Figure 7.1 shows the schematic arrangement of the two types of BJT along with a more accurate arrangement of the three regions of an *npn* BJT. The symbols for the *npn* and *pnp* devices are shown in parts (d) and (e) of the figure.

Figure 7.1
(a) An npn "textbook" BJT.
(b) A pnp "textbook" BJT.
(c) A more practical npn BJT. (d) Symbol for npn BJT. (e) Symbol for pnp BJT.

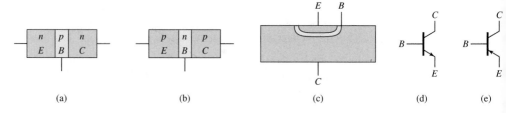

PRACTICAL Considerations

The two rectangular representations might be called "textbook" BJTs, since actual BJTs are not constructed with this geometry. These representations allow the operation of the device to be discussed more easily without regard to the more complex geometry of the practical device. Historically, the *rate-grown* BJT of the early 1960s approximated the textbook BJT quite closely. The advent of the planar process that led to the integrated circuit industry also led to the elimination of the rate-grown process. Although both the *npn* and *pnp* devices are fabricated, the *npn* is by far the most popular device due to its ease of fabrication. Consequently, in succeeding material we will emphasize this device.

The three regions are called the emitter, base, and collector. Each of the three regions connects to a wire, which allows electrical connections to be made. The base region is very narrow compared to the widths of the emitter and collector regions. In the textbook BJT, abrupt junctions are assumed to exist, and the cross-sectional area is assumed constant throughout the various regions.

The *active region* of the BJT is defined by a forward-biased base-emitter junction and a reverse-biased collector base junction. This region is of importance in linear or amplifying applications. Two other regions of importance in switching applications are the *cutoff* and *saturation* regions. Cutoff occurs when both junctions are reverse biased, and saturation occurs when both junctions are forward biased.

When introduced to the configuration of the BJT, an inquisitive person might wonder whether two back-to-back diodes, as shown in Fig. 7.2, might operate as a BJT. After all, the devices appear to make up an *npn* or a *pnp* geometry.

In fact, the diodes do not simulate BJT action. One reason is the difference in base region widths. The BJT has a very narrow base region that allows the collector voltage to influence the flow of current through the base region. The two diodes have an *n* region that consists of two separate segments, both of which are much wider than the BJT base width. The voltage connected to the simulated collector does not influence base current flow in the diodes in the same way collector voltage influences base current flow in the BJT.

On the other hand, it is possible for a BJT to be used as a diode. Figure 7.3 shows the BJT with a voltage source applied across the emitter-base regions. The collector terminal is floating, which means it is not connected to any voltage.

For the *pnp* device, all current injected into the emitter exits from the base. The same equation that describes the V-I characteristics of the diode applies to this case; that is, from Eq. (5.28)

$$I_E = I_{EO}\left(e^{qV_{EB}/kT} - 1\right) \tag{7.1}$$

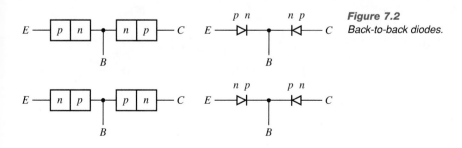

Figure 7.2
Back-to-back diodes.

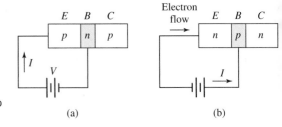

Figure 7.3
A BJT used as a diode: (a) pnp device, (b) npn device.

where I_{EO} is the emitter saturation current. In this circuit, the base current would equal the emitter current. A similar equation would apply to collector current if the collector-base junction were included in a loop with a voltage source and the emitter terminal were floating.

The behavior of the *npn* device is similar to that of the *pnp* device, except that the applied voltage is of the opposite polarity and the current flow is in the opposite direction.

A very significant behavior of a BJT is exhibited when the emitter-base junction is forward biased and the collector-base junction is reverse biased, as shown in Fig. 7.4 for an *npn* BJT. This is the active region of the device. The circuit shown is called the *common base configuration,* since both emitter and collector voltages are referenced to the base terminal.

In a single junction that is reverse biased, as is the collector-base junction, there is negligible current flow. This can be seen from Eq. (7.1), replacing V_{EB} with V_{CB}, which has a large negative value for a reverse bias. In effect, when a reverse bias is applied to a junction, a minute number of current carriers are available to conduct current.

The forward-biased emitter-base junction of Fig. 7.4 injects a large number of current carriers from emitter to base, which are now available to flow across the collector-base junction. In the *npn* BJT these carriers are free electrons.

If the concentration of electrons in the base region after doping is designated n_{bo}, the concentration of the carriers injected from the emitter to the base is

$$n(0) = n_{bo}e^{V_{EB}/V_T} \tag{7.2}$$

At the collector edge of the base region ($x = W$), the reverse-bias voltage at the collector keeps the concentration of carriers near zero, since

$$n(W) = n_{bo}e^{V_{CB}/V_T} \to 0$$

Figure 7.4
A BJT with one forward-biased junction and one reverse-biased junction.

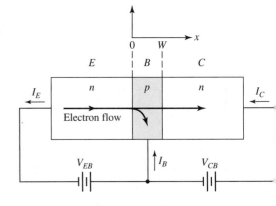

when V_{CB} is negative. A gradient of carriers is now present across the base region. The diffusion equation could be used to calculate this current. The gradient is

$$\frac{dn}{dx} = -\frac{n(0)}{W} = -n_{bo}\frac{e^{V_{BE}/V_T}}{W}$$

From Eq. (5.11), we can write the current as

$$I_E = qD_n\frac{dn}{dx}A = -qD_nAn_{bo}\frac{e^{V_{EB}/V_T}}{W} \tag{7.3}$$

The electrons flowing into the emitter and across the emitter-base junction now flow across the collector-base junction and out the collector terminal. Only a very small portion of the electrons flows out the base, generally less than 1% of the emitter current. Conventional current flows in the opposite direction to the electron flow, as indicated in Fig. 7.4 and by Eq. (7.3).

If not for the fact that the base region is so narrow, the electrons injected across the emitter-base junction could not be collected so effectively by the collector. It should be emphasized that this result is responsible for the amplifying properties of the BJT and explains why two back-to-back diodes cannot simulate the BJT in this mode of operation.

The emitter current across the emitter-base junction expressed by Eq. (7.3) is very similar to the diode equation of Eq. (7.1). This current is a strong function of forward junction voltage. The emitter current is determined by the emitter-base voltage, and the collector current is nearly equal to the emitter current. The constant of proportionality, α, is slightly less than unity, typically over 0.99 for low-power BJTs. Only a very small number of electrons leave the base region. From Kirchhoff's law, this current is equal in magnitude to $(1-\alpha)I_E$. We note that electrons injected from the emitter into the base of an *npn* BJT lead to electrons exiting from the collector and the base terminals. We then define input current (emitter current) as positive when exiting the emitter terminal, whereas current entering the base and collector terminals will be considered positive for the *npn* device.

There are two definitions of the parameter α. Both express a ratio of collector current to emitter current. An incremental α is defined as

$$\alpha = \frac{\Delta I_C}{\Delta I_E} \qquad \text{with} \quad V_C = \text{constant} \tag{7.4}$$

A dc α is defined as

$$\alpha_{dc} = \frac{I_C}{I_E} \qquad \text{with} \quad V_C = \text{constant}$$

Usually α_{dc} and α differ by less than 1% and generally can be assumed to have the same value.

7.1.2 COMMON-BASE CONFIGURATION

Figure 7.5 depicts an *npn* BJT connected for use as an amplifier in the common-base configuration using the normal symbol. The voltage sources V_{EB} and V_{CB} place the BJT in the active region.

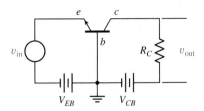

Figure 7.5
A BJT in the common-base configuration.

The dc current leaving the emitter terminal is established by the voltage V_{EB} that has a magnitude near 0.7 V for typical emitter currents in the mA range. The current entering the collector terminal is equal to αI_E and also forces the base current to equal $(1 - \alpha)I_E$. The base current is found from Kirchhoff's law, which requires the total current entering the BJT to sum to zero. With our definition of currents—that is, emitter current is positive when leaving the terminal and base and collector currents are positive when entering the terminals—this requirement can be expressed by

$$I_E = I_B + I_C \tag{7.5}$$

Solving for I_B gives

$$I_B = I_E - I_C = I_E - \alpha I_E = (1 - \alpha)I_E \tag{7.6}$$

Technically, dc calculations would use a model having a current generator with a strength of α_{dc}. Since this value differs little from α, no differentiation need be made. The emitter current is again expressed by the exponential variation given by Eq. (7.1). The V-I characteristics of the ideal transistor of Fig. 7.5 are shown in Fig. 7.6, along with the actual characteristics of a BJT. These actual BJT characteristics are indicated by solid lines and do not depart significantly from the curves for the ideal device, shown by the dashed lines.

The input curve is very slightly affected by the collector to base voltage. This effect is ignored in most applications. The output curves exhibit a slight upward slope with a change in V_{CB}. This effect is also ignored in most applications; however, there are a few situations in which this nonzero slope must be considered. More will be said about this situation when small-signal modeling is discussed.

In principle, α could be evaluated from the characteristic curves, but the accuracy of a graphical evaluation is generally no better than the approximation of this quantity by unity.

For dc calculations, the base-emitter diode would be replaced with a suitable dc equivalent circuit. This diode model could be as simple as a 0.6-V to 0.7-V voltage source.

Figure 7.6

Characteristic curves for a BJT in the common-base configuration: (a) input characteristics, (b) output characteristics.

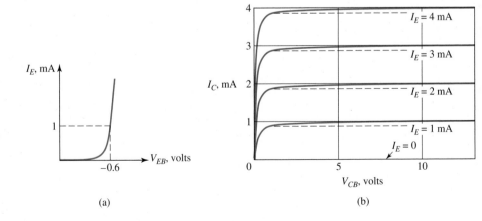

EXAMPLE 7.1

Assume that $V_{BE} = 0.7$ V and $\alpha = 0.99$ for the circuit of Fig. 7.7. Calculate the quiescent collector voltage, V_{CQ}.

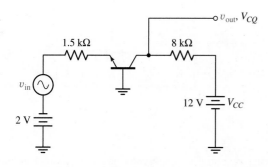

Figure 7.7
*Common-base circuit for
Example 7.1.*

SOLUTION The 2-V source in the base-emitter loop sets the dc emitter current at

$$I_E = \frac{2 - V_{BE}}{1.5} = \frac{2 - 0.7}{1.5} = 0.867 \text{ mA}$$

The collector current is αI_E or $0.99 I_E$, but only a 1% error is made if I_C is assumed to equal I_E. The quiescent collector voltage is approximately

$$V_{CQ} = V_{CC} - I_C R_C = 12 - 0.867 \times 8 = 5.06 \text{ V}$$

7.1.3 COMMON-EMITTER CONFIGURATION

The common-emitter configuration shown in Fig. 7.8 is the most commonly used config-uration. For this circuit the two quantities I_C and V_{CE} are the output quantities. The base current is the input parameter, and the emitter terminal is common to both input and output. We can write the output current I_C in terms of the input current I_B by noting that

$$I_E = I_B + I_C \tag{7.7}$$

and

$$I_C = \alpha I_E$$

Eliminating I_E results in

$$I_C = \frac{\alpha}{1 - \alpha} I_B = \beta I_B \tag{7.8}$$

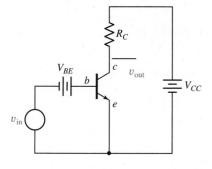

Figure 7.8
*The common-emitter
configuration.*

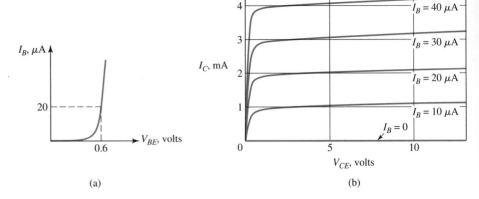

Figure 7.9
*Common-emitter
characteristics: (a) input
characteristics, (b) output
characteristics.*

where $\beta = \alpha/(1 - \alpha)$. The quantity β can be very large and is a strong function of α. For low-power BJTs, β may range from 80 to 300. This quantity is defined as the short-circuit current gain from base to collector. There are two different values of β: an incremental value and a dc value. These parameters are defined as

$$\beta \text{ (incremental)} = \frac{\Delta I_C}{\Delta I_B} \quad \text{with } V_{CE} \text{ constant} \tag{7.9}$$

and

$$\beta_{dc} = \frac{I_C}{I_B} \quad \text{with } V_{CE} \text{ constant} \tag{7.10}$$

The values of β and β_{dc} can differ significantly; however, to simplify further calculations in this chapter, they will be assumed to have equal values.

The characteristic curves for the common-emitter configuration are shown in Fig. 7.9. There are several differences between the common-base and common-emitter characteristics. For the common-emitter configuration:

1. I_C is small, but finite for zero values of input current. This is not apparent on the graph in Fig. 7.9(b).

2. I_C increases more with output voltage for a constant input current.

3. The slope of the curves increases at higher values of I_C.

The first point relates to the leakage current of the BJT, but this effect is often negligible for low-power silicon BJTs, and the collector current for zero base current can be assumed zero. In power stages, this leakage current must often be considered.

The second point can be modeled by a term involving the Early voltage when needed. The collector current is modified to include this effect by writing

$$I_C = \beta I_B \left(1 + \frac{V_{CE}}{V_A}\right) \tag{7.11}$$

where V_A is a constant referred to as the Early voltage of the device. This value will be constant for a given device and is typically in the range of 50–100 V. The term in parentheses in Eq. (7.11) leads to an upward slope of I_C as V_{CE} increases.

The third point is also accounted for by Eq. (7.11), which has a higher slope of I_C versus V_{CE} as βI_B increases.

The reciprocal of the slope of the V-I characteristics is defined as the *incremental resistance from collector to emitter*, designated r_{ce}. The slope of I_C with V_{CE} can be evaluated

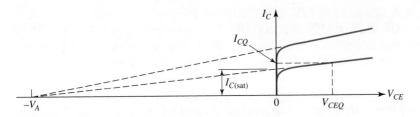

Figure 7.10
The Early voltage.

by differentiating Eq. (7.11) with respect to V_{CE}, which gives

$$\text{slope} = \frac{1}{r_{ce}} = \frac{\beta I_B}{V_A} \qquad (7.12)$$

The term βI_B is equal to the collector current that flows when $V_{CE} \approx 0$ and can be written as $I_{C(sat)}$, the collector current at saturation.

If a given V-I characteristic curve is extended into the negative voltage quadrant from $I_{C(sat)}$ to the point of intersection with the voltage axis, this voltage will equal $-V_A$. In Fig. 7.10, it is easy to see that

$$r_{ce} = \frac{V_A}{I_{C(sat)}}$$

The extension of the curve into the negative voltage quadrant for each value of I_B intersects at approximately $-V_A$. Higher currents lead to larger slopes and smaller values of collector-to-emitter resistance.

We generally know a quiescent collector-emitter voltage and a quiescent collector current for the BJT circuit. When this is the case, the resistance can also be calculated from

$$r_{ce} = \frac{V_A + V_{CEQ}}{I_{CQ}} \qquad (7.13)$$

Characteristic curves for the common-collector configuration or emitter follower are not usually plotted, since the common-emitter characteristics can be used for this configuration also.

EXAMPLE 7.2

In Fig. 7.11, $V_{CC} = 12$ V, $V_{EE} = -3$V, $R_B = 500$ kΩ, and $R_C = 10$ kΩ. If the BJT has values of $\beta = 180$ and $V_{BE} = 0.7$ V, calculate V_{CQ}, neglecting the Early effect.

SOLUTION Summing voltages around the base-emitter loop leads to

$$I_B R_B + V_{BE} = -V_{EE}$$

Substituting values into this equation and solving for I_B results in

$$I_B = \frac{2.3}{500} = 4.6 \ \mu\text{A}$$

The collector current is β times this value or

$$I_C = 180 \times 4.6 = 0.83 \text{ mA}$$

The collector voltage is

$$V_{CQ} = V_{CC} - I_C R_C = 12 - 10 \times 0.83 = 3.7 \text{ V}$$

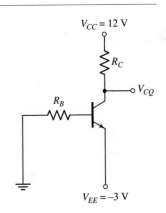

Figure 7.11
Common-emitter circuit for Example 7.2.

7.1.4 A QUALITATIVE DESCRIPTION OF BJT AMPLIFICATION

We are now in a position to qualitatively discuss the mechanisms involved in current and voltage amplification by BJTs. Consider the common-base amplifier of Fig. 7.12. Assuming that the sources V_{EB} and V_{CC} will bias the BJT in its active region, we can note the effects of v_{in}. The small ac voltage will modulate the gradient of carriers and, hence, the current injected across the emitter-base junction. According to Eq. (7.3) small changes in V_{EB} result in large current changes because of the exponential variation of current with voltage. Therefore, large emitter current changes are produced by the small input voltage changes. This ac current travels through the base region with very small loss due to recombination, since from Eq. (7.4)

$$i_c = \alpha i_e \text{(incremental equation)}$$

The collector current develops a voltage across R_C, which is the output voltage and can be much larger than v_{in}. The operation can be summarized as follows:

1. A small value of v_{in} produces large emitter current changes (exponential with voltage).
2. The current travels through the BJT to R_C, developing a large output voltage.

With the configuration of Fig. 7.12, voltage gains as high as 100–200 are obtained.

The current gain for the common-base stage is quite low, however, since the source current is i_e and the output current is αi_e. This current gain is equal to α, which is slightly less than unity. We should also note that the phase shift between input and output voltages is zero.

The operation of the common-emitter amplifier shown in Fig. 7.8 is quite similar to that of the common base. The input source is again applied across the emitter-base junction; therefore, the same magnitude of voltage gain is expected. There are two differences between the common-emitter and common-base configurations. The first is that when the input source goes positive, it is actually raising the emitter-base voltage and hence the emitter current. Collector current increases, dropping a larger voltage across R_C. The output voltage goes more negative when the input goes positive and is then 180° out of phase with the input voltage. The other difference is that the input source current is base current in the common-emitter configuration. This current is much smaller than either emitter or collector current, meaning there is a high current gain (β) in this configuration. Since both current and voltage gains for a common-emitter stage are high, the power gain is quite high. It is this fact that makes this stage the most widely used.

Figure 7.9(a) represents the input characteristics of the BJT in the common-emitter configuration. The input variables are base current and base-to-emitter voltage. If the dc conditions are $I_B = 20\ \mu A$ and $V_{BE} = +0.6$ V, a small variation of input voltage leads to large variations in I_B and even larger variations in I_C. The changes in I_C lead to a voltage variation across R_C that can be quite large. The output voltage change may be 5 V for an input change of -0.05 V, leading to a voltage gain of -100 V/V.

Figure 7.12
A common-base amplifier.

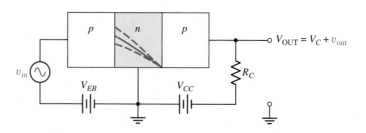

EXAMPLE 7.3

If the voltage sources v_{in} and V_{BE} of Fig. 7.8 deliver a current of

$$I_B = (5 + 1 \ \sin \ \omega t) \ \mu A$$

calculate the output voltage and the ac voltage gain of the circuit. For this example $V_{CC} = 5$ V and $R_C = 4$ kΩ. Assume that the BJT has a value of $\beta = \beta_{dc} = 80$. A more accurate set of input characteristics than that shown in Fig. 7.9(a) would be required to calculate the input voltage. Let us assume that a 1-μA change in current at a bias value of 5 μA leads to a peak voltage change of 0.006 V. Also assume that $V_A \gg V_{CE}$.

SOLUTION The current I_C is found from Eq. (7.10) [or from Eq. (7.11) with $V_{CE}/V_A \approx 0$] to be

$$I_C = \beta I_B = 80(5 + 1 \ \sin \ \omega t) \ \mu A = (0.4 + 0.08 \ \sin \ \omega t) \ mA$$

The total output voltage, including both dc and ac components, can be written as

$$V_{OUT} = V_{CC} - I_C R_C = 5 - 4(0.4 + 0.08 \ \sin \ \omega t)$$

$$= V_C + v_{out} = +3.4 - 0.32 \ \sin \ \omega t \ V$$

In order to find the ac or small-signal voltage gain, the ac input voltage of 0.006 V is used. We note that as I_B increases in the direction shown, V_{BE} also increases or becomes more positive, while V_{OUT} becomes more negative. The voltage gain is then

$$A_{MB} = \frac{\Delta V_{OUT}}{\Delta V_{BE}} = -\frac{0.32}{0.006} = -53.3 \ V/V$$

PRACTICE Problems

7.1 Calculate the ac output voltage in Example 7.3 if the base current delivered is changed to $I_B = (4 + 1.2 \sin \omega t) \ \mu A$. *Ans:* 0.38 V peak.
7.2 Calculate the voltage gain in Practice Problem 7.1 if the ac input voltage resulting from the current source input has a peak value of 0.0075 V. *Ans:* 51.2 V/V.

7.2 Graphical Analysis of the BJT

IMPORTANT Concepts

1. This section demonstrates a method of graphically analyzing a circuit that contains a nonlinear device such as a diode or a BJT.
2. The V-I characteristics for both input and output of the BJT must be given graphically to apply this method in calculating voltage gain.

If the characteristic curves for an electronic device are available, a load line can be used to analyze a simple circuit that contains the device. This is not a popular method of analysis, because characteristic curves are not readily available for each device used. The justification for considering this method is the additional insight gained relative to the operation of electronic devices. In general, an equivalent circuit approach to BJT design is much more effective than graphical analysis, but it is instructive to investigate the load-line analysis for the purpose of obtaining more information about the BJT.

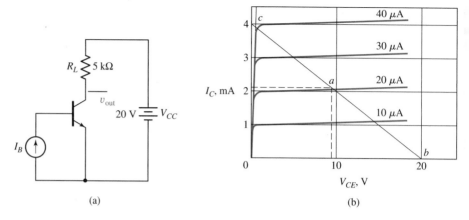

Figure 7.13
(a) Common-emitter amplifier
circuit. (b) Characteristics.

Consider the circuit of Fig. 7.13(a) with the corresponding V-I characteristics of Fig. 7.13(b). The load line is sketched on the V-I characteristics of the device with the slope $= -1/5$ kΩ. Actually, only one set of device characteristics is shown. An infinite number of curves could be drawn corresponding to other I_B values than those given. In effect, there are an infinite number of possible operating points. On the other hand, once I_B is specified, the operating point is uniquely determined. If I_B is 20 μA, the operating point is given by point a, and the output quantities are then found to be $V_{CE} = 9.5$ V and $I_C = 2.1$ mA. Now let us assume that I_B changes slowly to zero current. In this case, the operating point will move slowly to the right along the load line, reaching point b when $I_B = 0$. This point corresponds to the BJT being cut off, and V_{CE} is equal to the power supply voltage or $V_{CE} = V_{CC} = 20$ V.

If I_B is now slowly increased to 40 μA, point c on the load line will be reached. This point corresponds to saturation, and the BJT will have only a small voltage drop from collector to emitter. A typical value might be 0.1 V. The collector current is equal to approximately 4 mA. This value is found by making the approximation $V_{CE} = 0.1 \approx 0V$, which leads to

$$I_C = \frac{V_{CC} - V_{CE}}{5\ k\Omega} \approx \frac{20V}{5\ k\Omega} = 4\ \text{mA}$$

Now I_B can be increased beyond this value of 40 μA. If 70 μA of base current is used, the operating point does not change, since the $I_B = 70$ μA curve also passes through point c. In most BJTs, the base current can become 10 to 20 times the minimum current required for saturation without harming the BJT.

The common-emitter amplifier can then take on any output voltage ranging from approximately zero volts to the power supply voltage, depending on the input current. If the input consists of a dc current plus an ac signal component, then the output voltage will also consist of these two components. For example, if

$$I_B = (20 + 10\ \sin\ \omega t)\ \mu\text{A}$$

then the output voltage will be $(9.5 - 5\ \sin\ \omega t)$ V. The quiescent level (dc level with no ac input) will be $V_{CEQ} = 9.5$ V. As I_B swings to 30 μA (peak value), V_{CE} moves to 4.5 V; as I_B swings to 10 μA, V_{CE} swings to 14.5 V. An ac current of 10 μA peak enters the base of the BJT and causes a 5-V peak voltage output (inverted in phase). The voltage gain of the stage can again be found by referring to the input characteristics. The peak-to-peak input current swing may generate an input voltage swing from 0.54 V to 0.66 V, or a peak input voltage of 0.06 V. The voltage gain of the stage is then $-(5/0.06) = -83$ V/V.

The load-line approach has demonstrated a very important point that can be applied later; that is, the active region of a common-emitter amplifier with no emitter resistance is essentially equal to the power supply voltage. The maximum voltage is V_{CC} while the minimum value is approximately zero volts.

The characteristic curves for a BJT are not always available from the manufacturer. Of course, they can be generated by a curve tracer. However, the variation in characteristics from one BJT to another of the same type is generally so great that this approach is of little practical value. If a circuit is designed for production, it must exhibit the same performance when its BJTs are replaced with others of the same type. In order to achieve this constant performance as characteristics change, circuit configurations are used that are difficult to analyze using load-line techniques. We will consider these circuits later in this chapter.

PRACTICE Problem

7.3 What is the expression for output voltage in the circuit of Fig. 7.13(a) resulting from a base current of $I_B = (20 + 8 \sin \omega t) \, \mu A$? *Ans:* $V_C = 9.5 - 4 \sin \omega t$.

7.3 Discrete Circuit Biasing

IMPORTANT Concepts

1. It is important to bias a BJT stage to a point in the active region between the extremes of saturation and cutoff, because this allows an ac output to swing in both directions about the quiescent point.
2. A production circuit must be stable with respect to changes in β. This parameter varies greatly from one BJT to another of the same type. Similar circuits, having BJTs with different values of β, must perform similarly.
3. A common-emitter stage typically applies current to the base terminal to achieve the proper quiescent point.
4. The base-current bias method requires a single resistor between power supply and base to set the quiescent point. This method of bias does not lead to a practical circuit, since the quiescent output voltage is a strong function of BJT β.
5. The emitter-bias stage is a popular stage in practical discrete amplifiers.
6. The midpoint of the active region often represents a reasonable choice for V_{CQ}.
7. The stability with changes in β and temperature is quite good for the emitter-bias stage if the biasing resistors are chosen properly.
8. The emitter-bias stage can provide a very stable quiescent voltage as β changes over a wide range. Typically, a 100% increase in β leads to less than 0.5 V change in V_{CQ}.

7.3.1 GENERAL CONSIDERATIONS IN BIASING

Biasing is a very important aspect of linear amplifier design. In Section 7.2 it was found that the output voltage swing of the BJT is limited at one extreme by cutoff and at the other extreme by saturation. Cutoff occurs when the emitter-base voltage is insufficient to cause emitter current to flow. A further decrease in this voltage causes no corresponding change in collector current when the BJT is in cutoff; hence the input signal is not able to control the output signal. Saturation occurs when the collector-base junction becomes forward biased. A further increase of input signal does not increase collector current, and the output signal is not controlled by the input signal. In a linear circuit, it is important that neither of these boundaries be reached, since the output signal must be proportional to the input signal at all times. If an ac signal is to be amplified, the BJT must allow the output voltage to swing in both directions from the quiescent or no-signal value. Cutoff and saturation must be avoided, even at the extreme values of output voltage swing. The process by which the quiescent output voltage is caused to fall somewhere between the cutoff and saturated values is referred to as *biasing.*

Figure 7.13(b) demonstrates the points of cutoff and saturation for a common-emitter stage. A load-line is superimposed on the characteristic curves. Point c corresponds to saturation. A further increase of input current above 40 μA does not change the output voltage or current at this point. Point b corresponds to cutoff occurring for zero or negative input currents, where, again, the collector voltage or current is no longer controlled by the input current. A logical choice for the quiescent output voltage would be the center of the active region, corresponding to point a on the characteristics. The quiescent output voltage would then be 10 V, with the possibility of a positive swing to approximately 20 V (at cutoff) and a negative swing to 0 V (at saturation). In practice, distortion will take place for large swings, due to the unequal spacing of the curves, but this will be neglected for now. Since this quiescent level is exactly at the center of the active region, the maximum possible symmetrical output swing is allowed. Occasionally, for reasons to be considered in the following chapters, the quiescent point will be offset from the midpoint. However, for many practical cases the midpoint is a reasonable choice for the quiescent output voltage.

One of the first factors in selecting a bias network is the simplicity and economy of the scheme. Obviously, a regulated current source could be used to provide the bias current, but this method would be very expensive. In discrete circuits, biasing is often accomplished with a resistive network driven by the power-supply voltage. Resistors are the least expensive element in discrete circuits and are relatively precise. Of course, the circuit must include a power supply for other purposes, so using this source to supply bias current through resistors adds little to the circuit cost.

For integrated circuits, a resistor is more expensive than a BJT, because it occupies considerably more space on the chip. Consequently, IC bias schemes use as few resistors as possible along with additional BJT devices. In some bias configurations, individual devices must be matched with similar or equal β and $V_{BE(on)}$ values. Relatively good matching can be done for devices on an IC chip, but it is more difficult to match discrete devices. Bias schemes requiring matched βs are rarely used for discrete designs.

A second consideration in biasing relates to the stability of the output voltage with respect to changes in temperature or changes in β. A quiescent output voltage is selected to provide reasonable swings in both directions. This quiescent voltage is set far enough away from saturation or cutoff to result in linear operation as the signal voltage varies. As temperature or β changes, an inappropriate bias scheme may allow the quiescent output voltage to change drastically. This voltage may approach the saturation or cutoff boundary, allowing the signal voltage to drive the output to one of these extreme points. The resulting output signal would be highly distorted.

Temperature changes will occur as the ambient temperature changes and also as a result of changing power dissipation of the circuit. Production circuits must be designed to work properly over some specified temperature range, for example 0°C to 50°C.

Temperature also affects the β of a BJT, and β values will also differ from one BJT to another. If 10,000 identical systems are to be manufactured, there will be a β variation of BJTs from one system to the next. To achieve similar performance, the bias scheme must minimize the effect of this variation of β on quiescent voltages.

This section will discuss two biasing schemes, the base-bias and the emitter-bias configurations. The base-bias scheme is quite impractical for production circuits, whereas the emitter-bias scheme is a popular configuration in production circuits.

7.3.2 BASE-CURRENT BIAS

The circuit of Fig. 7.14 demonstrates one method of obtaining bias current. This method of bias is seldom used for production circuits, because it allows the quiescent level to shift drastically with changes in β or temperature. However, it serves to familiarize one with the

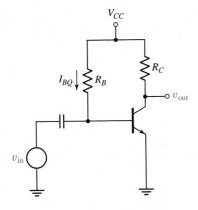

Figure 7.14
Base-current bias.

techniques involved in biasing a stage. The coupling capacitor isolates the source from bias considerations, since it appears as an open circuit for dc signals.

We can set the value of I_{BQ} by means of R_B if the voltage at the base of the BJT is known, since

$$I_{BQ} = \frac{V_{CC} - V_{BE}}{R_B} \tag{7.14}$$

Usually, V_{BE} is not known accurately, but it will be approximately 0.6–0.7 V for silicon BJTs. A particular value of V_{BE} can be used for all cases because of the sharpness of the input characteristics. Figure 7.15 shows the rapid increase of base current with applied voltage for silicon BJTs. The curve breaks so sharply that it leads to little error to assume that the vertical dashed line represents the input characteristics. Equation (7.14) reduces to

$$I_{BQ} = \frac{V_{CC} - 0.6}{R_B} \tag{7.15}$$

We note that for typical supply voltages, for example 10 V or greater, the voltage from base to emitter can be neglected with little error. Using the simpler equation

$$I_{BQ} = \frac{V_{CC}}{R_B} \tag{7.16}$$

generally results in an error smaller than 5%.

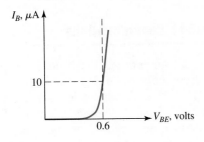

Figure 7.15
Input characteristics.

EXAMPLE 7.4

For the circuit of Fig. 7.14, assume that $R_C = 5$ kΩ, $\beta = 120$, and $V_{CC} = 20$ V. Select a value of R_B to bias a silicon BJT to the center of its active region.

SOLUTION One procedure is as follows.

1. Calculate V_{CEQ}

$$V_{CEQ} = \frac{V_{CC}}{2} = 10 \text{ V}$$

2. Calculate I_{CQ} to lead to this value of V_{CEQ}. Since $V_{CEQ} = V_{CC} - R_C I_{CQ}$, we solve for I_{CQ} as

$$I_{CQ} = \frac{20 - 10}{5 \text{ k}\Omega} = 2 \text{ mA}$$

3. Calculate I_{BQ} using $I_{CQ} = \beta I_{BQ}$. For this example

$$I_{BQ} = \frac{2}{120} = 16.7 \text{ } \mu\text{A}$$

4. Calculate R_B from $(V_{CC} - V_{BE})/I_{BQ} = R_B$, which gives

$$R_B = \frac{20 - 0.6}{16.7 \text{ } \mu A} = 1.16 \text{ M}\Omega$$

For the case where V_{BE} is assumed to be zero, a value of

$$R_B = 2\beta R_C = 1.2 \text{ M}\Omega$$

will set V_{CEQ} at $V_{CC}/2$. This can be checked by considering the value of I_{BQ} resulting from using this value of R_B. The base current is then approximately

$$I_{BQ} = \frac{V_{CC}}{2\beta R_C}$$

The collector current is

$$I_{CQ} = \beta I_{BQ} = \frac{V_{CC}}{2R_C}$$

which gives a quiescent collector voltage of

$$V_{CEQ} = V_{CC} - I_{CQ}R_C = V_{CC} - \frac{V_{CC}}{2} = \frac{V_{CC}}{2}$$

PRACTICAL Considerations

PRACTICE Problems

7.4 For the circuit of Example 7.4, select R_B to set $V_{CQ} = 16$ V. *Ans:* 2.91 MΩ.
7.5 For the circuit of Example 7.4, select R_B to set $V_{CQ} = 5$ V. *Ans:* 776 kΩ.

A serious problem with base current bias in production circuits is its poor stability with respect to changes in β. Consider the circuit of Fig. 7.14 with the values of R_C and R_B as calculated in Example 7.4. If $\beta = 120$, the quiescent collector voltage is 10 V. What happens if we replace the BJT with one having a β of 60? The quiescent voltage is then 15 V. If β is 200, the quiescent voltage becomes 3.3 V. The variation in quiescent voltage with change in β is too great to allow the base-current bias stage to be of much practical value. If 1000 circuits using this stage were to be sold, the variation in β that occurs from one transistor to another would cause some units to perform unacceptably. This parameter can easily vary by a factor of two from one device to another of the same type.

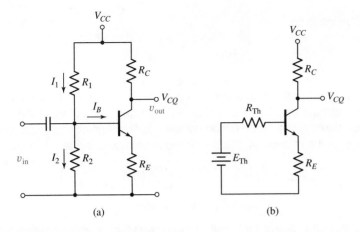

Figure 7.16
*Emitter bias: (a) actual circuit,
(b)equivalent dc circuit.*

7.3.3 EMITTER BIAS

Emitter bias is a popular method of bias because it leads to good stability with respect to changes in β and temperature. The circuit of Fig. 7.16 shows the emitter-bias scheme. This circuit employs current feedback to stabilize the temperature variations. Taking a Thévenin equivalent circuit of R_1, R_2, and V_{CC}, we obtain the circuit shown in Fig. 7.16(b). To find I_B, we write the loop equation

$$E_{Th} = R_{Th}I_B + V_{BE} + R_E I_E$$

We can also write

$$I_E = I_B + I_C = (\beta + 1)I_B \tag{7.17}$$

Solving these two equations for the quiescent value of I_B gives

$$I_{BQ} = \frac{E_{Th} - V_{BE}}{R_{Th} + (\beta + 1)R_E} \tag{7.18}$$

There are two methods of selecting R_1 and R_2 for the proper bias. Since Eq. (7.18) relates I_{BQ} to E_{Th} and R_{Th}, once I_{BQ} is known, E_{Th} and R_{Th} can be chosen to satisfy this equation. Then R_2 and R_1 can be found from the equations

$$\frac{R_2}{R_1 + R_2}V_{CC} = E_{Th} \quad \text{and} \quad \frac{R_1 R_2}{R_1 + R_2} = R_{Th}$$

A second method, which is simpler for design purposes, is to first select R_2 to be 10 to 20 times R_E to result in good stability and then to select R_1 to give the proper value of I_{BQ}.

PRACTICAL Considerations

We should note that the addition of R_E to the circuit decreases the size of the active region. The edge of the active region defined by cutoff remains the same. When the BJT turns off, the collector voltage becomes $V_{C\max} = +V_{CC}$. However, the edge of the region defined by saturation is no longer approximately zero volts. When the BJT saturates, the BJT can be approximated by a short circuit and V_C is calculated by voltage division to be

$$V_{C\min} = \frac{R_E}{R_E + R_C} V_{CC}$$

That is, the entire power-supply voltage drops across R_E and R_C when the BJT is saturated (assuming the saturation voltage of the BJT is zero). This minimum value of collector voltage might be 1 or 2 V or even greater in some applications. In order to allow an equal swing on both sides of the quiescent point, the quiescent voltage must be set at the midpoint of the two collector voltage extremes. This point is

$$V_{CQ} = \frac{V_{C\max} + V_{C\min}}{2} = \frac{V_{CC}}{2}\left(1 + \frac{R_E}{R_E + R_C}\right)$$

$$= \text{midpoint of active region} \tag{7.19}$$

When this condition is met, half of the power-supply voltage drops across the BJT and half drops across the sum of R_E and R_C. An example will demonstrate the second method of selecting R_2 and R_1 for a midpoint-bias voltage.

EXAMPLE 7.5

In the emitter-bias circuit of Fig. 7.16, $R_C = 10$ kΩ, $R_E = 1$ kΩ, and $V_{CC} = 20$ V. Given that $\beta = 80$ and $V_{BE} = 0.6$ V, find R_1 and R_2 to set V_{CQ} at the center of the active region.

SOLUTION We will first assume that $R_2 = 10R_E$ is a condition imposed on the circuit by stability requirements. It will be shown later that this is a reasonable choice of R_2 and gives a value of $R_2 = 10$ kΩ for this example. We now must find the desired value of I_{BQ}.

1. The center of the active region must be located. Using Eq. (7.19), we find that this value is

$$V_{CQ} = 10\left(1 + \frac{1}{11}\right) = 10.9 \text{ V}$$

2. The required collector current and base current can now be found to give this value of V_{CQ}:

$$I_{CQ} = \frac{V_{CC} - V_{CQ}}{R_C} = \frac{20 - 10.9}{10} = 0.91 \text{ mA}$$

$$I_{BQ} = \frac{I_{CQ}}{\beta} = \frac{0.91}{80} = 0.0114 \text{ mA}$$

3. We now select R_1 to give the calculated value of I_{BQ}. By noting in Fig. 7.16(a) that $I_1 = I_{BQ} + I_2$, we can find the necessary value of R_1:

$$I_2 = \frac{V_{BQ}}{R_2} = \frac{I_{EQ}R_E + V_{BE}}{R_2} = \frac{(I_{BQ} + I_{CQ})R_E + 0.6}{R_2}$$

$$= \frac{(0.92) \times 1 + 0.6}{10} = 0.152 \text{ mA}$$

Then

$$I_1 = I_{BQ} + I_2 = 0.011 + 0.152 = 0.163 \text{ mA}$$

Since the voltage across R_1 is $V_{CC} - V_{BQ}$, then

$$R_1 = \frac{V_{CC} - V_{BQ}}{I_1} = \frac{20 - 1.52}{0.163} = 113 \text{ k}\Omega$$

PRACTICAL Considerations

Generally, for design purposes the second method of selecting R_1 and R_2 is preferable to taking a Thévenin equivalent of the bias network. For analysis of a circuit having all values specified, the calculation is simplified by using the Thévenin equivalent.

If no other constraints are present, setting V_{CQ} at the midpoint of the active region is a reasonable choice. A constraint that may modify this choice is that of minimizing power-supply current. If an amplifier is to be powered by a battery and the output signal is relatively small, an operating point value of V_{CQ} should reduce current drain while allowing sufficient output peak voltage swing without reaching cutoff.

The emitter-bias circuit is popular in discrete circuit design because it has a very stable operating point even for large changes in β.

PRACTICE Problems

7.6 Rework Example 7.5 with $R_2 = 10$ kΩ and a desired value of $V_{CQ} = 14$ V. *Ans: $R_1 = 147$ kΩ.*
7.7 Rework Example 7.5 with $R_2 = 10$ kΩ and a desired value of $V_{CQ} = 7$ V. *Ans: $R_1 = 88$ kΩ.*

7.3.4 STABILITY WITH RESPECT TO CHANGES IN β

Variations in β can occur as a result of temperature changes. Furthermore, β variations will occur when BJTs are interchanged, resulting in the same effect as temperature variations of β. A figure for $\partial V_{CQ}/\partial T$, related to β, would not be very meaningful because it would not predict quiescent-point variations for interchanged BJTs. Usually, values of $\beta_{\min}$ and $\beta_{\max}$ are given by the manufacturer and all BJTs of the same type will have a β within this range. If the circuit is designed to operate properly when β assumes either of these two extreme values, the circuit is stable with respect to β changes, regardless of what produces these changes. Earlier we found that the base-current bias circuit is very unstable for β changes. We will now consider the emitter-bias stage.

Emitter-Bias Stage The equivalent circuit of Fig. 7.16(b) is an emitter-bias stage. For this circuit the collector voltage is given by

$$V_{CQ} = V_{CC} - I_{CQ} R_C = V_{CC} - \beta I_{BQ} R_C$$

$$= V_{CC} - \frac{\beta R_C [E_{Th} - V_{BE}]}{R_{Th} + (\beta + 1) R_E} \qquad (7.20)$$

If R_{Th} is much greater than $(\beta + 1) R_E$, the equation reduces to

$$V_{CQ} = V_{CC} - \frac{\beta R_C}{R_{Th}} [E_{Th} - V_{BE}]$$

In this case V_{CQ} decreases directly with increasing β. The circuit would be very unstable with respect to changes in β. The base-bias circuit, with $R_E = 0$, is a good example of this. On the other hand, if $(\beta + 1) R_E$ is much greater than R_{Th}, Eq. (7.20) becomes

$$V_{CQ} = V_{CC} - \frac{\beta R_C}{(\beta + 1) R_E} [E_{Th} - V_{BE}] = V_{CC} - \frac{\alpha R_C}{R_E} [E_{Th} - V_{BE}] \qquad (7.21)$$

Since α varies only 1% or 2% for very large variations in β, the collector voltage is quite stable in this case. Stability requirements can be satisfied for the emitter-bias stage if R_{Th}

is limited to a fraction of $(\beta + 1)R_E$. Some circuits requiring less parameter stability allow R_{Th} to become larger, but, in general, a reasonable upper limit on R_{Th} is $0.2(\beta + 1)R_E$.

We can consider another viewpoint to explain why the emitter-bias circuit is so much more stable than the base-bias stage. In the base-bias stage, the base current is constant at a given temperature at

$$I_{BQ} = \frac{V_{CC} - V_{BE}}{R_B}$$

Since collector current equals βI_B, if β is doubled, the collector current will also double. In the emitter-bias circuit the base current is

$$I_{BQ} = \frac{E_{Th} - V_{BE} - I_{EQ}R_E}{R_{Th}}$$

When β increases, I_{EQ} increases as $(\beta + 1)I_{BQ}$, but this decreases the value of I_{BQ} by raising the emitter voltage due to the increase in $I_{EQ}R_E$. I_{BQ} is lowered as β increases, which offsets the increase in I_{CQ} due to increased β.

EXAMPLE 7.6

In the emitter-bias circuit of Fig. 7.16(b), $R_{Th} = 10$ kΩ, $R_E = 1$ kΩ, $R_C = 5$ kΩ, $V_{CC} = +12$ V, $E_{Th} = 1.6$ V, and $V_{BE} = 0.5$ V. Calculate the quiescent collector voltage for $\beta = 80$. Compare this to the case for $\beta = 160$.

SOLUTION From Eq. (7.20) for $\beta = 80$,

$$V_{CQ} = 12 - \frac{80 \times 5 \times 10^3[1.6 - 0.5]}{10^4 + 8.1 \times 10^4} = 7.16 \text{ V}$$

If β is now changed to 160, the quiescent voltage becomes

$$V_{CQ} = 12 - \frac{160 \times 5 \times 10^3[1.6 - 0.5]}{10^4 + 16.1 \times 10^4} = 6.85 \text{ V}$$

The change in V_{CQ} due to a doubling of β is only 0.31 V. Recall that in the base-bias circuit, using a 20-V supply, the doubling of β from 60 to 120 changed V_{CQ} from 15 V to 10 V.

Sometimes it is advantageous to calculate the maximum quiescent-point change that can occur, assuming that all variables reach their extreme values at the same time. If the combination of variables so chosen to give the maximum possible shift allow a satisfactory design, then all other values of variables will result in a satisfactory design; this is referred to as *worst-case* design. As an example of this, consider the equation for collector voltage in the emitter-bias case,

$$V_{CQ} = V_{CC} - \frac{\beta R_C[E_{Th} - V_{BE}]}{R_{Th} + (\beta + 1)R_E} \tag{7.20}$$

Let us assume that $V_{CC} = 20$ V, $R_{Th} = 10$ kΩ, $R_E = 1$ kΩ, $R_C = 5$ kΩ, and $E_{Th} = 3.0$ V. The current gain, β, can vary from 100 to 200 (due either to temperature or interchanging of BJTs) and V_{BE} varies at a rate of -2mV/$^\circ$C, from 25°C to 75°C.

From an examination of Eq. (7.20), it is seen that as T increases, V_{CQ} will decrease due to a decrease in V_{BE}. V_{CQ} also decreases when β increases. Thus, the lower extreme of V_{CQ} will occur when β is at its maximum value and $T = 75^\circ$C. Since V_{BE} decreases by 2 mV/$^\circ$C,

$V_{BE} = 0.6 - (50)(0.002) = 0.5$ V at $T = 75°C$. The lower extreme of V_{CQ} is then

$$V_{CQ(min)} = 20 - \frac{200 \times 5 \times 10^3[3.0 - 0.5]}{10^4 + 201 \times 10^3} = 8.2 \text{ V}$$

The maximum value of V_{CQ} occurs when β is at its minimum and V_{BE} is maximum ($T = 25°C$). Therefore,

$$V_{CQ(max)} = 20 - \frac{100 \times 5 \times 10^3[3.0 - 0.6]}{10^4 + 101 \times 10^3} = 9.2 \text{ V}$$

The worst possible shift that can occur is a 1-V shift. All other combinations of β and V_{BE} will result in a quiescent collector voltage falling between 8.2 V and 9.2 V.

The concept of worst-case design is quite valuable in engineering practice. Circuit manufacturers can specify tolerance limits of important quantities based on worst-case calculations. Recognition of the importance of this type of consideration has resulted in this feature being included in several of the simulation programs for automatic circuit analysis. As the complexity of a circuit grows, worst-case design becomes more difficult to perform manually; thus, the inclusion of this feature in programs such as Spice is of great practical significance.

PRACTICAL Considerations

An interesting point can be discussed here in relation to the results of this section. If β can double and only cause a small bias-point shift (0.5–1.0 V), how accurately must β be known in order to design an emitter-bias circuit? Quite obviously, in Example 7.6, if β were estimated to be any number between 80 and 160, the value of V_{CQ} would be within 0.31 V of the desired value. The emitter-bias circuit causes the results of the design to depend more on element values than on BJT parameters. This point is very useful and demonstrates that β can be estimated very grossly and still be used to give accurate results with the emitter-bias circuit. One might then believe that the type of BJT used in an emitter-bias circuit is relatively unimportant. This conclusion is partially correct. For low-frequency, noncritical applications, several hundred types of BJTs can be used to satisfy a given design. We shall see later that the power requirements of the stage will impose additional constraints on the circuit. Furthermore, high frequency and switching performance of a circuit is usually limited by the BJT; hence special BJTs are required. Nevertheless, it is interesting to know that a great deal of design can be done with only a vague knowledge of the BJT characteristics.

PRACTICE Problem

7.8 Select R_1 to bias the stage of Fig. 7.16(a) to $V_{CQ} = 8$ V. Assume that $\beta = 100$, $V_{BE(on)} = 0.6$ V, $V_{CC} = 12$ V, $R_C = 4$ kΩ, $R_E = 0.5$ kΩ, and $R_2 = 8$ kΩ. Calculate the change in V_{CQ} if β now changes to 200. *Ans: $R_1 = 74$ kΩ, $\Delta V = -0.33$ V.*

7.4 Small-Signal or Linear Models

IMPORTANT Concepts

1. The equivalent circuit or model of the BJT, rather than the characteristic curves, is generally used in the design of amplifier circuits.
2. The incremental model of the BJT is valid only for output currents and voltages that are approximately linear functions of the input signal.

As a result of the difficulties in using characteristic curves, design methods based on characteristic curves have not been developed to the point of other design methods. For example, in linear amplifier design, the incremental or small-signal model of the BJT is the most popular method used.

Since a linear amplifier requires a linear input-output relationship, the voltage variation across the emitter-base junction of a BJT within the amplifier must be limited to small values. This limitation allows the exponential relationship of voltage and current in the diode to be approximated by a linear function over this small range of voltages. The model resulting from this approximation is called a small-signal or incremental model.

There is no output voltage level that can be used to define small-signal operation. The output voltage of a BJT stage might have a 2-V amplitude and still be considered a small signal. On the other hand, the voltage applied to the base-emitter junction must be quite small to lead to a linear current variation with this voltage. A rule of thumb is that the variation of voltage across this junction must be less than 0.1 V for this model to approximate linear behavior. Furthermore, the output voltage or current variations must not exceed the limits of the active region. When this occurrs, the operation becomes highly nonlinear.

7.4.1 THE COMMON-BASE CONFIGURATION

For most linear applications of the BJT using the common-base configuration, the small-signal model of Fig. 7.17 is appropriate. This model or equivalent circuit uses the small-signal diode model of Chapter 5 for the emitter-base diode. This resistance will be called r_e, since it appears in series with the emitter of the BJT.

In order to apply the small-signal model, the BJT must be placed in its active region, generally by one of the methods discussed in the last section.

Once the bias is established, the small-signal incremental model ignores the dc conditions of the circuit. Again, this model assumes that the applied signal voltage causes a variation of emitter-base voltage that is small compared to the bias voltage established across the emitter-base junction. In small-signal models, we will use lowercase characters to represent incremental variables such as voltage and current.

Since a small incremental current produces no corresponding incremental voltage change across a dc source, a short circuit replaces each dc voltage source in the incremental or small-signal model.

Use of the Equivalent Circuit   In earlier chapters, we discussed the design of amplifier circuits. We saw that three important quantities of interest in amplifier design are input impedance, output impedance, and voltage gain. In some applications, it is also important to know the current gain.

The equivalent circuit of Fig. 7.17 can be used to evaluate these quantities for the common-base stage. The diode resistance is determined by the bias current and, at room

Figure 7.17
Small-signal model for the BJT in the common-base configuration.

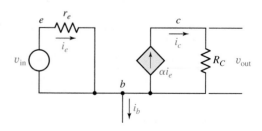

temperature, is

$$r_e = \frac{26}{I_E} \qquad (7.22)$$

The input impedance is obviously equal to r_e or

$$r_{in} = \frac{v_{in}}{i_{in}} = r_e \qquad (7.23)$$

The midband voltage gain is the ratio of the output voltage to the input voltage. The output voltage is given by

$$v_{out} = i_c R_C$$

In order to find the voltage gain, it is convenient to write v_{out} as a function of v_{in}. We can do this by noting that

$$i_c = \alpha i_e = \alpha \frac{v_{in}}{r_e}$$

Inserting this result into the equation for v_{out} and solving for voltage gain gives

$$A_{MB} = \frac{v_{out}}{v_{in}} = \frac{\alpha R_C}{r_e} \qquad (7.24)$$

For this configuration, the output impedance of the model is infinite; thus the output impedance of the overall circuit is

$$R_{out} = R_C \qquad (7.25)$$

The current gain is the ratio of current leaving the collector terminal (output current) to emitter current (input current) for the common base stage and is given by

$$A_i = \frac{i_c}{i_e} = \frac{\alpha i_e}{i_e} = \alpha \qquad (7.26)$$

The equivalent circuit used to evaluate these quantities is only a first-order model. However, it can be used with sufficient accuracy for a very high percentage of practical design problems. For large collector resistors the output impedance of the BJT may no longer be considered as an open circuit. In such cases, a large resistor may be added to the model from collector to base to account for this impedance.

PRACTICE Problems

7.9 Rework Example 7.7 if I_E and R_C are changed to 0.8 mA and 4 kΩ, respectively. *Ans:* 122 V/V.

7.10 Rework Example 7.7 if I_E and R_C are changed to 1.2 mA and 4 kΩ, respectively. *Ans:* 183 V/V.

7.11 Select R_C to give the circuit of Fig. 7.5(a) a voltage gain of 100 V/V. Assume that $I_E = 1$ mA. *Ans:* 2.63 kΩ.

EXAMPLE 7.7

Assume that the circuit of Fig. 7.5(a) is biased such that $I_E = 1$ mA. Given that $R_C = 5$ kΩ and $\alpha = 0.99$, find the input impedance, output impedance, and the midband voltage gain of the amplifier.

SOLUTION The equivalent circuit of Fig. 7.17 is used.
The input impedance equals r_e, which in this case is

$$r_{in} = r_e = \frac{26}{I_E} = 26 \ \Omega$$

The output impedance is seen to be

$$R_{out} = R_C = 5 \text{ k}\Omega$$

The voltage gain A_{MB} is found from Eq. (7.24) to be

$$A_{MB} = \frac{0.99 \times 5 \times 10^3}{26} = 190 \text{ V/V}$$

7.4.2 THE COMMON-EMITTER CONFIGURATION

A common-emitter amplifier is shown in Fig. 7.18. Basically, the only difference between the circuit shown in this figure and that of Fig. 7.5 is that, in Fig. 7.18, the input source has been moved around the emitter-base loop toward the base. The incremental input voltage is still applied across the emitter-base junction, but base current rather than emitter current now represents input current flowing from the source. If incremental base and collector currents are now assumed to be positive when entering these terminals, emitter current must be considered positive when exiting the emitter terminal. We can again write the relationship between input voltage and emitter current as

$$v_{\text{in}} = i_e r_e$$

Since input current is now base current rather than emitter current, a more appropriate form of this equation results by noting that

$$i_e = \frac{i_b}{1 - \alpha}$$

Since β is defined in terms of α by

$$\beta = \frac{\alpha}{1 - \alpha} \tag{7.27}$$

it can be shown that

$$\frac{1}{1 - \alpha} = \beta + 1 \tag{7.28}$$

We can then write v_{in} in terms of i_b as

$$v_{\text{in}} = i_b(\beta + 1)r_e$$

Solving for input impedance gives

$$r_{\text{in}} = \frac{v_{\text{in}}}{i_{\text{in}}} = \frac{v_{\text{in}}}{i_b} = (\beta + 1)r_e \tag{7.29}$$

Figure 7.18
A common-emitter amplifier and the equivalent circuit.

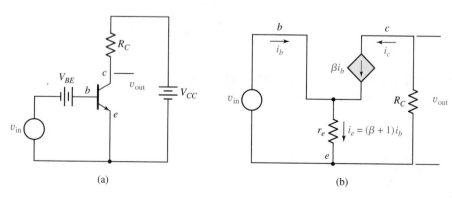

(a)   (b)

The output current is again collector current and can be written in terms of base current as

$$i_c = \alpha i_e = \frac{\alpha}{1 - \alpha} i_b = \beta i_b \tag{7.30}$$

The parameter β is the current gain from base to collector. Note that β is generally a very large number, typically in the range of 100 to 300 for low-power BJTs. If β is known and α is needed, Eq. (7.27) can be solved for α to result in

$$\alpha = \frac{\beta}{\beta + 1} \tag{7.31}$$

In the active region of the BJT, an incremental flow of current into the emitter results in base and collector currents out of their respective terminals. If incremental current flows out of the emitter, both base and collector currents flow into the device, for both *npn* and *pnp* BJTs. As a result of these directions, the equivalent circuit of Fig. 7.18(b) shows collector current flowing into the device when base current flows into the device. Although this might seem to be a trivial matter, it results in a phase inversion between input and output voltage of the common-emitter stage.

The voltage gain of the common-emitter stage can be found by writing

$$v_{\text{out}} = -i_c R_C = -\beta i_b R_C$$

The base current relates to input voltage as

$$i_b = \frac{v_{\text{in}}}{r_{\text{in}}}$$

From Eq. (7.29), the input impedance is

$$r_{\text{in}} = (\beta + 1) r_e$$

Utilizing these three equations allows the midband voltage gain to be written

$$A_{MB} = \frac{v_{\text{out}}}{v_{\text{in}}} = \frac{-\beta R_C}{(\beta + 1) r_e} = \frac{-\alpha R_C}{r_e} \tag{7.32}$$

This expression is similar to the equation for voltage gain in the common base stage except for the negative sign that precedes it. This sign reflects the fact that the output voltage is inverted in phase, referenced to the input voltage of the common emitter stage.

The output impedance of the circuit equals R_C in this circuit as well as the common base circuit.

Practical Modifications to the Common-Emitter Model

The equivalent circuit of Fig. 7.18(b) can be drawn in an alternate form that is easier to use. The element r_e appears as a common element to both the input and output loops. Since the current of the output loop is not affected by this resistance, it is reasonable to remove this element from the output loop. The effect of r_e on the input loop can be represented by placing a single resistance in this loop. The model of Fig. 7.19(a) shows an element equal to $(\beta + 1) r_e$ in the input loop with only a dependent source appearing between collector and emitter. The resistance $(\beta + 1) r_e$ is designated r_π. This model is often referred to as the hybrid-π equivalent circuit.

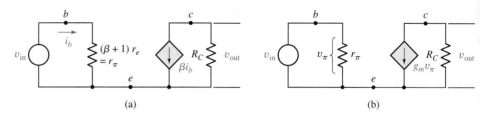

Figure 7.19
The hybrid-π, low-frequency model; (a) simple model; (b) modified nomenclature.

In checking the relationships between voltages and currents around the input and output loops, it is found that the same relationships hold for the models of Figs. 7.18(b) and 7.19(a). Physically, the circuit of Fig. 7.19(a) does not reflect the correct impedance levels of the actual BJT at the emitter. For the model of Fig. 7.18(b) and for the BJT itself, the base-to-emitter voltage is developed as emitter current flows through the emitter diode resistance r_e. The circuit of Fig. 7.19(a) implies that base current flows through the resistance $(\beta + 1)r_e$ to generate the voltage drop. Although this situation is not physically correct, the same V-I relationship results for either circuit; thus, both are acceptable models.

The dependent current generator can be written in terms of base current as βi_b or in terms of base-to-emitter voltage as $g_m v_\pi$. It is left to the student to show that these two sources are equivalent current generators if

$$g_m = \frac{\alpha}{r_e} \tag{7.33}$$

The parameter g_m is called the transconductance and has units of A/V or Siemens.

The equivalent circuits of Figs. 7.18 and 7.19 are sometimes called first-order models because they represent only the basic behavior of the BJT. There are two second-order mechanisms in the actual BJT that can be added to make the models more accurate: these are the ohmic base resistance, r_x, and the output impedance between collector and emitter, r_{ce}.

Because the base region is very narrow, the contact to the base and the intrinsic resistance of this narrow region lead to an element called the ohmic base resistance. The value of this parameter may range from 20 Ω to about 200 Ω. In low-frequency work, this resistance can often be neglected since the value of r_π appears in series with r_x and is generally much larger. This base resistance becomes more important in high-frequency work since the resistance r_π is shunted by a capacitance that lowers the impedance in series with r_x.

We recall that the V-I characteristics of the common-emitter stage sloped slightly upward. This effect leads to an incremental resistance from collector to emitter. In the incremental model, a resistance can be placed across the dependent collector current source to account for this effect. The resistance placed across the current source is designated r_{ce} and also relates to the Early voltage used in Eq. (7.11); this is the output resistance looking back into the collector terminal. It can be shown that this resistor is given by

$$r_{ce} = \frac{V_A}{\beta I_B} \tag{7.34}$$

Figure 7.20 shows a more accurate equivalent circuit for the common-emitter stage including r_x and r_{ce}. A typical value of r_{ce} is 40 kΩ. If this value is much larger than R_C,

Figure 7.20
An accurate model for the common-emitter BJT.

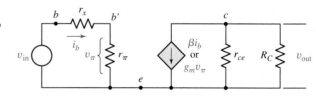

it is ignored. When an external resistor is placed in series with the emitter terminal, the effective output impedance is increased considerably over the value given by r_{ce}.

EXAMPLE 7.8

Assume that the circuit of Fig. 7.18(a) is biased such that $I_E = 1$ mA. Given that $R_C = 5$ kΩ and $\alpha = 0.99$, find the input impedance, output impedance, and the voltage gain of the amplifier.

SOLUTION From the equivalent circuit of Fig. 7.18(b) or Fig. 7.19, we see that the input impedance is

$$r_{in} = (\beta + 1)r_e \qquad (7.35)$$

which results in a value of

$$r_{in} = 26(100) = 2600 \ \Omega$$

The voltage gain is evaluated using Eq. (7.32) and is

$$A_{MB} = \frac{v_{out}}{v_{in}} = -\frac{\alpha R_C}{r_e} = -190 \text{ V/V}$$

The magnitude of the voltage gain is exactly the same as that in the common-base amplifier, but there is a phase inversion in the common-emitter stage.

It is to be expected that the voltage gain of the common-emitter and common-base amplifiers is the same, since the input voltage is applied across the same junction and the output is taken at the collector in both cases.

The output impedance is again 5 kΩ, neglecting r_{ce}, but the current gain in the common-emitter stage is quite high. This value is found by taking the ratio of the output or collector current to input or base current. This ratio is equal to β.

Because the common-emitter amplifier has a high current gain along with a high voltage gain, it is the most well-used stage of the three possible configurations.

7.4.3 THE EMITTER FOLLOWER (COMMON-COLLECTOR CONFIGURATION)

The emitter follower is very important in circuit design because it has a high input impedance and low output impedance. Even though its voltage gain is approximately unity, the impedance transformation properties make it a very useful circuit. Figure 7.21 shows the emitter-follower circuit along with an appropriate model.

The voltage gain can be found by writing the equations for v_{out} and v_{in}, which gives

$$v_{out} = (\beta + 1)i_b R_E$$

and

$$v_{in} = r_e(\beta + 1)i_b + R_E(\beta + 1)i_b$$

PRACTICE Problems

7.12 If an input signal source with a resistance of 1 kΩ drives the stage of Example 7.8, what is the overall voltage gain?
Ans: -137 V/V.

7.13 If an input signal source with a resistance of 1 kΩ drives the stage of Example 7.8, what value of R_C is needed to result in an overall midband voltage gain of -90 V/V?
Ans: 3.27 kΩ.

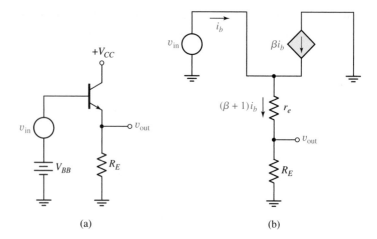

Figure 7.21
(a) An emitter follower.
(b) Model for the emitter
follower.

Solving these equations for v_{out}/v_{in} gives

$$A_{MB} = \frac{v_{out}}{v_{in}} = \frac{(\beta+1)R_E}{(\beta+1)r_e + (\beta+1)R_E} = \frac{R_E}{r_e + R_E} \qquad (7.36)$$

Since R_E is typically much larger than r_e, the voltage gain of the emitter follower is very close to unity. For most situations, it is appropriate to use

$$A_{MB} \approx 1$$

The input impedance is found to be approximately

$$r_{in} = \frac{v_{in}}{i_b} = (\beta+1)r_e + (\beta+1)R_E \qquad (7.37)$$

The output impedance can be found by shorting the input source and assuming that there is a test source at the output. The current through this output source is

$$i_{out} = i_{R_E} + i_e$$

where i_{R_E} is simply v_{out}/R_E. The emitter current is found by using the following equations:

1. $i_e = i_b(\beta+1)$
2. $v_{out} = i_e r_e = r_e(\beta+1)i_b$

Combining these equations leads to

$$\frac{v_{out}}{i_e} = r_e$$

and therefore

$$R_{out} = \frac{v_{out}}{i_{R_E} + i_e} = R_E \parallel r_e \qquad (7.38)$$

The output impedance of the BJT itself is just

$$r_{out} = r_e$$

It is obvious that the current gain is given by

$$A_i = \beta + 1 \tag{7.39}$$

The equations for voltage gain, current gain, and output impedance are accurate for most applications. The equation for input impedance results in errors as the input impedance approaches large values in the range of hundreds of kΩ, due to a high impedance that exists from base to collector. This junction forms a reverse-biased diode with very high resistance, perhaps 2 MΩ. This resistance appears in parallel with the input resistance calculated previously and can typically be neglected because it is so large. As the overall impedance becomes large, due to large values of R_E, the impedance from base to collector may need to be considered.

If a resistance R_B were placed in series with the input voltage source, the voltage gain would become

$$A_{MB} = \frac{v_{\text{out}}}{v_{\text{in}}} = \frac{(\beta+1)R_E}{R_B + (\beta+1)r_e + (\beta+1)R_E} \tag{7.40}$$

For typical values of R_E and R_B, the voltage gain would again be near unity.

The output impedance would now be

$$R_{\text{out}} = R_E \parallel \left(r_e + \frac{R_B}{\beta+1} \right) \tag{7.41}$$

Note that the emitter resistance r_e cannot be reflected to the input loop when calculating the output resistance. When this was done in the common-emitter circuit, the V-I relationship at the base was preserved. The calculation of output impedance involves the emitter current, and reflecting the resistance r_e to the base loop as $(\beta+1)r_e$ does not maintain the correct emitter-current relationship.

PRACTICE Problem

7.14 A generator resistance of 1 kΩ appears in series with the base lead in the emitter follower of Fig. 7.21. If $I_E = 2$ mA, $R_E = 3$ kΩ, and $\beta = 120$, find the midband voltage gain, the current gain, and the overall output impedance of the circuit. *Ans:* 0.993 V/V, 121 A/A, 21.1 Ω.

7.4.4 USING THE SMALL-SIGNAL MODELS WITH EXTERNAL RESISTORS

In most practical cases the BJT amplifier will include resistors other than those of the equivalent circuit. Most input sources will have some generator resistance, and many load resistors cannot be connected directly to a BJT terminal. For example, a microphone may have a generator impedance of 10 kΩ; this transducer will often supply the input signal to an amplifier. An audio speaker is a common load for certain amplifiers; this device may have an impedance of 40 Ω and cannot tolerate dc current. Consequently, the speaker must be coupled to the BJT output through a transformer or capacitor to eliminate the dc component of the BJT output signal.

These resistors can be added to the simple equivalent circuit for purposes of calculation. This will be demonstrated by the circuit of Fig. 7.22(a), which includes a generator or base resistance R_B, an external emitter resistance R_E, and a load resistance R_L. The emitter resistance R_E might be used to control the voltage gain. The capacitor does not allow the dc voltage appearing at the collector to reach the load.

The equivalent circuit shown in part (b) of the figure includes R_B in series with the base lead. The emitter resistance R_E is in series with r_e. In this particular equivalent circuit, r_e must be multiplied by $\beta + 1$; thus R_E must also be multiplied by this value.

We assume that the capacitor has negligible impedance to the midband signal frequencies of interest. The effective load seen by the collector current source is the parallel combination of R_C and R_L. This resistance value will be designated R_{eff}.

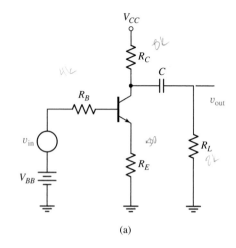

(a)

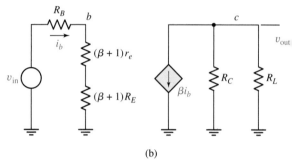

Figure 7.22
*Common emitter amplifier with
external resistors. (a) Actual
circuit. (b) Equivalent circuit.*

(b)

The equations for gain or impedance will be listed, but the derivations are left to the reader. The gain is given by

$$A_{MB} = \frac{-\beta R_{\text{eff}}}{R_B + (\beta + 1)(r_e + R_E)} \tag{7.42}$$

The impedance at point b (not including R_B) is approximately

$$r_{\text{in}} = (\beta + 1)r_e + (\beta + 1)R_E \tag{7.43}$$

If R_E is much larger than r_e and if $(\beta + 1)R_E$ is much larger than R_B, the gain is approximated by

$$A_{MB} = -\frac{R_{\text{eff}}}{R_E} \tag{7.44}$$

The input impedance is approximated by

$$r_{\text{in}} = (\beta + 1)R_E \tag{7.45}$$

The common-base and common-collector configurations can be handled equally well in this manner.

EXAMPLE 7.9

The circuit of Fig. 7.22 is biased such that $I_E = 0.5$ mA. The BJT has a value of $\beta = 180$. If $R_B = 4$ kΩ, $R_E = 500$ Ω, $R_C = 8$ kΩ, $R_L = 2$ kΩ, and $V_{CC} = 12$ V, calculate the voltage gain of the circuit.

Can the absolute value of voltage gain be increased to 6 by modifying the values of R_C and R_E without changing the dc emitter current? If so, select proper values.

SOLUTION The calculation of the voltage gain is done rather easily by using Eq. (7.42). The effective collector load is $R_{\text{eff}} = 1.6$ kΩ. The dynamic emitter resistance is

$$r_e = \frac{26}{0.5} = 52 \ \Omega$$

With these values the voltage gain is

$$A_{MB} = \frac{-180 \times 1.6}{4 + 181 \times 0.552} = -2.77 \text{ V/V}$$

This part of the example is an analysis problem. The second part is more closely related to design.

Equation (7.42) shows that in order to increase the absolute value of voltage gain, R_{eff} can be increased, R_E can be decreased, or a combination of both these actions can be effected. There is no single correct solution to this problem. It can be seen, however, that the voltage gain varies directly with R_{eff}. Attempts to increase R_{eff} by increasing R_C meet with little success. The maximum value of R_{eff} is 2 kΩ, which occurs when R_C increases to infinity. Not only does R_{eff} increase little with increasing R_C, the BJT would near saturation if R_C were increased toward 20 kΩ. Changing R_E appears to be a good approach to satisfying the voltage gain requirement. Assuming that R_{eff} remains constant, Eq. (7.42) can be solved for the value of R_E that leads to the correct voltage gain. The result is

$$R_E = \frac{\beta \times R_{\text{eff}}}{(\beta + 1) \times |A_{MB}|} - \frac{R_B}{(\beta + 1)} - r_e$$

$$= \frac{180 \times 1.6}{181 \times 6} - \frac{4}{181} - 0.052 = 0.191 \text{ k}\Omega$$

7.4.5 COMPARISON OF CONFIGURATIONS

Before ending this section, it will be useful to compare the performance of the three BJT configurations. In this comparison, shown in Table 7.1, a BJT with $\beta = 100$ is assumed. An emitter current of 1.3 mA is also assumed ($r_e = 20 \ \Omega$).

The common emitter stage can have moderate to high input impedance, high voltage gain, and high current gain. The common base stage has low input impedance, high voltage gain, and low current gain (approximately unity). The emitter follower has moderate to high input impedance, low voltage gain (approximately unity), and high current gain. An outstanding feature of the emitter follower is its low output impedance.

The common emitter stage offers the best combination of parameters and is the most commonly used stage in discrete design. The common base stage is used in applications that require a reasonable voltage gain with no phase inversion, especially if the input source resistance is small. We will later see that this stage has good high-frequency characteristics

Table 7.1   Comparison of Three BJT Configurations.

Configuration	R_{in} ($R_E = 0$)	R_{in} ($R_E = 1$ kΩ)	R_{out} ($R_C = 2$ kΩ)	A_{MB} ($R_E = 0$)	A_{MB} ($R_E = 1$ kΩ)	A_I ($R_E = 1$ kΩ)
Common emitter	2 kΩ	102 kΩ	2 kΩ	−100	−2	100
Common base	20 Ω	1.02 kΩ	2 kΩ	+100	+2	0.99
Common collector (emitter follower)	—	102 kΩ	20 Ω	—	+1	101

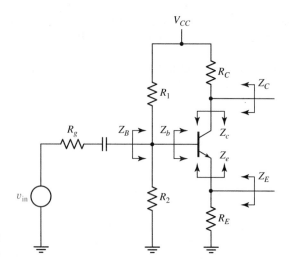

Figure 7.23
A BJT stage.

also. It is used often in integrated amplifier designs. The emitter follower is used to couple between a high source resistance and a small load resistance.

Several useful relationships can be summarized using the circuit of Fig. 7.23.

General Equations

$$\beta = \frac{i_c}{i_b} \qquad i_c = \beta i_b \qquad g_m = \frac{\alpha}{r_e}$$

$$r_e = \frac{26}{I_E(\text{mA})} \qquad r_{ce} = \frac{V_A}{\beta I_B} \qquad V_{BE}(\text{on}) = 0.7 \text{ V}$$

DC Bias Equations

$$E_{\text{Th}} = \frac{R_2}{R_1 + R_2} V_{CC}$$

$$R_{\text{Th}} = R_1 \parallel R_2$$

$$I_B = \frac{E_{\text{Th}} - 0.7}{R_{\text{Th}} + (\beta + 1)R_E}$$

Impedances Looking into the base:

$$Z_b = (\beta + 1)(R_E + r_e)$$

Overall input impedance:

$$Z_B = Z_b \parallel R_{\text{Th}}$$

Looking into the emitter:

$$Z_e = r_e + \frac{R_g \parallel R_{\text{Th}}}{\beta + 1} \approx r_e \approx \frac{1}{g_m}$$

Overall output impedance at the emitter:

$$Z_E = R_E \parallel Z_e$$

$I_E = (\beta + 1)I_B = \frac{I_e}{1-\alpha}$

Looking into the collector:
If R_E is bypassed with a capacitor,

$$Z_c = r_{ce}$$

If R_E is unbypassed,

$$Z_c = r_{ce}\left(1 + \frac{\beta R_E}{(\beta + 1)r_e + R_g + R_E}\right)$$

This value is generally much larger than r_{ce}.

Overall output impedance at the collector:
If R_E is bypassed with a capacitor,

$$Z_C = r_{ce} \parallel R_C$$

If R_E is unbypassed,

$$Z_C \approx R_C$$

Voltage Gain from Base to Collector

$$A_{BC} = -\frac{\beta R_C}{(\beta + 1)(r_e + R_E)} = -\frac{R_C}{R_E + 1/g_m} = -\frac{R_C}{r_e + R_E}$$

Voltage Gain from Base to Emitter

$$A_{BE} = \frac{R_E}{R_E + r_e} = \frac{R_E}{R_E + 1/g_m}$$

A thorough understanding of these equations allows BJT circuit analysis or design problems to be easily solved.

7.5 The BJT at High Frequencies

IMPORTANT Concepts

1. The major reactive effects in the BJT can be represented by depletion-region capacitances and a diffusion capacitance.
2. The Miller effect adversely affects the upper corner frequency of the common-emitter stage.
3. Several simplifications of the high-frequency BJT model allows manual analysis to be done.

At high frequencies certain effects in the BJT that have previously been neglected must be considered. In the hybrid-π circuit these effects can be taken into account by additional, small, shunt capacitors, even though the effects are due to more complicated mechanisms. At lower frequencies the capacitors become open circuits and can be neglected, reducing the equivalent circuit to the low-frequency hybrid-π of Fig. 7.19. The question of when these capacitive elements can be neglected must be answered in terms of the particular BJT of interest. For frequencies below a value called the β-cutoff frequency, f_β, the low-frequency circuit can normally be used with good accuracy. Near f_β and above, the reactive elements must be added to the equivalent circuit. The f_β of a BJT is, however, not an absolute measure of the BJT's frequency performance. One circuit configuration might allow the BJT to amplify at frequencies far above f_β, whereas another circuit arrangement would allow amplification only of frequencies lower than f_β.

7.5.1 REACTIVE EFFECTS

The reactive effects associated with the BJT are the same as those discussed in relation to a diode in Chapter 5. The charge that must be removed or supplied to the depletion region, when the voltage across the region is changed, gives rise to an effective depletion-region capacitance. There will be a depletion-region capacitance associated with the collector-base junction called C_μ and another associated with the emitter-base junction called $C_{b'e}$. From Eq. (5.32) these capacitances are seen to be a function of the voltages applied to the respective junctions; that is,

$$C_T = \frac{K_1}{(\psi_o - V)^m} \tag{7.46}$$

where K_1 is the zero-bias transition capacitance of the junction. Spice uses C_{JE} for the emitter-base and C_{JC} for the collector-base zero-bias values.

Under small-signal conditions, the dc junction voltage will be much larger than the ac component of voltage. The capacitance then will be determined by the dc bias voltage and will be unaffected by the signal voltage. For large-signal (switching) circuits an average value of capacitance can be calculated for each junction [see Eq. (5.39)].

Typical values of depletion-layer capacitance might range from 0.5 picofarad (1 pF $= 10^{-12}$ F) for very-high-frequency BJTs to 50 pF for low-frequency, high-power BJTs.

In addition to the depletion-layer capacitances, there is a diffusion capacitance that must be included in the equivalent circuit also. This capacitance appears across the emitter-base junction and accounts for the change in charge that must accompany the change in minority-carrier gradient when the junction voltage is modified. We will derive this diffusion capacitance here for the BJT.

The hole distribution in the base region of a *pnp* BJT is shown in Fig. 7.24 for a BJT in the active region. This figure assumes that the emitter is doped much more heavily than is the base region, and free electrons in the emitter region can be neglected. As the emitter-base junction voltage increases or decreases, the hole concentration at the junction does likewise, as shown by the dashed lines.

The total charge required to support the gradient of holes in the base region can be written as

$$Q_B = \frac{1}{2} q W p(0) A \tag{7.47}$$

Figure 7.24
Minority carrier distribution in pnp BJT base.

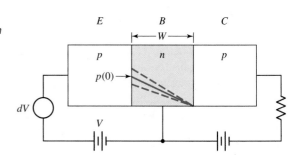

where W is the width of the effective base region and A is the cross-sectional area of the base region (assumed uniform). The concentration of holes injected from emitter into the base region is given by

$$p(0) = p_{bo}e^{qV_{EB}/kT} = p_{bo}e^{V_{EB}/V_T} \qquad (7.48)$$

Equation (7.47) can now be written as

$$Q_B = \frac{1}{2}qWAp_{bo}e^{V_{EB}/V_T} \qquad (7.49)$$

The capacitance, which is the change in charge with change in voltage, is then

$$C_\pi = \frac{dQ_B}{dV_{EB}} = \frac{qWA}{2V_T}p_{bo}e^{V_{EB}/V_T} \qquad (7.50)$$

In general the emitter-base bias voltage is unknown, but the emitter-bias current is known. It is more meaningful then to relate the diffusion capacitance to the dc emitter current rather than to the voltage. The emitter current can be written as (assuming only hole conduction)

$$I_E = \frac{qD_pA}{W}p_{bo}e^{V_{EB}/V_T} \qquad (7.51)$$

Using this equation to eliminate the exponential term from Eq. (7.50) we obtain

$$C_\pi = \frac{W^2}{2D_p} \times \frac{I_E}{V_T} = \frac{W^2}{2D_p} \times \frac{1}{r_e} \qquad (7.52)$$

The diffusion capacitance is seen to depend quite strongly on the base width of the BJT and also on the dc emitter current. A narrow base width is very important to high-frequency operation, since C_π must be made small. The impurity diffusion technique has permitted a great advancement in base-width control, which has resulted in higher-frequency BJTs. Unfortunately, the narrow base region introduces more ohmic resistance for the current flowing into the base contact and into the base region. This effect is modeled in the high-frequency circuit by the ohmic base resistance, r_x.

The high-frequency equivalent circuit can be constructed as shown in Fig. 7.25. This circuit includes the two depletion-layer capacitances, the diffusion capacitance, and the ohmic base resistance, as well as the low-frequency elements of the hybrid-π equivalent circuit. There are two immediate modifications that can be made to this circuit. Noting that C_π might range from 100 to several thousand picofarads for various BJTs, $C_{b'e}$ is neglected, since it will be so much smaller than C_π. The second modification is done as a matter of convenience. Rather than leave the current generator in terms of the current through r_π, we write its value in terms of the voltage v_π. Since this current is $i = v_\pi/(\beta+1)r_e$, the current generator will

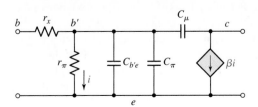

Figure 7.25
High-frequency BJT equivalent circuit.

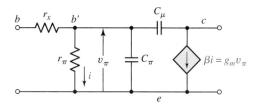

Figure 7.26
Simplified equivalent circuit.

have a value of $\beta i = (\alpha/r_e)v_\pi = g_m v_\pi$. The modified circuit is shown in Fig. 7.26. Both versions of the circuit (Figs. 7.25 and 7.26) are called the hybrid-π equivalent circuit.

As the frequency increases and the impedance of the capacitors becomes less, v_π is decreased and the collector current becomes less. We might note at this point that no output resistance is shown in the equivalent circuit. Even though r_{out} will be present, it is almost always negligible in the high-frequency circuit because the collector load resistance, R_C, will necessarily be small in discrete stages in order to achieve a large bandwidth.

It should be realized that the high-frequency equivalent circuit of Fig. 7.26 also applies to the common-emitter circuit with an external emitter resistor that is bypassed by a capacitor. In terms of ac signals the emitter is also at ground potential in this case.

Although these derivations were done for a "textbook" BJT, the basic concepts apply to diffused devices with nonuniform doping profiles, resulting in a similar equivalent circuit.

7.5.2 SIMPLIFICATION OF THE HIGH-FREQUENCY CIRCUIT

The circuit of Fig. 7.26 is quite difficult to analyze manually when a source and load are applied. It then becomes a three-loop problem with C_μ linking the input and output loops. It is quite time-consuming to analyze a three-loop problem manually. Thus, it seems reasonable to simplify the circuit, if such is possible with little loss of accuracy. It is appropriate to mention at this point that if the digital computer is used to analyze the network there is little need for further simplification. Analysis of a three-loop problem requires only slightly more time, and accuracy is guaranteed by programs such as pSpice©. For hand calculation, the time required to solve three simultaneous loop equations is considerably more than that required for the solution of two equations. Furthermore, the probability of error is much greater for the three-loop problem.

It is possible to simplify the circuit by reflecting the effect of C_μ to the input loop and to the output loop. This simplification was discussed as the Miller effect in Chapter 2. The current actually drawn by C_μ for a given applied input voltage can be found; then an additional element can be placed in the input loop to draw this same current. The troublesome element C_μ can then be removed with its effect now present due to the added element. The circuit becomes a two-loop problem that can more easily be analyzed. Before we conclude that this method of simplification is entirely valid we must note that C_μ has some direct loading effect on the output circuit. We saw in Chapter 2 that when the capacitor is multiplied and reflected to the input, a value of C_μ should also be reflected to the output. In narrowband applications wherein the resonant collector impedance is large or in some IC amplifier stages, it is essential to reflect the effects of C_μ to the output loop of the circuit. On the other hand, it has been shown that the loading on the output circuit is negligible for discrete video or wideband stages, since normal values of load resistance are quite small. Consequently, for the wideband amplifiers to be discussed we will reflect the effects of C_μ to the input circuit, noting that loading by C_μ on the output loop is negligible. In other words, the dominant pole of a discrete wideband amplifier is the pole caused by the input loop rather than the pole caused by the output loop.

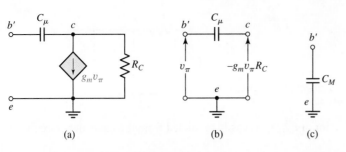

Figure 7.27
*Circuits used to find C_M:
(a) equivalent circuit to the right
of points b' and e, (b) voltage
appearing across C_μ,
(c) equivalent Miller
capacitance.*

Effect of C_μ on the Input Circuit When a voltage is applied to the input, a certain charge must be supplied by the source to charge C_μ. This charge can be calculated and the proper value of capacitance can be connected between b' and e to replace C_μ. This equivalent capacitance will be denoted by C_M. The circuit of Fig. 7.27 is helpful in finding the value of C_M when R_C is present.

When the voltage between b' and ground equals v_π, the collector voltage will be

$$v_c = -g_m v_\pi R_C \tag{7.53}$$

The total voltage across the capacitor is $v_{b'c} = v_{b'} - v_c$ or

$$v_{b'c} = v_\pi (1 + g_m R_C) \tag{7.54}$$

The voltage v_π causes a voltage of $v_\pi (1 + g_m R_C)$ to appear across the capacitor, because the collector voltage is dependent on the voltage v_π. The charge that must be supplied to C_μ is

$$Q_{b'c} = C_\mu v_{b'c} = C_\mu v_\pi (1 + g_m R_C) \tag{7.55}$$

This charge is exactly the same as that delivered to a capacitor across the points b' and e of value

$$C_M = C_\mu (1 + g_m R_C) \tag{7.56}$$

In terms of the input circuit, the equivalent circuit can then be drawn as shown in Fig. 7.28. Since the input loop is isolated from the output loop, the circuit is called unilateral. It is sometimes referred to as the hybrid-y equivalent circuit.

The next step in the development of the high-frequency circuit is to relate the element values to manufacturer's specified values or to known quantities. The circuit is quite useless from a design standpoint unless we know the element values.

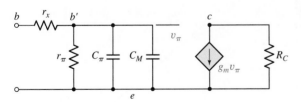

Figure 7.28
Unilateral equivalent circuit.

7.5.3 SHORT-CIRCUIT CURRENT GAIN

An examination of the behavior of the short-circuit current gain as frequency is varied is made by the manufacturer. From this examination, the β-cutoff frequency, f_β, or the current gain-bandwidth figure, f_t, can be specified. The measurements are taken with $R_C = 0$ and using a current source input, as shown in Fig. 7.29.

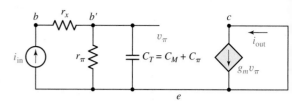

Figure 7.29
Measurement of short-circuit current gain.

The behavior of $i_{\text{out}}/i_{\text{in}}$ is easily predicted from the equivalent circuit. Since

$$v_\pi = i_{\text{in}} \frac{r_\pi}{1 + j\omega r_\pi C_T}$$

the output current is

$$i_{\text{out}} = g_m v_\pi = \frac{\alpha}{r_e} v_\pi = \frac{(\alpha/r_e)(\beta + 1)r_e}{1 + j\omega r_\pi C_T} i_{\text{in}} = \frac{\beta}{1 + j\omega r_\pi C_T} i_{\text{in}}$$

Because $R_C = 0$, the Miller capacitance is $C_M = C_\mu$, which can usually be neglected when compared to C_π. The short-circuit current gain is then

$$\frac{i_{\text{out}}}{i_{\text{in}}} = \frac{\beta}{1 + j\omega r_\pi C_\pi} \tag{7.57}$$

At low frequencies the current gain is β, whereas at high frequencies the gain drops asymptotically at -6 dB / octave. The graph of Fig. 7.30 shows this variation. The frequency where the gain is down 3 dB is defined as the β-cutoff frequency and is

$$f_\beta = \frac{1}{2\pi r_\pi (C_\pi + C_\mu)} \approx \frac{1}{2\pi r_\pi C_\pi} \tag{7.58}$$

The point in frequency at which the current gain drops to a value of unity (0 dB) is called f_t. From the curve, this value is seen to be β times f_β (Problem 7.42) or

$$f_t = \beta f_B \approx \frac{1}{2\pi r_e C_\pi} \tag{7.59}$$

Usually, f_t is the quantity supplied by the manufacturer rather than f_β. Knowing this value along with the value of r_e allows the calculation of C_π.

Figure 7.30
Short-circuit current gain as a function of frequency.

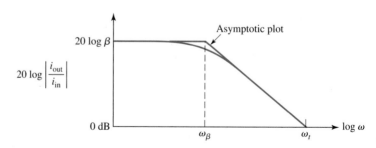

7.5.4 A PRACTICAL AMPLIFIER STAGE

In practice, a voltage source with a finite source resistance will be used and a nonzero value will be used for the load resistor R_C. The source resistance might result from a signal

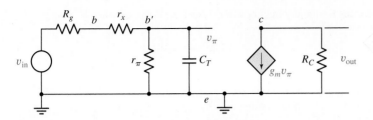

Figure 7.31
Equivalent circuit for practical amplifier.

generator or the output resistance of a previous stage. In any case, the circuit will appear as shown in Fig. 7.31. Since there is a finite value of R_C, the total capacitance in the input loop will be given by

$$C_T = C_\pi + C_M = C_\pi + (1 + g_m R_C)C_\mu \qquad (7.60)$$

The formal method of calculating the voltage gain is to write it as a function of frequency, and the corner frequency can then be found from this equation. We can start this method by writing

$$v_{out} = -g_m v_\pi R_C$$

The voltage v_π can be expressed in terms of v_{in} as

$$v_\pi = \frac{r_\pi}{(r_\pi + R_g + r_x)[1 + j\omega DC_\pi r_\pi \| (R_g + r_x)]} v_{in}$$

Substituting this expression into the previous equation gives a voltage gain of

$$A_o = \frac{-\beta R_C}{(r_\pi + r_x + R_g)[1 + j\omega C_T r_\pi \| (R_g + r_x)]} \qquad (7.61)$$

The corner frequency of the voltage gain is the frequency where the imaginary part of the denominator equals the unity real part or

$$\omega_2 = \frac{1}{C_T R_{Th}} \qquad (7.62)$$

where C_T is found by Eq. (7.60) and R_{Th} is the parallel combination of r_π and $R_g + r_x$.

Equation (7.61) can be arrived at more simply by evaluating the midband gain and then finding the corner frequency of v_π. Since the output current and voltage are proportional to v_π, the voltage gain will have the same corner frequency as v_π. The midband gain is

$$A_{MB} = \frac{-\beta R_C}{r_\pi + r_x + R_g} \qquad (7.63)$$

The corner frequency of v_π is found by taking a Thévenin equivalent of $R_g + r_x$ and r_π. The resulting circuit is shown in Fig. 7.32. The equivalent resistance is

$$R_{Th} = \frac{r_\pi(R_g + r_x)}{r_\pi + r_x + R_g}$$

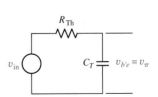

Figure 7.32
Equivalent circuit of input loop.

The corner frequency occurs when $1/\omega_2 C_T = R_{\text{Th}}$ or when

$$\omega_2 = \frac{1}{C_T \frac{r_\pi(R_g+r_x)}{r_\pi+R_g+r_x}} \tag{7.64}$$

This expression is equivalent to that given by Eq. (7.62).

Equation (7.61) can be written as

$$A_o = \frac{-\beta R_C}{[r_\pi + r_x + R_g][1 + j(\omega/\omega_2)]} \tag{7.65}$$

The source resistance affects both the midband gain and the upper corner frequency of the stage. When the input circuit is driven by a current source of infinite resistance, the time constant of v_π is simply

$$\tau = r_\pi C_T$$

As the source resistance is lowered to a finite value, the time constant is decreased to

$$\tau = \frac{r_\pi(R_g + r_x)}{r_\pi + r_x + R_g} C_T$$

The time constant of the circuit is the reciprocal of the radian bandwidth, so as τ decreases, the corner frequency increases.

PRACTICAL Considerations

For manual calculations, two constants are defined to simplify the calculations of upper corner frequency. The upper corner frequency of the common-emitter stage can be written as

$$f_2 = \frac{B}{D} f_\beta$$

where the D-factor is the ratio of C_T to C_π and B is called the broadband factor. This factor is given by

$$B = \frac{r_\pi + r_x + R_g}{r_x + R_g}$$

It can be shown that the D-factor can be approximated as

$$D = 1 + \omega_t R_C C_\mu$$

The Miller effect is accounted for by the D-factor, and the broadband factor accounts for the increase in bandwidth due to a finite generator resistance. Note that both factors depend on BJT specifications and amplifier component values. Calculation of f_2 is quite simple using these factors.

7.5.5 USE OF THE HIGH-FREQUENCY CIRCUIT

It is often mistakenly thought that a knowledge of the β-cutoff frequency, f_β, or the current gain-bandwidth product, f_t, is all that is required to predict the corner frequency. This is incorrect, as seen in the last two sections, due to

1. The Miller effect

2. The effect of the generator resistance on the input-circuit bandwidth

The hybrid-π circuit is quite useful not only in small-signal design, but also in switching-circuit design. Even though both r_π and C_π depend on the emitter current, the time constant and bandwidth are relatively independent of I_E. The product of r_π and C_π is independent of I_E, since r_π varies inversely, and C_π varies directly, with I_E.

The collector depletion-layer capacitance is given by the manufacturer and is measured at a specified voltage. In the amplifier circuit, the BJT might be used at a collector-base voltage different from that for which C_μ is given. For most high-frequency BJTs, which are manufactured by a diffusion process, the capacitance is approximated by

$$C_\mu = \frac{K_2}{(\psi_o - V_{CB})^{1/3}} \tag{7.66}$$

The constant K_2 can be evaluated at the specified voltage, and C_μ can then be calculated for the value of V_{CB} that is to be used in the circuit.

EXAMPLE 7.10

A BJT has the following parameters: $\alpha = 0.99$, $r_x = 200\ \Omega$, $C_\mu = 5$ pF at the collector-base voltage used for the stage, and $f_t = 10$ MHz. The BJT is used in the common-emitter configuration with $R_C = 1\ k\Omega$ and is driven by a source with a resistance of 500 Ω. Given that I_E is 1 mA, find the high-frequency 3-dB point (corner frequency) and the midband voltage gain.

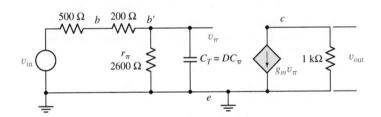

Figure 7.33
Equivalent circuit for
Example 7.10.

SOLUTION   The equivalent circuit is shown in Fig. 7.33. The midband gain is

$$A_{MB} = \frac{-\beta R_C}{r_\pi + r_x + R_g} = \frac{-99 \times 10^3}{2600 + 200 + 500} = -30 \text{ V/V}$$

The B-factor and D-factor can be found in order to calculate f_2:

$$B = \frac{r_\pi + r_x + R_g}{r_x + R_g} = \frac{3300}{700} = 4.71$$

$$D = 1 + R_C C_\mu \omega_t = 1 \times 10^3 \times 5 \times 10^{-12} \times 2\pi \times 10^7 = 1.31$$

The corner frequency is then

$$f_2 = \frac{B}{D} f_\beta = \frac{B}{D} \frac{f_t}{\beta} = \frac{4.71 \times 10^7}{1.31 \times 99} = 363 \text{ kHz}$$

If the B-factor and D-factor are not used, the total capacitance must be found along with R_{Th} as shown in Fig. 7.32. The diffusion capacitance is found from Eq. (7.59) to be

$$C_\pi = \frac{1}{2\pi r_e f_t} = 612 \text{ pF}$$

PRACTICE Problems

7.17 Rework Example 7.10 if R_C is changed to 500 Ω. *Ans:* −15 V/V, 410 kHz.
7.18 Rework Example 7.10 if $R_C = 500$ Ω, $C_\mu = 2$ pF, and $f_t = 120$ MHz. All other parameters remain the same. *Ans:* −15 V/V, 3.26 MHz.

The Miller capacitance is calculated from Eq. (7.56), giving

$$C_M = C_\mu(1 + g_m R_C) = 195 \text{ pF}$$

The total capacitance in the input loop is $C_T = 807$ pF.

The Thévenin resistance seen by the capacitance C_M is r_π (2600 Ω) in parallel with $R_g + r_x$ (700 Ω), which gives a value of $R_{Th} = 552$ Ω. The upper corner frequency is now given by

$$f_2 = \frac{1}{2\pi R_{Th}C_T} = 357 \text{ kHz}$$

Although this answer compares with the earlier answer, the use of the D-factor and B-factor is much easier.

DISCUSSION OF THE DEMONSTRATION PROBLEM

We will first consider the bias of the circuit for the demonstration problem, repeated here for convenience. If the maximum peak value of the output signal is to be 1 V, an incremental current of 5 mA must flow through R_L. The largest incremental current produced by the emitter follower equals the quiescent emitter current of this stage. This conclusion results because the maximum change in peak current occurs when the emitter current changes from its quiescent value to zero current. For this reason, the quiescent current of an output stage must always exceed the maximum required incremental current to the load. It therefore appears reasonable to set the emitter current of the output stage to 8 mA. The resistance R_{E2} can be selected as 1 kΩ, which leads to a quiescent emitter voltage on $Q2$ of 8 V.

We cannot bias $Q1$ yet because we do not have values of R_C and R_{E1}. Let us pick $R_C = 4$ kΩ. Since the base voltage of $Q2$ must equal about 8.7 V to establish an emitter voltage of 8 V, the collector current through R_C must lead to a 3.3-V drop across this resistance, which leads to a current of

$$I_{R_C} = \frac{3.3}{4} = 0.825 \text{ mA}$$

In order to select an appropriate value for R_{E1}, we recognize that a voltage gain from base to collector of $Q1$ must exceed the overall gain of the amplifier, because of the attenuation of

Amplifier for demonstration problem.

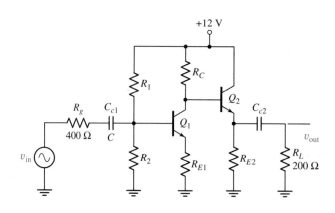

overall gain by input loop loading on the generator resistance. The gain of the output stage is near unity since this is an emitter follower. A more accurate expression of the base-to-emitter voltage gain of the output stage is

$$A_2 = \frac{R_{EQ}}{R_{EQ} + r_{e2}}$$

Taking $R_{EQ} = R_L \parallel R_{E2} = 167\ \Omega$ and $r_{e2} = 26/I_{E2} = 3.25\ \Omega$ leads to a value of $A_2 = 0.98$ V/V.

For an overall voltage gain of -22 V/V, a base-to-collector gain of perhaps -26 V/V is sufficient and establishes a constraint on R_{E1}, since

$$A_{BC} = -\frac{\alpha R_C}{r_{e1} + R_{E1}}$$

The incremental resistance, r_{e1}, is given by

$$r_{e1} = \frac{26}{I_{E1}} = \frac{26}{0.825} = 31.5\ \Omega$$

Equating A_{BC} to -26 and solving for R_{E1} results in $R_{E1} = 121\ \Omega$.

We will now choose the lower bias resistance, R_2, to equal 2.4 kΩ. For good stability, this resistance should not be much greater than 20 times R_{E1}. The dc base voltage of $Q1$ is calculated as

$$V_{B1} = R_{E1}I_{E1} + V_{BE(on)} = 0.825 \times 0.121 + 0.7 = 0.8\ \text{V}$$

Note here that I_{E1} was approximated as I_{C1}.

The dc current through R_2 is

$$I_2 = \frac{V_{B1}}{R_2} = \frac{0.8}{2.4} = 0.333\ \text{mA}$$

The base current of $Q1$ is, assuming $\beta = 180$

$$I_{B1} = \frac{I_{C1}}{\beta} = \frac{0.825}{180} = 0.0046\ \text{mA}$$

The total current that must be provided by R_1 is the sum of I_2 and I_{B1}, which is 0.338 mA. The resistance R_1 is now calculated as

$$R_1 = \frac{V_{CC} - V_{B1}}{I_1} = \frac{11.2}{0.338} = 33.1\ \text{k}\Omega$$

Having completed the bias design of the amplifier, we draw the midband incremental equivalent circuit of Fig. 7.34. Using $R_B = R_1 \parallel R_2$ and $R_{in1} = R_B \parallel [(\beta + 1)(R_{E1} + r_{e1})]$, we can calculate the midband gain as

$$A_{MBo} = \frac{R_{in1}}{R_{in1} + R_g} A_{BC} A_2 = \frac{2.07}{2.07 + 0.4} \times (-26) \times 0.98 = -21.4\ \text{V/V}$$

This selection of element values appears to satisfy the voltage gain specification.

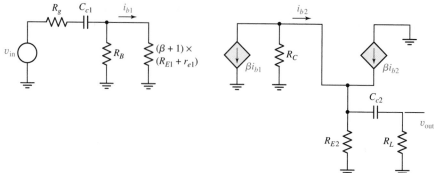

Figure 7.34
Equivalent circuit
for the demonstration
problem.

There will be two lower corner frequencies for this amplifier; one arising from the input loop and one arising from the output loop. The input loop corner frequency is

$$f_{low-in} = \frac{1}{2\pi C_{c1}(R_g + R_{in1})}$$

We will arbitrarily set this value to 6 Hz so it will have little effect on the overall corner frequency of 40 Hz; this requires a coupling capacitor of $C_{c1} = 10.7 \; \mu F$.

The output corner frequency is determined by the resistance seen by C_{c2}. This resistance is the sum of R_L and the output resistance looking into the emitter of $Q2$. This output resistance is

$$r_{out2} = R_{E2} \parallel \left(r_{e2} + \frac{R_{C1}}{(\beta + 1)} \right) = 25 \; \Omega$$

The output loop corner frequency is then

$$f_{low-out} = \frac{1}{2\pi C_{c2}(r_{out2} + R_L)}$$

We will set this value to 36 Hz, which requires a value of $C_{c2} = 19.8 \; \mu F$. Thus a pair of 20-μF capacitors could safely be used for this design.

SUMMARY

➤ The BJT transistor must be biased into the active region before it can amplify an incremental signal.

➤ The input signal modifies the base-emitter voltage, which results in a variation in collector and emitter current. The output signal is generated by channeling the collector or emitter current through a load resistance.

➤ The common-base configuration can have a high voltage gain, but the current gain is near unity. The common-emitter configuration can have high current gain and high voltage gain. The common-collector configuration or the emitter follower has a high current gain, but the voltage gain is near unity.

➤ The BJT can be biased into its active region using the emitter-bias scheme. This method results in good stability of operating point even if large β changes take place.

➤ After biasing the BJT properly, the incremental equivalent circuit can be used to calculate voltage gain, input impedance, output impedance, or current gain.

➤ The high-frequency equivalent circuit can be modified using the Miller effect to obtain a simplified, unilateral equivalent circuit.

PROBLEMS

SECTION 7.1.1 BJT BEHAVIOR

7.1 Plot Eq. (7.1) for $I_S = 10^{-15}$ A and $T = 300°$K. Use different scales for negative and positive currents.

7.2 Calculate the current flowing through an ideal silicon diode at room temperature if $I_S = 0.0001$ nA and a forward bias of 0.55 V is applied. Repeat for a forward bias of 0.75 V.

SECTION 7.1.2 COMMON-BASE CONFIGURATION

7.3 Given that the BJT in the figure has a value of $\alpha = 0.995$, find the output voltage, including dc and ac components, when

(a) $I_E = (1.0 + 0.5 \sin \omega t)$ mA

(b) $I_E = (2.0 + 0.5 \sin \omega t)$ mA

Compare the ac output voltages in (a) and (b) and explain any differences.

7.4 Repeat Problem 7.3 for $I_E = (2.0 + 2.0 \sin \omega t)$ mA.

7.5 Repeat Problem 7.3 for $I_E = (6.0 + 2.0 \sin \omega t)$ mA.

Figure P7.3

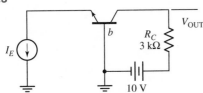

SECTION 7.1.3 COMMON-EMITTER CONFIGURATION

7.6 Given that the BJT in the figure has a value of $\beta = 150$, find the output voltage, including dc and ac components, when

(a) $I_B = (20 + 10 \sin \omega t)$ μA

(b) $I_B = (40 + 10 \sin \omega t)$ μA

Compare the ac output voltages in (a) and (b), and explain any differences.

Figure P7.6

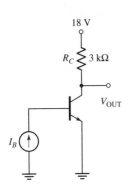

7.7 Repeat Problem 7.6 for $I_B = (15 + 15 \sin \omega t)$ μA.

7.8 Repeat Problem 7.6 for $I_B = (40 + 25 \sin \omega t)$ μA.

7.9 Assuming that the BJT in Problem 7.6 has a value of $\beta = 120$, find the output voltage and the ac voltage gain of the stage for $I_B = (20 + 5 \sin \omega t)$ μA. Sketch both input voltage and output voltage waveforms. Assume that the resulting ac input voltage has a peak value of 10 mV.

7.10 Repeat Problem 7.9, changing the load resistor from 3 kΩ to 4.5 kΩ. How does ac voltage gain vary with R_C?

7.11 Repeat Problem 7.9, changing the load resistor from 3 kΩ to 0.5 kΩ. How does ac voltage gain vary with R_C?

7.12 The short-circuit ac current gain from base to collector is defined as

$$\beta = \frac{\Delta I_C}{\Delta I_B} \qquad \text{with } V_{CE} \text{ constant}$$

Taking ΔI_B as 10 μA, use the characteristics of Fig. 7.9 to evaluate β at $V_{CE} = +10$ V and

(a) $I_C = 1$ mA

(b) $I_C = 2$ mA

(c) $I_C = 3$ mA

(d) $I_C = 4$ mA

7.13 Work Problem 7.12, given that I_C is 3 mA and

(a) $V_{CE} = 5$ V

(b) $V_{CE} = 10$ V

☆ **7.14** Assume that $\beta = 100$ and $V_{BE(on)} = 0.6$ V for the BJT shown. Assume that $V_A = \infty$.

(a) Find the quiescent output voltage and current.

(b) Sketch the collector characteristics for I_B of 0, 10 μA, 20 μA, 30 μA, and 40 μA.

(c) Draw the load line and locate the Q-point.

Now assume that $V_A = 60$ V.

(d) Calculate the value of r_{ce} at $I_C = 1.5$ mA.

Figure P7.14

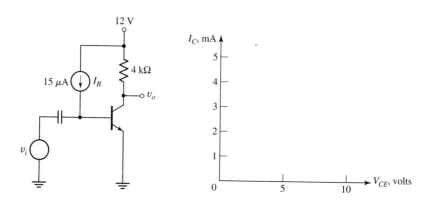

SECTION 7.2 GRAPHICAL ANALYSIS OF THE BJT

7.15 In the circuit shown in the figure, β is assumed to be 100. Find the output ac voltage, given that $I_B = (20 + 5 \sin \omega t)\ \mu$A.

If the dc β is assumed to equal 100 also, what is the quiescent (dc) voltage?

7.16 If I_B in Problem 7.15 is $(36 + A \sin \omega t)\ \mu$A, what is the maximum value that A can have and still keep the BJT from saturating at the peak of the input current swing? Assume that $\beta = \beta_{dc} = 100$.

7.17 If I_B in Problem 7.15 is $(6 + B \sin \omega t)\ \mu$A, what is the maximum value that B can have and still keep the BJT from cutting off at the negative peak of the input current swing?

Figure P7.15

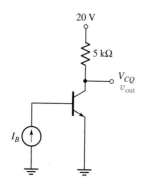

SECTION 7.3.2 BASE-CURRENT BIAS

D 7.18 In the base-current bias circuit of Fig. 7.14, $V_{CC} = 20$ V, $R_C = 5$ kΩ, $\beta = 100$, and $V_{BE} = 0.6$ V. Select R_B to set V_{CQ} at

(a) the midpoint of the active region

(b) 16 V

(c) 5 V

D 7.19 Repeat Problem 7.18 if β is changed to 200.

SECTION 7.3.3 EMITTER BIAS

D 7.20 Select R_1 for the emitter-bias circuit of Fig. 7.16 with $R_C = 8.2$ kΩ, $R_E = 1$ kΩ, $R_2 = 20$ kΩ, and $V_{CC} = 12$ V. Assume that $\beta = 100$ and $V_{BE} = 0.7$ V. Select R_1 to set V_{CQ} at the midpoint of the active region. Find the maximum symmetrical peak-to-peak output voltage that can be obtained before saturation or cutoff occurs.

D 7.21 Rework Problem 7.20 for $V_{CQ} = 4$ V.

D 7.22 Rework Problem 7.20 for $V_{CQ} = 9$ V.

D 7.23 The emitter-bias circuit of Fig. 7.16 has $R_C = 8.2$ kΩ, $R_E = 1$ kΩ, $V_{CC} = 20$ V, $\beta = 100$, and $V_{BE} = 0.7$ V. Choose reasonable values of R_1 and R_2 to set V_{CQ} to the midpoint of the active region.

D 7.24 An emitter-bias circuit (see Fig. 7.16) having $R_C = 6.8$ kΩ, $V_{CC} = 12$ V, and $R_E = 750$ Ω is to be capable of delivering a sinusoidal output signal of 3-V amplitude. Given that a BJT is used with $\beta = 60$ and $V_{BE} = 0.7$ V, find the upper and lower limits on V_{CQ} to satisfy the design. Choose reasonable values of R_1 and R_2 that will result in the lower of these two values.

SECTION 7.4.1 THE COMMON-BASE CONFIGURATION

7.25 Evaluate the midband voltage gain of the common-base stage shown. Assume that $\alpha = 0.995$ and $r_e = 20\Omega$.

Figure P7.25

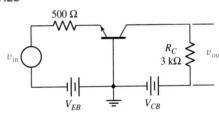

SECTION 7.4.2 AND 7.4.4 THE COMMON-EMITTER CONFIGURATION

7.26 Calculate the midband voltage gain of the circuit shown if $\alpha = 0.99$ and $I_E = 2$ mA. Calculate the input impedance of the stage. What will the maximum peak-to-peak output signal be when cutoff is reached?

Figure P7.26

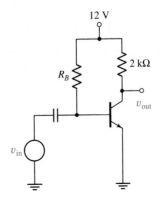

7.27 Repeat Problem 7.26 if I_E is changed to 1 mA.

7.28 If $\beta = 100$, calculate the midband voltage gain and sketch the output voltage including the dc component for the circuit shown.

Figure P7.28

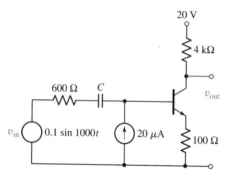

7.29 If I_{BQ} is reduced to 5 μA in Problem 7.28, calculate the midband voltage gain. Compare to the gain when $I_{BQ} = 30 \mu$A.

7.30 Calculate the midband voltage gain of the circuit shown if $\beta = 100$ and $I_E = 2$ mA. Calculate the input and output impedances of the circuit.

Figure P7.30

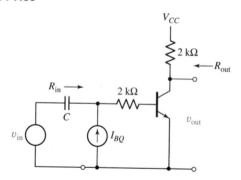

7.31 Repeat Problem 7.30 if a 100-Ω resistance is inserted between emitter and ground.

7.32 If $V_{BE(on)} = 0.6$ V, calculate the midband voltage gain for the circuit shown.

Figure P7.32

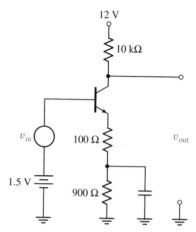

SECTION 7.4.3 THE EMITTER FOLLOWER

7.33 If $V_{BE(on)} = 0.6$ V and $\beta = 100$, calculate the midband voltage gain for the circuits shown.

Figure P7.33

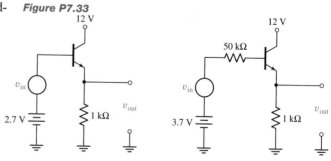

SECTION 7.5.1 REACTIVE EFFECTS

7.34 The small-signal depletion-layer capacitance for an abrupt junction is given by

$$C = K_1/\sqrt{\psi_o - V}$$

where K_1 is a constant, ψ_o is the barrier voltage, and V is the applied junction voltage (V has a positive value for forward-bias and a negative value for reverse bias). Given that the collector-base capacitance is 10 pF for a reverse bias of 6 V and that $\psi_o = 0.8V$, find the capacitance for a 12-V reverse bias.

7.35 Repeat Problem 7.34 for the case of a graded junction where $C = K_2/(\psi_o - V)^{1/3}$.

7.36 Using the formula given in Problem 7.34, evaluate C at 6-V reverse bias, assuming that $C = 10$ pF at 18 V.

7.37 Using the formula given in Problem 7.35, evaluate C at 6-V reverse bias, assuming that $C = 10$ pF at 18 V.

7.38 Given that ω_t is 10^8 rad/s, find the diffusion capacitance C_π for emitter currents of

(a) $I_E = 1$ mA

(a) $I_E = 5$ mA

7.39 Repeat Problem 7.38 for a BJT with $\omega_t = 2 \times 10^9$ rad/s.

SECTION 7.5.2–7.5.6 THE HIGH-FREQUENCY BJT STAGE

☆ **7.40** The BJT shown in the figure is an abrupt junction BJT with $\omega_t = 6.28 \times 10^9$ rad/s, $\beta = 70$, and $C_\mu = 20$ pF at $V_{CB} = 6$ V. Assuming that $\psi_o = 0.8$ V, find the bandwidth of v_{out}/i_{in} when $I_E = 1$ mA.

Figure P7.40

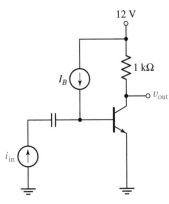

7.41 Repeat Problem 7.40, given that I_E is changed to 4 mA.

7.42 Assuming that f_β is defined as the frequency at which the short-circuit base-to-collector current gain is down 3 dB from the midband value and that f_t is defined as the frequency at which this gain is equal to unity, prove that $f_t = \beta \times f_\beta$.

☆ **7.43** When the diffusion capacitance decreases and can no longer be considered to be much greater than the depletionlayer capacitance, the formula for f_t is

$$f_t = \frac{1}{2\pi r_e(C_\pi + C_{b'e} + C_\mu)}$$

Explain why this quantity decreases at low emitter-bias currents.

7.44 The BJT in the circuit shown in the figure has the following parameters: $\beta = 100$, $r_x = 100\ \Omega$, $V_{BE} = 0.6$ V, $f_t = 10^7$ Hz, and $C_\mu = 10$ pF at the bias point used. Calculate the midband voltage gain and the upper corner frequency of the circuit. Neglect the loading effect of the bias resistors.

Figure P7.44

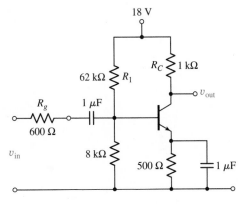

☆ **7.45** Repeat Problem 7.44, given that $f_t = 10^8$ Hz. Explain why the upper corner frequency does not increase by the same factor as f_t increases.

7.46 Repeat Problem 7.44, given that $f_t = 10^8$ Hz, $C_\mu = 5$ pF, and R_C is lowered to 200 Ω.

7.47 Repeat Problem 7.44, given that R_g is lowered to 200 Ω and R_C is lowered to 250 Ω.

☆ **7.48** Find the Miller-effect capacitances for Problems 7.44 and 7.45. Find C_π for both problems and compare to C_M. Calculate the D-factor for both cases.

7.49 Repeat Problem 7.44 for the case of $R_g = 3.6$ kΩ.

7.50 Repeat Problem 7.45 for the case of $R_g = 3.6$ kΩ.

7.51 Repeat Problem 7.45, assuming that R_1 is changed to 100 kΩ and that C_1 remains constant at 10 pF.

Integrated Circuit Design | 8

P rior to the early 1960s, all electronic circuits were constructed from discrete component circuits. Resistors, capacitors, inductors, and transistors were packaged individually to be assembled into a functioning electronic circuit on some carrier such as a printed circuit board (PCB). The discrete component circuit is still required in certain applications today, but the integrated circuit or IC is now the dominant form of implementation for an electronic system. There are major differences in the design principles used for the IC and the discrete circuit. In the next sections we explore many of these differences.

There are several classifications of the IC. One classification is based on the approximate number of components or elements contained in the circuit. Small-scale integration (SSI) refers to a circuit consisting of one to about 100 elements. The early logic ICs of the 1960s were primarily SSI circuits. Within a short time, medium-scale integration (MSI) became available. Circuits consisting of 100 to 1000 elements fall into this category. By the early 1970s, large-scale integration (LSI) was developed and used to create basic microprocessor chips and chips to control hand-held electronic calculators. Somewhere between 1000 and 10,000 elements may appear in an LSI circuit. Very-large-scale integration (VLSI) was necessary for the more advanced microprocessors that power the personal computer. This type of IC may contain between 10,000 and 1,000,000 elements. Systems that exceed 1,000,000 devices are often fabricated by ultra-large-scale integration (ULSI) methods.

Another method of classification of the IC relates to the type of function performed by the chip. The three broad categories here are digital, analog, and mixed-signal. The digital IC is used for logic and computer systems, and the analog IC is used in amplifiers, communication circuits such as mixers, and analog instrumentation. Mixed-signal circuits refer to those that process both analog and digital signals on the same chip. Analog-to-digital and digital-to-analog converters are examples of mixed-signal circuits.

8.1 Comparison of Integrated and Discrete Component Circuits

IMPORTANT Concepts

1. The fabrication processes used for the IC allow the creation of resistors, capacitors, diodes, MOSFETs, and BJTs.

257

2. The tolerances on component values are relatively low although ratios of resistors or capacitors can be made much more accurately.

3. IC components have several limitations that are not present with discrete components.

8.1.1 PHYSICAL STRUCTURE OF THE INTEGRATED CIRCUIT

The most popular type of integrated circuit is constructed in and on a silicon material. Silicon is an element that occurs very plentifully in nature, most notably in sand. However, in order to be useful in IC fabrication, the silicon is purified by removing as many impure atoms as possible. Typically, there may be 1 impurity atom for each 100,000,000,000 silicon atoms (one hundred billion). In addition to purification, the silicon is processed to have a very uniform lattice structure [Campbell, 1996].

The starting material for most fabrication processes is a round silicon wafer with a diameter ranging from about 1 inch to 15 inches. A round cylinder of fixed diameter results from the purifying process, and wafers of fixed, small thickness (20–30 mils) can be cut from the cylinder by a diamond-blade saw. The silicon material can also be doped during the purification/growing process to have a desired amount of resistivity.

The elements that are commonly created on (or in) the silicon are resistors, capacitors, diodes, bipolar transistors, and MOS transistors.

Resistors: Resistors can be created within the silicon material by doping the silicon by diffusion of impurities or by ion implantation of impurities into the material. The resistor can be connected to other elements of the circuit by depositing metal from the resistor ends to the elements.

Resistances with high conductivity can also be made from metal patterns deposited onto the chip surface. More recently, doped polysilicon is used rather than metal to create higher resistance values. Neither deposited resistors nor diffusion resistors can be created with accurate values. Typical tolerances on these resistors are 10% to 20%. However, resistance ratios can be controlled to less than 1%.

Another problem common to the fabrication of resistors on a chip is the parasitic capacitance problem. Since the deposited resistors are produced on a layer of oxide for insulation, a capacitance is formed between the resistor and other elements on the wafer, with the oxide layer acting as the dielectric. These parasitic capacitances can lead to lower performance levels, especially in high-frequency circuits.

A less popular method of resistor synthesis is the deposition of very thin films over thick layers of oxide. These resistors are called thin-film resistors and offer a wider range of values with fewer parasitics. On the other hand, processing steps and subsequent cost are added to the finished chip.

Because of the very small dimensions involved in IC resistors, individual resistor values are influenced by the proximity of other resistors. Matching of resistors often requires the placing of dummy resistors nearby to ensure that the same environment surrounds each resistor to be matched. Such a practice simply does not exist in discrete element circuit design.

Capacitors: Two popular methods exist in fabricating capacitors for an IC. One approach is to create a reverse-biased *pn*-junction. The depletion region capacitance is then used as the desired capacitor. The shortcoming of this method is the nonlinear variation of capacitance

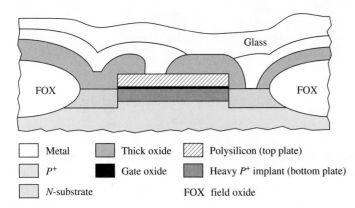

	Metal		Thick oxide		Polysilicon (top plate)
	P^+		Gate oxide		Heavy P^+ implant (bottom plate)
	N-substrate			FOX	field oxide

Figure 8.1
Cross section of typical capacitor.

with voltage across the junction. For large-signal applications, this type of capacitor can generate nonlinear distortion.

A more common method of creating a capacitor uses a doped silicon region as one plate, a thin oxide layer as the dielectric, and polysilicon or metal for the other plate. These capacitors show almost no variation of value as voltage across the device is varied. Unfortunately, the accuracy is comparable to resistor accuracy, and large values of capacitors cannot be created in a reasonable chip area. Figure 8.1 shows the cross section of a capacitor created in this manner.

Diodes: In theory, diodes are formed by creating an n region next to a p region and connecting leads to each region. In practice, other considerations make the fabrication of diodes slightly more complex. If a metal connection is made to a low-conductivity n material, not only will the material have high ohmic losses, but also a Schottky diode may form at this interface. Consequently, high conductivity regions are formed to make the metal contacts, as shown in Fig. 8.2.

The p^+ region is more heavily doped and makes a low-resistance connection to the deposited metal anode contact. The interface of the p^+ and n regions forms the junction of the diode. An n^+ region interfaces the n region but does not change the characteristics of the diode. It does, however, provide a low-resistance connection to the metal cathode connection. Such diodes can be formed in BJT or MOSFET processes.

It is also possible to use a portion of a BJT for a diode in BJT fabrication processes. Figure 8.3 shows three possibilities for this type of diode. The circuit of Fig. 8.3(a) uses the base-emitter junction with a floating collector. This diode does not have a very high reverse-breakdown voltage rating. Figure 8.3(b) shows a lower resistance diode that has emitter current for both input and output current. The configuration of Fig. 8.3(c) uses the base-collector junction while floating the emitter. This diode exhibits a higher reverse-breakdown voltage than that of Fig. 8.3(a).

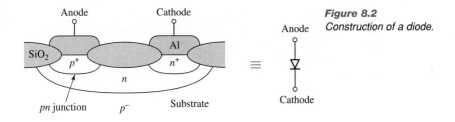

Figure 8.2
Construction of a diode.

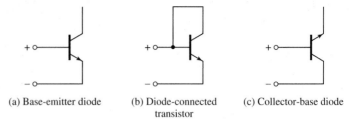

Figure 8.3
Diode connections for bipolar transistors.

(a) Base-emitter diode (b) Diode-connected transistor (c) Collector-base diode

Bipolar Junction Transistors: The structure of a single integrated BJT is shown in Fig. 8.4. The individual device shown is assumed to be isolated from all other components on the same substrate by junction isolation, as discussed in Section 5.5. The collector of the device is formed by the n-doped epitaxial layer, isolated from the p-type substrate that is required to connect to a more negative voltage than the collector. A reverse-biased diode is then formed with very little leakage current.

In the active region of operation, carriers are emitted from the emitter and flow vertically through the thin base region to the collector region. As carriers spread out, the collector region surrounds the base region to collect the carriers very effectively. Note that if the transistor is reversed, with carriers injected from collector to base by a forward bias, the emitter is too small to effectively collect these carriers. In the very early days of BJTs, a symmetrical device was made, but the resulting low β of the device limited its usefulness.

MOS and CMOS Devices: Because of the popularity of CMOS circuits, especially in digital applications, Fig. 8.5 shows the cross section of a chip that contains both a pMOS and an nMOS device. Although there are other configurations of devices, this figure represents a popular process. The lightly doped p-substrate is the starting material of the process. A slightly higher doped n-region forms the n-well. Within this n-well, the pMOS device is formed, while the accompanying nMOS device is formed in the substrate.

Figure 8.4
(a) Top view of a BJT. (b) Cross section of a BJT.

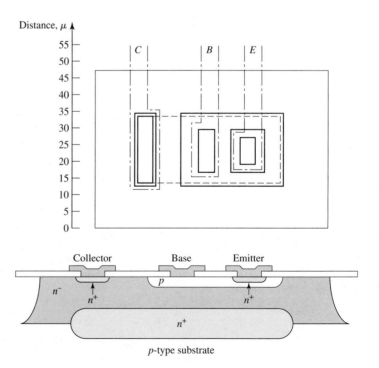

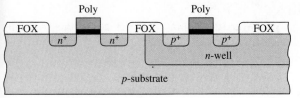

Figure 8.5
A CMOS circuit.

The insulation between the polysilicon gate terminals and the channels is a thin layer of oxide, and rather thick layers of oxide cover regions that do not have transistors underneath. The leftmost n^+ region is the source of the nMOS device, and the other n^+ region is the drain of this device. The rightmost p^+ region is the source of the pMOS device with the second p^+ region forming the drain. Metal layers are used to connect the devices properly to each other and to the power supply and ground.

Figure 8.6 shows the top view of a CMOS logic inverter. Note that the polysilicon splits to form both gates. The source and drain regions of the devices are formed in the silicon substrate by proper doping using ion implantation or other methods. Metal is deposited to connect the pMOS and nMOS drains and to run power and ground from IC pins to the devices. Typically, an oxide layer will appear between metal and substrate, and contacts must be made to the device by etching holes through the field oxide layer to the substrate. These contacts are then made by depositing metal that connects the proper region of the substrate to the metal conductors.

The width of the polysilicon gate region as it crosses the channel determines the gate length, L. The width of the channel determines the channel width, W. Both of these parameters are important in the design of MOSFET circuits. Channel lengths now vary down to 0.1 μ (micron), and lengths less than 1 μ are referred to as submicron devices.

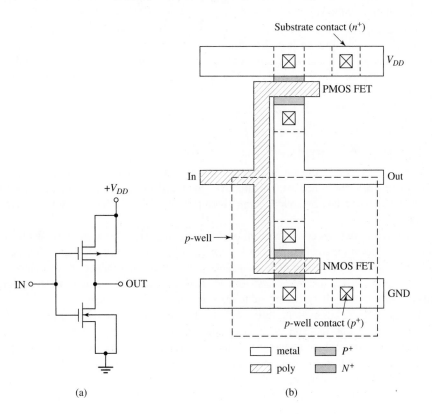

Figure 8.6
(a) Inverter schematic. (b) Top view of integrated circuit.

(a)   (b)

8.1.2 SIZE DIFFERENCES

Some microprocessor chips now contain over 100 million MOS devices. If this same microprocessor chip were implemented with discrete components containing two or three resistors per transistor stage, it might occupy one or two hundred cubic feet of volume. With the typical spacing of components used in a discrete circuit, this complex circuit might occupy a very large room rather than a small chip inside the personal computer. Size reductions by factors of 100,000 to 1,000,000 have resulted from the use of ICs. Items such as the personal computer and the hand-held electronic calculator would not be available if not for the IC.

There are three different types of limitation on a given integrated circuit. The first is the *core* limit, the second is the *pad* limit, and the third is the *pin* limit. A core-limited circuit is one that requires a small enough number of connections to circuits external to the chip that pads or pins do not limit the circuit. Instead, there is so much core circuitry that it is difficult to get the entire system to fit in the space available on the chip. Typically, digital VLSI systems contain so much circuitry that the core limit is approached before the pad or pin limit.

Pads are metal deposits, usually rectangular, that terminate conductors on the IC chip and provide space for the bonding of wires that connect to the pins of the chip. Pads are provided for all connections that must be made from the circuitry of the chip to external circuits or systems. A common practice is to place the pads around the periphery of the chip. The pads have minimum size restrictions, depending on the process, and if too many pads are required, it may be difficult to provide enough chip space to include all of these pads.

A third limitation is that of pins to be used for a given circuit. For many analog circuits such as op amps, the number of pins to be used may be specified by the company fabricating the amplifier. Many SSI or MSI logic circuits and op amps are fabricated on chips with somewhere between 12 and 24 pins. Integrated circuits classified as LSI, VLSI, or ULSI can use hundreds of pins.

When designing a circuit on a chip, different approaches may be taken depending on the limitations. For example, op amp circuits are generally pin limited to correspond to standard op amp chips. In such a case, it is possible to apply design methods that do not minimize circuit or core area. Many op amps contain several resistors because core space is not at a premium in these designs. Other chips, notably those containing large digital systems, may limit or eliminate the use of resistors since the IC may be core limited.

8.1.3 COST DIFFERENCES

Size reduction alone implies a price reduction of ICs over discrete circuits. The greatest factor in cost reduction, however, results from the mode in which the IC can be fabricated. Millions of devices can be created simultaneously on a large wafer during the fabrication process.

If the wafer contains complex circuits, the interconnections between individual components can be accomplished during fabrication, thereby minimizing packaging costs associated with discrete circuit components and wiring costs to interconnect these components. Although present-day discrete transistors are now created in this batch mode, packaging costs and especially wiring costs lead to large finished circuit costs.

A large discrete component digital computer with 1 megabyte of main memory might have cost $2,000,000 in 1966. Without accounting for inflationary effects, a comparable personal computer with 64 megabytes of main memory now costs around $1000. The cost reduction factor of ICs over discrete circuits may range from 1000 to 10,000.

It should be understood that the large reductions in cost only apply if large quantities of a chip are to be produced. The production costs of a chip are very high, but if many chips are to be fabricated, the production cost per chip can be very small. If only one simple electronic circuit is required, building a discrete component prototype is much less costly. On the other hand, if millions of these simple circuits or thousands of more complex circuits are to be produced and sold, the fabrication of IC chips becomes more cost effective.

8.1.4 COMPONENT DIFFERENCES

Discrete components such as resistors, capacitors, and inductors are available in a very large range of values. Very precise values can be obtained at the expense of a higher cost. Resistor values can extend to many $M\Omega$ and capacitor values to hundreds of μF. Metal core or air core inductors are available over a wide range of values, as also are transformers with a wide choice of turns ratios.

The IC chip becomes too large to be useful when the total resistance of resistors on the chip exceeds some value such as 100 kΩ. The same applies if the total capacitance exceeds 50–100 pF. Until recently, inductors could not be fabricated on a chip. Even now, integrated inductors with limited values only in the nH range are possible [Burghartz et al., 1998].

Another problem with IC components is the lack of precise control of values. Resistors or capacitors are typically fabricated with an absolute accuracy of around 20%. It is possible to create resistive ratios or capacitive ratios that approach a 0.1% accuracy, but absolute value control is very poor.

Because of the limited, inaccurate values of resistors and capacitors, the design philosophy for ICs differs from that of discrete component circuits. Before ICs were available, a designer attempted to minimize the number of transistors in a circuit. In fact, some companies estimated the component cost of a circuit by multiplying the number of transistors of the circuit by a cost per stage. This cost per stage was determined mainly by the transistor cost plus a small amount added to account for the passive component cost. The least expensive item in the discrete circuit is generally the resistor, which may cost a few cents.

In an IC, the greatest cost is often associated with the component that requires the most space. The BJT or MOSFET typically requires much less space than resistors or capacitors and is the least expensive component on the chip. An IC designer's attempt to minimize space or simplify the fabrication process for a given circuit generally results in much larger numbers of transistors and smaller numbers of resistors and capacitors than the discrete circuit version would contain. For example, a discrete circuit bistable flip-flop of 1960 vintage contained 2 transistors, 6 resistors, and 3 capacitors. A corresponding IC design had 18 transistors, 2 resistors, and no capacitors.

Another difference in design philosophies between discrete circuit and IC design is the concept of matching devices. Matching components in IC design can be better than in discrete design, but also presents unique problems as well. Components that are made from identical photographic masks can vary in size and value because of uneven etching rates influenced by nearby structures or by asymmetrical processes. Dummy components must often be included to achieve similar physical environments for all important components. Certain processes must be modified to result in symmetrical results. Metal conductors deposited on the IC chip can introduce parasitics or modify performance of the circuit if placed in certain areas of the chip; thus care must be taken in placing the metal depositions. Of course, these kinds of considerations are unnecessary in discrete circuit design.

Although discrete circuits can use matched devices such as differential stages, often single-ended stages can be designed to meet relatively demanding specifications. Discrete components can be produced with very tight tolerances to lead to circuit performance that falls within the accuracy of the specifications. Circuit design can be based on the absolute accuracy of key components such as resistors or capacitors. Although it may add to the price of a finished circuit, simple component-tuning methods can be incorporated into the production of critical discrete circuits.

Since the absolute values of resistors and capacitors created by standard processes for ICs cannot be determined with great accuracy, matching of components and devices is used to achieve acceptable performance. Differential stages are very common in IC design, since matched transistors and components are easy to create even if absolute values cannot be controlled accurately.

8.1.5 PERFORMANCE DIFFERENCES

A high-power output stage requires a large-area heat sink to keep the maximum temperature of the device below some acceptable limit. In 200–300 W audio amplifiers, the output transistors may dissipate 10–50 W. If this much power were dissipated in a small integrated circuit with no heat sink, the circuit would generally self-destruct due to the excessive temperature generated.

In high-power applications, large discrete power transistors, mounted on effective heat sinks, are used rather than integrated circuit chips. This area will probably remain the domain of discrete devices for some years to come.

It is possible to build power MOSFETs or BJTs by integrated circuit methods. However, these devices occupy a very large chip area, which makes thermal interfacing to the package more efficient. The large areas needed for power IC devices tend to make the size advantage of ICs over discrete circuits become insignificant.

8.1.6 SUMMARY OF DIFFERENCES

Table 8.1 summarizes several differences between discrete circuits and ICs.

Table 8.1 Differences Between Discrete Circuits and ICs

Discrete	Integrated
Large range of capacitor values	Limited capacitor values (50–100 pF)
Large range of resistor values	Limited resistor values (50 kΩ)
Large range of inductors	No inductors or very small values
Relatively precise component values	Imprecise component values
Values independent of environment	Values dependent on environment
Can design and construct circuit in a very short time	Fabrication of circuit takes much time
Single person can design and construct circuit	Many people involved in fabrication of circuit
Can easily modify design after construction	Cannot modify design after construction in most cases
Lower cost for small number of circuits	High cost for small number of circuits
Ability to construct circuits with high power outputs	Limited output power capabilities
High cost for large number of circuits	Low cost for large number of circuits
Occupies large volume	Occupies small volume
Higher dissipation per stage	Lower dissipation per stage
Design theory is well developed	More complex design considerations

8.2 Simulation in IC Design

IMPORTANT Concepts

1. The complexity of the IC dictates that simulation methods be used to verify performance before the chip is fabricated.
2. The BJT model for simulation has been available for many years and can result in accurate simulation results if characterized properly.
3. MOSFET models are undergoing an evolution to maintain accuracy as physical size becomes smaller. Many models lead to inaccurate simulation results.

In designing discrete-circuit systems, many circuits can be built without a thorough understanding of the details of operation. Simpler circuits can be constructed and tested to measure performance. If unacceptable, element values or configuration can be modified, perhaps several times, to achieve acceptable performance. Depending on circuit complexity, a relatively small time investment is necessary to produce an acceptable circuit.

For integrated circuits, the final design is "written in concrete"; that is, the fabricated circuit cannot generally be changed. If values are calculated incorrectly, or even if an element value deviation within a process tolerance is too great, the circuit may be worthless. The time from submitting the final design to the finished fabrication may be several weeks or several months. Furthermore, the cost of fabrication may be several thousands of dollars. Inaccurately determined performance cannot be tolerated in the design of ICs.

As a result of the importance of fabricating a working circuit on the first or second attempt, circuits are simulated carefully before the final design is complete. Although there are several circuit simulation programs available, those related to or based on the Spice program are rather popular at the present time. The next section will briefly discuss this program.

Spice, which is an acronym for Simulation Program with Integrated Circuit Emphasis, was developed in the early 1970s at the University of California, Berkeley. It has been revised several times, and the Berkeley version is public domain software. Other commercial versions of Spice have been independently developed and marketed as proprietary software. Examples of commercial versions of Spice are PSpice©, HSpice©, and ALLSpice©. All versions of Spice use similar syntax and computational algorithms. Spice has been widely accepted by the VLSI industry as the standard method of device-level simulation.

Spice uses a nodal analysis approach and can be used for nonlinear dc, nonlinear transient, and linear ac circuit solutions. It also includes a temperature analysis module. For nonlinear dc and transient analyses, the program accounts for the nonlinear effects of all devices as specified by model parameters. For ac analysis, the dc operating points of all devices are first determined, then the small-signal equivalent circuit for each device is obtained. The resulting analysis allows for no nonlinear behavior and could easily result in a signal with a magnitude that exceeds the limits of the active region. The user is expected to interpret the results correctly for the ac analysis. Although the transient analysis may take much longer, the results are accurate, whereas some ac results might require more scrutiny.

The first commercial personal computer-based Spice program was developed by MicroSim Corporation and has gained wide acceptance in both industry and academia. This version of Spice is called PSpice© and will be used in this section for illustrative examples.

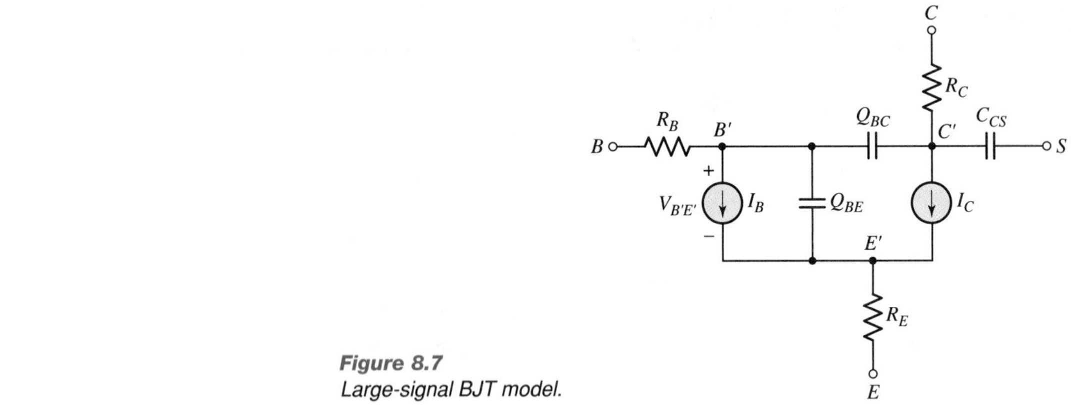

Figure 8.7
Large-signal BJT model.

8.2.1 THE BJT MODEL

Since we have previously discussed the major parameters of the BJT equivalent circuit, we will simply present the circuit models used in Spice programs. Figure 8.7 shows a large-signal BJT model based on the more complex Gummel–Poon model used in many versions of Spice. This model is used to calculate dc voltages and currents when the operating point, invoked by the .OP statement, is to be found. It is also used in transient analysis when the .TRAN statement is used.

The parameters R_B, R_E, and R_C represent the parasitic resistances at the base, emitter, and collector, respectively. The capacitors and current sources obey nonlinear equations involving the related voltages. The reader is referred to other references for definitions of these elements in terms of equations [Vladimirescu, 1994].

For ac analysis using the .AC statement, a piecewise linear equivalent circuit is used in all Spice models. This linearized circuit ignores active region boundaries. If a 1-V peak signal is applied to an amplifier that has a gain of 200 V/V, a real circuit will show hard clipping of the output voltage as saturation and cutoff are reached. A Spice simulation using ac analysis will indicate a 200-V peak output signal. Thus, if active region information is desired, either a transient analysis or a dc analysis that scans the input voltage over an appropriate range must be used.

The hybrid-π circuit of Fig. 8.8 shows the dominant elements of the model used for Spice. This circuit differs very little from that of Fig. 7.25, discussed in Section 7.5. The elements added to the equivalent circuit discussed in Section 7.5 are R_B, R_E, R_C, C_{cs}, r_μ, and r_o. For many applications R_B is replaced by r_x, and the other elements are negligible.

Figure 8.8
Small-signal, hybrid-π model of the BJT.

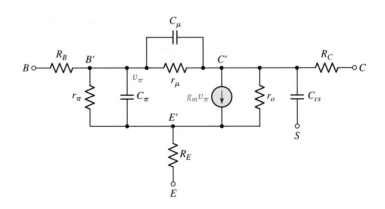

8.2.2 BIPOLAR SPICE MODEL PARAMETERS

Table 8.2 defines the Spice model parameters typically used for an *npn* bipolar transistor along with some typical values for an integrated circuit device.

Many of the Spice model parameters listed in Table 8.2 relate to second-order effects that are not represented by the model of Fig. 8.8. Most of these parameters have default values that are used in the analysis programs even when the user does not specify values. For less complex simulations, the user need only specify a small number of parameters.

Table 8.2   Bipolar Spice Model Parameters

Symbol	Parameter Value	Typical	Unit
IS	Reverse saturation current	2.28E-17	A
ISE	B-E leakage current	2.28E-14	A
ISC	B-C leakage current	2.28E-14	A
IKF	Corner for β_F high-current rolloff	0.02	A
IKR	Corner for β_R high-current rolloff	∞	A
IRB	Current where RB falls halfway to minimum value	∞	A
ITF	Coefficient for bias dependence of TF	0	A
VAF	Forward Early voltage	60	V
VAR	Reverse Early voltage	8	V
VTF	Coefficient for TF B-C voltage dependence	∞	V
VJE	B-E built-in potential	0.804	V
VJC	B-C built-in potential	0.334	V
VJS	C-S (Substrate) built-in potential	0.536	V
RB	Zero-bias base resistance	160	Ω
RBM	Minimum base resistance at high currents	RB	Ω
RE	Emitter ohmic resistance	4	Ω
RC	Collector ohmic resistance	240	Ω
CJE	Zero-bias B-E junction capacitance	0.157p	F
CJC	Zero-bias B-C junction capacitance	0.207p	F
CJS	Zero-bias C-S junction capacitance	1.06p	F
TF	Ideal forward transit time	0.157p	s
TR	Ideal reverse transit time	16n	s
BF	Forward current gain	100	
BR	Reverse current gain	1	
NF	Forward current emission coefficient	1	
NE	B-E leakage emission coefficient	1.5	
NR	Reverse current emission coefficient	1	
NC	B-C leakage emission coefficient	2	
MJE	B-E junction grading factor	0.31	
MJC	B-C junction grading factor	0.209	
MJS	C-S junction grading factor	0.269	
XTB	Beta temperature coefficient	0	
XTF	Coefficient for TF bias dependence		
XCJC	Fraction of CJC connected to base node	1	
PTF	Excess phase at $1/(2\pi\,\mathrm{TF})$ Hz	5	o
EG	Energy gap	1.1	eV
XTI	IS temperature exponent	12.8	

For more complex simulations, the second-order parameters can also be specified. A few examples will demonstrate the use of the Spice program.

EXAMPLE 8.1

Assume that an *npn* BJT is characterized by the Spice model given in Table 8.2. Using Spice, calculate and plot a set of collector characteristics (I_C vs. V_{CE}) for values of base current ranging from 0 to 20 μA in 5-μA increments. Graphically evaluate the forward Early voltage and the current gain, β, from the characteristic curves and compare to the model parameters.

SOLUTION Table 8.3 shows the netlist file to accomplish the specified plot. Figure 8.9 shows a plot of the results.

Table 8.3 Netlist for Example 8.1

```
EX8-1.CIR

VCE 2 0
IB 0 1
Q 1 2 1 0 0 NPNS
.MODEL NPNS NPN (BF=100 VAF=60 VAR=8 IS=2.28E-17
+ RB=160 RE=4 RC=240 IKF=0.02 TF=0.157N TR=16N
+ CJS=1.06P VJS=0.536 MJS=0.269 CJE=0.157P VJE=0.804
+ MJE=0.31 CJC=0.207P VJC=0.334 MJC=0.209 XTI=12.8 PTF=5)
.DC VCE 0 20 0.1 IB 0 20U 5U
.PROBE
.END
```

From this figure an incremental β at $V_{CE} = 5$ V is found to be

$$\beta = \left[\frac{\Delta I_C}{\Delta I_B}\right]_{V_{CE}=5} = \frac{1.35 - 0.9}{0.015 - 0.010} = 90$$

The forward Early voltage can be evaluated by projecting the collector characteristics to the zero current value, as shown in Fig. 8.10. This gives a value of $VAF = 60$ V. The values of β and VAF from Spice compare reasonably well with the model parameters.

Figure 8.9
Collector characteristics.

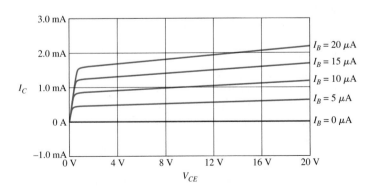

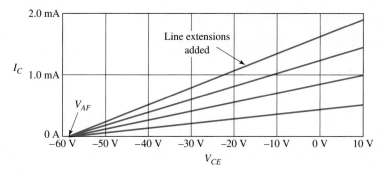

Figure 8.10
Graphical determination of
Early voltage.

EXAMPLE 8.2

Given the circuit of Fig. 8.11 with $\beta = 100$ and $V_{BE\,(on)} = 0.8$ V. Calculate the quiescent collector voltage and the small-signal voltage gain of the circuit. Simulate the circuit with Spice to verify these results.

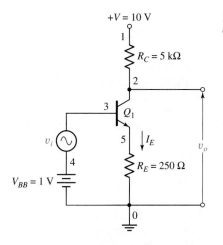

Figure 8.11
Circuit for Example 8.2.

SOLUTION The emitter current is calculated as

$$I_E = \frac{V_{BB} - V_{BE(on)}}{R_E} = \frac{0.2}{0.25} = 0.8 \text{ mA}$$

The value of r_e is then

$$r_e = \frac{26}{0.8} = 32.5 \ \Omega$$

The small-signal gain is found to be

$$A \approx \frac{-R_C}{r_e + R_E} = \frac{-5}{0.25 + 0.0325} = -17.7$$

The collector voltage is calculated to be

$$V_C = V_{CC} - R_C \alpha I_E = 10 - 5 \times 0.792 = 6.04 \text{ V}$$

The netlist for this circuit is shown in Table 8.4.

Table 8.4 Input Circuit File for Example 8.2

```
EX8-2.CIR

VCC 1 0 DC 10
VBB 4 0 DC 1
VAC 3 4 AC 0.1
RC 1 2 5K
RE 5 0 250
Q1 2 3 5 0 NPNS
.MODEL NPNS NPN (BF=100 VAF=60 VAR=8 IS=2.28E-17
+ RB=160 RE=4 RC=240 IKF=0.02 TF=0.157N TR=16N
+ CJS=1.06P VJS=0.536 MJS=0.269 CJE=0.157P VJE=0.804
+ MJE=0.31 CJC=0.207P VJC=0.334 MJC=0.209 XTI=12.8 PTF=5)
.OP
.AC DEC 10 100 10MEG
.PROBE
.END
```

The .OP option generates the output shown in Table 8.5.

Table 8.5 Operating Point Information from Spice Output File

Name Model	Q1 npns
IB	7.91E-06
IC	7.49E-04
VBE	8.11E-01
VBC	-5.25E+00
VCE	6.06E+00
BETADC	9.48E+01
GM	2.78E-02
RPI	3.27E+03
RX	1.60E+02
RO	7.88E+04
CBE	4.63E-12
CBC	1.16E-13
CBX	0.00E+00
CJS	5.39E-13
BETAAC	9.11E+01
FT	9.34E+08

The node voltage for the collector (node 2) gives a voltage value of 6.255 V. This result is somewhat higher than the 6.04 V calculated, because $V_{BE(on)}$ is found to be 0.811 V rather than the value of 0.8 V used in manual calculations. The Bode plot shows a midband gain of -16.8, which is a little lower than the calculated value. This gain also relates to the lower emitter current flowing in the simulation as a result of a higher value of $V_{BE(on)}$, which results in a larger value of r_e and a lower value of gain.

8.2.3 THE MOSFET MODEL

There is no single model that is used for the MOSFET as is the case for the BJT. There are several levels of models, and the newer subthreshold devices are modeled by several pages of equations, sometimes several hundreds of pages.

In the typical situation, a library file in Spice contains appropriate models for the MOS devices that will be used in the actual fabrication process. This model will account for doping levels, oxide thicknesses, and other parameters of the construction process. When simulation of a circuit is to be done, the MOS device specifications include important dimensions of the devices to be used. For example, the channel length and width, the areas of the source and drain regions, and modified perimeters of the source and drain regions are specified in the model call statement. The circuit is analyzed to find all steady-state voltages. With this information along with the process parameters and dimensions of the devices, all capacitance values are calculated and included in the circuit to be analyzed. The bandwidth can then be found from the simulation. If the bandwidth is unacceptable, device dimensions can be changed, within prescribed limits, or configurations can be changed to improve performance. An example will demonstrate a simple circuit simulation using a MOSFET.

EXAMPLE 8.3

Given the amplifier of Fig. 8.12. The model for the nMOSFET is in a file called ORBIT. LIB. Plot the frequency response of the voltage gain of the amplifier.

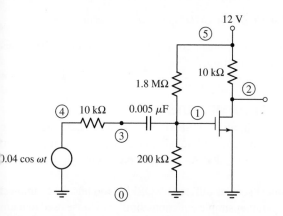

Figure 8.12
A MOSFET amplifier.

SOLUTION The Spice program for this circuit is shown in Table 8.6.

The dimensions of the source and drain for this device are assumed to be $200 \mu \times 5\mu$. The perimeter figure is found by adding the width and twice the length of the drain or source regions. The results of this simulation are $A_{MB} = 21.4$, $f_{high} = 10.1$ MHz, and $f_L = 169$ Hz. The overall frequency response is shown in Fig. 8.13. The values of capacitances for the MOSFET along with conductances calculated by Spice are tabulated in Table 8.7.

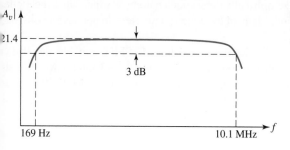

Figure 8.13
Frequency response of amplifier.

Table 8.6 Spice Program for Circuit of Example 8.3

```
EX8-3.CIR

VDD 5 0 DC 12
R3 5 2 10K
R1 5 1 1800K
R2 1 0 200K
C1 1 3 0.005U
R4 3 4 10K
VIN 0 4 AC 0.04
M1 2 1 0 0  N1 L=2U W=200U AS=1000P AD=1000P PS=210U PD=210U
.AC DEC 10 100 100MEG
.LIB ORBIT.LIB
.OP
.PROBE
.END
```

Table 8.7 MOSFET
Parameters Calculated
by Spice

Parameter	Value
CBS	2.43E-13
CBD	8.93E-14
CGSOV	5.50E-14
CGDOV	5.50E-14
CGBOV	5.52E-16
CGS	1.92E-13
CGD	0.0
CGB	0.0
GM	2.69E-3
GDS	1.92E-5
VT	0.765

The largest component of capacitance appears from gate to ground. Although there is capacitance from drain to ground, this value is considerably smaller than the capacitance at the input. The upper corner frequency can be calculated manually by considering the input circuit, which has an equivalent resistance of 8.5 kΩ and an input capacitance of

$$C_{in} = CGS + CGSOV + CGBOV + CGDOV(1 + A) = 1.48\,\text{pF}$$

Ignoring output effects leads to an approximate corner frequency of 11.3 MHz. This value is about 10% higher than the simulated value.

8.2.4 CMOS SPICE MODEL PARAMETERS

Table 8.8 lists the model parameters for simpler MOSFET models. We emphasize that the continual decrease in MOS geometries leads to additional effects that must be included in the model to produce accurate simulation results. As the geometries change, the models continue to evolve.

MOSFET versions of Spice support several different MOSFET models. The simplest model, designated Level 1, is based on rather simple equations similar to those used in manual analysis. It is primarily useful only for bias-point calculations and functional verification of a circuit's operation.

The Level 2 model provides a more complete description of device operation based on first- and second-order physical effects. It is more accurate in the transition region between the triode and active regions. It also includes a method of calculating the effective channel length and is more accurate than the Level 1 model for short channel lengths in the vicinity of 2 μ.

The Level 3 model adds several empirically determined parameters that cause the model to more closely fit observed operation. Many of the added parameters do not have an obvious relation to the device physics.

Other Spice models continue to be developed for the MOSFET, notably the BSIM models, developed at the University of California, Berkeley. The BSIM3 and BSIM4 models are especially useful for submicron gate lengths and for operation of the MOSFET in the weak inversion or subthreshold region. Most companies have developed proprietary models used in their respective fabrication processes.

Table 8.8   Spice MOSFET Model Parameters

Name	Description	Units	Default	Example
LEVEL	Model index		1	
VTO	Threshold voltage ($VSB = 0$)	V	0	0.8
KP	Transconductance parameter	A/V^2	2E-5	1E-3
GAMMA	Bulk threshold parameter	$V^{1/2}$	0	0.5
PHI	Surface potential	V	0.6	0.65
LAMBDA	Channel length model parameter	V^{-1}	0	1E-4
RD	Drain ohmic resistance	Ω	0	10
RS	Source ohmic resistance	Ω	0	10
RSH	D & S sheet resistance	Ω/sq.	0	30
CBD	Zero-bias B-D junction capacitance	F	0	5P
CBS	Zero-bias B-S junction capacitance	F	0	2P
CJ	Zero-bias bulk-bottom capacitance	F/m^2	0	1E-4
MJ	Bulk-junction grading coefficient		0.5	0.5
CJSW	Zero-bias sidewall capacitance	F/m^2	0	1E-9
MJSW	Bulk-junction sidewall gradient coefficient	0.33	0.3	
PB	Bulk-junction potential	V	1	0.8
IS	Bulk-junction saturation current	A	1E-14	1E-14
JS	Bulk-junction saturation current density	A/m^2	0	1E-8
CGDO	G-D overlap capacitance per channel width	F/m	0	4E-11
CGSO	G-S overlap capacitance per channel width	F/m	0	4E-11
CGBO	G-B overlap capacitance per channel width	F/m	0	4E-10
NSUB	Substrate doping	cm^{-3}	0	1E16
NSS	Surface state density	cm^{-2}	0	1E10
NFS	Fast surface state density	cm^{-2}	0	1E10
TOX	Oxide thickness	m	1E-7	1E-7
TPG	Type of gate material		1	1
XJ	Metallurgical junction depth	m	0	5E-7
LD	Lateral diffusion	m	0	2E-7
UO	Surface mobility	$cm^2/V - s$	600	600
UCRIT	Critical field for mobility degradation	V/cm	1E4	1E4
UEXP	Critical exponent in mobility degradation		0	0.1
VMAX	Maximum drift velocity	m/s	0	5E4
NEFF	Channel charge coefficient		1	4
XQC	Oxide share coefficient		1	0.5
Delta	Width effect on threshold voltage		0	1
THETA	Mobility modulation	V^{-1}	0	0.1
ETA	Static feedback coefficient		0	1
KAPPA	Saturation field factor		0.2	0.4

8.3 Approximation in BJT Circuit Design

The ability to approximate circuit performance is important as a design is initiated. Of several possible approaches to meeting a set of specifications, approximate analysis might eliminate unacceptable configurations with little loss of the designer's time. Furthermore, an understanding of the approximate behavior of the circuit allows improved performance to be obtained through intelligent change of element values or configuration.

The methods discussed in Chapters 6 and 7 allow approximate analysis of many less complex circuits. Parameters from a Spice simulation can be used in these approximate methods if available.

REFERENCES

Burghartz, J., Edelstein, D., Soyuer, M., Ainspan, H., and Jenkins, K. "RF circuit design aspects of spiral inductors on silicon," *IEEE International Solid-State Circuits Conference Digest of Technical Papers,* February 1998, pp. 246–247.

Campbell, S. A. *The Science and Engineering of Microelectronic Fabrication,* Oxford University Press, New York, 1996.

Vladimirescu, A. *The Spice Book,* Wiley, New York, 1994.

Integrated Circuit Design with the MOSFET

The MOSFET has become more popular in circuit design in the last decade, especially for mixed-signal circuits that combine both digital and analog circuits on a single chip. With the improvements in high-frequency performance, the MOSFET can compete with the BJT up to frequencies in the low GHz range. Of course, the heterojunction BJT can operate at much higher frequencies than this, approaching frequencies of 100 GHz.

This chapter will consider several configurations of amplifying stages used in MOSFET amplifier design. Each configuration is analyzed in terms of midband voltage gain, upper corner frequency, and the size of the output active region.

Design of MOSFET circuits is considerably different from the design of BJT circuits. For example, the dimensions of the channel are used as part of the design procedure in MOSFET circuits. The width of the channel affects the transconductance, the output resistance and capacitance, and the midband voltage gain of the stage. The specification of channel width is often one step in the design of a MOSFET amplifier stage. In digital circuits, the width and the length are generally of the same order of magnitude. In amplifier design, the dimension of channel width may be hundreds of times greater than the channel length to achieve high values of voltage gain. Such considerations are not necessary in BJT amplifier design, which provides one reason for covering BJT design in a separate chapter. The next chapter will consider the BJT.

One circuit that is very significant in IC design is the current mirror. This circuit not only provides bias for amplifier stages, it also provides an incremental resistive load for various amplifier stages. The first section of this chapter will discuss some basic configurations of this important circuit, before moving to various amplifier configurations.

DEMONSTRATION PROBLEM

The circuit shown is a common-source amplifier with a current mirror load. If $K_n = \mu_n C_{ox} W_n / 2L_n = 0.12$ mA/V^2, $K_p = \mu_p C_{ox} W_p / 2L_p = 0.1$ mA/V^2, $\lambda_n = \lambda_p = 0.01$ V^{-1}, and $V_{Tn} = -V_{Tp} = 1$ V, find V_1 to set $V_{DSQ1} = 2.5$ V. Calculate the midband voltage gain of the amplifier.

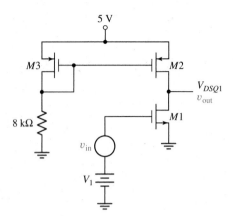

Amplifier for Demonstration Problem

This problem requires a knowledge of the dc current versus voltage relationships of the MOSFETs to find the bias voltage, V_1. A knowledge of the incremental equivalent circuit for the MOSFET is needed to calculate the midband voltage gain.

9.1 MOSFET Current Mirrors

IMPORTANT Concepts

1. A simple MOSFET current mirror is constructed in the same configuration as that of the BJT current mirror.
2. The ratio of output to input current can be determined by the aspect ratios of the two devices.
3. The Wilson current mirror can be used to achieve higher output impedance for the output device.

The BJT current mirror was developed in the 1960s for use in op amp circuits. As the MOSFET increased in capability, the MOSFET current mirror evolved from the BJT circuit. MOSFET current mirrors operate on similar principles to the BJT mirrors to be discussed in Chapter 10 and use similar configurations. One advantage of MOSFET current mirrors over BJT mirrors is that the MOS devices draw zero control current. The BJT stages exhibit small errors due to the finite base currents required. On the other hand, the matching of threshold voltages on MOS devices is generally not as good as the V_{BE} matching of bipolar devices. Since BJT mirrors require additional considerations beyond those of MOSFET mirrors, a more thorough discussion of this circuit appears in Chapter 10.

Figure 9.1 shows a simple nMOS current mirror consisting of two matched devices, $M1$ and $M2$. This current mirror may be used to create a constant bias current for an IC amplifier stage. Integrated circuits provide the capability of matching device characteristics quite closely, and the current mirror takes advantage of this capability. In the circuit of Fig. 9.1, the current I_o is intended to be equal to I_{in}. Although not shown in the figure, the external circuit through which I_o flows connects to the drain of $M2$. The current I_{in} equals the drain current of $M1$ whereas I_o is the drain current of $M2$.

Device $M1$ is in its active region, since drain current is flowing and the drain voltage equals the gate voltage. With V_{DS2} sufficiently positive to put $M2$ in the active region, the

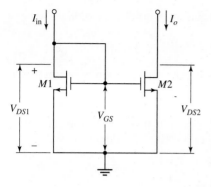

Figure 9.1
A simple nMOS current mirror (sink).

ratio of output current, I_o, to input current, I_{in}, can be expressed as

$$\frac{I_o}{I_{in}} = \left[\frac{L_1 W_2}{L_2 W_1}\right]\left[\frac{V_{GS} - V_{T2}}{V_{GS} - V_{T1}}\right]^2\left[\frac{1 + \lambda_2(V_{DS2} - V_{DSP2})}{1 + \lambda_1(V_{DS1} - V_{DSP1})}\right]\left[\frac{\mu_2 C_{ox2}}{\mu_1 C_{ox1}}\right] \quad (9.1)$$

In ICs, it is possible to match devices so that $\mu_1 C_{ox1} = \mu_2 C_{ox2}$, $V_{T1} = V_{T2}$, and $\lambda_1 \approx \lambda_2$. With these conditions satisfied, Eq. (9.1) can be written as

$$\frac{I_o}{I_{in}} = \left[\frac{L_1 W_2}{L_2 W_1}\right]\left[\frac{1 + \lambda(V_{DS2} - V_{DSP2})}{1 + \lambda(V_{DS1} - V_{DSP1})}\right] \quad (9.2)$$

As a result of the connection between gate and source of $M1$, $V_{DS1} = V_{GS}$ and, since $V_{DSP1} = V_{GS} - V_{T1}$, we can simplify the denominator of Eq. (9.2) further. We can express $[1 + \lambda(V_{DS1} - V_{DSP1})]$ as

$$[1 + \lambda(V_{GS} - V_{GS} + V_{T1})] = 1 + \lambda V_{T1}$$

Finally, if we limit the output voltage such that $V_{DS2} = V_{DS1}$, Eq. (9.2) reduces to

$$\frac{I_o}{I_{in}} = \frac{L_1 W_2}{L_2 W_1} \quad (9.3)$$

This equation indicates that in the simple MOSFET current mirror, the ratio of I_o to I_{in} may be scaled to any desired value by scaling the aspect ratios (W/L) of the devices.

There are three effects that cause the MOSFET current mirror performance to differ from that predicted by Eq. (9.3). These are:

1. Channel length modulation as V_{DS2} changes, as predicted by Eq. (9.2)
2. Threshold voltage mismatch
3. Imperfect geometrical matching

EXAMPLE 9.1

Assume that a matched pair of MOSFETs are used in the current mirror of Fig. 9.1 with values of $\lambda = 0.032$ V^{-1}, $\mu C_{ox} = 70$ μA/V^2, $W/2L = 10$, and $V_T = 0.9$ V. If a 5-V source in series with a resistor, R, is connected to the drain of $M1$ to create the input current, calculate the value of R needed to create an input current of 100 μA. Calculate the output current when $V_{DS2} = 3$ V.

SOLUTION From the equation for drain current in the active region we write

$$I_{D1} = \frac{\mu C_{ox} W}{2L} [V_{GS} - V_T]^2 [1 + \lambda(V_{DS1} - V_{DSP1})]$$

Since $V_{DS1} = V_{GS}$ and $V_{DSP1} = V_{GS} - V_T$, this equation can be expressed as

$$I_{D1} = 100 = 700 [V_{GS} - 0.9]^2 [1 + 0.032 \times 0.9]$$

We can now solve for the value of V_{GS} to result in the specified drain current. This value is $V_{GS} = 1.27$ V. The resistance needed to create 100 μA of drain current is

$$R = \frac{5 - V_{DS1}}{I_{D1}} = \frac{5 - 1.27}{0.1} = 37.3 \text{ k}\Omega$$

The output current is calculated from

$$I_{D2} = \frac{\mu C_{ox} W}{2L} [V_{GS} - V_T]^2 [1 + \lambda(V_{DS2} - V_{DSP2})]$$

In this case, $V_{DS2} - V_{DSP2} = V_{DS2} - (V_{GS} - V_T) = 3 - 1.27 + 0.9 = 2.63$ V. The output current is then $I_o = 104$ μA.

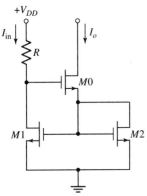

Figure 9.2
nMOS version of the Wilson current mirror.

A MOSFET version of the BJT Wilson current mirror, discussed in Chapter 10, may be used to reduce the output compliance error of the mirror. Figure 9.2 illustrates such an nMOS mirror.

For matched devices, the current I_{in} may be expressed as

$$I_{in} = \frac{V_{DD} - 2V_{GS}}{R} \tag{9.4}$$

where V_{GS} is the gate-to-source voltage of all three devices. The output current is then

$$I_o = I_{in} \left[\frac{L_1 W_2}{L_2 W_1} \right] \tag{9.5}$$

The voltage at the drain of $M2$ remains constant at a value of V_{GS}, keeping I_o constant, as the output voltage at the drain of $M0$ varies over a large range. This stage has a higher output impedance than that of the simple current mirror because it has a source impedance. Device $M2$ presents an impedance between drain and ground that equals $1/g_{m2}$, thus increasing the output impedance of $M0$. A higher output impedance leads to a smaller current change with output voltage.

It is also possible to create current mirror sources with almost identical performance to these sinks by using pMOS devices. Figure 9.3 shows a simple mirror and a Wilson-type mirror using pMOS devices. Although more complex current mirrors are sometimes used

PRACTICE **Problems**

9.1 For the finished design of Example 9.1, calculate the output current when $V_{DS2} = 4$ V. *Ans: 107 μA.*
9.2 At what voltage must V_{DS2} be in the current mirror of Example 9.1 to cause an output current of 102 μA? *Ans: 2.37 V.*
9.3 Using the device parameters of Example 9.1, select the resistance R of this example to lead to an output current of 180 μA when $V_{DS2} = 4$ V. *Ans: R = 22.3 kΩ.*

Figure 9.3
Current sources: (a) simple mirror, (b) Wilson mirror.

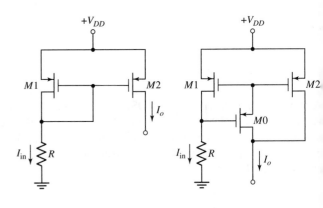

in critical circuit design, the remainder of this chapter will apply the simple mirror of this section.

9.2 Amplifier Configurations for MOSFET Integrated Circuits

IMPORTANT Concepts

1. Most MOSFET IC amplifier stages do not use a resistor as the load for the amplifying stage.
2. The load is generally created by one or more other MOS devices. These devices are called active loads.
3. Active load stages can lead to very high voltage gains.
4. The ratio of channel width to channel length, called the aspect ratio, is quite significant in MOSFET circuit design.

In amplifier design, MOS circuits have become increasingly important, but not to the extent of CMOS digital circuits. Many of the MOSFET designs are nMOS or pMOS circuits rather than CMOS circuits, but may be referred to as CMOS designs, because an established CMOS process is used to create the design even though no devices appear in the complementary or CMOS configuration.

Whereas digital circuit design almost eliminates the need for resistors or capacitors, analog circuit design may use resistors or capacitors in the design of amplifiers. It is useful to minimize the need for these elements, since they may occupy spaces in which tens or hundreds of MOS devices may fit. In the next few subsections we will discuss MOSFET amplifier configurations. Before proceeding to these amplifiers, some useful relations for the MOSFET will be tabulated. These are written in terms of an nMOS device in Table 9.1 and were derived in Chapter 6.

PRACTICE Problem

9.4 A MOSFET has values of $g_m = 500\ \mu\text{A/V}$ and $r_{ds} = 130\ \text{k}\Omega$ at $I_D = 120\ \mu\text{A}$. Approximate the values of g_m and r_{ds} for this device at $I_D = 1\ \text{mA}$. *Ans:* $g_m = 1.44\ \text{mA/V}$, $r_{ds} = 15.6\ \text{k}\Omega$.

9.2.1 SIMPLE AMPLIFIER STAGES

A simple stage that can produce a controlled voltage gain is the diode-connected load stage, shown in Fig. 9.4. The numbers in circles represent the node numbers to be used in a later

Table 9.1   Useful Relations for an nMOS Device

DC Equations

$V_{\text{eff}} = V_{GS} - V_T$	Positive for triode or active region
	Zero or negative for subthreshold or cutoff
$V_{DSP} = V_{GS} - V_T = V_{\text{eff}}$	Drain-source pinchoff voltage
	Borders triode and active region
$I_D = \frac{\mu C_{\text{ox}} W}{L}\left[(V_{GS} - V_T)V_{DS} - \frac{V_{DS}^2}{2}\right]$	Triode region
$I_D = \frac{\mu C_{\text{ox}} W}{2L}[V_{GS} - V_T]^2[1 + \lambda(V_{DS} - V_{\text{eff}})]$	Active region

Small-Signal Equations

$g_m = \sqrt{2\mu C_{\text{ox}}(W/L)I_D} = \frac{2I_D}{V_{\text{eff}}}$	Transconductance
$r_{ds} = \frac{1}{\lambda I_{DP}}$	Drain-source resistance

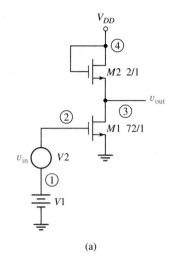

(a)

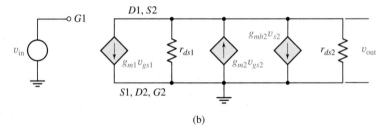

Figure 9.4
(a) An amplifier with a diode-connected load.
(b) Equivalent circuit.

(b)

simulation. This amplifier configuration demonstrates several important points relating to general MOSFET IC amplifier design. The transconductance term, g_{mb2}, accounts for the body effect as the source to body voltage changes in device $M2$.

The amplifier can be treated as a gain stage, $M1$, driving a load impedance where the load impedance is that looking into the source of $M2$. Figure 9.5 shows the model for $M2$ used to calculate the impedance at the source terminal. The impedance, R_{S2}, looking into $S2$ can be found by assuming that a voltage, v_{s2}, is applied to $S2$. The resulting current is calculated and divided into v_{s2} to find R_{S2}.

With the gate voltage of $M2$ tied to ground, the gate-to-source voltage reduces to $v_{gs2} = v_{g2} - v_{s2} = -v_{s2}$. When v_{s2} is applied to the source of $M2$, the current into this terminal is

$$i_{s2} = g_{m2}v_{s2} + g_{mb2}v_{s2} + \frac{v_{s2}}{r_{ds2}}$$

Figure 9.5
Circuit used for impedance calculation.

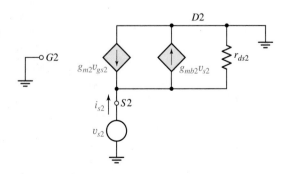

The impedance looking into this terminal is then

$$R_{S2} = \frac{v_{s2}}{i_{s2}} = \frac{1}{g_{m2} + g_{mb2} + g_{ds2}} \tag{9.6}$$

where $g_{ds2} = 1/r_{ds2}$.

The midband voltage gain is found by recognizing that device $M1$ sees this load in parallel with its own output impedance, r_{ds1}. This parallel resistance is the output resistance, R_{out}, which can be expressed as

$$R_{\text{out}} = r_{ds1} \parallel R_{S2} = \frac{1}{g_{m2} + g_{mb2} + g_{ds2} + g_{ds1}} \tag{9.7}$$

The midband voltage gain is then

$$A_{MB} = -g_{m1} R_{\text{out}} = -\frac{g_{m1}}{g_{m2} + g_{mb2} + g_{ds2} + g_{ds1}} \tag{9.8}$$

Typically, the largest term in the denominator will be g_{m2}, which allows the gain to be approximated by

$$A_{MB} \approx -\frac{g_{m1}}{g_{m2}} \tag{9.9}$$

For example, a 1-μ device ($L = 1\mu$) may have values of $g_m = 0.2$ mA/V, $g_{mb2} = 0.022$ mA/V, and $g_{ds} = 0.001$ mA/V, which leads to a 12% error in the approximation of Eq. (9.9).

Examination of Eq. (9.9) shows that, for higher gains, g_{m1} must be much larger than g_{m2}, which can be accomplished by controlling the aspect ratios (W/L) of the two devices. The approximate variation of gain with device scaling can be derived from a consideration of the equation for transconductance in the active region. This equation is found from Table 9.1 and is

$$g_m = \sqrt{2\mu_n C_{\text{ox}} (W/L) I_D} \tag{9.10}$$

Since the drain currents of both devices are equal, the voltage gain approximation of Eq. (9.9) can now be written as

$$A_{MB} \approx -\frac{\sqrt{2\mu_n C_{\text{ox}} (W_1/L_1) I_D}}{\sqrt{2\mu_n C_{\text{ox}} (W_2/L_2) I_D}} = -\left[\frac{(W_1/L_1)}{(W_2/L_2)} \right]^{1/2} \tag{9.11}$$

It can be seen from Eq. (9.11) that the magnitude of the voltage gain varies as the square root of the ratio of aspect ratios. If a gain of -6 is required, both devices can have the same channel length, with device $M1$ using a channel width that is 36 times that of $M2$.

We recognize that the actual gain may be slightly lower than the value indicated by Eq. (9.11), but the approximation is sufficient to begin a design that can be "tweaked" during simulation.

One consideration that becomes more important as power-supply voltages are lowered in value is that of *headroom*. This term is used to indicate how much of the output voltage swing cannot be used if serious distortion is to be avoided. If a stage that uses a 5-V dc power supply has a headroom of 1.2 V, then the maximum usable output voltage range is $5 - 1.2 = 3.8$ V. For the circuit of Fig. 9.4, the output can be driven to within a few tenths of a volt of ground by a positive input signal. The drain voltage can swing to its pinchoff

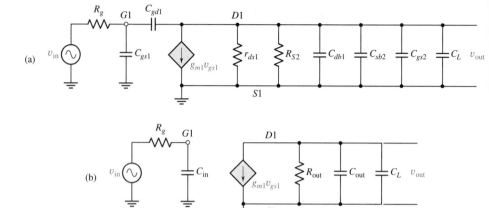

Figure 9.6
(a) High-frequency model of
diode-connected load stage.
(b) Simplified high-frequency
model.

value, which is $V_{GS1} - V_{Tn} = V_{eff1}$. This value is often near 200–300 mV. When the output swings positive, the load device must continue to conduct current; thus the gate-to-source voltage must equal or exceed the threshold voltage plus the effective voltage. The headroom for this stage is then a few tenths of a volt, perhaps 0.3 V, plus the threshold voltage of $M2$. This headroom voltage is approximately V_{GS1}.

The upper corner frequency of an amplifier stage is often significant, but depends on generator resistance and load capacitance as well as on the stage itself. Figure 9.6(a) shows the high-frequency model for this stage including a load capacitance and a generator resistance. Capacitor C_{gd} is gate-to-drain capacitance, C_{gs} is gate-to-source capacitance, C_{db} is drain-to-bulk capacitance, and C_{sb} is the source-to-bulk capacitance. The capacitance C_{gd1} bridges the input and output nodes and can be reflected to the input and output terminals using the Miller effect. The input capacitance of Fig. 9.6(b) is $C_{in} = C_{gs1} + (1 + |A_{MB}|)C_{gd1}$, and the output capacitance is $C_{out} = C_{gd1} + C_{db1} + C_{sb2} + C_{gs2}$.

From Eq. (9.7), the output resistance seen by the output capacitance is

$$R_{out} = \frac{1}{g_{m2} + g_{ds1} + g_{mb2} + g_{ds2}} \approx \frac{1}{g_{m2} + g_{mb2}} \qquad (9.12)$$

The amplifying stage has two upper corner frequencies; one caused by the input circuit and one caused by the output circuit. These frequencies are calculated by

$$f_{in-high} = \frac{1}{2\pi R_g C_{in}} \qquad (9.13)$$

$$f_{out-high} = \frac{1}{2\pi R_{out}(C_{out} + C_L)} \qquad (9.14)$$

Depending on circuit values, these frequencies may be widely or narrowly separated. If widely separated—for example, if the higher one is at least 5 times the lower frequency—then the lower frequency approximates the overall upper corner frequency, f_{2o}. If the two frequencies are different by less than a factor of five, the method of Chapter 3 must be used to calculate the overall upper corner frequency.

The circuit of Fig. 9.4, implemented on a 0.5-μ process, but with gate lengths of 1 μ, is simulated by PSpice© to demonstrate several of the points made in this section. The Spice netlist file is shown in Table 9.2. A 5-V power supply is used.

Table 9.2 Spice Netlist File for Amplifier with Diode-Connected Load

```
CH9.CIR

V1 1 0 1.0V
V2 2 1 AC 0.005V
V3 4 0 5V
M1 3 2 0 0 N L=1U W=72U AD=360P AS=360P PD=82U PS=82U
M2 4 4 3 0 N L=1U W=2U AD=10P AS=10P PD=12U PS=12U
.AC DEC 10 100 1000MEG
.OP
.PROBE
.LIB C5X.LIB
.END
```

This program uses a model for the MOSFETs named C5X.LIB.

PRACTICAL Considerations

In specifying a MOSFET device for analysis by Spice, the length and width of the channel are first specified. The surface area of the drain and source follow; then the external perimeters of the drain and source are specified. For device $M1$, the channel length in microns is 1 and the width is 72. The area of the drain is determined by W_1 and the layout length of the drain region. For this device, the width is 72; thus the length of the drain region must be 5 to result in an area of $72 \times 5 = 360$ square microns. The perimeter used for the drain is not equal to $2W_1 + 2L_D$. Rather, it is the perimeter of the drain region minus the width of the channel: $PD = W_1 + 2L_D = 72 + 10 = 82$. Certain capacitances are based on the external perimeter of the associated region, but the capacitance associated with the side of the region that abuts the channel is accounted for in a separate calculation. The same considerations apply to the source terminal also.

The circuit is first simulated with neither source resistance nor load capacitance. The results of this simulation are $A_{MBsim} = -7.15$ V/V, $f_{2o-sim} = 202.1$ MHz, with a headroom of about 2.5 V. The headroom is found by doing a dc scan of the input voltage and watching the output voltage for departures from linearity.

From calculation the midband voltage gain is found to be

$$|A_{MB}| \approx \left[\frac{(W_1/L_1)}{(W_2/L_2)} \right]^{1/2} = \sqrt{36} = 6 \text{ V/V}$$

The capacitance values given by the Spice simulation are

$$C_{in} = C_{gs1} + (1 + |A_{MB}|)C_{gd1} = 168 + (1 + 6)21.6 = 319.2 \text{ fF}$$

and

$$C_{out} = C_{gd1} + C_{db1} + C_{sb2} + C_{gs2} = 21.6 + 141 + 7.5 + 5 = 175.1 \text{ fF}$$

The output resistance, calculated from Eq. (9.12) using parameters from the simulation, is $R_{out} = 4.35$ kΩ. The upper corner frequency with no source resistance and no load

Table 9.3 Summary of Simulation Results for the Diode-Connected Load Stage

$A_{MBsim} = -7.15$ V/V		
C_L, pF	R_g, kΩ	f_{2o-sim}, MHz
0	0	202
1	0	30.1
0	339	1.52
1	339	1.48

capacitance is then

$$f_{2o} = \frac{1}{2\pi R_{out} C_{out}} = \frac{1}{2\pi \times 4350 \times 175.1 \times 10^{-15}} = 209 \text{ MHz}$$

When a load capacitance of 1 pF is added across the output terminals, the simulation shows an upper corner frequency of 30 MHz, and the calculation leads to a value of 31.1 MHz.

If a rather large signal generator resistance of 339 kΩ, which will also be used in succeeding circuits for comparison purposes, is added to the circuit along with the 1-pF output capacitance, the simulated value of upper corner frequency is 1.48 MHz. The upper corner frequency is lowered from 30 MHz to 1.48 MHz, due to the generator resistance and input capacitance. The overall upper corner frequency must then be caused primarily by the input circuit; thus, we calculate a value of

$$f_{2o} = \frac{1}{2\pi R_g C_{in}} = \frac{1}{2\pi \times 339 \times 10^3 \times 319.2 \times 10^{-15}} = 1.47 \text{ MHz}$$

If a generator resistance that equals the output resistance of this stage is used—that is, $R_g = 4.35$ kΩ—the upper corner frequency is 83.7 MHz without C_L and 26.1 MHz with $C_L = 1$ pF added. Table 9.3 summarizes these results for the diode-connected load stage.

PRACTICAL Considerations

Several practical points relating to IC design can be based on this analysis.

1. Although the load is not a resistor, the same techniques used in the analysis of discrete circuits are still valid. The equivalent resistance of the MOSFET load is first found and substituted for the load resistance.

2. The aspect ratio is important in determining the performance of the circuit. Very large aspect ratios often result in analog design, while high-frequency digital circuits generally keep this ratio at a low value to minimize capacitance and required *real estate* or chip volume. Schematics of analog MOSFET circuits label the width and length of each device near the device, as shown in Fig. 9.4.

3. Approximate results are useful to provide a starting point for circuit simulations that are a necessity before an IC chip is laid out. Fabrication runs are very expensive, and mistakes must be avoided to minimize cost. Thus, the simulation step is never omitted in the IC design process. This step will use parameters for the MOS devices that are based on the actual process to be used in fabrication.

4. Headroom may be an important consideration, since IC chips often use low-voltage dc supplies.

There are some disadvantages to the stage discussed in this section that limit its usefulness. Larger channel areas lead to increased capacitance, so the bandwidth of high-gain stages will be less than that of lower-gain stages for the diode-connected load stage. On the other hand, it is a simple stage that has a low output impedance and is, therefore, not affected to a great extent by load capacitance.

In the following chapter we will see that the BJT never uses the configuration of Fig. 9.4, since the impedance looking into the emitter is r_e. This load impedance is too low to achieve a significant voltage gain. We will also see that the load impedance cannot be increased by scaling the areas of a BJT, since r_e is not a function of area. This difference is very significant in BJT design and MOSFET design, the latter of which uses aspect ratio as a critical design parameter.

The load device of Fig. 9.4 can be replaced by a pMOS device, as shown in Fig. 9.7. This configuration eliminates the body effect of the load device and increases the resistance due to a smaller value of μ_p compared to μ_n. The value of μ_n is about three times that of μ_p. The approximation of Eq. (9.9) becomes more accurate, and the voltage gain can be written as

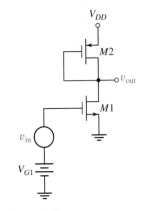

Figure 9.7
A diode-connected pMOS load.

$$|A_{MB}| = \left[\frac{\mu_n C_{oxn}(W_1/L_1)}{\mu_p C_{oxp}(W_2/L_2)}\right]^{1/2} \approx \left[\frac{3(W_1/L_1)}{(W_2/L_2)}\right]^{1/2} \tag{9.15}$$

To approximate the upper corner frequency of the diode-connected pMOS stage, the capacitances of Fig. 9.8 are added. It is generally true that $C_{ds} \ll C_{db}$, giving a total capacitance from output to ground of

$$C_{out} = C_{db1} + C_{gd1} + C_{db2} + C_{gs2}$$

The output resistance can be found as

$$R_{out} = \frac{1}{g_{m2} + g_{ds2} + g_{ds1}} \approx \frac{1}{g_{m2}}$$

It is left as an exercise for the reader to derive this expression for output impedance.

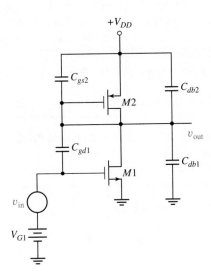

Figure 9.8
Parasitic capacitances determining the upper 3-dB frequency.

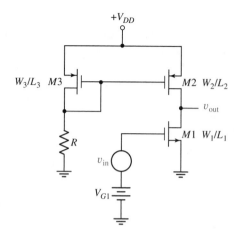

Figure 9.9
A MOSFET stage with an active current source load.

PRACTICE **Problem**

9.5 Two n-devices are used in a diode-connected amplifier stage similar to that of Fig. 9.4(a). If $L_1 = L_2 = 0.6\ \mu$, calculate W_1/W_2 for a midband voltage gain of -4.3 V/V. *Ans:* 18.5.

The upper corner frequency is then

$$f_{2o} = \frac{1}{2\pi C_{out} R_{out}} \tag{9.16}$$

This value is approximately

$$f_{2o} = \frac{g_{m2}}{2\pi (C_{db1} + C_{db2} + C_{gs2} + C_{gd1})} \tag{9.17}$$

In many cases, the load capacitance or input capacitance to the next stage will lower f_{2o} from this value, as seen earlier in this section.

9.2.2 ACTIVE LOAD STAGE

A stage used often in IC design is the active load stage. Typically, the active load is a current source, often based on the current mirror. A simple MOSFET active load stage appears in Fig. 9.9.

There is a compelling reason to use active load stages in IC design. The device that acts as the load for the amplifying stage can present a large incremental load while allowing a reasonable drain current to flow. This large incremental load leads to a high gain for the stage, while the dc current can lead to acceptable values of transconductance and bias current. If a fixed resistor with a large value were used to achieve high gain, the dc drop across this element would be prohibitive for normal bias currents.

PRACTICAL **Considerations**

A simple current mirror might provide 100 μA of current to an amplifying stage while presenting 60 kΩ of incremental resistance to the stage. If a simple 60-kΩ resistance replaced the current mirror, the same incremental resistance would prevail, but the dc voltage drop across the resistor would be

$$V_R = 0.1 \times 60 = 6\ \text{V}$$

The dc power supply for many MOSFET IC amplifiers is 5 V or less; thus, using a simple resistor is out of the question.

The difference between an active load and a simple resistor becomes even more apparent when a more complex current mirror is used to produce a resistive load of several hundred kilohms or greater. The dc drop across a several hundred kilohm resistor, conducting a typical bias current, might be over 100 V.

The pMOS current mirror of Fig. 9.9 provides bias current to the amplifying device, $M1$. This device also sees an incremental resistive load that equals

$$R_{\text{out}} = r_{ds1} \parallel r_{ds2} = \frac{1}{g_{ds1} + g_{ds2}} \tag{9.18}$$

The midband voltage gain is now

$$A_{MB} = -g_{m1} R_{\text{out}} \tag{9.19}$$

Since the values of r_{ds} can be high for both devices, the voltage gain for a single stage can also be reasonably high.

The output voltage can be driven positive to the point that $M2$ reaches its pinchoff value for V_{DS}. This voltage may be $V_{DD} - V_{\text{eff2}} = V_{DD} - 0.4$ V. In the negative direction, the output can be driven to the point that $M1$ reaches pinchoff, which is equal to $V_{\text{eff1}} = 0.4$ V. The headroom is perhaps 0.6–0.8 V.

The output capacitance for the amplifier of Fig. 9.9 is

$$C_{\text{out}} = C_{db1} + C_{db2} + C_{gd1} + C_{gd2}$$

which gives an upper corner frequency of

$$f_{2o} = \frac{(g_{ds1} + g_{ds2})}{2\pi(C_{db1} + C_{db2} + C_{gd1} + C_{gd2})} \tag{9.20}$$

Again, we emphasize that a load capacitance or input capacitance of a following stage will lower this value.

EXAMPLE 9.2

The current mirror of Fig. 9.10 supplies a current of 50 μA to the amplifying stage. The dc output voltage is adjusted by voltage source, V_1, to be 2.4 V. If $g_{m1} = 0.19$ mA/V, $g_{ds1} = 0.95$ μA/V, $g_{ds2} = 2$ μA/V, $C_{db1} = 11.0$ fF, $C_{db2} = 32.0$ fF, $C_{gd1} = 1.5$ fF, and $C_{gd2} = 4.5$ fF,

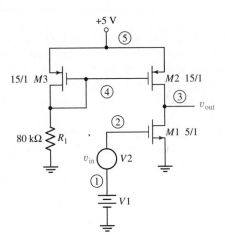

Figure 9.10
Amplifier for Example 9.2.

(a) Calculate the midband voltage gain, A_{MB}.

(b) Calculate the upper corner frequency, f_{2o}.

(c) Verify results with a Spice simulation.

SOLUTION The output resistance for this stage is

$$R_{\text{out}} = \frac{1}{g_{ds1} + g_{ds2}} = \frac{10^6}{2.95} = 339 \text{ k}\Omega$$

Using Eq. (9.19) allows the voltage gain to be found as

$$A_{MB} = -g_{m1} R_{\text{out}} = -0.19 \times 10^{-3} \times 339 \times 10^3 = -64.4 \text{ V/V}$$

The output capacitance is

$$C_{\text{out}} = C_{db1} + C_{db2} + C_{gd1} + C_{gd2} = 11.0 + 32.0 + 1.5 + 4.5 = 49 \text{ fF}$$

From Eq. (9.20), this gives an upper corner frequency of

$$f_{2o} = \frac{1}{2\pi C_{\text{out}} R_{\text{out}}} = \frac{1}{2\pi \times 49 \times 10^{-15} \times 339 \times 10^3} = 9.58 \text{ MHz}$$

The Spice netlist file is listed in Table 9.4. Corresponding node and element numbers are shown in Fig. 9.10.

Table 9.4 Spice Netlist File for Example 9.2

```
EX9-2.CIR

R1 0 4 80K
V1 1 0 1.275V
V2 2 1 AC 0.005V
V3 5 0 5V
M1 3 2 0 0 N L=1U W=5U AD=25P AS=25P PD=15U PS=15U
M2 3 4 5 5 P L=1U W=15U AD=75P AS=75P PD=25U PS=25U
M3 4 4 5 5 P L=1U W=15U AD=75P AS=75P PD=25U PS=25U
.AC DEC 10 100 10MEG
.OP
.PROBE
.LIB C5X.LIB
.END
```

The results of this simulation are $A_{MB\text{sim}} = -63.6$ V/V and $f_{2o-\text{sim}} = 8.53$ MHz, which compare well with the calculated results. A dc scan on the input voltage allows the active region of the output voltage to be evaluated. For this circuit, the active region extends from 0.4 V to 4.6 V, which also agrees well with theory.

Adding Generator Impedance and Load Impedance Only the output capacitance was used in developing Eq. (9.20) for upper corner frequency. As mentioned in connection with the diode-connected load stage, there are two other considerations that must be made for the practical circuit. One is the additional capacitance added between output and ground due to the input capacitance of the following stage or from the output pad of an IC chip. The second is the generator resistance or output resistance of the previous stage that drives the input capacitance of this stage. Both loops must be considered to result in an accurate upper corner frequency.

Let us suppose that the output of the amplifier of Fig. 9.10 is brought to an output pad and pin of an IC chip. The capacitance associated with the pad might exceed 1 pF. If the simulation is repeated with a 1-pF load added, the upper corner frequency drops from 8.53 MHz to 431 kHz, a very large drop.

If this stage were loaded by a comparable stage rather than by 1 pF, the corner frequency might drop by a factor of two rather than a factor of 20. In order to avoid this drop, a buffer stage may be added to drive the 1-pF load without the large reduction in corner frequency. Another possibility is to increase the values of W for all the devices. Although this increases the capacitance, the output resistance decreases and g_m increases to result in a comparable voltage gain. The effect of the load capacitance on upper corner frequency will now be much less.

As mentioned earlier, the stage of Fig. 9.10 is driven by a perfect voltage generator with zero output resistance. If this stage were driven by an identical stage, the output impedance of the first stage would become the generator resistance for the second stage. To demonstrate the effect of generator resistance on upper corner frequency, a resistance of 339 kΩ is used as a generator resistance for the circuit of Fig. 9.10. This particular value of resistance will be used in succeeding examples for comparison purposes. No external load capacitance is used in this simulation. The simulated upper corner frequency is lowered in this situation from a value of 8.35 MHz to 3.08 MHz.

This value can be calculated by noting that the input circuit will now cause a corner frequency determined by the generator resistance and the input capacitance. The input capacitance equals the sum of C_{gs1}, C_{gb1}, and the Miller effect capacitance $(1 + |A_{MB}|)C_{gd1}$. From the output file of the simulation, these values are

$$|A_{MB}| = 63.6 \text{ V/V} \qquad C_{gs1} = 13.35 \text{ fF} \qquad C_{gd1} = 1.5 \text{ fF} \qquad C_{gb1} = 0.3 \text{ fF}$$

The total input capacitance resulting is approximately 111 fF. With a value of $R_g = 339$ kΩ for generator resistance, this adds a corner frequency of

$$f_{\text{in-high}} = \frac{1}{2\pi C_{\text{in}} R_g} = 4.23 \text{ MHz}$$

The amplifier now has an input corner frequency of 4.23 MHz and an output corner frequency of 8.53 MHz. We use the method of Chapter 3 to calculate the overall upper corner frequency when two single-pole upper corner frequencies make up the amplifier response. The result is a calculated overall corner frequency of $f_{2o-\text{calc}} = 3.56$ MHz. This value exceeds the simulated value of 3.08 MHz by 15%.

If a generator resistance of 339 kΩ and a 1-pF load capacitance are both added to the amplifier, the new upper corner frequency is found from simulation to be 399 kHz. Table 9.5 summarizes the results of this simulation.

The active current source load provides a method to achieve high midband voltage gains without the large discrepancy in size required by the diode-connected load amplifier.

Table 9.5   Summary of Simulation Results for the Current-Source Load Stage

$A_{MB\text{sim}} = -63.6$ V/V		
C_L, pF	R_g, kΩ	$f_{2o-\text{sim}}$, MHz
0	0	8.53
1	0	0.43
0	339	3.08
1	339	0.40

PRACTICE Problems

9.6 Work Example 9.2 for a generator resistance of $R_g = 50$ kΩ.
9.7 Work Example 9.2 for a generator resistance of $R_g = 50$ kΩ and an input capacitance to a following stage of $C_{in2} = 100$ fF.
9.8 The drain current of the current mirror output stage of Fig. 9.10 is increased to 100 μA. Assuming that the capacitances remain constant, calculate the new midband voltage gain and upper corner frequency for the amplifier.
Ans: $A_{MB} = -45.4$ V/V, $f_{2o} = 19.2$ MHz.
9.9 Work Example 9.3 if a midband voltage gain of 180 V/V is required.
Ans: $W = 457 \mu$.

It suffers from poor frequency response when a large capacitive load, much larger than the output capacitance of the stage, is present.

We can note that the midband voltage gain given by Eq. (9.19) will be increased if R_{out} is increased. This resistance can be increased by using a more complex current mirror to make the output resistance much larger than the output resistance of the amplifying stage. While this increases the midband voltage gain, the upper corner frequency due to the output loop, calculated from Eq. (9.20), decreases by the same factor. If this corner frequency is the dominant one, midband voltage gain and bandwidth can be exchanged directly by varying the output resistance of the current mirror.

PRACTICAL Considerations

Although the upper corner frequency of the active load stage is significantly affected by a 1-pF capacitance, it is possible to minimize this problem. The channel width of the output stage can be increased considerably to result in an increase in output capacitance and a corresponding decrease in output resistance. The upper corner frequency with no external load can be approximately equal to that of the smaller device. However, when a 1-pF load capacitance is added, the percentage increase of capacitance is much less for the large device than for the small device. The effect on upper corner frequency is then much less for the larger device.

The effect of channel width, W, on the voltage gain can be seen in the following example.

EXAMPLE 9.3

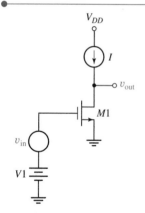

Figure 9.11
Circuit for Example 9.3.

In the circuit of Fig. 9.11, the current source generates 100 μA of current and has an incremental output resistance of $r_{cs} = 100$ kΩ. The device M1 is a $1 = \mu$ gate length device with $\mu C_{ox} = 0.06$ mA/V^2 and $\lambda = 0.03$ V^{-1}. Find the channel width to result in a midband voltage gain of 100 V/V.

SOLUTION The midband voltage gain of this stage is

$$A_{MB} = -g_m R_{out}$$

where

$$R_{out} = r_{cs} \| r_{ds1}$$

The value of r_{ds1} can be approximated by assuming that the drain pinchoff current is approximately equal to 100 μA. The result is

$$r_{ds1} = \frac{1}{\lambda I_D} = \frac{1}{0.03 \times 0.1} = 333 \text{ k}\Omega$$

Using the 100-kΩ impedance of the current source, the output resistance is found to be

$$R_{out} = 100 \| 333 = 76.9 \text{ k}\Omega$$

Recalling that

$$g_m = \sqrt{2\mu C_{ox}(W/L)I_D}$$

we can write the magnitude of the voltage gain as

$$|A_{MB}| = g_m R_{out} = \sqrt{2 \times 0.06 \times (W/L) \times 0.1} \times 76.9 = 100 \text{ V/V}$$

Solving this equation for the ratio of W/L leads to a value of 141 μ for W.

9.2.3 SOURCE FOLLOWER WITH ACTIVE LOAD

The source follower provides a buffer stage, but the midband voltage gain is low, even less than the value of unity approached by the BJT emitter-follower stage. The bandwidth is quite high for both the emitter-follower and the source-follower stages. Figure 9.12 demonstrates a source follower with a current mirror load.

The device $M2$ presents a resistance of r_{ds2} between the source of $M1$ and ground. In addition, device $M1$ presents a resistance of r_{ds1} in parallel with $1/g_{mb1}$ to the dc power supply, which is also ground for incremental signals. Again we note that the body effect in $M1$ must be included, since the source-to-substrate voltage of this device varies with the output signal. In fact, it equals the output signal.

The circuit of Fig. 9.12(b) is redrawn in Fig. 9.13 and the pertinent parasitic capacitances are added. The current source strength, $g_{m1}v_{gs1}$, can be written as $g_{m1}(v_{in} - v_{out})$. In the equivalent circuit, this current can be generated by two separate sources, as shown in Fig. 9.13. The source terminal now appears at the top of the figure while the drain terminal is grounded.

The two current sources $g_{m1}v_{out}$ and $g_{mb1}v_{out}$ can be converted to conductances g_{m1} and g_{mb1}, respectively. As a voltage appears at $S1$, the currents through these conductances equal the values that would be generated by the sources. Figure 9.14 shows an alternate equivalent circuit that is used to find the voltage gain as a function of frequency. This circuit results from taking a Thévenin equivalent of the current source, $g_{m1}v_{in}$, and the parallel resistance, R_{out}.

The circuit of Fig. 9.14 is analyzed to find that

$$A(j\omega) = g_{m1} R_{out} \frac{1 + j\omega C_{gs1}/g_{m1}}{1 + j\omega(C_{gs1} + C_{out})R_{out}} \tag{9.21}$$

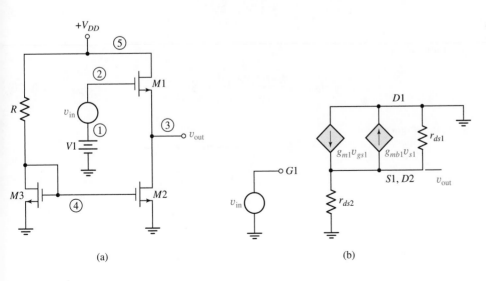

(a) (b)

Figure 9.12
(a) A source-follower stage with current source load.
(b) Equivalent circuit.

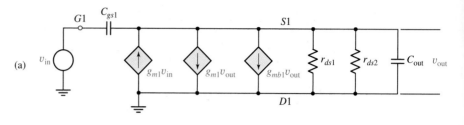

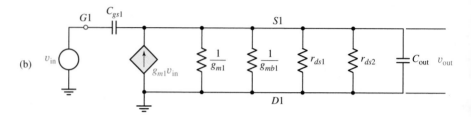

Figure 9.13
*(a) The high-frequency
source-follower equivalent
circuit. (b) Using
transconductances
for two current sources.*

where

$$R_{\text{out}} = \frac{1}{g_{m1} + g_{mb1} + g_{ds1} + g_{ds2}} \tag{9.22}$$

and

$$C_{\text{out}} = C_{sb1} + C_{db2} + C_{gd2} \approx C_{sb1} + C_{db2} \tag{9.23}$$

The midband gain can be evaluated from Eqs. (9.21) and (9.22) to be

$$A_{MB} = \frac{g_{m1}}{g_{m1} + g_{mb1} + g_{ds1} + g_{ds2}} \tag{9.24}$$

In many submicron processes, the value of the denominator of Eq. (9.24) might equal 1.15 to 1.2 times g_{m1}. This leads to values of midband gain ranging from about 0.8 to 0.9 V/V.

The transfer function for voltage gain as a function of frequency shows a zero at

$$f_{\text{zero}} = \frac{g_{m1}}{2\pi C_{gs1}} \tag{9.25}$$

and a pole at

$$f_{\text{pole}} = \frac{g_{m1} + g_{mb1} + g_{ds1} + g_{ds2}}{2\pi (C_{gs1} + C_{\text{out}})} \tag{9.26}$$

Figure 9.14
*Alternate equivalent circuit of
the source follower.*

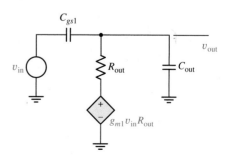

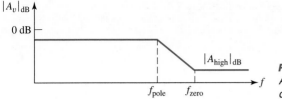

Figure 9.15
*Asymptotic frequency response
of source follower.*

Typically, the zero frequency is higher than the pole frequency, and the asymptotic frequency response appears as in Fig. 9.15. At high frequencies the capacitor C_{gs1} feeds through from input directly to output, causing a gain of

$$A_{high} = \frac{C_{gs1}}{C_{gs1} + C_{out}} \tag{9.27}$$

As we consider the large-signal operation of the source follower, we see that the output voltage can be driven positive to a value of $V_{DD} - V_{eff1} - V_{T1}$. The negative voltage can reach V_{eff2} before $M2$ leaves the active region. The headroom is then $V_{T1} + V_{eff1} + V_{eff2}$.

EXAMPLE 9.4

In the source follower and current mirror of Fig. 9.12, a dc drain current of 50 μA passes through $M1$ and $M2$. The following parameters apply to the devices of the circuit at this bias point: $g_{m1} = 199$ μA/V, $g_{mb1} = 44$ μA/V, $g_{ds1} = 1$ μA/V, $g_{ds2} = 1.6$ μA/V, $C_{gs1} = 7.1$ fF, $C_{sb1} = 12.6$ fF, $C_{db2} = 12.7$ fF, and $C_{gd2} = 1.5$ fF. Calculate the midband voltage gain, the pole frequency or upper corner frequency, and the zero frequency. Also calculate the high-frequency gain due to capacitor feedthrough of C_{gs1}. Simulate this circuit with Spice.

SOLUTION The midband voltage gain is calculated from Eq. (9.24) as

$$A_{MB} = \frac{g_{m1}}{g_{m1} + g_{mb1} + g_{ds1} + g_{ds2}} = \frac{199}{245.6} = 0.810 \text{ V/V}$$

The pole frequency that will be near the upper corner frequency is found from Eq. (9.26) to be

$$f_{pole} = \frac{g_{m1} + g_{mb1} + g_{ds1} + g_{ds2}}{2\pi(C_{gs1} + C_{out})} = \frac{245.6 \times 10^{-6}}{2\pi \times 33.9 \times 10^{-15}} = 1.15 \text{ GHz}$$

From Eq. (9.25), the zero frequency is calculated to be

$$f_{zero} = \frac{g_{m1}}{2\pi C_{gs1}} = \frac{199 \times 10^{-6}}{2\pi \times 7.1 \times 10^{-15}} = 4.46 \text{ GHz}$$

The high-frequency gain is

$$A_{high} = \frac{C_{gs1}}{C_{gs1} + C_{out}} = \frac{7.1}{33.9} = 0.209 \text{ V/V}$$

Table 9.6 Spice File for Example 9.4

```
EX9-4.CIR

R1 4 5 75K
V1 1 0 3.0V
V2 2 1 AC 0.1V
V3 5 0 5V
M1 5 2 3 0 N L=1U W=5U AD=25P AS=25P PD=15U PS=15U
M2 3 4 0 0 N L=1U W=5U AD=25P AS=25P PD=15U PS=15U
M3 4 4 0 0 N L=1U W=5U AD=25P AS=25P PD=15U PS=15U
.AC DEC 10 100 10000MEG
.OP
.PROBE
.LIB C5X.LIB
.END
```

The Spice simulation file is listed in Table 9.6.

The frequency response resulting from this simulation is shown in Fig. 9.16. The simulated midband gain is 0.811 V/V, which is very close to the calculated value of 0.810 V/V. The upper corner frequency from the simulation is 1.01 GHz, compared to a calculated value of 1.15 GHz. The simulated zero frequency is 6.15 GHz, and the calculated value is 4.46 MHz. The high-frequency gain had a simulated value of 0.16 V/V, whereas the calculated value was 0.21 V/V.

A dc scan on the input voltage shows an output active region extending from approximately 0.32 to 3.03 V. The lower voltage is approximately $V_{\text{eff}2}$, and the upper voltage should approximate $V_{DD} - V_{\text{eff}1} - V_{T1}$.

Figure 9.16
Simulated frequency response for source follower.

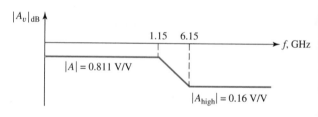

We note that the upper corner frequency for the source follower is 1.01 GHz, compared to 8.53 MHz for the comparable-sized current source load stage of Example 9.2. The primary reason for this is the difference in the output resistance of the two stages. For the current source load stage, the output resistance is 339 kΩ, and the source-follower output resistance is about 4 kΩ. In order to compare this stage when driven by a signal generator resistance, a 339-kΩ resistance is inserted in the input lead. This simulation leads to an overall upper corner frequency of 101.5 MHz. The corresponding value for the current source load of Example 9.2 is 3.08 MHz.

With a capacitive load of 1 pF and a generator resistance of 339 kΩ, a Spice simulation shows an upper corner frequency of 37 MHz. Although this value is much lower than the unloaded value, it is considerably more than the corresponding frequency of the current source load stage, which was 399 kHz. Table 9.7 summarizes these results.

PRACTICAL Considerations

As noted earlier, capacitive loading can be a serious problem for an output stage that connects to an IC pin, which will connect to the input of another circuit. Because of the high output resistance of configurations such as the active load stage, a small capacitance of a pF or less will decrease the upper corner frequency significantly. The source-follower stage is often inserted between a high-gain stage and the output pin of a chip to serve as a buffer stage. This buffer presents a rather low capacitance to the preceding stage, since there is no Miller multiplication of C_{gd} and, as seen from the simulation, a 1-pF load capacitance can result in a reasonably high upper corner frequency.

PRACTICE Problems

9.10 For the source follower of Fig. 9.12, derive the output resistance looking into the source terminal of $M1$.
9.11 If g_{ds1} and g_{ds2} are considered negligible for the source follower of Example 9.4, what will the midband voltage gain be?
Ans: 0.819 V/V.

Table 9.7   Summary of Simulation Results for Source Follower

$A_{MBsim} = 0.811$ V/V

C_L, pF	R_g, kΩ	f_{2o-sim}, MHz
0	0	1,010
1	0	36.6
0	339	101.6
1	339	37.2

9.2.4 THE CASCODE CONNECTION

The cascode connection has been used for many years in high-frequency BJT circuits. It has become more important in IC MOSFET design in recent years. We will discuss the MOSFET cascode circuit in this section. Figure 9.17(a) shows the basic cascode circuit with a current source load.

The equivalent circuit for the upper device would include a current generator of value $g_{m2}v_{gs2}$. For this connection, $v_{gs2} = -v_{s2}$; thus this current generator reverses the direction of current flow and removes the negative sign to lead to the value of $g_{m2}v_{s2}$, shown in Fig. 9.17(b).

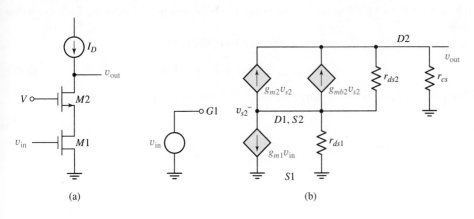

Figure 9.17
(a) The MOSFET cascode amplifier stage. (b) The equivalent circuit.

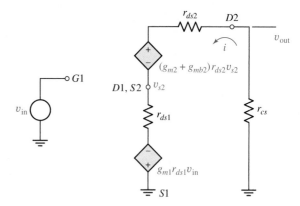

Figure 9.18
An alternate equivalent circuit for the cascode.

The resistance r_{cs} is the output resistance of the current source. For a simple source, this value would be r_{ds}. For a Wilson mirror, it would be considerably larger. The equivalent circuit can be converted to a Thévenin circuit, as shown in Fig. 9.18. The output voltage is

$$v_{out} = -ir_{cs}$$

where i is the incremental drain current. Note that the upper voltage source is proportional to the source voltage of $M2$, v_{s2}. This voltage can be expressed as

$$v_{s2} = -g_{m1}r_{ds1}v_{in} + ir_{ds1}$$

If the transconductance of $M2$ is added to the body effect transconductance—that is,

$$g_{t2} = g_{m2} + g_{mb2}$$

the midband voltage gain can be found to be

$$A_{MB} = -\frac{(g_{m1}r_{ds1} + g_{t2}g_{m1}r_{ds1}r_{ds2})r_{cs}}{r_{ds1} + r_{ds2} + r_{cs} + g_{t2}r_{ds1}r_{ds2}} \tag{9.28}$$

This expression represents the product of two relatively large gains. As bias current decreases, the r_{ds} terms increase faster than the g_m terms decrease. Gains of several thousand can be obtained with this circuit. However, as r_{ds} increases, the high-frequency response decreases, since the output resistance must drive the parasitic capacitance at the output of the stage.

A differential version of this stage is often used as an input stage in MOSFET op amp chips. A slightly different version of the cascode is the folded cascode that replaces $M2$ by a pMOS device with a drain connection to ground. We will not consider the folded cascode stage.

The headroom of the cascode circuit can be found to be

$$V_{headroom} = V_{eff1} + V_{eff2} - V_{eff3} \tag{9.29}$$

The value of V_{eff3} will be negative for the pMOS device of a simple current mirror.

Figure 9.19
High-frequency model of the cascode.

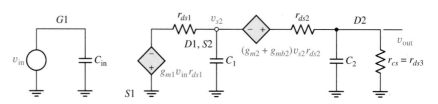

The upper corner frequency of the cascode can be found from the equivalent circuit, shown in Fig. 9.19. The analysis of this circuit is somewhat complex. An accurate value of the upper corner frequency can be found by simulation, but it can be approximated rather easily from the equivalent circuit. We first note that with no signal generator resistance, the input capacitance will not influence the bandwidth. We next note that the corner frequency due to the capacitance C_1 will be affected by the parallel combination of r_{ds1} and the impedance looking into the source of $M2$. This impedance will be much less than r_{ds1} and can be approximated by

$$R_{S2} = \frac{1}{g_{m2} + g_{mb2}}$$

This corner frequency will be quite large, at least compared to that caused by C_2. If the current source is a simple current mirror, its output impedance will be $r_{cs} = r_{ds3}$. The resistance seen by C_2 is then $r_{ds3} \parallel R_{out2}$, where R_{out2} is the impedance looking into the drain of $M2$. This latter impedance will be quite large since the source resistance of $M2$ equals r_{ds1}. This source resistance increases the output resistance of $M2$ significantly; thus, the resistance seen by C_2 is approximately r_{ds3}. The upper corner frequency of the cascode can then be approximated by

$$f_{2o} \approx \frac{1}{2\pi r_{ds3} C_2} \tag{9.30}$$

where $C_2 = C_{db2} + C_{gd2} + C_{db3} + C_{gd3}$.

EXAMPLE 9.5

The cascode circuit of Fig. 9.20 has a dc drain current for all transistors of 50 μA. This current is supplied by a simple current mirror with $M3$ as the output device. With the bias voltages shown, the parameters are $g_{m1} = 181$ μA/V, $g_{m2} = 195$ μA/V, $g_{ds1} = 5.87$ μA/V, $g_{mb2} = 57.1$ μA/V, $g_{ds2} = 0.939$ μA/V, $g_{ds3} = 3.76$ μA/V, $C_{db2} = 9.8$ fF, $C_{gd2} = 1.5$ fF,

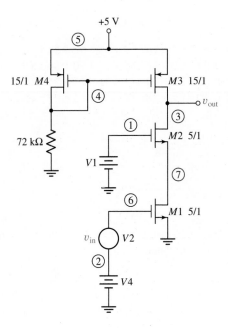

Figure 9.20
Cascode circuit for Example 9.5.

$C_{db3} = 40.9$ fF, and $C_{gd3} = 4.5$ fF. Calculate the midband voltage gain and the approximate upper corner frequency of this cascode stage. Verify these results with a Spice simulation.

SOLUTION The midband gain is calculated from Eq. (9.28).

$$A_{MB} = -\frac{(g_{m1}r_{ds1} + g_{t2}g_{m1}r_{ds1}r_{ds2})r_{ds3}}{r_{ds1} + r_{ds2} + r_{ds3} + g_{t2}r_{ds1}r_{ds2}} = -46.8 \text{ V/V}$$

We note that this gain can be approximated by

$$A_{MB} = -g_{m1}r_{ds3} = -\frac{181}{3.76} = -48.1 \text{ V/V}$$

Which assumes that the incremental current generated by device $M1$, $g_{m1}v_{in}$, flows through $M2$ to develop the output voltage across r_{ds3}.

This low value of voltage gain could be increased sharply by increasing the output impedance of the current mirror to something much greater than r_{ds3}.

The upper corner frequency is approximated by Eq. (9.30) as

$$f_{2o} = \frac{1}{2\pi r_{ds3}C_2} = \frac{1}{2\pi \times 56.7 \times 10^{-15} \times 266 \times 10^3} = 10.6 \text{ MHz}$$

The Spice netlist is included in Table 9.8.

Table 9.8 Spice Netlist for Example 9.5

```
EX9-5.CIR

R1 0 4 72K
V1 1 0 2.1V
V2 2 6 AC 0.005V
V3 5 0 5V
V4 2 0 1.3V
M1 7 6 0 0 N L=1U W=5U AD=25P AS=25P PD=15U PS=15U
M2 3 1 7 0 N L=1U W=5U AD=25P AS=25P PD=15U PS=15U
M3 3 4 5 5 P L=1U W=15U AD=75P AS=75P PD=25U PS=25U
M4 4 4 5 5 P L=1U W=15U AD=75P AS=75P PD=25U PS=25U
.AC DEC 10 100 100MEG
.OP
.PROBE
.LIB C5X.LIB
.END
```

The results of the simulation are $A_{MBsim} = -46.9$ V/V and $f_{2o-sim} = 9.65$ MHz. The calculated gain is very close to the simulated value, and the upper corner frequency is calculated to be about 10% higher than the simulated value.

In order to demonstrate the effect of current on gain and bandwidth, the dc drain current of the circuit is dropped from 50 μA to 10.7 μA. For this situation, the simulated results

Table 9.9   Summary of Simulation Results for
Cascode Amplifier

$A_{MB\mathrm{sim}} = -46.9$ V/V		
C_L, pF	R_g, kΩ	$f_{2o-\mathrm{sim}}$, MHz
0	0	9.65
1	0	0.55
0	339	8.95
1	339	0.55

are $A_{MB\mathrm{sim}} = -94.1$ V/V and $f_{2o-\mathrm{sim}} = 2.52$ MHz. The gain has approximately doubled, while the bandwidth has dropped by a factor of about 4.

The effect on voltage gain can be seen from the approximation

$$A_{MB} = -g_{m1}r_{ds3}$$

Since g_m varies as the square root of drain current and r_{ds} varies approximately inversely with current, the voltage gain at a new current, I_{D2}, compares to the voltage gain at a current of I_{D1} as

$$A_{MB}(I_{D2}) = \sqrt{\frac{I_{D1}}{I_{D2}}} A_{MB}(I_{D1})$$

For this circuit, the new gain at 10.7 μA would be approximated as

$$A_{MB}(10.7) = \sqrt{\frac{50}{10.7}} A_{MB}(50) = 2.16 \times (-46.8) = -101 \text{ V/V}$$

The bandwidth decrease is due to the increase in output resistance when drain current is lowered.

If a signal generator resistance is present, a second corner frequency is added at the input due to C_{in} of Fig. 9.19. When a 339-kΩ generator resistance is added to the simulation of Table 9.8, the new upper corner frequency is 8.95 MHz. This value represents a relatively small change from the original value of 9.65 MHz, which implies that C_{in} is rather small. Thus we see one of the advantages of the cascode connection; that is, the input capacitance is small because the Miller effect is minimized by a small voltage gain from gate to drain of $M1$.

If a load capacitance of 1 pF is added but no generator resistance is present, the upper corner frequency is lowered to 552 kHz. Adding the generator resistance to this circuit leads to no further reduction of the upper corner frequency.

The headroom for the cascode circuit is simulated to be about 0.94 V. Using simulated values of the effective voltages of Eq. (9.29) gives a calculated value of

$$V_{\mathrm{headroom}} = 0.47 + 0.45 + 0.41 = 1.33 \text{ V}$$

These results are summarized in Table 9.9.

9.2.5 THE ACTIVE CASCODE AMPLIFIER

An amplifier that behaves much like the cascode circuit, but offers slightly more headroom and requires one less bias voltage source, is the active cascode stage of Fig. 9.21. It also

PRACTICE Problem

9.12 Rework Example 9.5 if the current from the mirror is changed to 75 μA.

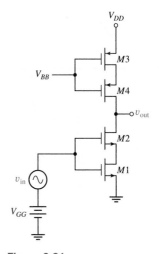

Figure 9.21
*An active cascode stage with
a partial cascode load.*

exhibits high gain, although it has a lower bandwidth than the cascode stage with a current mirror load. The only difference between this circuit and the cascode circuit is that the gates of both nMOS amplifying stages, $M1$ and $M2$, are tied together and driven by the input signal. The load is formed by the two pMOS devices, $M3$ and $M4$, which again connect both gates together and also to a bias source. The connection of $M3$ and $M4$ is called a partial cascode stage.

The partial cascode can be found to present an output impedance, looking into the drain of $M4$, of

$$R_{o4} = r_{ds3} + r_{ds4} + g_{t4} r_{ds3} r_{ds4} \tag{9.31}$$

where $g_{t4} = g_{m4} + g_{mb4}$.

The midband voltage gain is evaluated as

$$A_{MB} = -\left(\frac{g_{m1}}{g_{ds1}} + \frac{g_{m2}}{g_{ds2}} + \frac{g_{m1} g_{t2}}{g_{ds1} g_{ds2}} \frac{R_{o4}}{R_{o4} + R_{o2}} \right) \tag{9.32}$$

where

$$R_{o2} = r_{ds1} + r_{ds2} + g_{t2} r_{ds3} r_{ds4} \tag{9.33}$$

and $g_{t2} = g_{m2} + g_{mb2}$.

The device $M2$ is chosen to have a larger aspect ratio than that of $M1$, and the same is true of $M4$ compared to $M3$. Thus the outer devices, $M1$ and $M3$, are forced to operate in the triode region or near the edge of the triode and the active region.

The gain expression can be reduced by using aspect ratios of the pMOS devices that are three times the corresponding nMOS devices, which will result in the following approximate transconductance relationships:

$$g_{t2} = g_{t4} \qquad g_{ds1} = g_{ds3} \qquad g_{ds2} = g_{ds4}$$

With some simple approximations, the voltage gain can be expressed as

$$A_{MB} \approx -\frac{g_{t2} g_{m1}}{2 g_{ds1} g_{ds2}} \tag{9.34}$$

The output impedance of the active cascode for comparable devices is considerably greater than that of the cascode. Of course, additional devices could be used in the current mirror load of the cascode to increase the output impedance. For a four-transistor circuit, the active cascode has higher gain, smaller headroom, and fewer required bias voltages. The disadvantage is a smaller bandwidth.

The headroom is

$$V_{\text{headroom}} = V_{\text{eff1}} + V_{\text{eff3}} \tag{9.35}$$

which may be 0.4 to 0.5 V.

Table 9.10   Spice Netlist File for Active Cascode Amplifier

```
ACTCAS.CIR

V1 1 0 1.333V
V2 5 0 5V
V3 4 1 AC 0.001V
V4 7 0 3.52V
M1 3 4 0 0 N L=1U W=5U AD=25P AS=25P PD=15U PS=15U
M2 2 4 3 0 N L=1U W=40U AD=200P AS=200P PD=50U PS=50U
M3 6 7 5 5 P L=1U W=15U AD=75P AS=75P PD=25U PS=25U
M4 2 7 6 5 P L=1U W=120U AD=600P AS=600P PD=130U PS=130U
.AC DEC 10 100 100MEG
.OP
.PROBE
.LIB C5X.LIB
.END
```

A Spice netlist file of the active cascode with drain currents equal to about 50 μA is shown in Table 9.10.

The results of this simulation are indicated in Table 9.11.

Both the cascode and active cascode will show an increased gain and decreased bandwidth as drain current decreases. For the circuits of Figs. 9.17(a) and 9.21, the drain currents are decreased to approximately 10 μA. The new simulated results for the cascode circuit with no additional loading are $A_{MBsim} = -70.9$ V/V and $f_{2o-sim} = 3.03$ MHz. The results for the active cascode are $A_{MBsim} = -624$ V/V and $f_{2o-sim} = 63$ kHz.

All of the stages discussed in this chapter require a voltage bias on the gate of the amplifying stage. Voltage reference circuits can be constructed to provide the necessary voltage bias; however, these sources will not be considered here.

Table 9.11   Summary of Results for Active Cascode Amplifier

$A_{MBsim} = -291$ V/V		
C_L, pF	R_g, kΩ	f_{2o-sim}, kHz
0	0	249
1	0	69.5
0	339	87.2
1	339	45.9

DISCUSSION OF DEMONSTRATION PROBLEM

The amplifier of the Demonstration Problem is repeated here for convenience. The first part of the problem consists of solving for the current through the 8-kΩ resistance at the current mirror input, which will allow the output current from $M2$ to be equated to the current into $M1$ with an output voltage of 2.5 V.

Amplifier for Demonstration
Problem

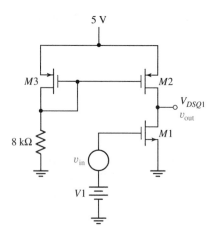

The drain current of $M3$ can be written as

$$I_{D3} = K_p[V_{GS3} - V_{Tp}]^2[1 - \lambda(V_{DS3} - V_{DSP3})]$$

We note that $V_{GS3} = V_{G3} - V_{DD} = 8I_{D3} - 5$ V and $V_{DS3} - V_{DSP3} = V_{Tp} = -1$ V. Writing the right side of the drain current equation in terms of I_{D3} leads to a quadratic equation. The solution of this equation results in $I_{D3} = 0.2888$ mA and $V_{G3} = 2.31$ V. The pinchoff current of both $M2$ and $M3$ is

$$I_{DP} = K_p[V_{GS} - V_{Tp}]^2 = 0.1 \times [-1.69]^2 = 0.2856 \text{ mA}$$

The pinchoff voltage is $V_{DSP} = -1.69$ V.

We now equate the current, I_{D2}, to the current I_{D1}, which allows the equation

$$I_{DSP2}[1 - \lambda(V_{DS2} - V_{DSP2})] = I_{DSP1}[1 + \lambda(V_{DS1} - V_{DSP1})]$$

to be written. The left side of this equation is I_{DS2} and the right side is I_{DS1}. Noting that $I_{DS2} = 0.1[-1.69]^2$ leads to

$$I_{DS2} = 0.1[-1.69]^2[1 - 0.01(-2.5 + 1.69)] = 0.2879 \text{ mA}$$

This value is equated to the right side of the equation, which is

$$I_{DS1} = 0.2879 = 0.12[V_{GS1} - 1]^2[1 + 0.01(1.035 - 0.01V_{GS1})]$$

Solving for V_{GS1} gives

$$V_{GS1} = 2.54 \text{ V}$$

When this voltage is applied to the gate of $M1$, a current of 0.2879 mA and a voltage of $V_{DSQ1} = 2.5$V result.

The incremental values of r_{ds1} and r_{ds2} can now be found from

$$r_{ds} = \frac{1}{\lambda I_{DSP}}$$

The values of r_{ds1} and r_{ds2} are equal and are both 350 kΩ. The resistance between output and ground is the parallel combination of these two resistances or 175 kΩ.

The midband voltage gain is

$$A_{MB} = -g_{m1} R_{out}$$

The value of g_m is found to be

$$g_m = \sqrt{2\mu C_{ox}(W/L)I_D}$$

For $M1$, the transconductance is 0.37 mA/V, which gives a gain of

$$A_{MB} = -0.37 \times 175 = -64.7 \text{ V/V}$$

SUMMARY

➤ The MOSFET amplifier designed for implementation as an IC reduces to a minimum the number of resistors required. Active loads often replace the resistors used in discrete MOSFET amplifiers.

➤ The current mirror is used extensively in IC design. It serves as a bias current source as well as an active load.

➤ The individual amplifier stage must not be analyzed as an isolated stage. The driving source resistance that may arise from a previous stage and the load capacitance that may arise from a following stage or an IC output pad must be included in the analysis of frequency response.

➤ The source follower has a low voltage gain but a very good frequency response.

➤ The cascode and partial cascode connections provide rather efficient amplifying configurations.

PROBLEMS

SECTION 9.1 MOSFET CURRENT MIRRORS

9.1 If $\mu W C_{ox}/2L = 50 \ \mu\text{A/V}^2$, $V_T = 0.8$ V, and $\lambda = 0.02$, calculate the drain current in the circuit shown.

Figure P9.1

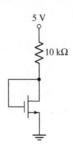

5 V

10 kΩ

9.2 For the circuit of Problem 9.1, calculate the value of pinchoff current, I_{DP}, and pinchoff voltage, V_{DSP}.

9.3 If $\mu W C_{ox}/2L = 80\mu A/V^2$, $V_T = -1.0$ V, and $\lambda = 0.02$, calculate the drain current in the circuit shown.

Figure P9.3

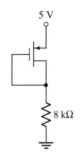

5 V

8 kΩ

9.4 For the circuit of Problem 9.3, calculate the value of pinchoff current, I_{DP}, and pinchoff voltage, V_{DSP}.

9.5 If $\mu W C_{ox}/2L = 50$ $\mu A/V^2$, $V_T = 0.8$ V, and $\lambda = 0.02$, calculate the ratio I_o/I_{in} for the circuit of Fig. 9.1 when $I_{in} = 72$ μA and $V_{DS2} = 6$ V. Assume matched MOSFETs.

☆ **9.6** In Fig. 9.1, the current source, I_{in}, is generated by a 12-V voltage source in series with a 60-kΩ resistor. Matched devices are used, except for V_{T2}, with values as given in Problem 9.5.

(a) Calculate the current, I_{in}.

(b) Calculate V_{GS}.

(c) If I_o/I_{in} is to be unity when $V_{DS2} = 8$ V, what is the required value of V_{T2}?

(d) With the value of V_{T2} in part (c), what is I_o/I_{in} when $V_{DS2} = 16$ V?

9.7 Derive the output impedance, looking into the drain, for the circuit shown. Express the result in terms of r_{ds}, g_m, and R_S. Neglect body effect for this derivation. Compare this impedance to that resulting when $R_S = 0$.

Figure P9.7

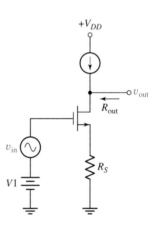

$+V_{DD}$

v_{out}

R_{out}

v_{in}

R_S

V1

SECTION 9.2.1 SIMPLE AMPLIFIER STAGES

9.8 What is the approximate midband voltage gain for the circuit of Fig. 9.4(a) if the dimensions of $M1$ are $W_1 = 24$ μ and $L_1 = 1$ μ and the dimensions of $M2$ are $W_2 = 6$ μ and $L_2 = 2$ μ?

9.9 What should the value of W_1 be in Fig. 9.4(a) to lead to a midband voltage gain of -42 V/V? The length of the channel of $M1$ is $L_1 = 1$ μ, and the dimensions of $M2$ are $W_2 = 2$ μ and $L_2 = 2$ μ.

D 9.10 If the circuit of Fig. 9.4(a) is to have a midband voltage gain of approximately -4.8 V/V, to what value should W_1 be changed?

9.11 What is the upper corner frequency of the amplifier of Fig. 9.4(a) if $C_{out} = 98$ fF, $g_{m2} = 0.4$ mA/V, and $g_{mb2} = 0.05$ mA/V? Assume that the load capacitance is 50 fF.

SECTION 9.2.2 ACTIVE LOAD STAGE

9.12 In the active load stage of Fig. 9.10, the current from the mirror is increased from 50 μA to 100 μA. What would you now expect the midband voltage gain to be? Explain your calculations.

☆ **9.13** Rework the Demonstration Problem if all parameters remain the same except $\lambda_p = 0.024$ V^{-1} and $\lambda_n = 0.02$ V^{-1}.

☆ **9.14** Rework the Demonstration Problem if all parameters remain the same except $K_n = 0.16$ mA/V^2.

9.15 In the circuit of Fig. 9.10, a resistance of value 400 kΩ is placed in series with the input signal generator. Calculate the value of load capacitance that causes the input upper corner frequency to equal the output upper corner frequency. What is the overall upper corner frequency? Use the capacitance values from the circuit simulation.

9.16 In the circuit of Fig. 9.10, a resistance is placed in series with the input signal generator. Calculate the value of this resistor to cause the input upper corner frequency to equal the output upper corner frequency. Assume a zero value for load capacitance. Use the capacitance values from the circuit simulation. What is the overall upper corner frequency?

☆ **D 9.17** In the circuit shown, the current source has infinite impedance and can be adjusted to supply whatever current is required to keep $M1$ in the active region as V_{GS} is changed. If $\lambda = 0.03$ V^{-1}, $V_T = 1.02$ V, and $\mu C_{ox} W/2L = 0.2$ mA/V^2,

(a) Calculate A_{MB} when $V_{GS} = 2$ V.

(b) Find the value of V_{GS} required to double this voltage gain. Assume that $V_{DS} = 4$ V for both cases.

Figure P9.17

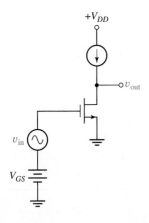

D 9.18 If the current source in Problem 9.17 is fixed at a value of 100 μA, what must V_{GS} be to result in $V_{DS} = 4$ V? What is the value of A_{MB} for this bias?

9.19 Rework Example 9.3 for a required midband voltage gain of -76 V/V.

9.20 In the circuit of Example 9.3, the drain current is lowered to 50 μA. Approximate the new value of midband voltage gain. Explain this approximation.

D 9.21 Using devices with $L = 1$ μ, $|V_T| = 1$ V, $\lambda = 0.02$ V^{-1}, and $\mu C_{ox} = 0.06$ mA/V^2, design an active load stage with a simple current mirror load to have a midband gain of -120 V/V.

SECTION 9.2.3 SOURCE FOLLOWER WITH ACTIVE LOAD

9.22 Rework Example 9.4 if all parameters remain the same except g_m, which becomes $g_{m1} = 300$ mA/V.

9.23 Find the overall upper corner frequency of the source follower of Example 9.4 if a load capacitance of 0.6 pF is added.

9.24 Find the overall upper corner frequency of the source follower of Example 9.4 if a load capacitance of 0.6 pF is added and a generator resistance of 100 kΩ is present. Calculate the midband voltage gain.

SECTION 9.2.4 THE CASCODE CONNECTION

9.25 In Example 9.5, using the cascode circuit of Fig. 9.20, the current from the mirror is increased from 50 μA to 100 μA. What would you now expect the midband voltage gain to be? Explain your calculations.

SECTION 9.2.5 THE ACTIVE CASCODE AMPLIFIER

9.26 Explain why the midband voltage gain increases with decreasing drain current in the active cascode amplifier.

Integrated Circuit Design with the BJT

10

As mentioned in Chapter 8, design of electronic circuits on a chip can be considerably different from discrete circuit design. Some analog circuits do not require high component densities, whereas others may pack a great deal of circuitry into a small chip space. For low-density circuits, more resistors and small capacitors may be used, but for high-density circuits, these elements must be minimized.

This chapter will consider circuits that replace resistors and capacitors with additional BJTs, just as Chapter 9 considered replacing these elements with additional MOSFETs. The organization of this chapter is similar to that of Chapter 9. The BJT current mirror, which is quite popular in biasing of IC amplifiers, will be discussed first. The chapter will then proceed to active load amplifier stages and other single-stage amplifier configurations in order to lay a foundation for the important op amp chip to be discussed in Chapter 11.

DEMONSTRATION PROBLEM

A typical problem that uses the principles appearing in this chapter is shown in the amplifier circuit. For all devices $\beta = 80$, $|V_{BE(on)}| = 0.7$ V, and $|V_A| = 62$ V (Early voltage). Assume that $C_\mu = 2$ pF, $C_\pi = 20$ pF, and $C_{cs} = 2$ pF for all devices. The dc voltage at the output is 4 V. Calculate the overall midband voltage gain and upper corner frequency of the amplifier.

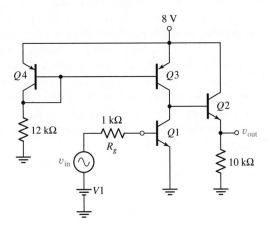

BJT amplifier for Demonstration Problem.

In order to solve this problem, the following questions must be answered.

1. What determines the output current of a current mirror?
2. What determines the voltage gain when the load consists of the output stage of a current mirror?
3. What determines the upper corner frequencies of a common-emitter stage and an emitter follower?

These questions will be answered in general terms throughout this chapter.

10.1 Integrated Circuit Biasing with Current Mirrors

IMPORTANT Concepts

1. The simple current mirror is often used on integrated circuit chips to provide bias current for amplifier stages.
2. The output impedance of current mirrors can be increased by certain modifications. A higher output impedance leads to a better approximation of a true current source.

The popular emitter-bias scheme of Fig. 7.16 for discrete BJT stages often uses a large emitter bypass capacitor to achieve high ac gain. Discrete amplifier stages may also use relatively large interstage coupling capacitors. The unavailability of large capacitors and inductors in IC designs requires that special techniques be used to establish bias currents for integrated amplifiers. Differential stages and complex feedback circuits are often used to obtain the correct bias in the IC amplifier. Although differential amplifiers will be discussed in the next chapter, a significant bias scheme used in differential BJT and other IC amplifier stages will be considered in the following paragraphs.

10.1.1 THE SIMPLE CURRENT MIRROR

The simple current mirror of Fig. 10.1 represents a popular method of creating a constant current bias for differential stages. The concept of the current mirror was developed specifically for analog integrated circuit biasing and is a good example of a circuit that takes advantage of the excellent matching characteristics that are possible in integrated circuits. In the circuit of Fig. 10.1, the current I_o is intended to be equal to I_{in}. Although not shown in the figure, the external circuit through which I_o flows connects to the collector of $Q2$.

In Chapter 9 we discussed the specific operation of the MOSFET current mirror. The following material will expand on the use of current mirrors and apply this material to the BJT version of this circuit. It is useful to discuss the more generalized concept of a current mirror and to introduce some appropriate terminology. Figure 10.2 illustrates a block diagram representation of a current mirror where the input or reference current and output currents are shown. Current mirrors can be designed to serve as sinks or sources, as indicated in the figure.

The general function of the current mirror is to reproduce or mirror the input or reference current to the output while allowing the output voltage to assume any value within some specified range. The current mirror can also be designed to generate an output current that

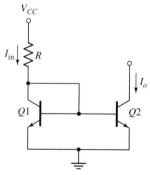

Figure 10.1
Current mirror bias stage.

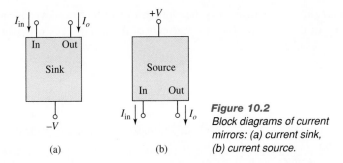

Figure 10.2
Block diagrams of current mirrors: (a) current sink, (b) current source.

equals the input current multiplied by a scale factor K. The output current can be expressed as a function of input current as

$$I_o = K I_{in} \tag{10.1}$$

where K can be equal to, less than, or greater than unity. This constant can be established accurately and will not vary with temperature.

Current Source Operating Voltage Range Figure 10.3(a) shows an ideal or theoretical current sink with a practical sink indicated in Fig. 10.3(b). The voltage at node A in the theoretical sink can be tied to any voltage above or below ground without affecting the value of I. On the other hand, the practical circuit of Fig. 10.3(b) requires that the transistor remain in the active region to provide an output current of

$$I = I_C = \alpha I_E = \alpha \frac{V_E}{R} = \alpha \frac{V_B - V_{BE}}{R} \tag{10.2}$$

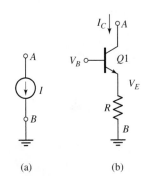

Figure 10.3
Current sink circuits: (a) ideal sink, (b) practical sink.

The collector voltage must exceed the voltage V_B at all times for active region operation. The upper limit on collector voltage is determined by the breakdown voltage of the transistor. The output voltage must then satisfy

$$V_B < V_C < (V_E + BV_{CE}) = (V_B - 0.7 + BV_{CE}) \tag{10.3}$$

where BV_{CE} is the breakdown voltage from collector to emitter of the transistor. The voltage range over which the current source operates within a prescribed accuracy is called the *output voltage compliance range* or the *output compliance*.

Current Mirror Analysis The current mirror is again shown in Fig. 10.4. If devices $Q1$ and $Q2$ are assumed to be matched devices, we can write

$$I_{E1} = I_{E2} = I_{EO}e^{V_{BE}/V_T} \tag{10.4}$$

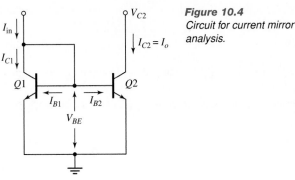

Figure 10.4
Circuit for current mirror analysis.

where $V_T = kT/q$, $I_{EO} = AJ_{EO}$, A is the emitter area of the two devices, and J_{EO} is the current density of the emitters. The base currents of the two devices will also be identical and can be expressed as

$$I_{B1} = I_{B2} = \frac{I_{EO}}{\beta + 1} e^{V_{BE}/V_T}$$

(10.5)

Device $Q1$ operates in the active region, but near saturation by virtue of the collector-base connection. This configuration is called a diode-connected transistor. The collector current of $Q1$ is β times the base current or

$$I_{C1} = \beta I_{B1} = \frac{\beta}{\beta + 1} I_{EO} e^{V_{BE}/V_T}$$

(10.6)

Whereas device $Q1$ is constrained so that $V_{CE} = V_{BE(\text{on})}$ by the connection between base and collector, device $Q2$ does not have this constraint. The collector voltage for $Q2$ will be determined by the external circuit that connects to this collector.

If we limit the voltage V_{C2} to small values relative to the Early voltage, I_{C2} is approximately equal to I_{C1}. For integrated circuit designs, the voltage required at the output of the current mirror is generally small, often making this approximation valid.

The input current to the mirror is slightly larger than the collector current and is expressed as

$$I_{\text{in}} = I_{C1} + 2I_B$$

(10.7)

Since $I_o = I_{C2} = I_{C1} = \beta I_B$, we can write Eq. (10.7) as

$$I_{\text{in}} = \beta I_B + 2I_B = (\beta + 2)I_B$$

(10.8)

Relating I_{in} to I_o results in

$$I_o = \frac{\beta}{\beta + 2} I_{\text{in}}$$

(10.9)

For typical values of β these two currents are essentially equal. Thus, a desired bias current, I_o, is generated by creating the desired value for I_{in}.

The current I_{in} is normally established by connecting a resistance, R_1, between a voltage source V_{CC} and the collector of $Q1$ to set I_{in} to

$$I_{\text{in}} = \frac{V_{CC} - V_{BE}}{R_1}$$

(10.10)

Control of collector current for $Q2$ is then accomplished by choosing proper values of V_{CC} and R_1.

If V_{C2} becomes larger, the Early effect of Eq. (7.11) must be considered. This equation is

$$I_C = \beta I_B \left(1 + \frac{V_{CE}}{V_A}\right)$$

(10.11)

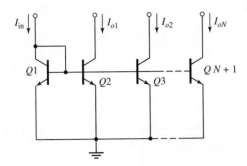

Figure 10.5
Multiple output current mirror.

A more accurate expression for the output current is

$$I_{C2} = \frac{\beta\left(1 + \frac{V_{C2}}{V_A}\right)}{2 + \beta\left(1 + \frac{V_{C1}}{V_A}\right)} I_{in} \tag{10.12}$$

where V_A is the Early voltage. This equation can be used to find the voltage compliance range of the current mirror.

Figure 10.5 shows a multiple output current mirror. It can be shown that the output current for each identical device in Fig. 10.5 is

$$I_o = \frac{\beta}{\beta + N + 1} I_{in} \tag{10.13}$$

where N is the number of output devices.

The preceding analysis of the current mirror has assumed equal transistor sizes. The output currents can be scaled by changing the relative areas of the output BJTs compared to the diode-connected BJT. The ratio of output current to input current scales directly with the ratio of emitter-base junction area of the output device to that of the input device.

The current sinks can be turned into current sources by using *pnp* transistors and a power supply of opposite polarity. The output devices can also be scaled in area to make I_o larger or smaller than I_{in}. The schematic of Fig. 10.6 indicates a multiple output current mirror

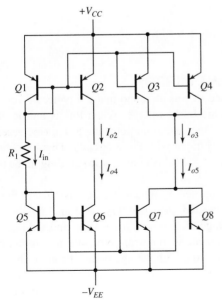

Figure 10.6
Multiple output sources and sinks.

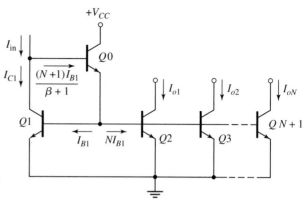

Figure 10.7
Improved multiple output current mirror.

that includes both sources and sinks. Although parallel transistors are shown to indicate higher current output devices, single devices with larger areas would be implemented on the chip.

10.1.2 A CURRENT MIRROR WITH REDUCED ERROR

The difference between output current in a multiple output current mirror and the input current can become quite large if N is large. One simple method of avoiding this problem is to use an emitter follower to drive the bases of all devices in the mirror, as shown in Fig. 10.7. The emitter follower, $Q0$, has a current gain from base to collector of $\beta + 1$, which reduces the difference between I_o and I_{in} to

$$I_{in} - I_o = \frac{N+1}{\beta+1} I_B \tag{10.14}$$

The output current for each device is

$$I_o = \frac{I_{in}}{1 + \frac{N+1}{\beta(\beta+1)}} \tag{10.15}$$

10.1.3 THE WILSON CURRENT MIRROR

In the simple current mirrors discussed, it was assumed that the collector voltage of the output stage was small compared to the Early voltage. When this is untrue, the output current will not remain constant, but will increase as output voltage (V_{CE}) increases. In other words, the output compliance range is limited with these circuits. This limitation occurs because the output impedance of $Q2$ in Fig. 10.4 is relatively low, falling in the tens of $k\Omega$ range.

An improved current mirror was proposed by Wilson and is illustrated in Fig. 10.8.

The Wilson current mirror is connected such that $V_{CB2} = 0$ and $V_{BE1} = V_{BE2}$. The device $Q1$ has a collector-emitter voltage of $V_{CE1} = V_{BE1} + V_{BE0}$, and $Q2$ has a value of $V_{CE2} = V_{BE1}$. Both $Q1$ and $Q2$ now operate with a near-zero collector-emitter bias, even though the collector of $Q0$ might feed into a high voltage point. It can be shown that the output impedance of the Wilson mirror is increased by a factor of approximately $\beta/2$ over the simple mirror. This higher impedance translates into a higher output compliance range.

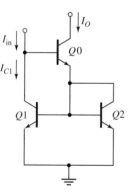

Figure 10.8
Wilson current mirror.

This circuit also reduces the difference between input and output current as a result of the emitter-follower stage.

PRACTICAL Considerations

The following chapter introduces the significant IC op amp. The op amp is a very high gain amplifier that requires several amplifying stages. Each amplifying stage requires a bias current. This chip then requires multiple bias current sources. In practice, these sources can be implemented by a current mirror with multiple output stages having properly scaled emitter areas. A single resistor along with multiple mirror stages require far less chip real estate than would the use of separate bias circuits for each amplifying element.

10.2 High-Gain Stages Using Active Loads

IMPORTANT Concepts

1. An active load may consist of the collector-to-emitter circuit of a transistor biased into its active region. This device replaces the passive resistor often used in the collector of a gain stage.
2. The incremental output resistance, looking into the collector terminal of the passive stage, can be large, leading to a high voltage gain. The dc voltage drop across the active load is quite low.
3. The active load often takes the form of a current source.

In order to achieve high voltage gains and eliminate load resistors, active loads are used in BJT IC amplifiers just as they are in MOSFET stages. In a conventional common-emitter stage, the gain is limited by the size of the collector resistance. The midband voltage gain of a common-emitter stage is given by

$$A_{MB} = -\frac{\alpha R_C}{(r_e + R_E)}$$

It would be possible to increase this voltage gain by increasing R_C; however, making R_C large can lead to some serious problems. A large collector load requires a low quiescent collector current to result in proper bias. This situation may lead to lower values of β, since current gain in a silicon transistor typically falls at low levels of emitter current. In order to achieve a voltage gain of 1000 V/V, a collector load of perhaps 100–200 kΩ might be required. The low collector current needed for proper bias, perhaps a few microamps, would lead to a low value of β and a very high value of r_e. The desired high voltage gain may not be achievable under these conditions.

A solution to this problem would result if the collector load presented a low resistance to dc signals but presented a high incremental resistance. This combination of impedances can result in a stable operating point along with a high gain. An ideal element to use for

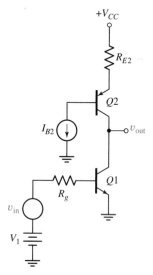

Figure 10.9
A transistor stage with an active load.

the collector load of a transistor is another transistor. This device present a low dc and high incremental impedance, and it is a simple element to implement on a chip.

10.2.1 A CURRENT SOURCE LOAD

The circuit of Fig. 10.9 demonstrates one type of BJT active load. The transistor $Q1$ is the amplifying element with $Q2$ acting as the load. Transistor $Q1$ looks into the collector of $Q2$. The incremental output impedance at the collector of a transistor having an emitter resistance in the low $k\Omega$ range can easily exceed 500 $k\Omega$. With such a high impedance, this transistor approximates a current source.

The dc collector currents of both transistors are equal in magnitude. This magnitude can be set to a value that leads to a reasonable value of β. Since $Q2$ has a very high output impedance, the midband voltage gain will be determined primarily by the collector-to-emitter resistance of $Q1$ and can be calculated from

$$A_{MB} = \frac{-\beta_1 r_{ce1}}{R_g + r_{\pi 1}} \tag{10.16}$$

where r_{ce1} is the output impedance of $Q1$. If the generator resistance, R_g is negligible, this equation reduces to

$$A_{MB} = -\frac{r_{ce1}}{r_{e1}} = -\frac{(V_A + V_{CQ1})/I_C}{V_T/I_E} \approx -\frac{V_A + V_{CQ1}}{V_T} \tag{10.17}$$

For an Early voltage of $V_A = 80$ V and $V_T = 0.026$ V, a small-signal voltage gain exceeding -3000 V/V could result. In a normal application, this stage would drive a second stage. The input impedance of the second stage will load the output impedance of the first stage, further lowering the gain. Depending on the input impedance of the second stage and the impedance of the active load stage, the gain magnitude may still exceed 1000 V/V.

The concept of an active load that presents a large incremental resistance while allowing a large dc quiescent current is important in integrated circuit design. It can be extended to FET amplifiers or hybrid bi-FET amplifiers with an FET amplifying stage and an active BJT load.

In addition to the current source load just considered, the current mirror stage can also be used to provide the active load of a differential stage. This topic is discussed in the next subsection.

EXAMPLE **10.1**

Assume that the active load in the circuit of Fig. 10.10 has an infinite output impedance, $V_{CQ1} = 4$ V, and $V_{EB2} = 0.7$ V. The Early voltage of the amplifying device is 68 V and $\beta = 150$. Calculate the midband voltage gain of this stage.

SOLUTION In order to apply Eq. (10.16), the output impedance and r_π for $Q1$ must be found. The quiescent emitter current of $Q2$ is determined by the base-emitter circuit of $Q2$. The emitter voltage of $Q2$ is 0.7 V higher than the base voltage or 8.7 V. The emitter current

Figure 10.10
An active load amplifier.

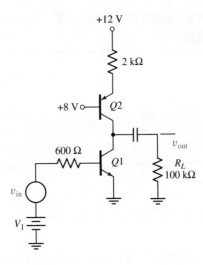

is then

$$I_{E2} = \frac{12 - 8.7}{2} = 1.65 \text{ mA}$$

The emitter current of $Q1$ will be very near to that of $Q2$. Since $I_{C1} \approx I_{E1}$, the output impedance can be found from Eq. (7.13) to be

$$r_{\text{out1}} = r_{ce1} = \frac{V_A + V_{CQ1}}{I_{C1}} = \frac{68 + 4}{1.65} = 43.6 \text{ k}\Omega$$

The value of $r_{\pi 1}$ is

$$r_{\pi 1} = (\beta + 1)r_e = 151 \times \frac{26}{1.65} = 2.38 \text{ k}\Omega$$

The load impedance consists of r_{ce1} in parallel with $100 \text{ k}\Omega$ or $R_{\text{out}} = 30.4 \text{ k}\Omega$. The voltage gain is

$$A_{MB} = \frac{-\beta R_{\text{out}}}{R_g + r_{\pi 1}} = -\frac{150 \times 30.4}{0.6 + 2.38} = -1530 \text{ V/V}$$

PRACTICAL Considerations

Another component that can be used to create a collector load with high incremental or ac impedance and low dc impedance is a transformer. This element generally has only a few ohms of dc primary resistance while presenting an ac primary resistance of $n^2 R_L$, where n is the turns ratio from primary to secondary and R_L is the resistive load connected across the secondary terminals.

For example, if $n = 5$ and $R_L = 8 \ \Omega$, the ac primary resistance is $25 \times 8 = 200 \ \Omega$. The dc resistance of the primary may be 3 Ω; thus, the ac resistance is much greater than the dc resistance.

Although transformers are used in high-power amplifiers, they cannot be fabricated by standard IC processes. Thus the active load developed by a biased transistor remains the most popular load device for IC processes.

PRACTICE Problems

10.4 Assume that the active load in the circuit of Fig. 10.10 is replaced by a simple current mirror with an output current of 1.65 mA. The output stage of the current mirror has an Early voltage of $V_A = 60$ V. Calculate the midband voltage gain of the amplifier. Approximate r_{ce2} as V_A/I_C. *Ans:* −833 V/V.

10.5 Repeat Practice Problem 10.4 if the 100-kΩ load is removed from the circuit. *Ans:* −999 V/V.

10.3 Amplifier Configurations In BJT Integrated Circuits

The current mirror serves as an active load for several important BJT IC amplifier stages to be considered in the following paragraphs.

10.3.1 THE CURRENT MIRROR LOAD

A rather simple configuration for an amplifying stage is shown in Fig. 10.11. In this stage, the output impedance of the current mirror is not large enough to be negligible as it was in the circuit of Fig. 10.9. Thus, the analysis will have to account for this element.

In Chapter 7, the high-frequency response of a discrete circuit was considered. Typically, this value was determined by the input circuit, including the Miller effect capacitance. The collector load resistance in a discrete stage is usually small enough that the output circuit does not affect the upper corner frequency. In the circuit of Fig. 10.11, as in most IC amplifier stages, the output impedance is very high compared to the discrete stage. For this circuit, the output impedance of the amplifier consists of the output impedance of $Q2$ in parallel with that of $Q1$. This value will generally be several tens of $k\Omega$.

The equivalent circuit of the amplifier of Fig. 10.11 is indicated in Fig. 10.12. The value of R_{out} is

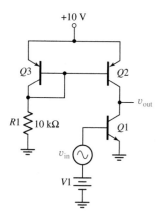

Figure 10.11
A common-emitter stage with current mirror active load.

$$R_{out} = r_{o1} \| r_{o2} = r_{ce1} \| r_{ce2} \tag{10.18}$$

The capacitance in parallel with R_{out} is approximately

$$C_{out} = C_{\mu1} + C_{\mu2} + C_{cs1} + C_{cs2} \tag{10.19}$$

In this equation, $C_{\mu1}$ and $C_{\mu2}$ are the collector-to-base junction capacitances, and C_{cs1} and C_{cs2} are the collector-to-substrate capacitances of the respective transistors. If no generator resistance is present, $C_{\mu1}$ will also appear in parallel with the output terminal and ground. When R_g is present, we will still approximate the output capacitance with the same

Figure 10.12
Equivalent circuit of the amplifier in Fig. 10.11.

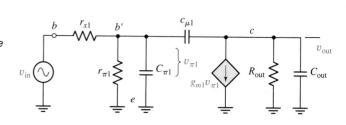

equation, although feedback effects between the output and the bases of $Q1$ and $Q2$ actually modify the value slightly.

The midband gain is easy to evaluate as

$$A_{MB} = \frac{-\beta_1 R_{out}}{r_{x1} + r_{\pi 1}} \tag{10.20}$$

The upper corner frequency is now more difficult to evaluate than that of the discrete circuit, which often has a low value of collector load resistance. In the discrete circuit, the input loop generally determines the overall upper corner frequency of the circuit. Although the Miller effect will be much larger in the IC stage, lowering the upper corner frequency of the input loop, the corner frequency of the output loop will also be smaller due to the large value of R_{out}. Both frequencies may influence the overall upper corner frequency of the amplifier.

The calculation of upper corner frequency begins by reflecting the bridging capacitance, C_μ, to both the input and the output. The value reflected to the input side, across terminals b' and e, is

$$(1 - A_{b'c1})C_{\mu 1} \tag{10.21}$$

as in the discrete circuit amplifier. Thus, the total input capacitance in parallel with $r_{\pi 1}$ is

$$C_{in} = C_{\pi 1} + (1 - A_{b'c1})C_{\mu 1} \tag{10.22}$$

The upper corner frequency resulting from the input circuit of this stage is

$$f_{in-high} = \frac{1}{2\pi C_{in} R_{eq}} \tag{10.23}$$

where $R_{eq} = r_{x1} \| r_{\pi 1}$.

The upper corner frequency resulting from the output side of the stage is

$$f_{out-high} = \frac{1}{2\pi C_{out} R_{out}} \tag{10.24}$$

The actual overall upper corner frequency, f_{2o}, must be found using the method of Chapter 3 for a two-pole response. An example will demonstrate these points.

EXAMPLE 10.2

Assume that the circuit of Fig. 10.11 is biased so that the collector currents of $Q1$ and $Q2$ have a magnitude of 1.14 mA. The parameters for $Q1$ are $\beta = 160$, $r_{x1} = 10\ \Omega$, $r_{ce1} = 68\ k\Omega$, $C_{\pi 1} = 20\ pF$, and $C_{\mu 1} = 2.1\ pF$. For device $Q2$, the necessary parameters are $r_{ce2} = 21\ k\Omega$ and $C_{\mu 2} = 3.1\ pF$. Each device has a value of $C_{cs1} = C_{cs2} = 2.5\ pF$. In this circuit, the power supply is 10 V and $R1 = 10\ k\Omega$.

Calculate the midband voltage gain and the upper corner frequency for this amplifier stage. Do a Spice simulation using 2N3904 (*npn*) and 2N3905 (*pnp*) transistors.

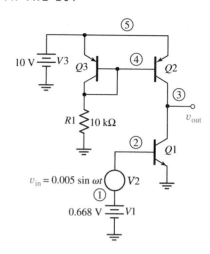

Figure 10.13
Schematic for Windows Spice
simulation.

SOLUTION The midband voltage gain can be calculated from Eq. (10.20) after evaluating $r_{\pi 1}$ and R_{out}. These resistances are

$$r_{\pi 1} = (\beta + 1)r_{e1} = 161 \times \frac{26}{1.14} = 3672 \ \Omega$$

and

$$R_{out} = r_{ce1} \parallel r_{ce2} = 68 \parallel 21 = 16 \ k\Omega$$

The midband gain is then

$$A_{MB} = -\frac{\beta R_{out}}{r_{x1} + r_{\pi 1}} = -\frac{160 \times 16,000}{10 + 3672} = -695 \ V/V$$

The upper corner frequency is found from a consideration of the two poles caused by the input circuit and the output circuit. The corner frequency of the input circuit is

$$f_{in-high} = \frac{1}{2\pi C_{in} R_{eq}} = \frac{1}{2\pi (C_{\pi 1} + [1 - A_{b'c}]C_{\mu 1})(r_{x1} \parallel r_{\pi 1})}$$

Table 10.1 Spice
Netlist File for Example
10.2

$$= \frac{1}{2\pi \times 1481 \times 10^{-12} \times 10} = 10.7 \ MHz$$

Since $r_{\pi 1} \gg r_{x1}$, the value of 10 Ω was used for R_{eq}. In addition, the midband gain was used to approximate $A_{b'c}$.

The corner frequency of the output circuit is

```
EX10-2.CIR


R1 0 4 10K
V1 1 0 0.668V
V2 2 1 AC 0.005V
V3 5 0 10V
Q1 3 2 0 0 Q2N3904
Q2 3 4 5 5 Q2N3905
Q3 4 4 5 5 Q2N3905
.AC DEC 100 100 1G
.OP
.PROBE
.LIB BIPOLAR.LIB
.END
```

$$f_{out-high} = \frac{1}{2\pi C_{out} R_{out}} = \frac{1}{2\pi \times (2.1 + 3.1 + 2.5 + 2.5) \times 10^{-12} \times 16,000} = 975 \ kHz$$

The input corner frequency is much higher than the output corner frequency; consequently, the latter value approximates the overall corner frequency. The result is a value of $f_{2o} = 975$ kHz.

The schematic for a Windows Spice simulation is shown in Fig. 10.13 with node and element numbers added. The Spice netlist file used to simulate this circuit is shown in Table 10.1.

Note that the connections for the BJT correspond to collector node, base node, emitter node, and substrate node. Normally, the substrate for a *pnp* device connects to the positive power supply voltage. The substrate for the *npn* device connects to the most negative power supply rail, which is often the ground terminal.

The results of the simulation are $A_{MBsim} = -685$ V/V and $f_{2o-sim} = 900$ kHz. The calculated and simulated values of midband voltage gain agree within 2%, and the upper corner frequency values are within 10%.

Table 10.2   Summary of Results for the Common-Emitter Current-Source Load Stage

C_L, pF	R_g, kΩ	A_{MBcal}, V/V	A_{MBsim}, V/V	f_{2o-cal}, kHz	f_{2o-sim}, kHz
0	0	−696	−685	975	900
10	0	−696	−685	480	493
0	10	−187	−186	40	39
10	10	−187	−186	39	38

In order to evaluate the effects of a signal generator resistance and a larger capacitive load, a 10-kΩ resistance was inserted in the base lead of $Q1$, which led to a simulated midband gain of −186 V/V and an upper corner frequency of 39 kHz. If a 10-pF capacitor is placed across the output and no generator resistance is used, the voltage gain returns to the value of −685 V/V, but the upper corner frequency is lowered to 480 kHz. Adding the generator resistance while the capacitor loads the output gives $A_{MB} = -186$ V/V and $f_{2o} = 37.8$ kHz. The comparison of calculated and simulated values for the different loading conditions is given in Table 10.2.

The calculated results are designated A_{MBcal} and f_{2o-cal}, and the simulated values are designated A_{MBsim} and f_{2o-sim}. The task of calculating values for Table 10.2 is left to the student. Note that adding a 10-kΩ generator resistance lowers the upper corner frequency by a large factor, approximately 25, whereas adding 10 pF to the output lowers this frequency by a factor of about two.

PRACTICAL Considerations

For a BJT with a high value of current gain-bandwidth product, f_t, the current source load stage has a relatively low upper corner frequency. In Example 10.2, the value of f_t for these BJTs is 300–400 MHz. The resulting upper corner frequency is just 975 kHz. It should be recognized that this type of stage is used in op amp chips that will ultimately have a very low upper corner frequency. When used in an amplifier, the op amp chip will use feedback to improve the overall upper corner frequency. Thus, the low value of f_{2o} for an individual gain stage is not significant in these applications.

10.3.2 THE EMITTER FOLLOWER

A stage that can be used to minimize the adverse effect on frequency response caused by a generator resistance is the emitter follower. Although this stage has a voltage gain near unity, it can be driven by a higher voltage gain stage while the emitter follower can drive a low impedance load. A typical stage is shown in Fig. 10.14. The output stage of the *npn* current mirror, $Q2$, serves as a high impedance load for the emitter follower, $Q1$. An equivalent circuit that represents the emitter follower of Fig. 10.14 is indicated in Fig. 10.15. For this circuit, $g_{m1} = \alpha_1/r_e \approx 1/r_e$.

PRACTICE Problem

10.6 Calculate the value of R_g that lowers the upper corner frequency of the circuit in Example 10.2 to 100 kHz. Assume a 10-pF load capacitance. Calculate the midband voltage gain. *Ans:* $R_g = 1.43$ kΩ, $A_{MB} = -501$ V/V.

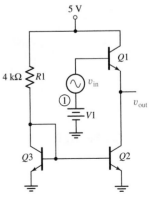

Figure 10.14
An emitter follower.

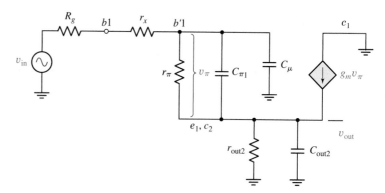

Figure 10.15
Equivalent circuit for the emitter follower.

This circuit can be analyzed to result in a voltage gain of

$$A = \frac{\frac{C_{\pi 1}}{\Pi C(R_g + r_{x1})}\left(j\omega + \frac{g_{m1}r_{\pi 1} + 1}{r_{\pi 1}C_{\pi 1}}\right)}{-\omega^2 + bj\omega + d} \tag{10.25}$$

where

$$\Pi C = C_{out2}C_{\mu 1} + C_{out2}C_{\pi 1} + C_{\pi 1}C_{\mu 1} \tag{10.26}$$

$$b = \frac{1}{\Pi C}\left(\frac{C_{\pi 1} + C_{out2}}{R_g + r_{x1}} + \frac{C_{\pi 1} + C_{\mu 1}(1 + g_{m1}r_{ce2})}{r_{ce2}} + \frac{C_{out2} + C_{\mu 1}}{r_{\pi 1}}\right) \tag{10.27}$$

and

$$d = \frac{R_g + r_{x1} + r_{\pi 1} + r_{ce2}(g_{m1}r_{\pi 1} + 1)}{\Pi C(R_g + r_{x1})r_{\pi 1}r_{ce2}} \tag{10.28}$$

Note that the output capacitance of $Q2$ can be approximated as the sum of $C_{\mu 2}$ and C_{cs2}.
The midband voltage gain is found from Eq. (10.25) by letting $\omega \to 0$. This value is

$$A_{MB} = \frac{(1 + g_{m1}r_{\pi 1})r_{ce2}}{R_g + r_{x1} + r_{\pi 1} + (1 + g_{m1}r_{\pi 1})r_{ce2}} \tag{10.29}$$

This gain is very near unity for typical element values.

The bandwidth is more difficult to calculate since the response has one zero and two poles. The zero for the circuit of Fig. 10.15 is typically larger than the lowest frequency pole. If these frequencies canceled, the larger pole would determine the corner frequency. Since they do not cancel, the overall upper corner frequency is expected to be smaller than the larger pole frequency. An accurate calculation can be made from Eq. (10.25) when the parameters are known.

EXAMPLE 10.3

The emitter-follower circuit of Fig. 10.14 is biased so that $I_{C1} = 1.08$ mA. The value of β_1 is 155, and the ohmic base resistance is 10 Ω. The collector-base depletion capacitance for both $Q1$ and $Q2$ is 2.5 pF as also is the collector-to-substrate capacitance. The value of f_t is approximately 300 MHz, and the Early voltage is 75 V for the transistors.

Calculate

. The midband voltage gain of the circuit

. The approximate upper corner frequency of the circuit

imulate the operation of this stage using Spice to find the voltage gain and upper corner requency.

OLUTION The value of r_e is found as

$$r_e = \frac{26}{I_E} \approx \frac{26}{1.09} = 24 \ \Omega$$

Jsing this value, the diffusion capacitance or base-to-emitter capacitance can be found as

$$C_{\pi 1} = \frac{1}{2\pi r_e f_t} = \frac{1}{2\pi \times 24 \times 3 \times 10^8} = 22 \text{ pF}$$

The output resistance of $Q2$ is calculated from

$$r_{ce2} \approx \frac{V_A}{I_C} = \frac{75}{1.09} = 68.8 \text{ k}\Omega$$

his equation neglects the voltage V_{CEQ2}. The value of $r_{\pi 1}$ is $(\beta + 1)r_e = 3744 \ \Omega$. All lement values in the equivalent circuit of Fig. 10.15 are now known.

Substituting element values into Eq. (10.25) results in

$$A = \frac{1.239 \times 10^{10} \left(1 + j\frac{\omega}{1.89 \times 10^9}\right)}{-\omega^2 + j\omega 1.581 \times 10^{10} + 2.348 \times 10^{19}}$$

The midband gain of this circuit is $A_{MB} = 0.9993$ V/V. The zero frequency is 301 MHz. The two pole frequencies are 264 MHz and 2.28 GHz. The gain expression can also be vritten

$$A = \frac{0.9993 \left(j\frac{f}{301 \ MHz} + 1\right)}{\left(j\frac{f}{264 \ MHz} + 1\right) \left(j\frac{f}{2.28 \ GHz} + 1\right)}$$

Jsing iterative methods, the upper corner frequency is found to be 1.67 GHz. Although the voltage gain is only unity, the upper corner frequency is much higher than the common-emitter amplifier with a current mirror load.

The schematic for simulation is shown in Fig. 10.16 with node and element numbers idded. The Spice netlist file is shown in Table 10.3.

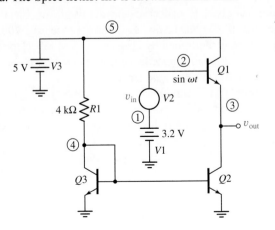

Figure 10.16
Schematic for simulation of the emitter follower.

Table 10.3
Spice Netlist File for Example 10.3

EX10-3.CIR

R1 5 4 4K
V1 1 0 3.2 V
V2 2 1 AC 1V
V3 5 0 5V
Q1 5 2 3 0 Q2N3904
Q2 3 4 0 0 Q2N3904
Q3 4 4 0 0 Q2N3904
.AC DEC 100 100 10G
.OP
.PROBE
.LIB BIPOLAR.LIB
.END

The results of the simulation are $A_{MBsim} = 0.9993$ V/V and $f_{2o-sim} = 2.56$ GHz. Whereas the midband voltage gain compares well to the calculated value, the simulated upper corner frequency is somewhat higher than the calculated value.

Table 10.4 Summary of Results for the Emitter-Follower Stage

C_L, pF	R_g, kΩ	A_{MBcal}, V/V	A_{MBsim}, V/V	f_{2o-cal}, MHz	f_{2o-sim}, MHz
0	0	0.9993	0.9993	1670	2560
10	0	0.9996	0.9993	420	420
0	10	0.9982	0.9975	6.56	6.61
10	10	0.9987	0.9975	6.85	6.87

It is again useful to consider the effects of adding a large generator resistance or a load capacitance to the emitter-follower stage. Table 10.4 summarizes the results of simulations for different combinations of source resistance and load capacitance.

Note that the insertion of a 10-kΩ generator resistance lowers the upper corner frequency more than the addition of the 10-pF load capacitance. Note also that the addition of the 10-pF capacitance to the circuit that includes a 10-kΩ generator resistor leads to a higher upper corner frequency rather than a lower value. This result is from a shift in the lower pole frequency to a value that exceeds that of the zero frequency when the load capacitance is added. The added value of C_L when no generator resistance is present does not have the same effect.

Surprisingly, it is possible for the two poles to become complex, depending on element values. When this occurs, the frequency response can exhibit a peak and the step response can exhibit ringing.

PRACTICE Problem

10.7 An emitter follower has a midband voltage gain of 0.99. The voltage gain has a zero at 500 MHz, one pole at 200 MHz, and another pole at 1.6 MHz. Find the upper corner frequency of the voltage gain. *Ans: 236 MHz.*

PRACTICAL Considerations

The emitter follower, like the source follower, is often used as a buffer to interface between a high voltage gain stage and a low impedance load. The emitter follower loads the preceding stage only slightly, but provides a high current to the load with an accompanying high upper corner frequency.

10.3.3 THE CASCODE AMPLIFIER STAGE

One of the problems with the common-emitter stage using an active load is the Miller effect. This stage has a high voltage gain from base to collector. The circuit of Fig. 10.11 with the values of Example 10.2 has an inverting voltage gain, A_{MB}, that approaches -700 V/V. The base-collector junction capacitance is multiplied by $(1 + |A_{MB}|)$ and reflected to the input loop. This capacitance adds to the diffusion capacitance from point b' to point e and decreases the upper corner frequency to a relatively small value.

The cascode amplifier stage of Fig. 10.17 minimizes the capacitance reflected to the input. In this circuit, the input capacitance is primarily composed of the diffusion capacitance of $Q1$. The gain from base to collector of $Q1$ is quite low since the collector load of this device consists of the impedance looking into the emitter of $Q2$. This impedance is approximately equal to the base-emitter diode resistance of $Q2$, which is

$$r_{e2} = \frac{26}{I_{E2}}$$

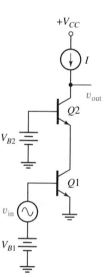

Figure 10.17
A cascode stage.

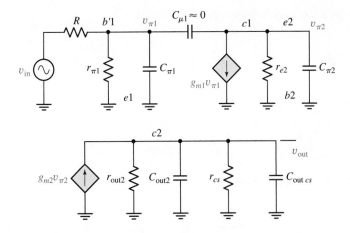

Figure 10.18
Equivalent circuit of the cascode stage.

The upper device passes the incremental signal current of $Q1$ to its collector and develops a large voltage across the current source impedance. There is no Miller multiplication of capacitance from the input of $Q2$ (emitter) to the output (collector), since the gain is noninverting and negligible capacitance exists between emitter and collector. Thus, the cascode stage essentially eliminates Miller effect capacitance and its resulting effect on upper corner frequency.

A high-frequency equivalent circuit of this stage is shown in Fig. 10.18. The resistance R includes any generator resistance and the base resistance, r_{x1} of $Q1$. The resistance r_{cs} is the output resistance of the current source. The output capacitance is the sum of $C_{\mu2}$, C_{cs2}, and any capacitance at the current source output. The resistance r_{out2} can be quite large, since $Q2$ sees a large emitter resistance looking into the collector of $Q1$. This emitter load leads to negative feedback that increases the output resistance of $Q2$.

The midband voltage gain is calculated from the equivalent circuit of Fig. 10.18 after eliminating the capacitors. This gain is found rather easily by noting that the input current to $Q1$ is

$$i_{b1} = \frac{v_{in}}{R + r_{\pi1}} \tag{10.30}$$

This current will be multiplied by β_1 to become collector current in $Q1$. This current also equals the emitter current of $Q2$. The emitter current of $Q2$ is multiplied by α_2 to become collector current of $Q2$. The output voltage is then

$$v_{out} = i_{c2} \times R_3 \tag{10.31}$$

where $R_3 = r_{out2} \parallel r_{cs}$. This resistance could be very large if the current source resistance, r_{cs}, is large. The value of r_{out2} will be high since the emitter of $Q2$ sees a resistance of r_{ce1}. Combining this information results in a midband voltage gain of

$$A_{MB} = \frac{-\beta_1 \alpha_2 R_3}{R + r_{\pi1}} \tag{10.32}$$

If no generator resistance is present and if $R_3 = 100$ kΩ, this midband voltage gain might exceed 5000 V/V.

The gain as a function of frequency can be found as

$$A = A_{MB} \frac{1}{1 + j\omega C_{\pi 1}(r_{\pi 1} \| R)} \frac{1}{1 + j\omega r_{e2}C_{\pi 2}} \frac{1}{1 + j\omega R_3 C_{out}} \qquad (10.33)$$

Typically, the corner frequency of the second frequency term in Eq. (10.33), that is,

$$f_2 = \frac{1}{2\pi r_{e2} C_{\pi 2}}$$

is much higher than that of the first term,

$$f_{in-high} = \frac{1}{2\pi (r_{\pi 1} \| R)C_{\pi 1}}$$

especially if R is large compared to $r_{\pi 1}$. In this case, since $r_{e1} \approx r_{e2}$ as a result of equal emitter currents, then $r_{e2} \ll (\beta + 1)r_{e1}$. For hand analysis of the cascode circuit, the second term in the expression for gain is often neglected.

The gain can then be written as

$$A = A_{MB} \frac{1}{1 + j \frac{f}{f_{in-high}}} \frac{1}{1 + j \frac{f}{f_{out-high}}} \qquad (10.34)$$

where

$$f_{out-high} = \frac{1}{2\pi C_{out} R_3} \qquad (10.35)$$

and $f_{in-high}$ was defined previously.

The capacitance C_{out} is the sum of the current source output capacitance and the output capacitance of $Q2$, giving

$$C_{out} = C_{out2} + C_{outcs}$$

If a current mirror with output stage $Q3$ generates the collector bias current for $Q1$ and $Q2$, the output capacitance is

$$C_{out} = C_{\mu 2} + C_{\mu 3} + C_{cs2} + C_{cs3}$$

EXAMPLE 10.4

The cascode circuit of Fig. 10.19 is driven by a current mirror with an output current of 0.38 mA. Devices $Q1$ and $Q2$ have values of $\beta = 140$ and $r_x = 10 \ \Omega$. The capacitor values are $C_{\pi 1} = C_{\pi 2} = 10.8 \ \text{pF}$, $C_{\mu 1} = C_{\mu 2} = C_{cs2} = 2.5 \ \text{pF}$, and $C_{\mu 3} = C_{cs3} = 5 \ \text{pF}$. The output impedance of the current mirror is 52.2 kΩ.

Calculate the midband voltage gain and the upper corner frequency of the circuit. Do a Spice simulation and compare the measured to the simulated results.

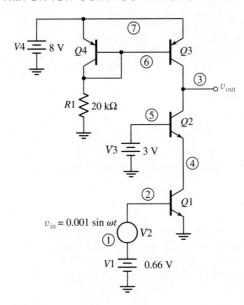

Figure 10.19
*Schematic for simulation of
cascode amplifier.*

SOLUTION The value of r_π for $Q1$ and $Q2$ is calculated to be

$$r_\pi = (\beta + 1)r_e = 141 \times \frac{26}{0.38} = 9647 \ \Omega$$

Assuming that r_{out2} is very large, the value of R_3 is approximated by $r_{cs} = 52.2$ kΩ. From Eq. (10.32), the midband voltage gain is

$$A_{MB} = \frac{-\beta_1 \alpha_2 R_3}{r_{\pi 1}} \approx \frac{-\beta_1 r_{cs}}{r_{\pi 1}} = \frac{-141 \times 52.2}{9.647} = -763 \ \text{V/V}$$

Table 10.5   Spice Netlist
File for Example 10.4.

```
EX10-3.CIR

R1 6 0 20K
V1 1 0 0.66V
V2 2 1 AC 0.001V
V3 5 0 3V
V4 7 0 8V
Q1 4 2 0 0 Q2N3904
Q2 3 5 4 0 Q2N3904
Q3 3 6 7 7 Q2N3905
Q4 6 6 7 7 Q2N3905
.AC DEC 100 100 100Meg
.OP
.PROBE
.LIB BIPOLAR.LIB
.END
```

The corner frequency of the input stage is

$$f_{in-high} = \frac{1}{2\pi(r_{\pi 1} \| R)C_{\pi 1}} = \frac{1}{2\pi \times 10 \times 10.8 \times 10^{-12}} = 1.47 \text{ GHz}$$

The upper corner frequency due to the output circuit is

$$f_{out-high} = \frac{1}{2\pi R_3 C_{out}} = \frac{1}{2\pi \times 52,200 \times 15 \times 10^{-12}} = 203 \text{ kHz}$$

In this equation, the output capacitance was taken as

$$C_{out} = C_{\mu 2} + C_{\mu 3} + C_{cs2} + C_{cs3} = 2.5 + 5 + 2.5 + 5 = 15 \text{ pF}$$

The overall upper corner frequency is $f_{2o} = 203$ kHz.

The Spice netlist file for this circuit is shown in Table 10.5 using the node and element numbers of Fig. 10.19.

The results of the simulation are $A_{MBsim} = -758$ V/V and $f_{2o-sim} = 205$ kHz.

Table 10.6 Summary of Results for the Cascode Stage

C_L, pF	R_g, kΩ	A_{MBcal}, V/V	A_{MBsim}, V/V	f_{2o-cal}, kHz	f_{2o-sim}, kHz
0	0	−763	−758	203	205
10	0	−763	−758	121	123
0	10	−372	−368	201	202
10	10	−372	−368	121	123

PRACTICE Problem

10.8 If a very large generator resistance is inserted in the circuit of Example 10.4, what is the lower limit on $f_{in-high}$?
Ans: 1.53 MHz.

Additional calculations and simulations were done using a 10-pF load capacitor and/o a 10-kΩ generator resistance. These results are summarized in Table 10.6.

We observe that the insertion of a 10-kΩ generator resistance has little effect on the upper corner frequency. This is to be expected as a result of the minimization of the Mille effect. The input capacitance is small enough that it has little effect on the overall uppe corner frequency even with the larger generator resistance in the input loop. In the active load stage of Fig. 10.13 considered earlier, insertion of a 10-kΩ generator resistance lowered the upper corner frequency by a factor of about 25.

DISCUSSION OF THE DEMONSTRATION PROBLEM

The amplifier circuit for the demonstration problem is repeated here. In order to determine the voltage gain of the common-emitter stage, the values of r_π and r_{out} must be determined. These depend on the output current of the current stage. Using Eq. (10.9), this current can be approximated as

$$I_o = \frac{\beta}{\beta+2}I_{in} = \frac{80}{82} \times \frac{8-0.7}{12} = 0.59 \text{ mA}$$

This current is the collector current of $Q1$ and approximates the emitter current of this device.

The resistance r_{e1} is then

$$r_{e1} = \frac{26}{0.59} = 44.1 \ \Omega$$

which leads to

$$r_{\pi 1} = (\beta + 1)r_{e1} = 81 \times 44.1 = 3569 \ \Omega$$

The output impedances of $Q1$ and $Q3$ are next calculated to be

$$r_{ce1} = r_{ce3} = \frac{V_A}{I_C} = \frac{62}{0.59} = 105 \ \text{k}\Omega$$

The voltage gain of $Q1$ can now be calculated after noting that the load for this stage is made up of the parallel combination of r_{ce1}, r_{ce3}, and the input impedance of the emitter follower. The emitter-follower input resistance is approximately $R_{in2} = (\beta + 1)R_{E2} = 810 \ \text{k}\Omega$, neglecting r_{e2}. The voltage gain of this stage is

$$A_{MB1} = -\frac{\beta(r_{ce1} \| r_{ce3} \| R_{in2})}{R_g + r_{\pi 1}} = -\frac{80 \times 49.3}{1 + 3.57} = -863 \ \text{V/V}$$

The voltage gain of the emitter follower is

$$A_{MB2} = \frac{R_{E2}}{R_{E2} + r_{e2}}$$

Since the voltage across R_{E2} is 4 V, the current through the emitter of $Q2$ is 0.4 mA, which results in $r_{e2} = 65 \ \Omega$ and $A_{MB2} = 0.994$. The overall midband gain is the product of A_{MB1} and A_{MB2}. This product is

$$A_{MB} = -863 \times 0.994 = -858 \ \text{V/V}$$

Because the upper corner frequency of the emitter follower is much greater than that of the common-source stage, the overall upper corner frequency will be equal to that of the common-source stage. This stage will have an upper corner frequency due to the input loop and another due to the output loop.

The output capacitance is calculated by Eq. (10.19) to be

$$C_{out} = C_{\mu 1} + C_{\mu 3} + C_{cs1} + C_{cs3} = 8 \ \text{pF}$$

The output resistance has previously been found as 49.3 kΩ, giving an upper corner frequency of

$$f_{\text{out-high}} = \frac{1}{2\pi C_{\text{out}} R_{\text{out}}} = 404 \text{ kHz}$$

The input capacitance is calculated from Eq. (10.22). This value is

$$C_{\text{in}1} = C_{\pi 1} + (1 + |A_{b'c}|)C_{\mu 1} = 20 + 1106 \times 2 = 2232 \text{ pF}$$

This capacitance sees an equivalent resistance of $R_{\text{eq}} = R_g \| r_{\pi 1} = 1 \| 3.57 = 781 \ \Omega$. The input loop corner frequency is

$$f_{\text{in-high}} = \frac{1}{2\pi C_{\text{in}} R_{\text{eq}}} = 91.3 \text{ kHz}$$

The two upper corner frequencies cause an overall upper corner frequency of 87 kHz.

SUMMARY

➤ Current mirrors are used to provide bias current for some IC amplifier stages and can also be used as active loads.

➤ Many IC amplifier stages use active loads to achieve high voltage gains. The high incremental resistance of an active load stage results in a high voltage gain but may limit the upper 3-dB frequency of the circuit.

➤ The emitter follower with an active load provides a voltage gain of approximately unity and a very high upper 3-dB frequency. This stage can drive a large capacitive load.

➤ The cascode stage minimizes the Miller effect capacitance at the input and provides a high voltage gain.

PROBLEMS

SECTION 10.1.1 THE SIMPLE CURRENT MIRROR

D 10.1 For the simple current mirror of Fig. 10.1, assume that $\beta_1 = \beta_2 = 100$, $V_{BE1} = V_{BE2} = 0.6$ V, $V_{CC} = 5$ V, and $V_A = \infty$. Select R to result in $I_o = 1.00$ mA.

D 10.2 If $V_A = 50$ V in Problem 10.1, select R to result in $I_o = 1.00$ mA when $V_{C2} = 4$ V.

☆ **10.3** If I_o can vary by ±5% in the mirror of Problem 10.2, determine the voltage compliance of the output circuit.

10.4 What is the percentage variation in output current in the mirror of Problem 10.2 as V_{C2} varies from 2 V to 6 V? What is the incremental output resistance of the mirror?

SECTION 10.1.2 A CURRENT MIRROR WITH REDUCED ERROR

10.5 Calculate the ratio I_o/I_{in} for the multiple current source circuit of Fig. 10.5, assuming $N = 4$, $\beta = 120$, and equal sizes for all transistors.

☆ **10.6** If $Q1$ has an emitter area that is 1/4 the size of the four current sinks it drives in Fig. 10.5, what is the ratio of output current of one sink to input current for $\beta = 120$?

☆ **10.7** If $Q1$ has an emitter area that is four times the size of the four current sinks it drives in Fig. 10.5, what is the ratio of output current of one sink to input current for $\beta = 120$?

SECTION 10.1.3 THE WILSON CURRENT MIRROR

10.8 A simple current mirror sinks 1.00 mA with an output voltage of 4 V. The current increases by 10% at an output voltage of 9 V, resulting in a voltage compliance of 5 V. This current mirror is now replaced by a Wilson current mirror designed to sink 1.00 mA with an output voltage of 4 V. The Wilson circuit has a voltage compliance of 64 V. Calculate the output impedance of both circuits.

10.9 In the Wilson current mirror of Fig. 10.8, derive an expression for incremental resistance seen looking into the collector/base of $Q2$.

☆ **10.10** In the Wilson current mirror of Fig. 10.8, derive an expression for output impedance of the circuit in terms of β, r_{e0}, r_{e2}, r_{ce0}, and any other necessary parameters.

SECTION 10.2.1 A CURRENT SOURCE LOAD

10.11 In the circuit shown, $\beta_1 = 210$, $\beta_2 = 90$, $V_{BE1} = -V_{BE2} = 0.68$ V, and $r_{out1} = 35$ kΩ. The quiescent output voltage is 5 V. Calculate the midband voltage gain of the circuit.

Figure P10.11

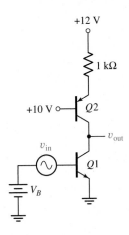

+12 V

1 kΩ

+10 V ○—| $Q2$

v_{out}

v_{in}

$Q1$

V_B

10.12 Repeat Problem 10.11 if a 10-kΩ resistor is inserted in the base lead of $Q1$.

10.13 In the circuit of Problem 10.11, $V_{A2} = 56$ V. Calculate the output resistance, r_{out2}, looking into the collector of $Q2$. Compare this to the output resistance of $Q1$, $r_{out1} = 35$ kΩ. Is it reasonable to assume that $r_{out2} = \infty$ in calculating the voltage gain?

10.14 For both transistors of the circuit, assume that $\beta = 100$ and $r_{ce} = 60$ kΩ.

(a) If both emitter currents in (a) of the figure are 0.8 mA, calculate the midband voltage gain of the amplifier.

(b) After adding a 100-Ω resistor to the emitter of $Q2$, the following data were taken for the circuit in (b) of the figure:

V_{C2}, V	3	4	5	6	7
I_{C2}, mA	0.792	0.796	0.800	0.804	0.808

If $V_{CQ1} = 5$ V in the circuit of (c), calculate the midband voltage gain.

Figure P10.14

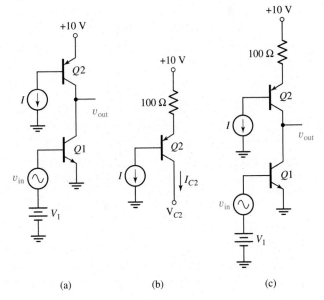

+10 V

$Q2$

I

v_{out}

$Q1$

v_{in}

V_1

(a)

+10 V

100 Ω

$Q2$

I

I_{C2}

V_{C2}

(b)

+10 V

100 Ω

$Q2$

v_{out}

I

$Q1$

v_{in}

V_1

(c)

☆ **10.15** Derive an expression for the incremental output impedance for circuit (b) in Problem 10.14.

☆ **D 10.16** Select the emitter resistance of $Q1$ in Fig 10.10 to lead to a midband gain of -1200 V/V. Assum that β and V_{EB2} remain at the values given in Exampl 10.1 and V_1 can be changed to any appropriate value t maintain $V_{CQ1} = 4$ V.

SECTION 10.3.1 THE CURRENT MIRROR LOAD

10.17 For the circuit shown, $\beta_1 = 200$, $\beta_2 = 100$, $|V_{A1}| = 60$ V, $|V_{A2}| = 40$ V, and $V_{BE1} = -V_{BE2} = -V_{BE3} = 0.7$ V. Assume that the quiescent output voltage is 5 V. Calculate the midband voltage gain of the circuit.

Figure P10.17

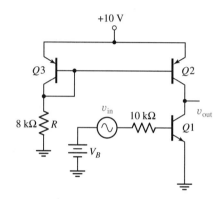

10.18 If R of the current mirror in Problem 10.17 is changed to 20 kΩ and V_B is adjusted to keep $V_{CQ1} = 5$ V, calculate the midband voltage gain of the circuit.

10.19 If the simple current mirror of Problem 10.17 is replaced by a Wilson current mirror with an output impedance of 420 kΩ, calculate the midband voltage gain of the stage. Assume that the output current of the mirror remains equal to the value in Problem 10.17.

10.20 In the circuit of Problem 10.17 at the bias point used, the diffusion capacitance of $Q1$ is $C_\pi = 20$ pF, $C_{\mu 1} = 2$ pF, $C_{\mu 2} = 4$ pF, $C_{cs1} = 3$ pF, and $C_{cs2} = 3.5$ pF. Calculate the upper corner frequency of the circuit.

10.21 Repeat Problem 10.20 if the source resistance is changed to 1 kΩ. Assume that $r_x = 50\Omega$.

☆ **D 10.22** In Problem 10.20, how large can a load capacitance be to result in a 10% reduction in bandwidth compared to the case of no load capacitance?

SECTION 10.3.2 THE EMITTER FOLLOWER

10.23 The three transistors of Fig. 10.16 are identical with $V_{BE} = 0.68$ V, $\beta = 180$, $r_x = 100\ \Omega$, $r_\pi = 2.4$ kΩ, and $r_{out} = 40$ kΩ. If a 100-kΩ generator resistance is inserted in series with $V2$, keeping the bias current constant, what is the midband voltage gain?

10.24 In Problem 10.23, the capacitors $C_\mu = 3$ pF, $C_{\pi 1} = 30$ pF, and $C_{out2} = 5$ pF. Calculate the upper corner frequency of the voltage gain.

SECTION 10.3.3 THE CASCODE AMPLIFIER STAGE

10.25 In Fig. 10.19, the resistor $R1$ is changed from $20\,\text{k}\Omega$ to $10\,\text{k}\Omega$. Calculate the new midband voltage gain. Assume that the output resistance of the current source, r_{cs}, changes from $52.2\,\text{k}\Omega$ to $70\,\text{k}\Omega$.

11 | The Differential Stage and the Op Amp

Very few circuits have had an impact on the electronics field as great as that of the op amp circuit. This name is a shortened version of operational amplifier, a term that has a rather specific meaning. Before IC chips became available, operational amplifiers were expensive high-performance amplifiers used primarily in a system called an analog computer. The analog computer was a real-time simulator that could simulate physical systems governed by differential equations. Examples of such systems are circuits, mechanical systems, and chemical systems. One popular use of the analog computer was the simulation of automatic control systems. The control of a space vehicle had to be designed on paper before test flights took place. The test flights could then be used to fine-tune the design. The analog computer was far more efficient than the digital computer of the day in simulating differential equations and was far less costly.

The op amp was an important component of the analog computer that allowed several mathematical operations to be performed with electronic circuits. The op amp could be used to create a summing circuit, a difference circuit, a weighting circuit (amplifier), or an integrating circuit. Using these capabilities, complex differential systems can be created with an output voltage that represents the physical output variable of the simulated system. The time variable may be scaled, but for many simulations, the output represents a real-time solution.

One of the key features of an op amp is the differential input to the amplifier. This input allows differences to be formed, and also allows the creation of a virtual ground or virtual short across the input terminals. This virtual short is used in summing several current signals into a node without affecting the other input current signals. These signals are then summed and easily converted into an output voltage.

The virtual ground also allows perfect integration by the op amp using an additional resistor and capacitor. This feature is essential in the simulation of differential equations.

Some engineers mistakenly refer to other amplifiers as op amps, but unless these amplifiers possess the capability to create a virtual ground and do mathematical operations, this is a misnomer.

We will begin this chapter with a discussion of the differential stage, which is included as the input stage for every op amp manufactured. The overall configuration and operation of a typical op amp will then be considered. A treatment of more advanced op amps will conclude the chapter.

PRACTICAL Considerations

In the late 1950s, the op amp was implemented as a modular circuit and sold by such companies as Burr-Brown and Philbrick for a few hundred dollars each. These amplifiers were very high-performance units and formed the basis of many analog computer systems. In 1964, Fairchild Semiconductor Corporation introduced the first integrated circuit op amp, the 702. The more popular 709 was introduced the following year, followed by the LM101 and the 741 in 1967. The first two IC op amps, the 702 and 709, required an external RC circuit for stability purposes. The LM101 and 741 included this network on the chip. The 741 has probably sold more units than any other analog chip in history.

DEMONSTRATION PROBLEM

An op amp is implemented by three amplifier stages as shown. The overall voltage gain of the op amp is a three-pole response given by

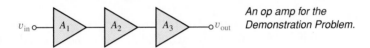

An op amp for the Demonstration Problem.

$$A = \frac{10,000}{\left(1 + j\frac{f}{10,000}\right)\left(1 + j\frac{f}{80,000}\right)\left(1 + j\frac{f}{4\times10^6}\right)}$$

Will this circuit be stable in the unity-gain configuration ($F = 1$)? If not, calculate the frequency at which oscillations may occur. To what value must the 10-kHz pole be changed to make the circuit marginally stable for $F = 1$?

An understanding of stability considerations for three-pole amplifiers is necessary to complete this problem.

11.1 The Differential Amplifier

IMPORTANT Concepts

1. The differential stage allows a low-drift, dc amplifier to be constructed.
2. The integrated circuit differential stage provides the basis for the popular operational amplifier or op amp chip.

11.1.1 THE BASIC DIFFERENTIAL PAIR

The differential pair, differential stage, or differential amplifier is very important in the electronics field. Virtually every op amp chip includes a differential pair as the input stage. Two of the major advantages of the differential stage are its immunity to temperature effects and its ability to amplify dc signals.

The differential amplifier uses a pair of identical stages connected in a configuration that allows the temperature drift of one stage to cancel that of the other stage.

The basic differential stage consists of two balanced amplifiers as shown in Fig. 11.1. The two devices can be connected and operated in several different configurations. The mode of operation most often used with the differential amplifier is the *differential input–double-ended output* mode. Differential input refers to a situation wherein the voltage appearing

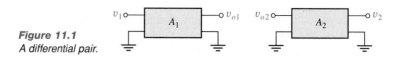

Figure 11.1
A differential pair.

across the input of one stage is equal in magnitude but opposite in polarity to the voltage appearing across the input of the other stage. Double-ended output refers to the fact that the output voltage is taken as the difference in voltage between the output voltage of each stage. In single-stage amplifiers, the output voltage appears between the circuit output and ground and is called a *single-ended* output. In Fig. 11.1, the double-ended output voltage is

$$v_{\text{out}} = v_{o1} - v_{o2}$$

or

$$v_{\text{out}} = v_{o2} - v_{o1}$$

depending on the choice of output terminals.

If this differential pair used a single-ended output, it could be taken between the output of stage 1 and ground or the output of stage 2 and ground.

A second input mode that could be used is the *common mode*. If the same signal is applied to both inputs, the circuit is said to operate in common mode. If the amplifier stages have exactly equal gains, the signals v_{o1} and v_{o2} will be equal. The double-ended output signal would then be zero in the ideal differential stage operating in common mode. In practice the two amplifier gains will not be identical; thus, the common-mode output signal will have a small value. The common-mode gain A_{CM} is defined as

$$A_{CM} = \frac{v_{o1} - v_{o2}}{v_{CM}} = A_1 - A_2 \qquad \text{(11.1}$$

where v_{CM} is the common-mode voltage applied to both inputs and A_1 and A_2 are the voltage gains of the two stages.

The differential gain applies when the input signals v_1 and v_2 are not equal, and the gain in this case is

$$A_D = \frac{v_{o1} - v_{o2}}{v_1 - v_2} = \frac{A_1 v_1 - A_2 v_2}{v_1 - v_2} = \frac{v_{\text{out}}}{v_1 - v_2} \qquad \text{(11.2}$$

where A_D is the differential voltage gain of the amplifier with a double-ended output. If the double-ended output had been defined as $v_{o2} - v_{o1}$ rather than $v_{o1} - v_{o2}$, only the algebraic sign of A_D would change.

In the general case, both common-mode and differential signals will be applied to the amplifier. We can demonstrate the significance of this situation by supposing that we have a differential pair with a differential gain of 100 and a common-mode gain of 0.01. If we apply values to the inputs of $v_1 = 5$ mV and $v_2 = -5$ mV, the double-ended output would be

$$v_{\text{out}} = v_{o2} - v_{o1} = A_D(v_1 - v_2) = 100 \times 0.01 = 1 \text{ V}$$

If the inputs are now changed to values of $v_1 = 6.005$ V and $v_2 = 5.995$ V, the differential input voltage remains at 10 mV. However, the common-mode input voltage is now 6 V. Using superposition, the output voltage can be calculated from

$$v_{\text{out}} = A_D(v_1 - v_2) + A_{CM}v_{CM} \qquad \text{(11.3}$$

The common-mode input voltage can be found from

$$v_{CM} = \frac{v_1 + v_2}{2} \tag{11.4}$$

In this example the common-mode input voltage is 6 V; thus the double-ended output is

$$v_{\text{out}} = 100 \times 0.01 + 0.01 \times 6 = 1.06 \text{ V}$$

We note that even though the differential input signal remains constant, the change in common-mode input leads to a slight change in output voltage.

PRACTICAL Considerations

Both inputs to a differential amplifier may have different voltages applied to them, but still contain a common-mode input voltage. For example, if 2.02 V is applied to input 1 and 1.98 V to input 2, the common-mode voltage is found by averaging these two values to be $v_{CM} = 2$ V. The differential input is $v_2 - v_1 = 1.98 - 2.02 = -0.04$ V. This situation arises when the differential inputs are driven by preceding stages that have an output consisting of a dc bias voltage and an incremental signal voltage.

In the ideal situation with perfectly symmetrical stages, the common-mode input would lead to zero output; thus a measure of the asymmetry of the differential pair is the common-mode rejection ratio, defined as

$$CMRR = 20 \log \frac{|A_D|}{|A_{CM}|} \text{ dB} \tag{11.5}$$

For the preceding example, this value would be

$$CMRR = 20 \log \frac{100}{0.01} = 80 \text{ dB}$$

The larger the *CMRR*, the smaller is the effect of A_{CM} on the output voltage compared to A_D.

One great advantage of the differential stage is in the cancellation of drift at the double-ended output. Temperature drifts in each stage are often common-mode signals. For example, the change in forward voltage across the base-emitter junction with constant current is about -2 mV/°C. As the temperature changes, each junction voltage changes by the same amount. These changes can be represented as equal voltage signals applied to the two inputs. If the stages are closely matched, very little output drift will be noted. The integrated differential amplifier can perform considerably better than its discrete counterpart, since component matching is more accurate and a relatively uniform temperature prevails throughout the chip. Power-supply noise is also a common-mode signal and has little effect on the output signal.

As previously mentioned, the mode of operation most often used with the differential amplifier is the differential input, double-ended output mode. In this configuration, the input voltage applied to one stage should be equal in magnitude but opposite in polarity to the voltage applied to the other stage. One method of obtaining these equal magnitude, opposite polarity signals using only a single input source is shown in Fig. 11.2.

If the input resistances of both stages are equal, half of v_{in} will drop across each stage. Whereas the voltage from terminal a to terminal b of stage 1 represents a voltage drop, the voltage across the corresponding terminals of stage 2 represents a rise in voltage.

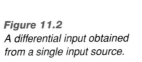

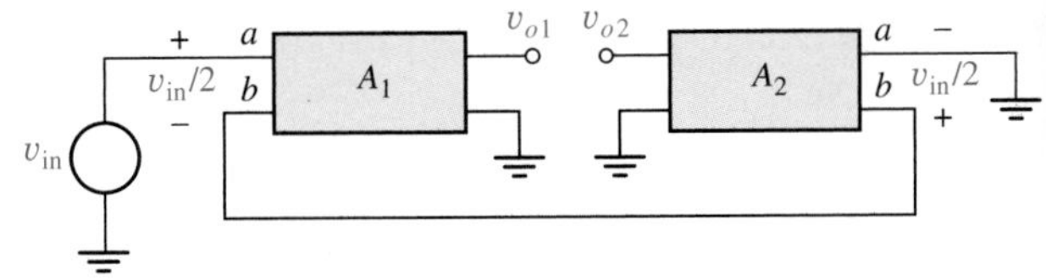

Figure 11.2
*A differential input obtained
from a single input source.*

We can write

$$v_1 = \frac{v_{\text{in}}}{2} \quad \text{and} \quad v_2 = -\frac{v_{\text{in}}}{2}$$

PRACTICE Problems

11.1 A differential pair such as that of Fig. 11.1 has values of $A_1 = 106.1$ and $A_2 = 106.3$. If a double-ended output is taken, calculate the common-mode gain of the amplifier. *Ans: $A_{CM} = -0.2$ V/V.*

11.2 Calculate the differential gain of the amplifier of Drill Problem 11.1. *Ans: $A_D = 106.2$ V/V.*

11.3 Calculate the *CMRR* for the amplifier of Practice Problem 11.1. *Ans: CMRR = 54.5 dB.*

Each stage has the same magnitude of input voltage but is opposite in polarity to that of the other stage. From Eq. (11.2) we can write the differential gain as

$$A_D = \frac{v_{\text{out}}}{v_{\text{in}}} = \frac{v_{o1} - v_{o2}}{v_1 - v_2} = \frac{v_{o1} - v_{o2}}{v_{\text{in}}/2 - (-v_{\text{in}}/2)}$$

$$= \frac{A_1 \frac{v_{\text{in}}}{2} + A_2 \frac{v_{\text{in}}}{2}}{\frac{v_{\text{in}}}{2} + \frac{v_{\text{in}}}{2}} = \frac{A_1 + A_2}{2} \tag{11.6}$$

11.1.2 THE BJT DIFFERENTIAL PAIR

Two versions of a BJT differential pair are shown in Fig. 11.3. The circuit of Fig. 11.3(a) has no external resistances in the input loop and is more typical of an integrated circuit stage. The circuit of Fig. 11.3(b) typifies a discrete differential pair.

Small-Signal Voltage Gain Figure 11.4 represents a simple equivalent circuit for the BJT differential pair. After drawing the equivalent circuit, the steps in calculating the single-ended or double-ended voltage gain of a differential pair are

1. Calculate the input current as $i_{\text{in}} = v_{\text{in}}/R_{\text{in}}$.
2. Note that the base currents are related to i_{in} by $i_{b1} = i_{\text{in}}$ and $i_{b2} = -i_{\text{in}}$.
3. Calculate collector currents as $i_{c1} = \beta_1 i_{b1}$ and $i_{c2} = \beta_2 i_{b2}$.
4. Calculate collector voltages from $v_{o1} = -i_{c1} R_{\text{Ceq1}}$ and $v_{o2} = -i_{c2} R_{\text{Ceq2}}$, where

$$R_{\text{Ceq}} = R_C \| r_{\text{out}} \tag{11.7}$$

With these values, the single-ended or double-ended voltage gain can be found.

Figure 11.3
(a) A simple differential pair.
(b) A differential pair with additional resistors.

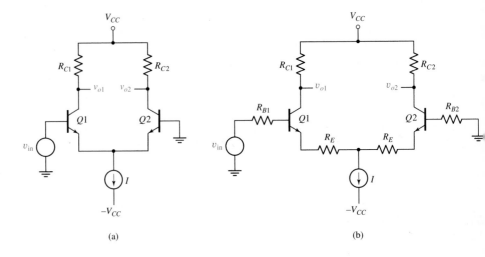

(a)

(b)

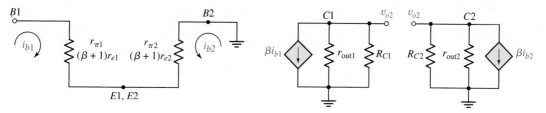

Figure 11.4
Equivalent circuit for the BJT differential pair of Fig. 11.3(a).

In the normal situation, we assume perfectly symmetrical pairs with $\beta_1 = \beta_2 = \beta$ and $R_{C1} = R_{C2} = R_C$. We also assume that the bias current I splits equally between the two stages, giving $I_{E1} = I_{E2} = I/2$. With these conditions, the steps in calculating the voltage gain for the circuit of Fig. 11.3(a) lead to the following:

1. The total input resistance in the input loop is

$$R_{\text{in}} = 2(\beta + 1)r_e \tag{11.8}$$

so

$$i_{\text{in}} = \frac{v_{\text{in}}}{2(\beta + 1)r_e}$$

2. The base currents are then

$$i_{b1} = \frac{v_{\text{in}}}{2(\beta + 1)r_e}$$

and

$$i_{b2} = \frac{-v_{\text{in}}}{2(\beta + 1)r_e}$$

3. The collector currents are

$$i_{c1} = \beta \frac{v_{\text{in}}}{2(\beta + 1)r_e}$$

and

$$i_{c2} = -\beta \frac{v_{\text{in}}}{2(\beta + 1)r_e}$$

4. The single-ended output voltages are then

$$v_{o1} = -\beta \frac{v_{\text{in}}}{2(\beta + 1)r_e} R_{C\text{eq}}$$

and

$$v_{o2} = \beta \frac{v_{\text{in}}}{2(\beta + 1)r_e} R_{C\text{eq}}$$

The single-ended gain of the first stage of Fig. 11.3(a) is

$$A_{S1} = \frac{v_{o1}}{v_{\text{in}}} = \frac{-\beta R_{C\text{eq}}}{2(\beta + 1)r_e} \tag{11.9}$$

The single-ended gain of the second stage equals this value in magnitude but shows no phase inversion and, therefore, has a positive algebraic sign.

Taking v_{out} as $v_{o2} - v_{o1}$, the double-ended differential gain is

$$A_D = \frac{v_{o2} - v_{o1}}{v_{\text{in}}} = \frac{\frac{\beta v_{\text{in}}}{2(\beta+1)r_e} R_{C\text{eq}} - \left(\frac{-\beta v_{\text{in}}}{2(\beta+1)r_e} R_{C\text{eq}}\right)}{v_{\text{in}}} = \frac{\beta R_{C\text{eq}}}{(\beta+1)r_e} \qquad \textbf{(11.10)}$$

This differential double-ended voltage gain is equal in magnitude to the voltage gain of a single transistor amplifier with a load of R_C and no external emitter resistance. One might question the validity of using two stages to achieve the same voltage gain as that of a single-stage circuit. The advantages of decreased temperature drift and good power-supply rejection in critical applications are typically far more important than using an extra device, especially for IC chips.

The method used to calculate the voltage gain of the differential pair is easily extended to other configurations. The simple but important case of additional resistance in the input loop can be treated with little effort. Figure 11.3(b) shows a circuit with two additional emitter resistors and two base resistors.

The only change from the previous case is in the input resistance. If $R_{B1} = R_{B2} = R_B$, the input resistance is now

$$R_{\text{in}} = 2R_B + 2(\beta+1)R_E + 2(\beta+1)r_e \qquad \textbf{(11.11)}$$

After applying the suggested steps in calculating voltage gain, only the denominator of Eq. (11.10) is modified, to result in a gain of

$$A_D = \frac{\beta R_{C\text{eq}}}{(\beta+1)R_E + (\beta+1)r_e + R_B} \qquad \textbf{(11.12)}$$

We have noted that the voltage gain of a perfectly matched differential stage with two equal resistances in both base circuits is equal to that of a single stage with a single base resistance. This is because the arrangement of the differential pair attenuates the input signal of each stage by a factor of one-half, cutting the gain of each stage by one-half. The double-ended output arrangement causes the two single-ended gains to be additive, thus restoring the gain to the value of a single stage.

It should be mentioned that the output impedance of each transistor looking into the collector, r_{out}, can be rather high compared to the external collector resistance, R_C, especially for the discrete differential pair with external emitter resistance, R_E. In many cases, the value of $R_{C\text{eq}}$ is then assumed to equal R_C.

It may not be obvious why two separate equal resistors are used in the base circuit of the differential pair shown in Fig. 11.3(b). We must recognize that the calculation of voltage gain is based on perfect matching, which implies that the dc collector currents through the two stages are equal. Since collector current is αI_E, the emitter currents of both stages must also be equal. This will only occur if the dc bias current source sees equal resistances looking into both emitters. The impedances seen by this source depend on the resistance from each base to ground; thus, two equal base resistors provide better matching of emitter currents, which then leads to equal values of r_e for the two stages. Equal collector currents also result in equal values of quiescent collector voltages. Since the double-ended output voltage is the difference of the collector voltages, the quiescent output voltage will be zero for perfect matching.

An example will demonstrate several points that must be considered in the design of the differential stage.

EXAMPLE 11.1

In the circuit of Fig. 11.3(b), the dc bias current is 4 mA. If $\alpha = 0.993$, $R_{B1} = R_{B2} = 1,000\ \Omega$, $R_E = 30\ \Omega$, $R_C = 1.6\ \text{k}\Omega$, $V_{CC} = 10\ \text{V}$, and $V_{BE(\text{on})} = 0.7\ \text{V}$,

(a) Calculate the dc collector currents.

(b) Calculate the dc or quiescent collector voltages.

(c) Calculate the maximum peak value of v_{out} before serious distortion results.

(d) Calculate the incremental differential voltage gain of the circuit defined as

$$A_D = \frac{v_{\text{out}}}{v_{\text{in}}} = \frac{v_{o2} - v_{o1}}{v_{\text{in}}}$$

(e) If the base resistor of $Q2$ is changed to $R_{B2} = 400\ \Omega$, calculate the dc collector current through each device.

SOLUTION The answer to part (a) is found by assuming perfect matching of $Q1$ and $Q2$ along with equal values of R_B. The dc bias current will then split equally, giving $I_{E1} = I_{E2} = 2\ \text{mA}$. The collector currents are very near this value, given by $I_C = \alpha I_E$. In this case, $I_C = 1.986\ \text{mA}$.

(b) The quiescent collector voltages will equal

$$V_{CC} - I_C R_C = 10 - 1.986 \times 1.6 = 6.82\ \text{V}$$

(c) Each BJT can swing to a maximum collector voltage of 10 V at cutoff and a minimum voltage of approximately 0 V (the dc level of the bases) at saturation. This means that the positive peak of the collector voltage is $10 - 6.82 = 3.18$ V and the negative peak is 6.82 V. Serious distortion will then be limited by cutoff with a peak collector voltage swing of 3.18 V. The output voltage is the difference of the two collector voltages; thus, as one collector swings 3.18 V toward cutoff, the other collector moves in the opposite direction by an equal amount. The peak output voltage is then twice the limit of the collector voltage or 6.36 V. Note that the peak-to-peak output voltage before hitting cutoff of a transistor is twice this value or 11.72 V. Of course, nonlinear distortion may limit the swing to a smaller value before cutoff is reached.

(d) Equation (11.12) can be used to calculate the differential voltage gain once r_e and β are found. Figure 11.5 shows the equivalent incremental input circuit for this stage. For an emitter current of 2 mA, the value of r_e is

$$r_e = \frac{26}{I_E} = \frac{26}{2} = 13\ \Omega$$

The value of β is given by

$$\beta = \frac{\alpha}{1 - \alpha} = \frac{0.993}{0.007} = 142$$

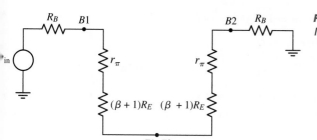

Figure 11.5
Incremental input circuit.

Applying the gain equation and assuming $r_{out} \gg 1.6$ kΩ gives

$$A_D = \frac{142 \times 1600}{143 \times (13 + 30) + 1000} = 31.8 \text{ V/V}$$

(e) This part of the example is intended to demonstrate the problem that occurs when the differential pair is not balanced. The dc equivalent circuit of Fig. 11.6 pertains to this situation. The voltage at the node above the dc current source can be found from

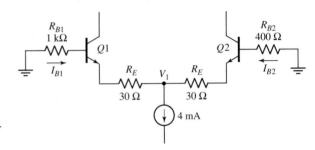

Figure 11.6
DC equivalent circuit for part (e).

$$V_1 = -[I_{B1}R_{B1} + V_{BE(on)} + (\beta + 1)I_{B1}R_E]$$

Summing the drops through $Q2$ rather than $Q1$ gives

$$V_1 = -[I_{B2}R_{B2} + V_{BE(on)} + (\beta + 1)I_{B2}R_E]$$

These expressions can be equated with I_{B1} and I_{B2} as the only unknowns. A second equation that must be satisfied results from summing the current into the same node above the bias current source, leading to

$$(\beta + 1)I_{B1} + (\beta + 1)I_{B2} = 4 \text{ mA}$$

These two equations are solved to result in

$$I_{B1} = 13.1 \ \mu\text{A} \quad \text{and} \quad I_{B2} = 14.8 \ \mu\text{A}$$

The corresponding emitter and collector currents are

$$I_{E1} = 1.88 \text{ mA}, \quad I_{E2} = 2.12 \text{ mA}$$

$$I_{C1} = 1.86 \text{ mA}, \quad I_{C2} = 2.10 \text{ mA}$$

The two quiescent collector voltages are no longer equal, resulting in a nonzero quiescent output voltage. The collector voltages are

$$V_{CQ2} = 10 - 1.6 \times 2.1 = 6.64 \text{ V}$$

and

$$V_{CQ1} = 10 - 1.6 \times 1.86 = 7.02 \text{ V}$$

The quiescent output voltage is now

$$V_{outQ} = V_{CQ2} - V_{CQ1} = 6.64 - 7.02 = -0.38 \text{ V}$$

We will later see that this nonzero quiescent voltage may have some serious consequences when this stage is followed by additional gain stages in an op amp, creating a finite output voltage offset when the inputs are shorted together.

The previous calculations for the differential pair assume there is no load across the output terminals. If this circuit drives a succeeding stage, the input impedance to this following stage will load the differential pair. Example 11.2 considers this situation.

EXAMPLE 11.2

If the matched circuit of Example 11.1 has a 10-kΩ resistor placed across the output terminals, calculate the differential voltage gain of the pair.

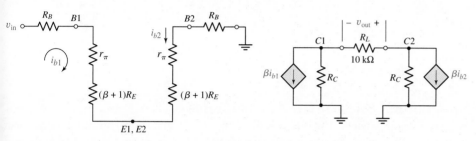

Figure 11.7
Equivalent circuit for Example 11.2.

SOLUTION The equivalent circuit, assuming r_{out} is very large, is shown in Fig. 11.7. One method of solution converts the dependent current sources to the dependent voltage sources of Fig. 11.8. From this equivalent circuit, it is apparent that the open circuit output voltage is attenuated by the resistance across the output terminals. The new voltage gain becomes

$$A_D(\text{loaded}) = A_D \frac{R_L}{R_L + 2R_C} \qquad (11.13)$$

For this example, the loaded gain is calculated as

$$A_D(\text{loaded}) = 31.8 \times \frac{10}{10 + 3.2} = 24.1 \text{ V/V}$$

A second approach to this problem results from noting that when the collector of one BJT is driven positive by the input signal, the other collector moves an equal voltage in the opposite direction. Each end of the load resistor, R_L, is driven in equal but opposite directions. The midpoint of the resistor is always at 0 V, for incremental signals. To calculate the loaded voltage gain, a resistance of $R_L/2$ can be placed in parallel with each collector resistance to give an equivalent collector resistance of

$$R_{Ceq} = R_C \parallel R_L/2 \qquad (11.14)$$

Equation (11.12) can now be used to calculate the loaded differential gain. For this example, the equivalent collector resistance becomes

$$R_{Ceq} = 1.6 \parallel 5 = 1.21 \text{ k}\Omega$$

Using Eq. (11.12) again results in a loaded gain of 24.1 V/V.

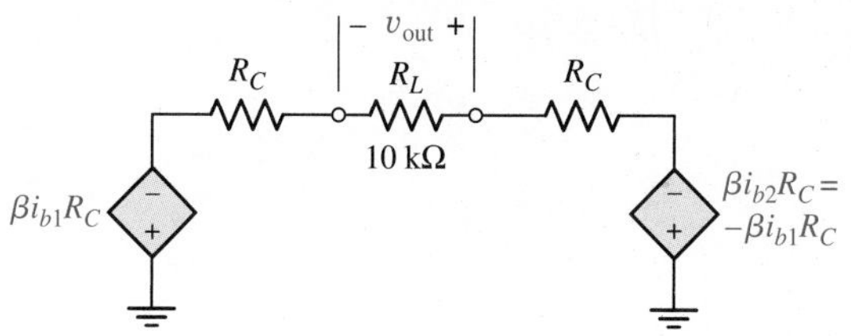

Figure 11.8
Equivalent output circuit for Example 11.2.

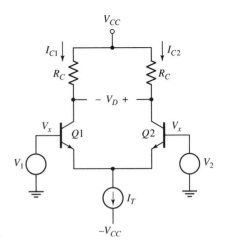

Figure 11.9
A simple differential amplifier.

An Alternate Voltage Gain Calculation

Figure 11.9 shows a differential amplifier using *npn* transistors. The variables in this figure are large-signal quantities rather than incremental quantities. Both transistors are biased by a current source, I_T, called the *tail current*. Input signals V_1 and V_2 are applied as shown, giving a differential input signal of

$$V_x = V_1 - V_2 \tag{11.15}$$

The voltages V_1 and V_2 can consist of both dc and incremental components. Since there are no external resistors in the input loop, the differential voltage must also be

$$V_x = V_{BE1} - V_{BE2} \tag{11.16}$$

The two collector currents can be approximated as

$$I_{C1} = I_{EO}e^{V_{BE1}/V_T} \tag{11.17}$$

$$I_{C2} = I_{EO}e^{V_{BE2}/V_T} \tag{11.18}$$

where $V_T = kT/q$, assuming the devices are matched.

The double-ended or differential output voltage can be written

$$V_D = V_{C2} - V_{C1} = (I_{C1} - I_{C2})R_{Ceq} = I_D R_{Ceq} \tag{11.19}$$

where

$$I_D = I_{C1} - I_{C2} \tag{11.20}$$

and

$$R_{Ceq} = R_C \parallel r_{out}$$

We also note that the sum of the collector currents can be approximated as

$$I_{C1} + I_{C2} = I_T \tag{11.21}$$

The ratio of I_D to I_T is

$$\frac{I_D}{I_T} = \frac{I_{C1} - I_{C2}}{I_{C1} + I_{C2}} \tag{11.22}$$

From Eq. (11.16)

$$V_{BE1} = V_x + V_{BE2} \tag{11.23}$$

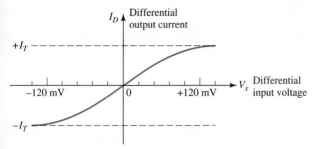

Figure 11.10
*Output current vs. input voltage
for differential pair.*

allowing us to use Eq. (11.17) to write

$$I_{C1} = I_{EO}e^{(V_x+V_{BE2})/V_T} = I_{EO}e^{V_x/V_T}e^{V_{BE2}/V_T} \qquad (11.24)$$

Substituting Eqs. (11.18) and (11.23) into Eq. (11.22) leads to

$$\frac{I_D}{I_T} = \frac{e^{V_x/V_T} - 1}{e^{V_x/V_T} + 1} = \tanh \frac{V_x}{2V_T} \qquad (11.25)$$

where tanh is the hyperbolic tangent defined by

$$\tanh y = \frac{e^y - e^{-y}}{e^y + e^{-y}}$$

The collector current difference can then be expressed as

$$I_D = I_T \tanh \frac{V_x}{2V_T} \qquad (11.26)$$

Equation (11.26) is graphed in Fig. 11.10. For small values of differential input voltage, the curve is quite linear. For larger values, above about 100 mV, the differential current approaches I_T. This corresponds to one device approaching cutoff while all of the tail current flows through the other device. Equation (11.26) can also be used to calculate a voltage gain for the differential pair. The double-ended output voltage is given by $V_D = I_D R_{Ceq}$ or

$$R_{Ceq}I_T \tanh \frac{V_x}{2V_T} \qquad (11.27)$$

For small arguments, a tanh function can be approximated by that argument; that is, $\tanh x \approx x$ for small values of x, which allows the output voltage to be written as

$$V_D \approx \frac{R_{Ceq}I_T V_x}{2V_T} \qquad \text{for} \quad V_x \ll V_T \qquad (11.28)$$

The incremental voltage gain can then be expressed as

$$A_D = \frac{V_D}{V_x} = R_{Ceq}\frac{I_T}{2V_T} \qquad (11.29)$$

Since this equation only applies for small values of V_x, we can assume that I_D is near zero and the emitter currents are approximately equal. Each emitter current will then be approximately $I_T/2$, and Eq. (11.29) can be written

$$A_D = \frac{V_D}{V_x} = R_{Ceq}\frac{I_{E1}}{V_T} = R_{Ceq}\frac{I_{E2}}{V_T} \qquad (11.30)$$

The transconductance of a common emitter stage is given by

$$g_m = \frac{i_c}{v_\pi} = \frac{\alpha I_E}{V_T} \approx \frac{I_E}{V_T} = \frac{1}{r_e} \qquad (11.31)$$

Using this expression in Eq. (11.30) leads to

$$A_D = \frac{V_D}{V_x} = g_m R_{Ceq} \tag{11.32}$$

11.4 In the circuit of Fig. 11.3(a), the value of I is 2 mA. If the collector resistors have values of 10 kΩ and $\beta = 100$ for both devices, calculate the magnitude of the differential voltage gain. *Ans:* 384.6 V/V.

11.5 In Fig. 11.3, the outputs of the stage of part (a) feed the differential inputs of the stage in part (b) after removing the input source and base resistances of the stage in part (b). The value of I in both stages is 2 mA. The value of all collector resistors is 10 kΩ and $R_E = 100$ Ω. Calculate the magnitude of the overall double-ended voltage gain. Assume that $\beta = 100$ for all devices. *Ans:* 16,740 V/V.

This equation can be compared to the earlier equation for gain developed from the equivalent circuit, Eq. (11.10), which is

$$A_D = \frac{\beta R_{Ceq}}{(\beta + 1) r_e} \tag{11.33}$$

With the approximations $\beta/(\beta + 1) \approx 1$ and $g_m \approx 1/r_e$, the two expressions are seen to be equivalent.

PRACTICAL Considerations

The expression for differential gain of the circuit in Fig. 11.9, $A_D = g_m R_{Ceq}$, can be extended to the case that external resistors appear in series with the two emitters. If R_E is inserted in series with each emitter, the same expression can be applied if the value of g_m is modified properly. This value should be changed from

$$g_m = \frac{1}{r_e}$$

to a value of

$$g_m = \frac{1}{r_e + R_E}$$

In the equivalent circuit, R_E appears in series with r_e; thus, replacing r_e by $r_e + R_E$ is reasonable.

11.1.3 THE MOSFET DIFFERENTIAL PAIR

Figure 11.11 shows a simple MOSFET differential pair. The current source again determines the quiescent drain voltages. Assuming matched devices, both device source currents equal one-half of the current generated by the current source. The drain currents also equal this value. The quiescent drain voltages are then

$$V_{D1} = V_{DD} - \frac{I}{2} R_1 \tag{11.34}$$

Figure 11.11
A MOSFET differential pair.

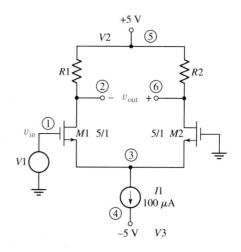

and

$$V_{D2} = V_{DD} - \frac{I}{2}R_2 \qquad \text{(11.35)}$$

The voltage at the sources of the devices can be found by using the active drain current equation. Assuming $\lambda = 0$, this is

$$I_D = \frac{\mu C_{ox} W}{2L}(V_{GS} - V_T)^2$$

where V_T is the threshold voltage of the devices. Since $I_D = I/2$, this equation allows V_{GS} to be found as

$$V_{GS} = V_T + \sqrt{LI/(\mu C_{ox} W)} \qquad \text{(11.36)}$$

In this circuit, the dc voltage at the gates is 0 V, resulting in

$$V_{SQ} = -\left[V_T + \sqrt{LI/(\mu C_{ox} W)}\right] \qquad \text{(11.37)}$$

Voltage Gain of the MOSFET Pair   Typically, the drain resistors are chosen to be equal. We will assume that $R_1 = R_2 = R_D$ in the following discussion. The input resistance at midband frequencies is assumed to be infinite; thus the input signal current is zero. The incremental equivalent circuit of the pair is indicated in Fig. 11.12, neglecting the body effect. Half of the input voltage drops across the gate and source of each device, generating incremental drain voltages of

$$v_{o1} = -g_m R_{Deq} \frac{v_{in}}{2} \qquad \text{(11.38)}$$

and

$$v_{o2} = g_m R_{Deq} \frac{v_{in}}{2} \qquad \text{(11.39)}$$

where $R_{Deq} = R_D \parallel r_{out}$.

The differential gain is then

$$A_D = \frac{v_{o2} - v_{o1}}{v_{in}} = g_m R_{Deq} \qquad \text{(11.40)}$$

The value of r_{out} is not equal to r_{ds}. Because each source sees an incremental resistance looking into the source terminal of the other device, the value of r_{out} will be higher than r_{ds}.

Figure 11.12
Equivalent circuit for the MOSFET differential pair.

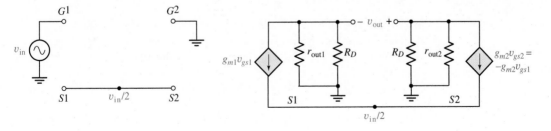

EXAMPLE 11.3

The devices of the differential pair of Fig. 11.11 have matched values of $V_T = 0.82$ V and $\mu C_{ox} W/2L = 200$ μA/V^2. The drain resistors are 50 kΩ and are much smaller than the output resistances of the devices. The current source $I = 100$ μA and the power supply voltage $V_{DD} = 5$ V. Calculate the quiescent drain voltages, the quiescent source voltages, and the incremental differential voltage gain.

SOLUTION The dc current is again assumed to split equally to give drain currents of 50 μA, which leads to quiescent drain voltages of

$$V_{DQ} = V_{DD} - I_D R_D = 5 - 0.05 \times 50 = 2.5 \text{ V}$$

Equation (11.37) can now be applied to find the source voltage, which gives

$$V_{SQ} = -\left[V_T + \sqrt{LI/(\mu C_{\text{ox}} W)}\right] = -\left[0.82 + \sqrt{100/400}\right] = -1.32 \text{ V}$$

In order to use Eq. (11.40) for the differential gain, we must first evaluate g_m. From Table 8.1 this can be found from

$$g_m = \sqrt{2\mu C_{\text{ox}}(W/L)I_D}$$

Using values from this example leads to $g_m = 200$ μA/V. The differential gain is

$$A_D = g_m R_{D\text{eq}} = 0.2 \times 50 = 10 \text{ V/V}$$

The upper corner frequency of the differential stage, when driven by a low-impedance signal source, is determined by the output circuit. It is easy to show that

$$f_{\text{high}} = \frac{1}{2\pi C_{\text{out}} R_{D\text{eq}}} \tag{11.41}$$

where $C_{\text{out}} \approx C_{db} + C_{gd}$.

EXAMPLE 11.4

Do a Spice simulation of the circuit of Example 11.3 using the model for the MOSFET from the Spice examples of Chapter 9. Use $L = 1$ μ and $W = 5$ μ. Determine the upper corner frequency of the differential pair and compare this to the calculated value, using appropriate element values from the output file from the simulation.

SOLUTION The Spice netlist file for this simulation is shown in Table 11.1.

The device parameters from the output file are $g_m = 176$ μA/V, $r_{ds} = 1.35$ MΩ, $C_{db} = 11$ fF, and $C_{gd} = 1.5$ fF. The calculated value of differential gain is

$$A_D = g_m R_{D\text{eq}} = 0.176 \times 50 = 8.8 \text{ V/V}$$

The upper corner frequency is calculated to be

$$f_{\text{high}} = \frac{1}{2\pi C_{\text{out}} R_{D\text{eq}}} = \frac{1}{2\pi \times 12.5 \times 10^{-15} \times 50,000} = 255 \text{ MHz}$$

The simulation gives a differential gain of 8.3 V/V and an upper corner frequency of 264 MHz.

Table 11.1   Spice netlist file for Example 11.4

```
EX11-4.CIR

R1 2 5 50K
R2 6 5 50K
I1 3 4 100U
V1 1 0 AC 0.05V
V2 5 0 5V
V3 0 4 5V
M1 2 1 3 0 N L=1U W=5U  AD=25P AS=25P PD=15U PS=15U
M2 6 0 3 0 N L=1U W=5U  AD=25P AS=25P PD=15U PS=15U
.AC DEC 10 100 1G
.OP
.PROBE
.LIB C5X.LIB
.END
```

PRACTICE Problem

11.6 Rework Example 11.3 if $V_T = 0.9$ V and $\mu C_{ox} W/2L = 400 \ \mu\text{A/V}^2$. Assume that the output resistances of the MOSFETs equal 280 kΩ. *Ans:*
$V_{DQ} = 2.5$ V,
$V_{SQ} = -1.25$ V, $g_m = 283$ μA/V, $A_D = 12$ V/V.

11.2 A Typical Op Amp Architecture and Specifications

IMPORTANT Concepts

1. Many op amps include three amplifying stages in cascade. The first stage is always a high-gain differential stage. The second stage typically has a moderate value of voltage gain. The last stage is often a buffer stage with high current gain and a voltage gain that is near unity.

2. The high-frequency poles of each stage introduce phase shift at higher frequencies that may cause the op amp to oscillate.

3. The upper corner frequency of the first stage is often reduced by adding capacitance at the output of this stage. This eliminates the possibility of oscillation but leads to a lower bandwidth for the op amp.

Although the technology used to implement op amps has changed considerably over the years, the basic architecture has remained remarkably constant. In this section we will first discuss that architecture and some approaches to its implementation. After this topic is considered, the subject will turn to methods of specifying amplifier performance.

The architecture of many op amps appears as shown in Fig. 11.13. The first section is a differential amplifier required by all op amps to allow a virtual ground and the implementation of mathematical operations. This stage is generally designed to have a very high differential voltage gain, perhaps several thousand. The bandwidth is generally rather low as a result of the high voltage gain. In some cases, this differential amplifier will use an active load that will also convert the double-ended output of the differential stage to a single-ended output that will drive the second voltage amplifier.

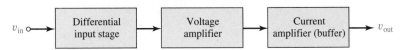

Figure 11.13
Architecture of typical op amp.

The second amplifier of Fig. 11.13 is a voltage amplifier with a more modest voltage gain that may range between 10 and 500 V/V. Often this stage is used to compensate the op amp, a topic that will be discussed later in this chapter. For now it is sufficient to say that this stage will probably be an inverting stage that can multiply the apparent value of some capacitance placed between the input and output of the stage. In this case the Miller effect is used to advantage. As a result of the Miller effect, leading to a large capacitive load presented to the input stage, this stage will have a very low upper corner frequency, perhaps 10–20 Hz.

The last stage is the output stage. It may be nothing more than an emitter follower or a source follower that has a large current amplification along with a voltage gain that is near unity. This stage will have a very high upper corner frequency.

11.2.1 THE HIGH-GAIN DIFFERENTIAL STAGE

The normal way to achieve high voltage gains is to use an active load for the amplifying stages, as demonstrated in Chapters 9 and 10. The incremental resistance presented to the amplifying stage is very high while the dc voltage across the active load is small. One popular choice for an active differential stage load is the current mirror. This load provides very high voltage gain and also converts the double-ended output signal to a single-ended signal referenced to ground. Figure 11.14 shows a block diagram of such an arrangement.

With no input signal applied to the differential stage, the tail current splits equally between I_{diff1} and I_{diff2}. The input current to the mirror equals this value of $I_{\text{tail}}/2$. This value is also mirrored to the output of the mirror, giving $I_{\text{out}} = I_{\text{tail}}/2$. We will assume that the voltage between the current mirror output and the second differential stage is approximately zero, although this assumption is unnecessary to achieve the correct result. The current to the resistance R is now zero.

When a signal is applied to the differential input, it may increase the current I_{diff1} by a peak value of ΔI. The input current to the mirror now becomes

$$I_{\text{in}} = \frac{I_{\text{tail}}}{2} + \Delta I$$

The output current from the mirror also equals this value. However, the input signal to the differential stage will decrease I_{diff2} by the same amount that I_{diff1} increases. Thus, we can write

$$I_{\text{diff2}} = \frac{I_{\text{tail}}}{2} - \Delta I$$

Figure 11.14
A current mirror active load for a differential stage.

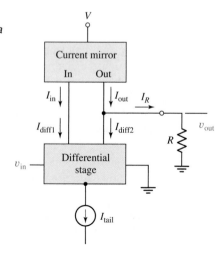

The current to the resistance R increases from its quiescent value of zero to

$$I_R = I_{out} - I_{diff2} = 2\Delta I$$

The incremental output voltage resulting is then

$$v_{out} = 2R\Delta I \qquad (11.42)$$

When an incremental input signal is applied to the differential pair, half of this voltage will drop across each base-emitter junction of the pair. The result is equal incremental differential stage currents in the two output devices, but they will be in opposite directions. An incremental input signal, v_{in}, will produce incremental currents of

$$i_{diff1} = \Delta I = \frac{g_m v_{in}}{2} \qquad (11.43)$$

and

$$i_{diff2} = -\Delta I = -\frac{g_m v_{in}}{2} \qquad (11.44)$$

where g_m is the transconductance of devices 1 and 2. Assuming negligibly large output resistances of the current mirror and the differential stage, the incremental output voltage becomes

$$v_{out} = 2i_{diff1} R = g_m R v_{in}$$

with a resulting midband voltage gain of

$$A_{MB} = g_m R \qquad (11.45)$$

If the output resistances of the mirror and differential stage are significant, the voltage gain can be found by combining these resistances in parallel with the load resistance to form $R_{eff} = R_{out} \parallel R$. This resistance then replaces R in Eq. (11.45). The load resistance may, in fact, be the incremental input resistance of the following stage. Very large values of voltage gain can result from this configuration.

For the MOSFET, the transconductance can be found from simulation or calculated from Table 9.1. For the BJT, the transconductance is given by $g_m = \alpha/r_e \approx 1/r_e$.

Although this expression is the same as that for the differential gain of a resistive load stage, given by Eqs. (11.10) or (11.40), two significant points should be made. First of all, the impedance R can be much greater than any resistive load that can be used in a differential stage. Large values of R in the differential stage would cause saturation of the stages for reasonable values of tail currents. In addition, large values of R are more difficult to fabricate on an IC chip. The current mirror solves this problem. The second point is that the output voltage of the differential pair with a current mirror load is a single-ended output, which can be applied to a following simple amplifier stage. However, the rejection of common-mode variables caused by temperature change is still in effect with the current mirror stage. If a resistive load differential stage must provide a single-ended output, the gain drops by a factor of 2, compared to the double-ended output, and common-mode rejection no longer takes place.

EXAMPLE 11.5

The BJT differential stage of Fig. 11.15 drives a current mirror load. The input impedance to the following stage is 100 kΩ. The output resistance of the current mirror is 80 kΩ, and the output resistance of stage two of the differential pair is 150 kΩ. The tail current is 3.6 mA. Calculate the midband voltage gain of the circuit, v_{out}/v_{in}.

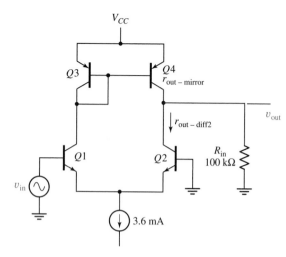

Figure 11.15
Circuit for Example 11.5.

PRACTICE Problems

11.7 Rework Example 11.5 if I_{tail} is lowered to 1 mA. Assume that the output impedance of the current mirror is now 130 kΩ, but all other parameters remain the same. Is it reasonable to expect the output impedance of the mirror to increase? *Ans: A_{MB} = 709 V/V; yes, output impedance varies inversely with collector current.*

11.8 Rework Example 11.5 if the current mirror is replaced by a Wilson current mirror with an output impedance of 500 kΩ. *Ans: A_{MB} = 3709 V/V.*

SOLUTION The gain can be calculated from Eq. (11.45) after finding both R_{eff} and g_m. The value of R_{eff} is given by

$$R_{eff} = r_{out-diff2} \parallel r_{out-mirror} \parallel R_{in} = 150 \parallel 80 \parallel 100 = 34.3 \text{ k}\Omega$$

The transconductance is $g_m = \alpha/r_e \approx 1/r_e$ where $r_e = 26/(I_{tail}/2)$. For this circuit

$$r_e = \frac{26}{1.8} = 14.4 \ \Omega$$

giving a transconductance of $g_m = 1/14.4 = 0.0692$ A/V. The voltage gain is then

$$A_{MB} = g_m R_{eff} = 0.0692 \times 34,300 = 2374 \text{ V/V}$$

This is a considerably higher value of gain than could be achieved with a resistive load differential stage.

11.2.2 THE SECOND AMPLIFIER STAGE

Before the purpose of the second amplifier stage can be fully understood, a discussion on circuit stability must take place. A more complete coverage of this topic is included in Chapter 12.

Feedback and Stability of the Op Amp The op amp is used in a feedback configuration for essentially all amplifying applications. The inverting and noninverting configurations were considered as early as Chapter 4 in this text. Because of nonideal or parasitic effects, it is possible for the feedback amplifier to exhibit unstable behavior. Oscillations at the output of the amplifier can exist, which have no relationship to the applied input signal. This, of course, negates the desired linear operation of the amplifier. It is necessary to eliminate any undesired oscillation signal from the amplifier.

The open-loop voltage gain of a three-stage amplifier such as that of Fig. 11.13 may be represented by a gain function of

$$A = \frac{A_{MB}}{\left(1 + j\frac{\omega}{P_1}\right)\left(1 + j\frac{\omega}{P_2}\right)\left(1 + j\frac{\omega}{P_3}\right)} \qquad \text{(11.46)}$$

where P_1, P_2, and P_3 are the dominant poles of gain stages 1, 2, and 3, respectively. The quantity A_{MB} is the low-frequency or midband gain of the amplifier.

When this op amp is used in a negative feedback configuration such as that shown in Fig. 11.16, the loop gain must be analyzed to see whether the conditions for oscillation occur. The feedback factor for this circuit has been evaluated in Chapter 4 and is

$$F = \frac{R_2}{R_2 + R_F} \tag{11.47}$$

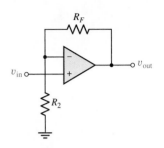

Figure 11.16
A noninverting stage.

In order to check the stability of this circuit, a zero-volt input signal (short circuit) can be applied to the noninverting input terminal. The loop gain from inverting terminal to output and back to inverting terminal can then be found with the noninverting terminal shorted to ground. This loop gain can be expressed as

$$AF = \frac{-A_{MB}F}{\left(1 + j\frac{\omega}{P_1}\right)\left(1 + j\frac{\omega}{P_2}\right)\left(1 + j\frac{\omega}{P_3}\right)} \tag{11.48}$$

Carrying out the multiplication of the denominator factors leads to the expression

$$AF = \frac{-A_{MB}F}{-j\frac{\omega^3}{P_1 P_2 P_3} + j\omega\left(\frac{1}{P_1} + \frac{1}{P_2} + \frac{1}{P_3}\right) - \omega^2\left(\frac{P_1 + P_2 + P_3}{P_1 P_2 P_3}\right) + 1} \tag{11.49}$$

Oscillations can occur around a feedback loop if the following two conditions are met:

$$(1) \text{ The angle of } AF \text{ is } 0° \text{ or some multiple of } 360° \tag{11.50}$$

$$(2) \ |AF| \geq 1 \tag{11.51}$$

In words, the loop gain AF can cause oscillations only if it has a $0°$ (or $360°$) phase shift and the magnitude of AF is unity or greater.

For this op amp stage, the first condition can only occur if the imaginary part of the denominator goes to zero and the denominator becomes a negative, real number. Setting the imaginary part of the denominator equal to zero results in a potential oscillation frequency that satisfies

$$\omega^2 = P_1 P_2 + P_2 P_3 + P_1 P_3 \tag{11.52}$$

With this value of ω^2, the imaginary part of the denominator goes to zero, leaving only the real part. Substituting for the ω^2 term in the remaining real part leads to a loop gain of

$$AF = \frac{-A_{MB}F}{-\left(2 + \frac{P_1 + P_2}{P_3} + \frac{P_1 + P_3}{P_2} + \frac{P_2 + P_3}{P_1}\right)} \tag{11.53}$$

The negative signs of both numerator and denominator cancel, so that the sign of the loop gain is positive at this potential oscillation frequency; that is, the angle of the loop gain is zero. If we want the amplifier to be stable, we must ensure that the magnitude of the denominator of Eq. (11.53) is larger than the magnitude of the numerator to result in $|AF| < 1$. We should also be interested in doing *worst-case* design; that is, we want this magnitude to be less than unity for the largest value of numerator that can occur. The midband gain will have some very large maximum value for a given design. The feedback factor has a practical upper limit of unity when a unity gain buffer stage is used. Thus, the upper limit of the numerator magnitude is $A_{MB\text{max}}$. The magnitude of the denominator must now be made larger than this value.

If the three gain stages are designed with no shunt capacitances added to limit the upper corner frequencies, typical values might be in the hundreds of kHz to hundreds of MHz range. For ease of calculation, let us assume that the op amp is fabricated with values of $P_1 = 10^6$ rad/s, $P_2 = 2 \times 10^6$ rad/s, and $P_3 = 3 \times 10^6$ rad/s. For these values, the magnitude of the denominator of AF, from Eq. (11.53), equals 10. In order to avoid instability, the maximum gain would have to be limited to values less than 10. For a three-stage op amp, a practical value of A_{MBmax} might be 300,000. Thus, we would expect oscillations to occur for a unity-gain configuration of this op amp.

One possibility to achieve normal op amp gains while maintaining a stable circuit is to intentionally modify one of the three upper corner frequencies of the op amp. For example, suppose we limited P_1 to 10 rad/s while keeping the same values of P_2 and P_3. For these values, the magnitude of the denominator of Eq. (11.53) equals 5×10^5. For the maximum numerator value of 300,000, resulting from the unity-gain configuration, the magnitude of AF would now be 0.6, ensuring that oscillations do not occur.

The gain of the op amp stage from noninverting input to output would then be given by

$$A = \frac{A_{MB}}{\left(1 + j\frac{\omega}{10}\right)\left(1 + j\frac{\omega}{2 \times 10^6}\right)\left(1 + j\frac{\omega}{3 \times 10^6}\right)}$$

The frequency response of this gain is sketched in Fig. 11.17 for a midband gain of 300,000. We note that the magnitude of this gain has fallen below a value of unity (0 dB) before the second upper corner frequency of 2×10^6 is reached. Normally, engineers do not destroy bandwidth of a stage intentionally; however, in this case it is necessary to stabilize the operation of the op amp in the feedback configuration.

Lowering the upper corner frequency of one stage is not a trivial matter. Although it may not need to be reduced to 10 rad/s in a practical op amp, it is often lowered to 10–100 Hz. A capacitor must be added to the appropriate point in the stage to drop this frequency, but a relatively large capacitor is required. In the early days of the IC op amp, the two terminals between which a capacitor was to be added were connected to two external pins of the chip. A discrete capacitor of sufficient value was then added externally.

In 1967, the capacitor was added to the IC chip using the Miller effect to multiply the capacitor value. Returning to Fig. 11.13, it is seen that a capacitor can be added between input and output of the second stage. This capacitor is typically of value 30 pF that, due to the Miller effect, is multiplied by a factor of $(A_2 + 1)$ where $-A_2$ is the gain of the second stage. The Miller capacitance is reflected to the input of the second stage and loads the output of the first stage. This large effective capacitance, driven by the large output impedance of the first stage, produces a very low upper corner frequency for the first stage.

In a 741 op amp design, the gain A_2 is approximately 400 V/V. With a 30-pF capacitance bridging input to output, the effective capacitance at the input is about 0.012 μF. This creates a bandwidth for the op amp of 10 Hz or 62.8 rad/s; thus, the output impedance of the previous stage must be near 1.33 MΩ.

This method of solving the instability problem is referred to as *dominant pole compensation*. The lower value of pole frequency is seen from Fig. 11.17 to dominate the amplifier performance up to frequencies above the useful range of gain.

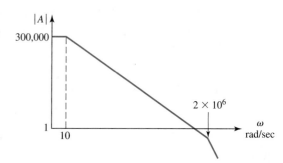

Figure 11.17
Op amp frequency response with $P_1 = 10$ rad/s.

In many op amps, the bandwidth reduction for compensation may be smaller. For lower gain op amps, the capacitance value required will also be reduced. Since FET devices typically have lower transconductances than bipolar devices, placing FET devices on the input will reduce overall gain, relax compensation requirements, and result in a better overall bandwidth. Processes that allow the fabrication of bipolar and FET devices on the same chip have become very popular in the construction of higher bandwidth op amps. These processes lead to BiFET, BiMOS, or BiCMOS op amps.

EXAMPLE 11.6

The architecture of Fig. 11.13 is used for an op amp. The first stage has a midband voltage gain of 1200 V/V, a dominant pole at $2\pi \times 96$ krad/s, and an impedance between output and ground (including R_{in2}) of 112 kΩ. The second stage has a midband voltage gain of -130 V/V and a dominant pole at $2\pi \times 1$ Mrad/s. The third stage has a midband voltage gain of 0.94 V/V and a bandwidth of $2\pi \times 10$ Mrad/s. Select the smallest capacitor that can be used to bridge the input and output nodes of the second amplifier to cause stability for the unity-gain configuration.

SOLUTION When this circuit is connected as an op amp in the unity-gain configuration, the gain around the feedback loop has a midband magnitude of

$$A_{MB}F = 1200 \times 130 \times 0.94 \times 1 = 146,640$$

Using $P_1 = 2\pi \times 96$ krad/s, $P_2 = 2\pi \times 10^6$ rad/s, and $P_3 = 2\pi \times 10^7$ rad/s, the denominator of Eq. (11.53) has an uncompensated value of

$$2 + 0.11 + 10.10 + 114.6 = 126.81$$

The loop gain at the possible oscillation frequency is then

$$|AF| = \frac{146,640}{126.81} = 1156$$

This circuit is certain to oscillate with this high value of loop gain. In order to drop the magnitude to unity by changing P_1, the new value of P_1 is found by solving

$$AF = \frac{-A_{MB}F}{-\left(2 + \frac{P_1' + P_2}{P_3} + \frac{P_1' + P_3}{P_2} + \frac{P_2 + P_3}{P_1'}\right)}$$

for P_1' after equating AF to unity, and using the specified values of $A_{MB}F$, P_2, and P_3. This leads to a value of

$$P_1' = 2\pi \times 75.1 \text{ rad/s}$$

This value can be approximated by recognizing that the last term of the denominator will dominate the numeric value when P_1' is small. Approximating the denominator by this last term results in

$$P_1' = \frac{P_2 + P_3}{A_{MB}F}$$

Now that the required value of the pole is known, we turn to the equation that will result in this value. The pole P_1' will be determined as

$$P_1' = \frac{1}{R_{out1}C_{in2}} = \frac{1}{112,000C_{in2}}$$

Solving for C_{in2} gives a value of 18.92 nF. The value that must be inserted to bridge input and output of the second amplifier is found from the Miller effect equation. With a gain of $A_2 = -130$ V/V for this stage, we can write

$$C_{in2} = (|A_2| + 1)C = 131C$$

Which leads to a value of $C = 144$ pF.

From a practical standpoint, this value is larger than desirable for an IC chip; thus, the gain of the differential stage might be lowered while raising the gain of the second amplifier to reduce the required capacitance.

A major purpose of the second amplifier stage is to provide Miller effect capacitance for compensation purposes.

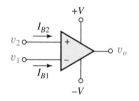

Figure 11.18
An op amp diagram.

11.2.3 OP AMP SPECIFICATIONS

To this point, we have ignored certain imperfections of the op amp to develop the underlying basis of operation. Because transistors will require finite voltages and currents to bias stages into the active region and will not be matched perfectly, nonideal effects may influence the performance of the stage. Some specifications that are important for design purposes are defined in this subsection. The diagram of Fig. 11.18 is used to define various terms used in these definitions.

Input Offset Voltage (V_{OS}): Mismatch of the transistors in the differential input stage leads to a finite output dc voltage when both inputs are shorted to ground. This finite output voltage is called the *output offset voltage*. A slight voltage mismatch in the differential pair is amplified by succeeding stages to create a larger voltage at the output. Inaccurate biasing of later stages also contributes to the output offset. Inaccuracies in later stages are amplified by smaller factors than are early stage inaccuracies.

The *input offset voltage* is the voltage that must be applied across the differential input terminals to cause the output voltage of the op amp to equal zero. Theoretically, this voltage could be found by measuring the output voltage when the input terminals are shorted, then dividing this value by the gain of the op amp. In practice, this may not be possible because the gain may not be known or the output offset may exceed the size of the active region. The value of V_{OS} is typically a few millivolts for monolithic or IC op amps.

Input Offset Voltage Drift (TCV_{OS}): Temperature changes affect certain parameters of the transistors, leading to a drift in the output dc offset voltage with temperature. In BJT devices, the voltage across the base-to-emitter junction for a constant emitter current drops by approximately 2 mV/°C. In MOSFET devices, the threshold voltage changes with temperature, leading to drain current variations predicted by

$$I_D = \frac{\mu C_{ox} W}{2L} [V_{GS} - V_T]^2$$

Small drifts of voltage in early stages will be amplified by following stages to produce relatively large drifts in output voltage. Because the output dc signal may not exist within the active region of the op amp, the drift is again referred to the input.

The *input offset voltage drift* is defined as the change in V_{OS} for a 1°C change in temperature (near room temperature). A value of 10–20 μV/°C is typical for IC op amps.

Input Bias Current (I_B): A BJT differential stage, such as that of Fig. 11.3(a), will require a finite amount of base current for biasing purposes, even if both inputs are grounded.

The *input bias current* of an op amp is defined as the average value of bias current into each input with the output driven to zero. The two bias currents are generally slightly different, so I_B is

$$I_B = \frac{(I_{B1} + I_{B2})}{2} \tag{11.54}$$

The input bias current for the MOSFET is very small compared to the BJT. In most cases, this value is zero for a MOSFET differential input pair.

Input Offset Current (I_{OS}): The *input offset current* is the difference between the two input bias currents when the output is at zero volts. This parameter is

$$I_{OS} = |I_{B1} - I_{B2}| \tag{11.55}$$

Common-Mode Input Voltage Range (CMVR): The voltage range over which the inputs can be simultaneously driven without causing deterioration of op amp performance is called the *common-mode voltage range* or *CMVR*. In most op amps, the *CMVR* is a few volts less than the rail-to-rail value of the power supplies. In many applications, both inputs are forced to move together due to the virtual short between the input terminals when negative feedback is used.

Common-Mode Rejection Ratio (CMRR): The ratio of input common-mode voltage to change in input offset voltage is called the *common-mode rejection range* or *CMRR*. An equivalent definition is the ratio of differential voltage gain to common-mode voltage gain. IC op amps range from 80 to 100 dB for the *CMRR*. This parameter was mentioned earlier in the chapter and is a measure of the mismatch of incremental gain from each of the two inputs to output. If the incremental gains from each input to output were equal, the *CMRR* would be infinite.

Power-Supply Rejection Ratio (PSRR): The *power-supply rejection ratio* or *PSRR* is the ratio of change in the input offset voltage to a unit change in one of the power-supply voltages. An op amp with two power supplies requires that a PSRR be specified for each power supply.

An example will demonstrate the use of several of these specifications.

EXAMPLE 11.7

In the op amp of Fig. 11.19, the offset voltage is specified as $V_{OS} = \pm 2$ mV, the input bias current is $I_B = 1$ μA, and the input offset current is $I_{OS} = \pm 0.25$ μA. Calculate the worst-case output offset voltage.

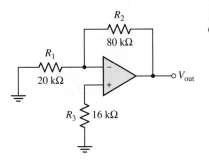

Figure 11.19
Circuit for Example 11.7.

SOLUTION In a given circuit, some of the offset components may be of such a polarity to cause cancellation of another component. A second circuit might have components that are additive. In a worst-case analysis, all offset components are taken with polarities that lead to the largest possible offset error. This provides an upper limit on the error that might occur if several thousand identical amplifiers were constructed.

A second point that is somewhat subtle is that the output offset due to each effect can be calculated by assuming an ideal op amp that is driven by the input offset value. For this amplifying stage, it is appropriate to use superposition, calculating the output offset voltage due to each parameter, then adding the results to get the worst-case value.

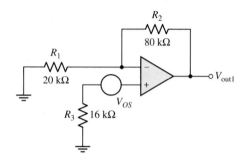

Figure 11.20
Calculation of output offset
voltage.

The circuit of Fig. 11.20 shows an offset voltage in series with the noninverting terminal of the amplifier. Since the op amp is assumed to be ideal, the input impedance at this terminal is infinite. No current flows through the 16-kΩ resistance, so the voltage at the noninverting terminal equals the input offset voltage. The gain of the noninverting stage multiplied by the input offset voltage leads to the output offset voltage, which is expressed as

$$V_{\text{out1}} = V_{OS}\left(1 + \frac{R_2}{R_1}\right)$$

where V_{out1} is the output offset voltage due to input offset voltage. This leads to a value of

$$V_{\text{out1}} = \pm 2 \text{ mV} \times \left(1 + \frac{80}{20}\right) = \pm 10 \text{ mV}$$

The output voltage due to bias currents can be found from the circuit of Fig. 11.21. Assuming an output voltage near ground potential, the voltage at the inverting terminal due to bias current I_{B1} is

$$V^- = -I_{B1} R_1 \| R_2$$

The voltage at the noninverting terminal is

$$V^+ = -I_{B2} R_3$$

Figure 11.21
Calculation of effect of input
offset current.

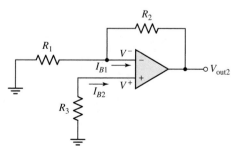

The difference of these voltages will be amplified by the gain of the op amp. In this case, the value of R_3 was selected to equal $R_1 \parallel R_2$, leading to a difference voltage of zero. There will be no output offset voltage due to bias currents as a result of this choice of R_3.

The offset effect of the input offset current can also be calculated from the circuit of Fig. 11.21. The output voltage can be expressed as

$$V_{out2} = A(V^+ - V^-)$$

Writing

$$V^+ = -(I_{B1} \pm I_{OS})R_3$$

and

$$V^- = V_{out2}\,\frac{R_1}{R_1 + R_2} - I_{B1} \times R_1 \parallel R_2$$

and noting that $R_3 = R_1 \parallel R_2$ leads to

$$V_{out2} = \pm R_2 I_{OS}$$

The output offset voltage due to input offset current for this circuit is

$$V_{out2} = \pm I_{OS}\,R_2 = \pm 0.25\ \mu\text{A} \times 80\ \text{k}\Omega = \pm 20\ \text{mV}$$

The worst-case output offset voltage for this stage is then $V_{out} = \pm 30$ mV.

11.3 A Practical Op Amp

IMPORTANT Concepts

1. The 741 op amp is an example of an op amp that uses the typical architecture discussed in the preceding section.
2. Several additional features are included in the 741 op amp to make it a practical amplifier.

11.3.1 THE 741 OP AMP

The 741 op amp was developed in 1967 and became one of the most widely used designs in op amps. It was originally implemented with bipolar technology and later in CMOS. Although many improved op amps are now available, this design provides a good vehicle for discussing the general op amp architecture of Fig. 11.13. A simplified schematic/block diagram is shown in Fig. 11.22.

As in the general architecture, the circuit consists of three stages of gain, the first being a differential stage ($Q1$, $Q2$), the second an inverting voltage amplifier ($Q5$), and the third a buffer stage. Because the first stage typically provides a high voltage gain, its performance will largely determine key parameters of the op amp, including input offset voltage and current. This differential stage uses a current mirror ($Q3$, $Q4$) to provide an active load that results in a very high voltage gain for this stage. The capacitor C provides dominant-pole compensation by taking advantage of the Miller effect to increase the apparent value of this capacitor.

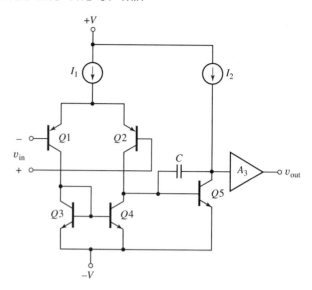

Figure 11.22
*Simplified diagram
of 741 op amp.*

Slew Rate The presence of the compensation capacitor in the circuit of Fig. 11.22 will not only limit the bandwidth, but will also limit the rate at which the op amp can respond to a large applied input signal. The maximum rate at which the output voltage of an op amp can change is called the *slew rate*. This value is determined by the rate of change of the critical node voltage inside the op amp. In the case of the 741 op amp, the critical node is the output of the differential stage at its connection to the capacitor C. When a large input signal exceeding 60–100 mV is applied to the differential stage, the input device $Q1$ will be shut off so that all of the tail current will be directed through $Q2$ to charge the capacitor. Further increase in input signal will not provide additional charging current. Figure 11.23 shows this case.

Since $Q3$ and $Q4$ are not conducting when $Q1$ is in cutoff, the total tail current is available to charge the capacitor. Device $Q5$ has a high voltage gain; hence V_{C5} will change

Figure 11.23
*Slew rate equivalent
of 741 op amp.*

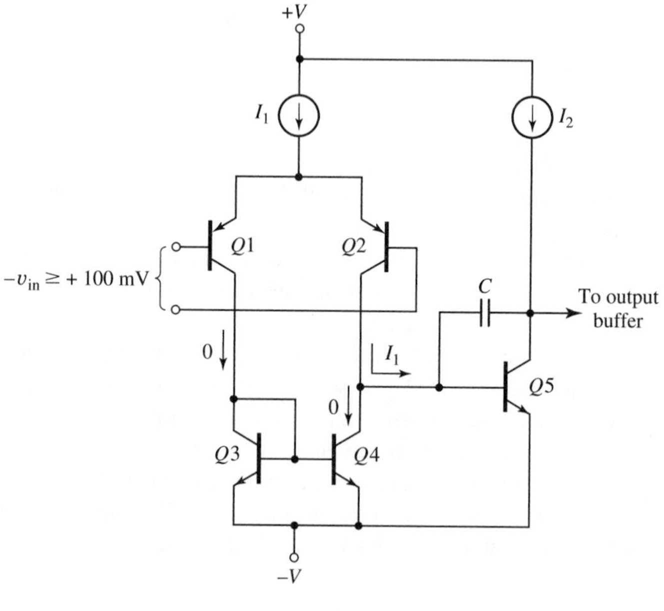

much more than V_{B5}. The collector voltage change of $Q5$ is then approximately equal to the capacitor voltage change. This rate of change can be expressed as

$$\frac{dV_C}{dt} = \frac{I_1}{C} = \text{slew rate at } Q5 \text{ output} \qquad (11.56)$$

Since the output of $Q5$ is passed to the op amp output through a unity voltage gain buffer, the output slew rate will equal the slew rate of the $Q5$ output. The 741 op amp may use a tail current of 30 μA. For a 30-pF capacitor, the slew rate of this op amp is about 1 V/μs.

Since slew rate is an important performance parameter for an op amp, efforts are generally made to increase it as much as possible. At first thought, it would appear that merely increasing the tail current of the input differential pair would help and would lead to an increased value of tail current I_1 in Eq. (11.56). However, in bipolar design, the transconductance of the input devices also increases directly with tail current, making the gain of the op amp higher. This higher gain requires a larger compensation capacitor, which increases the value of C in Eq. (11.56). This method does not often lead to significant improvement in slew rate.

If slew rate is to be increased by increasing the tail current, the voltage gain of the input stages can be limited by methods such as emitter degeneration. The increased tail current is not offset by a larger required value of compensating capacitor, and large improvements in slew rate can be effected. On the other hand, MOSFET designs with less inherent voltage gains require smaller capacitors and often have superior slew rates to bipolar designs.

Other Features of the 741   Figure 11.24 shows a complete schematic of a bipolar version of the 741.

Darlington Connection: The output of the input differential stage must work into a high input impedance to achieve a high voltage gain. A double emitter follower, called the Darlington connection, is used to produce a high input resistance to the second gain stage.

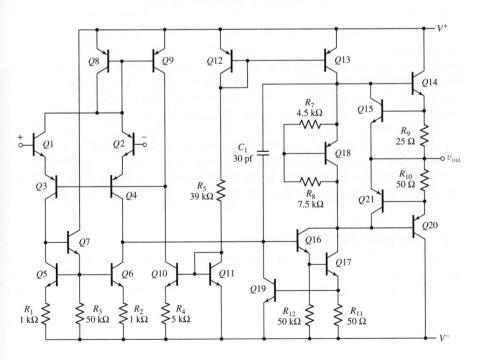

Figure 11.24
Schematic of the 741 op amp.

The Darlington connection is made up of $Q16$ and $Q17$. The collectors of both transistors are connected together, while the emitter of $Q16$ drives the base of $Q17$. Ignoring the small current drawn by resistor R_{12}, the current gain from the base of $Q16$ to the emitter of $Q17$ is $(\beta + 1)^2$. The impedance seen looking into the base of $Q16$ will then equal $(\beta + 1)^2$ times the impedance connected to the emitter of $Q17$. The resistance R_{11} makes up the emitter load of $Q17$. The Darlington connection is popular in situations that require a high input impedance.

npn Input Transistors and Biasing: In the original bipolar IC process used to fabricate the 741 op amp circuit, both *pnp* and *npn* devices were available. However, the *pnp* devices were implemented as *lateral* transistors whereas the *npn* devices were *vertical* transistors. The lateral *pnp* devices exhibited very poor current gain and frequency characteristics. If these devices had been used for the input stages, as shown in the diagram of Fig. 11.22, a very low input impedance and voltage gain would result. Adding the *npn* devices, $Q1$ and $Q2$ of Fig. 11.24, leads to superior first-stage performance in terms of voltage gain, bandwidth, and reduced input bias currents. The differential stage has a tail current set up by $Q10$ and mirrored through $Q9$ to $Q8$.

Output Stage: The output stage of the 741 op amp is a complementary emitter follower consisting of $Q14$ and $Q20$. The *npn* device conducts for positive output signals, and the *pnp* device conducts for negative output signals. Crossover distortion is minimized by biasing both bases to a point near the point of conduction. The level shift circuit consisting of device $Q18$ and resistors R_7 and R_8 provide this bias. Devices $Q15$ and $Q21$ provide short-circuit protection for the output devices. If a high current load is inadvertently placed across the output, the increased voltage drops across R_9, and R_{10} will forward bias devices $Q15$ and $Q21$ to rob the base current of the output devices. This puts the output devices into a less conductive state and minimizes power dissipation in these devices.

PRACTICAL Considerations

Although the nominal specifications of the 741 op amp show a midband gain of 200,000 V/V and a bandwidth of 5 Hz for a *GBW* = 1 MHz, these values may vary by as much as 100% from one op amp to another. A *GBW* of 2 MHz would not be unusual for this device. Such variations would be difficult to tolerate in design problems except for the fact that the op amp always uses feedback to create an amplifier with very stable voltage gains. This topic will be covered in detail in the following chapter.

Whereas the gain of an amplifier is established accurately by the ratio of the feedback resistors, the bandwidth may vary over a large range. This variation is often unimportant as long as it exceeds the required values. If it is significant, a capacitor placed across the feedback resistor can limit the bandwidth to the desired value.

11.3.2 BICMOS OP AMP DESIGN

Figure 11.25 shows the schematic of an op amp that also follows the architecture of the simplified op amp of Fig. 11.22 but employs a combination of bipolar and MOS devices. Using bipolar devices only for emitter followers allows the fabrication to be implemented on a standard p-well CMOS process. Devices $M1$ and $M2$ make up the differential input stage, and devices $M3$ and $M4$ provide a current mirror load with double-ended output to

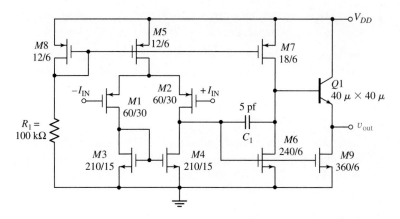

Figure 11.25
A simple BiCMOS op amp.

single-ended output conversion. Device $M6$ provides a common-source voltage gain stage with an active load provided by $M7$. The output stage of the amplifier is an emitter follower $Q1$ biased with a *pull-down* current source, $M9$.

The tail current for the input stage as well as the active load current for the second stage are provided by current mirror devices $M5$ and $M7$ slaved to reference device $M8$. The reference current for $M8$ is controlled by resistor R_1.

The compensating capacitor is only 5 pF, leading to a better slew rate and a higher bandwidth than the 741 bipolar op amp.

DISCUSSION OF THE DEMONSTRATION PROBLEM

The possible oscillation frequency for the op amp is found by solving Eq. (11.52)

$$\omega^2 = P_1 P_2 + P_2 P_3 + P_1 P_3 = (2\pi)^2 (8 \times 10^8 + 3.2 \times 10^{11} + 4 \times 10^{10})$$

which leads to a possible oscillation frequency of 600.7 kHz.

We must now check to see whether the magnitude of AF is one or greater. Equation (11.53) can be used to find that this magnitude is much greater than unity; thus, oscillations should occur.

This same equation can be used to find what value the first-stage pole must be lowered to in order to have marginal stability. This is done by finding the value of P_1 in Eq. (11.53) that leads to a magnitude of exactly unity. We use this information to write

$$A_{MB} F = 2 + \frac{P_1 + P_2}{P_3} + \frac{P_1 + P_3}{P_2} + \frac{P_2 + P_3}{P_1}$$

When P_1 is lowered, the last term of this expression will be the dominant term. The value of P_1 can be approximated by solving

$$A_{MB} F = 10{,}000 = \frac{P_2 + P_3}{P_1} = \frac{2\pi \times 4.08 \times 10^6}{P_1}$$

Which leads to a value for the pole frequency of $f_{\text{high1}} = 408$ Hz. The loop gain will be unity for this pole frequency, leading to marginal stability.

SUMMARY

➤ The differential stage is required for the input stage of an op amp. This stage allows a signal applied to one input to be amplified and inverted, while a signal applied to a second input will be amplified with no inversion.

➤ The differential gain of an op amp is very high, often in the range of 200,000–300,000 V/V. The bandwidth of the op amp is quite small in the range of tens of Hz.

➤ Many op amps are designed with a high-gain differential input stage, a moderate-gain second stage, and a low-voltage gain, high-current gain output stage.

➤ The op amp will be unstable in the unity-gain configuration unless bandwidth of one of the stages is reduced. Typically, the bandwidth of the first stage is reduced by adding capacitance to the output of this stage.

➤ Specifications for the op amp include differential gain, common-mode gain, common-mode rejection ratio, input offset voltage and current, and power-supply rejection ratio.

PROBLEMS

SECTION 11.1.1 THE BASIC DIFFERENTIAL PAIR

11.1 A differential pair such as that of Fig. 11.1 has values of $A_1 = 98.1$ and $A_2 = 98.2$. If a double-ended output is taken, calculate the common-mode gain and the differential gain of the amplifier.

11.2 Calculate the *CMRR* for the amplifier of Problem 11.1.

11.3 The *CMRR* of a differential amplifier is 78 dB. If the differential gain is 460 V/V, what is the common-mode gain of the amplifier?

11.4 If the differential gain of an amplifier is 820 V/V, what must the *CMRR* be to have a common-mode gain of 0.1 V/V?

SECTION 11.1.2 THE BJT DIFFERENTIAL PAIR

11.5 Calculate the differential gain for the circuit shown, assuming both devices have $\beta = 100$.

Figure P11.5

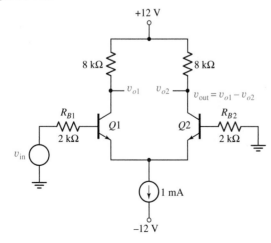

11.6 Find the quiescent output voltages V_{CQ1} and V_{CQ2} in the circuit of Problem 11.5. What is the largest peak value of v_{out} before saturation or cutoff is reached for the bias of this circuit?

11.7 In the circuit of Problem 11.5, the base resistors are changed to $R_{B1} = 3\,\text{k}\Omega$ and $R_{B2} = 4\,\text{k}\Omega$. Assuming that $V_{BE(on)} = 0.7$ V and $\beta_1 = \beta_2 = 100$, calculate

(a) The emitter currents of each transistor

(b) The largest peak value of v_{out} before serious distortion

(c) The differential voltage gain

11.8 Repeat Problem 11.7 if $R_{B1} = 2\,\text{k}\Omega$ and $R_{B2} = 4\,\text{k}\Omega$.

11.9 Calculate the double-ended voltage gain of the differential stage shown. Assume that $\beta = 100$ for both devices.

Figure P11.9

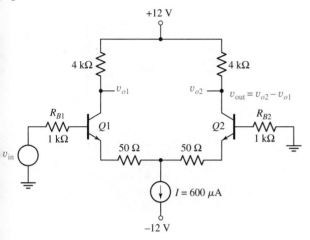

11.10 If R_{B1} is changed to 2 kΩ in the circuit of Problem 11.9, calculate I_{E1} and I_{E2}. Assume that $V_{BE1} = V_{BE2} = 0.7$ V. Calculate the single-ended voltage gain v_{o1}/v_{in} and compare to the value v_{o2}/v_{in}.

☆ **11.11** Evaluate g_m for the devices of Problem 11.5. Calculate the differential gain using Eq. (11.32), assuming the base resistors are shorted. If the base resistors are not shorted, modify your value of g_m appropriately to calculate the gain using Eq. (11.32).

☆ **11.12** Evaluate g_m for the devices of Problem 11.9, assuming the base resistors are shorted. Now modify your value of g_m to include the effects of the 50-Ω emitter resistors to calculate the gain using Eq. (11.32), again assuming the base resistors are shorted.

☆ **11.13** Assume that $\beta = 100$ for both devices of the figure and also assume that the dc bias current splits equally between the two emitters.

(a) The output voltage can be defined in one of two ways, either (1) $v_{out} = v_{o1} - v_{o2}$ or (2) $v_{out} = v_{o2} - v_{o1}$. Which definition results in a positive value of A_D?

(b) What is the value of quiescent collector voltage, V_{Q1}?

(c) What is the value of dc output voltage?

(d) What is the value of R_{in}?

(e) What is the value of $v_{out} = v_{o2} - v_{o1}$?

(f) If a 10-kΩ resistor is connected between the two collector terminals, what is the value of v_{out}?

Figure P11.13

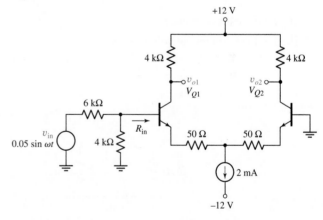

D 11.14 Change the value of the 10-kΩ resistor in Problem 11.13(f) to lead to an output voltage with a peak value of 0.75 V.

11.15 Assume that $\beta = 100$, $V_{CE(sat)} = 0$ V, and the dc bias current splits equally between emitters in the circuit shown.

Figure P11.15

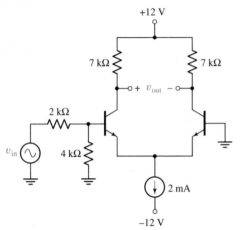

D 11.17 Assume that $\beta = 120$ for both transistors in the figure.

Figure P11.17

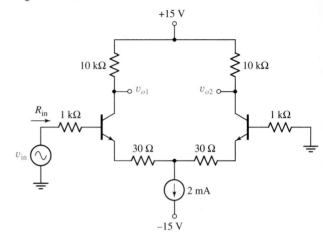

(a) Calculate the differential voltage gain for the circuit.

(b) If v_{in} is increased, what is the largest value that v_{out} can reach before serious distortion occurs (saturation or cutoff)?

☆ **11.16** Assume that $\beta = 100$ and $V_{BE(on)} = 0.7$ V for all transistors shown.

(a) Find the bias current I.

(b) Find the differential voltage gain

$$A_D = \frac{v_{o1} - v_{o2}}{v_{in}}$$

Figure P11.16

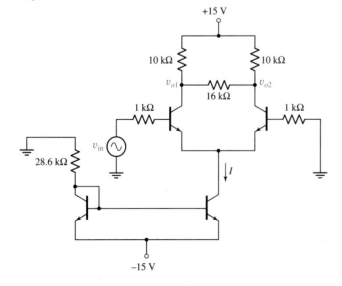

(a) Calculate the input resistance, R_{in}.

(b) Find the differential voltage gain

$$A_D = \frac{v_{o1} - v_{o2}}{v_{in}}$$

(c) If a 20-kΩ resistance is connected between the two collectors, calculate the differential voltage gain.

☆ **11.18** Assume that $\beta = 200$ for each transistor shown.

(a) Calculate $A_o = v_{out}/v_{in}$ for the amplifier.

(b) Will v_{out} be in phase or 180° out of phase with V_{in}?

Figure P11.18

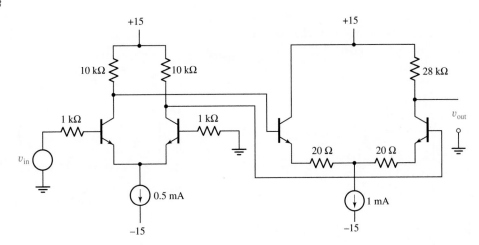

SECTION 11.1.3 THE MOSFET DIFFERENTIAL PAIR

11.19 Assume that $\mu C_{ox} W/2L = 600 \ \mu A/V^2$, $V_T = 0.8$ V, and $\lambda = 0$ for the circuit shown.

(a) If R_1 and R_2 have zero values, calculate the midband differential voltage gain, A_D.

(b) If $R_2 = 0$ and $R_1 = 200 \ \Omega$, calculate A_D, neglecting the body effect.

(c) If $R_2 = 10 \ k\Omega$ and $R_1 = 200 \ \Omega$, calculate A_D. What effect does R_2 have on the voltage gain?

(d) Calculate the dc voltage at the MOSFET source terminals.

11.20 Rework Problem 11.19 if the value of the current generator is increased to 4 mA.

Figure P11.19

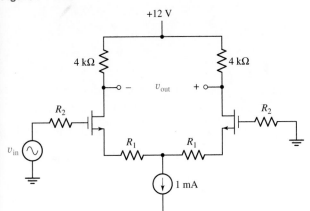

SECTION 11.2.1 THE HIGH-GAIN DIFFERENTIAL STAGE

11.21 Rework Example 11.5 if a Wilson current mirror with an output resistance of 1 MΩ replaces the current mirror load.

11.22 The output resistance of the current mirror stage in the circuit shown is 60 kΩ, and the output resistance of the differential stage is 100 kΩ. If $g_m = 400\ \mu$A/V for the differential devices, calculate the midband voltage gain at the output, neglecting the body effect.

Figure P11.22

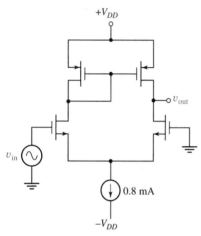

11.23 If the output impedance of the current mirror in the figure is $r_{out} = 100$ kΩ, as is the output impedance of the differential stage, calculate the voltage gain v_{out}/v_{in}.

Figure P11.23

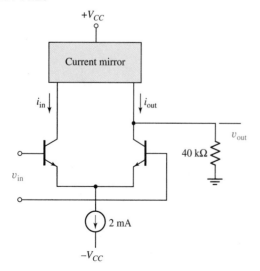

11.24 Assume that both devices in the circuit shown have values of $\beta = 150$ and the mirror has an infinite output impedance. Calculate the output voltage and express this voltage as a function of time.

Figure P11.24

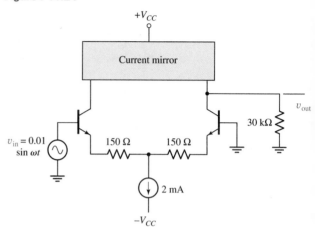

D 11.25 Select new emitter resistors in Problem 11.24 to result in a voltage gain of 300 V/V.

SECTION 11.2.2 THE SECOND AMPLIFIER STAGE

11.26 The three upper corner frequencies of an op amp are $P_1 = 10^5$ rad/s, $P_2 = 10^6$ rad/s, and $P_3 = 10^7$ rad/s. The maximum gain of the op amp is 100,000 V/V. Will this circuit be stable in the unity feedback configuration? If not, to what value must P_1 be lowered to lead to marginal stability?

11.27 The three upper corner frequencies of an op amp are $P_1 = 10^6$ rad/s, $P_2 = 10^6$ rad/s, and $P_3 = 10^7$ rad/s. The maximum gain of the op amp is 200,000 V/V. Will this circuit be stable in the unity feedback configuration? If not, to what value must P_1 be lowered to lead to marginal stability?

11.28 Given the idealized op amp shown in the figure in which $Q1$ and $Q2$ are perfectly matched and have infinite output impedance. Assume that the amplifier stage has infinite input impedance, no offset, and $A = 100$ V/V.

(a) Calculate the gain of the differential pair.

(b) If $Q3$ has $V_{BE} = 0.7$ V, estimate the input offset voltage of the op amp (dc input voltage to cause zero output voltage).

(c) If R_2 is changed to 120 kΩ while R_1 remains unchanged, calculate the new input offset (assume the same V_{BE3} as in part b).

11.29 Given the circuit shown with the following parameters for both *npn* and *pnp* devices: $I_S = 2.28 \times 10^{-17}$ A, $V_{AF} = 60$ V, and $\beta = 100$. Use Spice to determine

(a) I_1 and I_2 (dc bias currents)

(b) v_2/v_1 (small-signal value)

(c) v_3/v_1 (small-signal value)

(d) The dominant pole in the gain function produced by capacitor C

Figure P11.29

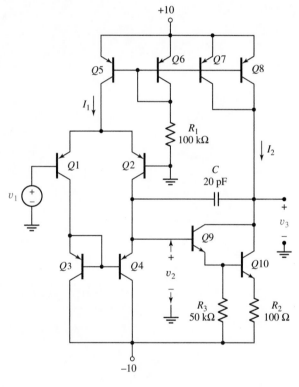

Figure P11.28

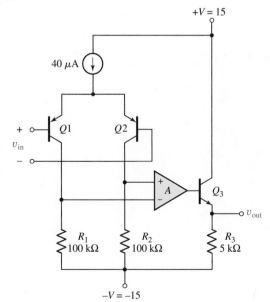

SECTION 11.2.3 OP AMP SPECIFICATIONS

11.30 For the inverting amplifier shown, the op amp is ideal with the exception that $V_{OS} = \pm 2$ mV. Find the worst-case value of output offset voltage.

Figure P11.30

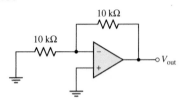

11.31 For the inverting amplifier shown, the op amp is ideal with the exception that $I_{B1} = I_{B2} = 1\ \mu A$. Find the worst-case value of output offset voltage.

Figure P11.31

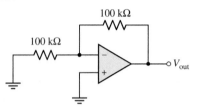

11.32 In the circuit of Problem 11.31, the connection from noninverting input to ground is replaced with a resistance R_3. Calculate the value of R_3 to result in an output offset voltage of 0 V.

Feedback Amplifiers | 12

Active devices are generally characterized by inaccurately defined device parameters and temperature variable parameters. For example, the transistor current gain from base to collector is highly variable from one transistor to another of the same type, and also depends on temperature. Likewise, the transconductance of the FET depends strongly on device, drain current, and temperature. It follows then that most amplifiers using these devices will not exhibit a constant, accurately specified gain unless the design minimizes the influence of device parameters on amplifier gain. On the other hand, passive networks containing resistors, capacitors, and inductors can usually be constructed with arbitrary accuracy for most engineering applications. The passive circuit, however, has no capability of power gain.

It is possible to combine the accuracy of passive circuits with the power gain of active circuits to produce stable amplifiers with precise gain figures. The overall accuracy may not reach the level of the passive component accuracy and the overall gain may not attain the maximum possible value, but a compromise in accuracy and gain can be effected by the use of negative feedback. Many production circuits require accuracy and reproducibility; thus, feedback is almost universally applied in commercial amplifier circuits.

There are other advantages that can be obtained in feedback amplifiers that are sometimes more important than the consideration of accuracy. For example, negative feedback can increase bandwidth, control input and output impedances, and lower distortion in the output signal due to nonlinearities or noise introduced by elements of the amplifier. Major disadvantages of feedback amplifiers are additional complexity, lower gain, and an increased tendency for the amplifier to oscillate and thereby generate unwanted output signals.

12.1 The Ideal Feedback Amplifier

12.2 The Practical Voltage Feedback Amplifier

12.3 Stability of Feedback Systems

DEMONSTRATION PROBLEM

In the amplifier shown, assume that $R_1 \parallel R_2 = R_3 \parallel R_4 = 100$ kΩ and that R_1, R_2, R_3, and R_4 bias devices $M1$, $M2$, and $M3$ properly. Also assume that $g_{m1} = g_{m2} = g_{m3} = g_m = 3$ mA/V, $r_{ds1} = r_{ds2} = r_{ds3} = r_{ds} = 80$ kΩ, and $g_{mb3} = 0.4$ mA/V.

(a) Calculate the midband voltage gain of the feedback amplifier.

(b) Calculate the impedance R'_{in-f} seen by the input signal generator.

(c) Calculate the output impedance, R_{out-f}.

(d) If the upper corner frequency of $M2$ is 160 kHz and this value is much smaller than the corner frequencies of $M1$ and $M3$, what is the bandwidth of the amplifier?

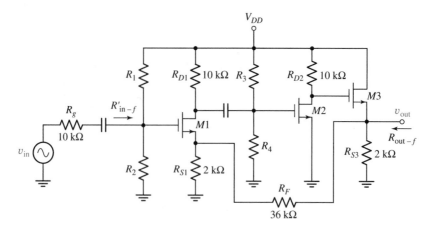

Feedback amplifier
for Demonstration
Problem.

A knowledge of the type of feedback applied in this circuit is necessary in order to work this problem. The ideal feedback equations along with corrections for nonideal impedance conditions must be applied to make these calculations. The effect of feedback on voltage gain, input impedance, output impedance, and bandwidth will be considered to solve this problem at the end of the chapter.

12.1 The Ideal Feedback Amplifier

IMPORTANT Concepts

1. An ideal negative feedback amplifier applies a difference signal to the amplifier input. The difference signal is formed by subtracting a signal that is proportional to the output from the input signal.
2. Although the feedback amplifier uses additional circuitry and is more complex, it leads to several advantages over nonfeedback amplifiers.
3. The feedback amplifier can increase the bandwidth, increase or decrease the input and output impedances, improve the signal-to-noise ratio, and make the performance of the circuit very stable.

To introduce the feedback concept we will assume that we are dealing with an ideal amplifier of gain A that has infinite input impedance and zero output impedance. This amplifier is considered to have no delay or phase shift between input and output signals. Figure 12.1 shows one possible configuration for a feedback amplifier. We will consider voltage relationships in this amplifier, but later we will see that currents in feedback amplifiers may also be of interest.

In Fig. 12.1, the ideal amplifier gain is defined by

$$A = \frac{v_{\text{out}}}{v_{ia}}$$

where v_{out} is the output voltage and v_{ia} is the input voltage to the amplifying element. The

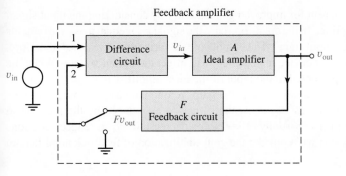

Feedback amplifier

Figure 12.1
An ideal negative-feedback amplifier.

overall gain of the feedback circuit is defined by

$$G = \frac{v_{\text{out}}}{v_{\text{in}}}$$

where v_{in} is the input voltage to the overall feedback amplifier.

The difference circuit is constructed so that the output signal is the algebraic difference between the signals at input 1 and input 2. If input 2 is grounded via the switch, the signal reaching the amplifier input is

$$v_{ia} = v_{\text{in}} - 0 = v_{\text{in}}$$

In this instance, the overall open-loop gain G_o is equal to the ideal amplifier gain A; that is,

$$G_o = \frac{v_{\text{out}}}{v_{\text{in}}} = \frac{v_{\text{out}}}{v_{ia}} = A$$

The *open-loop gain* refers to the forward gain when the feedback loop is not connected.

When the switch connects the difference circuit to the output of the feedback circuit, closing the feedback loop, a portion of the output signal is subtracted from the input signal. The gain or transfer function of the feedback circuit is designated as F and is normally less than unity. The signal reaching the ideal amplifier is reduced to

$$v_{ia} = v_{\text{in}} - F v_{\text{out}}$$

and the output voltage will consequently be smaller than the open-loop value. The output voltage is

$$v_{\text{out}} = A(v_{\text{in}} - F v_{\text{out}}) = G v_{\text{in}}$$

where G is the *closed-loop gain*. Solving for this value of gain G gives

$$G = \frac{v_{\text{out}}}{v_{\text{in}}} = \frac{A}{1 + AF} \tag{12.1}$$

The closed-loop gain G is always less than the open-loop gain G_o for the negative feedback circuit.

The product AF is the gain around the entire feedback loop. This quantity is called the *loop gain* or the *return ratio*. This latter term implies that a signal applied to the amplifier input will be returned to this same point, having been multiplied by AF. Throughout this chapter, we will refer to AF as the return ratio.

We can visualize the reason for improved gain stability by considering the input signal presented to the ideal amplifier. In the open-loop case the signal v_{ia} is equal to v_{in} and remains constant regardless of the output signal. The output voltage varies directly with A; therefore, temperature variations in A or device replacement will affect the output voltage. When the loop is closed, the signal presented to the input of the ideal amplifier is $v_{ia} = v_{in} - Fv_{out}$. The output voltage is again given by Av_{ia}. If the gain were to increase, the output voltage would also increase. However, as v_{out} rises, the signal v_{ia} is reduced by the increase in Fv_{out}; thus, the increase in A is partially offset by the decrease in v_{ia} to reduce variations in v_{out}. The following section will consider the gain stabilization of feedback amplifiers in a quantitative manner.

12.1.1 GAIN STABILITY

The gain stability can be discussed most readily if related to the sensitivity function, which compares the fractional change of a dependent variable to the fractional change of an independent variable. The sensitivity of v_{out} with respect to ideal amplifier gain A is defined as

$$S_{v_{out}, A} = \frac{\partial v_{out}/v_{out}}{\partial A/A} = \frac{A}{v_{out}} \frac{\partial v_{out}}{\partial A} \tag{12.2}$$

If the dependent variable exhibits a direct variation with the independent variable, a sensitivity of unity will result. Thus a 20% variation of the independent variable leads to a 20% variation of the dependent variable.

In an open-loop amplifier the sensitivity of output voltage with respect to gain is

$$\text{(Open loop) } S_{v_{out}, A} = \frac{A}{v_{out}} \frac{\partial v_{out}}{\partial A} = \frac{A}{Av_{in}} v_{in} = 1$$

This figure can be compared to that of the closed-loop amplifier. Since

$$v_{out} = \frac{A}{1 + AF} v_{in}$$

then

$$\frac{\partial v_{out}}{\partial A} = \frac{v_{in}}{(1 + AF)^2}$$

and

$$\text{(Closed loop) } S_{v_{out}, A} = \frac{A(1 + AF)}{Av_{in}} \frac{v_{in}}{(1 + AF)^2} = \frac{1}{1 + AF}$$

The fractional change in v_{out} is now less than the fractional change in A by a factor of $(1 + AF)$. For example, if $1 + AF = 100$, a 10% change in A results in an approximate change in v_{out} of only 0.1%.

The sensitivity of G with respect to A is equal to the closed-loop value of $S_{v_{out}, A}$; that is,

$$S_{G, A} = \frac{1}{1 + AF} \tag{12.3}$$

The overall gain becomes very stable as $(1 + AF)$ approaches large values.

We might consider an interesting limiting case in the calculation of gain stability. If AF is much greater than unity, the gain equation is

$$G = \frac{A}{1 + AF} \approx \frac{A}{AF} = \frac{1}{F} \tag{12.4}$$

Under this condition, the overall amplifier gain depends only on the feedback factor and is independent of the ideal amplifier gain A. The feedback amplifier gain is insensitive to changes in A. Overall gain now depends on F, but this feedback factor is generally determined by passive components with almost arbitrary precision. Off-the-shelf resistors and capacitors are accurate enough for many applications, whereas precise, low temperature drift elements can be secured for highly critical amplifiers.

We note that sensitivity of G with respect to A decreases as $(1 + AF)$ takes on larger values. The overall gain G simultaneously decreases with increases in $(1 + AF)$. Thus, gain stability is achieved at the expense of overall gain.

We can also calculate the gain sensitivity with respect to changes in the feedback factor F. This value is found to be

$$S_{G,F} = \frac{-AF}{1 + AF} \tag{12.5}$$

For large values of AF, this sensitivity approaches unity.

Numerically, the sensitivity of G with respect to F is greater than the sensitivity of G with respect to A. However, as temperature changes occur, G may be affected equally by changes in A and changes in F. The total derivative of G with respect to temperature T is

$$\frac{dG}{dT} = \frac{\partial G}{\partial A}\frac{dA}{dT} + \frac{\partial G}{\partial F}\frac{dF}{dT}$$

This derivative can also be expressed as

$$\frac{dG}{dT} = \frac{G}{A}S_{G,A}\frac{dA}{dT} + \frac{G}{F}S_{G,F}\frac{dF}{dT} \tag{12.6}$$

The sensitivity of G with respect to temperature is a meaningful quantity and can be written as

$$S_{G,T} = \frac{T}{G}\frac{dG}{dT} = S_{G,A}S_{A,T} + S_{G,F}S_{F,T} \tag{12.7}$$

Although $S_{G,F}$ is greater than $S_{G,A}$, the change in A with temperature, $S_{A,T}$, is generally much greater than the corresponding change in F, given by $S_{F,T}$. Thus, gain changes may be as significant as feedback factor changes.

PRACTICAL Considerations

From a practical standpoint it is often convenient to relate variations in G to the *temperature coefficients* of A and F. Temperature coefficients are generally specified by the manufacturer and can be used directly in calculating temperature effects. The temperature coefficient of G relates the fractional change in G to the change in T (rather than the fractional change in T); thus,

$$TC_G = \frac{1}{G}\frac{dG}{dT} = \frac{1}{T}S_{G,T} \tag{12.8}$$

Using Eqs. (12.6) and (12.7) and the definition of temperature coefficient allows us to write

$$TC_G = S_{G,A}\,TC_A + S_{G,F}\,TC_F \tag{12.9}$$

This equation can also be expressed as

$$TC_G = \frac{1}{1 + AF}\, TC_A - \frac{AF}{1 + AF}\, TC_F$$

The magnitude of the sensitivity of G with respect to F is larger than the sensitivity of G with respect to A by a factor of AF. Since TC_F usually depends on passive components whereas TC_A is a function of active devices, TC_F is much smaller than TC_A. Each term in Eq. (12.9) may contribute a significant amount to the total temperature coefficient of G. An example will demonstrate the relative importance of these terms.

EXAMPLE 12.1

PRACTICE Problems

12.1 For an ideal feedback amplifier, calculate the value of $S_{G,A}$ if $A = 1000$ V/V and $F = 0.042$ V/V. What is the closed-loop gain of the amplifier? *Ans:* $S_{G,A} = 0.0233$, $G = 23.3$ V/V.
12.2 Rework Example 12.1 if $TC_A = 5000$ ppm/°C and $TC_F = \pm40$ ppm/°C. *Ans:* 89.1 ppm/°C.

A feedback amplifier is constructed such that $F = 1/100$ and $A = 10{,}000$. The temperature coefficient of the feedback network is ±25 ppm/°C (parts per million per degree centigrade) and TC_A is 3000 ppm/°C. All quantities are room temperature values. Calculate the closed-loop gain and the maximum temperature coefficient of this gain.

SOLUTION The closed-loop gain is calculated from Eq. (12.1) to be $G = 99$. We note that the gain can be approximated with 1% accuracy by $G = 1/F$. The sensitivities are found from Eqs. (12.3) and (12.5). These values are $S_{G,A} = 9.9 \times 10^{-3}$ and $S_{G,F} = -9.9 \times 10^{-1}$. From Eq. (12.9) the temperature coefficient of closed-loop gain is

$$TC_G = (9.9 \times 10^{-3})(3 \times 10^3) + (0.99)(25) = 54.5 \text{ ppm/°C}$$

Changes in the feedback factor and the ideal gain A with temperature contribute almost equally to TC_G in this example. Note that the negative value of TC_F was used to find the maximum or worst-case value of TC_G.

12.1.2 SIGNAL-TO-NOISE RATIO

A signal containing distortion or noise is often characterized by the signal-to-noise ratio (*SNR*). This ratio can be expressed in terms of rms voltages or powers. Let us assume that we have an input voltage that can be expressed as

$$v_{\text{in}} = v_s + v_n$$

where v_s is the pure signal of interest and v_n is a noise component resulting from any of the possible electronic noise sources. The noise component could be introduced by resistors in the circuit, by pickup of extraneous signals, or by the active device used to generate the signal v_s. The noise component v_n may contain white noise, harmonic or nonlinear distortion, or an extraneous signal of almost any shape. Obviously, it may often be difficult to measure v_n in view of the possible complex makeup of this component. Assuming that v_n can be measured allows us to express the *signal-to-noise ratio* of v_{in} as

$$SNR = \frac{v_s}{v_n} \tag{12.10}$$

If v_{in} is now presented to a feedback amplifier that is assumed to add no noise to the amplified signal, the output voltage also consists of a noise component and the pure signal component. Both signal and noise components of v_{in} are amplified equally, leading to the same *SNR* for both v_{in} and v_{out}. We conclude here that the *SNR* of an input signal is not improved by noiseless amplification.

Most amplifiers of significant gain will introduce a noise component into the signal that is being amplified. This noise may be white noise, $1/f$ noise, nonlinear distortion, or extraneous noise picked up from unwanted sources (60-Hz hum and rf signals are examples of common extraneous pickup). In general, the *SNR* at the amplifier output is decreased over that measured at the input. In fact, because some of the noise components introduced within the amplifier increase with signal levels, the noise generated by the amplifier is often the dominant noise term.

If we designate the noise term introduced within the amplifier by v_{an}, the output signal of an open-loop amplifier becomes

$$v_{\text{out}} = Av_{\text{in}} + v_{an} = Av_s + Av_n + v_{an}$$

To examine the effect of feedback on amplifier noise we will neglect the noise contained in the input signal. In this case the *SNR* at the amplifier output is

$$SNR = \frac{Av_s}{v_{an}}$$

If feedback is now applied, the output signal decreases from the value without feedback. Since nonlinear noise introduced by the amplifier depends on the magnitude of the output voltage, a meaningful comparison of the open-loop and closed-loop amplifiers can only be made for equal output voltages. Hence, after feedback is applied we assume that the input signal is increased in magnitude, without affecting the input *SNR*, until the output signal equals the open-loop value. The noise introduced by the amplifier will again equal v_{an}, but a portion of the output noise will be fed back and subtracted from the input, resulting in a lower value of output noise v_{gn}.

The noise fed back to the input is Fv_{gn}. We can solve for v_{gn} since

$$v_{gn} = v_{an} - AFv_{gn}$$

giving

$$v_{gn} = \frac{v_{an}}{1 + AF} \tag{12.11}$$

The noise at the closed-loop amplifier output is less than the open-loop value by the same factor as the reduction in gain. Again neglecting noise in the input signal and recalling that this signal has been increased enough to yield an output voltage of Av_s, we find that

$$SNR = \frac{Av_s}{v_{an}} (1 + AF) \tag{12.12}$$

The signal-to-noise ratio of the feedback amplifier is better than that of the open-loop amplifier by the factor $(1 + AF)$.

PRACTICAL Considerations

At this point we must examine the validity of the assumption that the input signal can be increased without increasing the *SNR* of the input signal. Basically, the gain of the closed-loop amplifier is less than the open-loop value. To obtain equal output signals in the two cases requires that a preamplifier be inserted between the input signal and the feedback amplifier. This preamplifier will in general introduce noise that will degrade the *SNR* of the signal. Thus, the improvement in *SNR* indicated by Eq. (12.12) is overly optimistic and represents the maximum possible improvement.

If the noise at the output is primarily nonlinear distortion or a component that results from large output-signal swings, the preamplifier noise may be negligible. Equation (12.12) may then accurately describe the *SNR* of the amplifier. On the other hand, if the noise is introduced primarily within the preamplifier, the *SNR* may be improved only slightly over the open-loop amplifier. Typical of this case is the preamplifier that requires a very high input impedance leading to greater values of resistor noise and extraneous pickup. One must be rather careful when applying feedback to improve *SNR*, and give prime consideration to the points at which noise is introduced within the amplifier stages.

12.1.3 BANDWIDTH IMPROVEMENT

Certain configurations of feedback amplifier lead to considerable improvements in bandwidth over the open-loop amplifier. For the negative feedback amplifier of Fig. 12.1, let us assume that the ideal amplifier has a radian bandwidth of ω_2 and approaches a rolloff of -6 dB/octave above this value. The open-loop gain can then be written as

$$A = \frac{A_{MB}}{1 + j\omega/\omega_2}$$

where A_{MB} is the midband gain of the amplifier. The gain-bandwidth product is defined as $GBW = A_{MB} \times \omega_2$.

When the loop is closed we can apply Eq. (12.1) to calculate the gain G as a function of frequency, which gives

$$G = \frac{A}{1 + AF} = \frac{\frac{A_{MB}}{1 + j(\omega/\omega_2)}}{1 + \frac{A_{MB}F}{1 + j(\omega/\omega_2)}}$$

After manipulation, we can write this gain expression as

$$G = \frac{A_{MB}}{1 + A_{MB}F} \frac{1}{1 + j[\omega/\omega_2(1 + A_{MB}F)]}$$

$$= \frac{G_{MB}}{1 + j[\omega/\omega_2(1 + A_{MB}F)]} = \frac{G_{MB}}{1 + j\,\omega/\omega_{2f}} \tag{12.13}$$

where

$$G_{MB} = \frac{A_{MB}}{1 + A_{MB}F} \tag{12.14}$$

The bandwidth with feedback is

$$\omega_{2f} = \omega_2(1 + A_{MB}F) \tag{12.15}$$

The gain with feedback is a factor of $1 + A_{MB}F$ less than the gain without feedback, while the bandwidth increases over the open-loop bandwidth by this same factor. The gain-bandwidth product is then constant as the loop is closed, allowing gain and bandwidth to be interchanged directly by feedback. Figure 12.2 compares the frequency responses of the open-loop and closed-loop amplifiers.

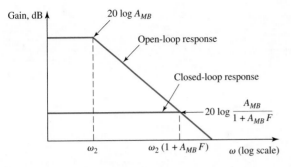

Figure 12.2
Frequency responses of open-loop and closed-loop amplifiers.

The extension of bandwidth indicated by Eq. (12.15) applies only to the ideal feedback stage of Fig. 12.1. In practical cases, F may also be frequency dependent and loading may not be negligible. Generally, the bandwidth given by Eq. (12.15) should be considered as an upper limit approached in few cases. An outstanding example of an amplifier that very nearly approximates the ideal case is the op amp considered in Chapter 4.

Not all feedback configurations will extend the voltage bandwidth of the amplifier. An example of this occurs in a MOSFET amplifier with an unbypassed resistance from source terminal to ground. The source resistance introduces negative feedback, which reduces the gain of the stage, but bandwidth may not be improved. In this case, however, the feedback voltage developed at the source does not exhibit the same frequency dependence as does the output signal. The feedback voltage is constant with frequency, while the capacitance that shunts the drain load resistance leads to a frequency-dependent output voltage. A rather intuitive criterion for increasing the bandwidth with negative feedback is that both the signal fed back to the input and the output voltage must exhibit similar frequency dependence.

EXAMPLE 12.2

The amplifying element used in a voltage feedback configuration is

$$A = \frac{1000}{1 + j\frac{f}{22\text{ kHz}}}$$

If the feedback factor is $F = 0.012$ V/V, calculate the midband voltage gain and the bandwidth of the feedback amplifier.

SOLUTION Since the midband gain of the amplifying element is 1000 V/V, the midband gain of the feedback amplifier is

$$G_{MB} = \frac{A_{MB}}{1 + A_{MB}F} = \frac{1000}{1 + 1000 \times 0.012} = 77 \text{ V/V}$$

The bandwidth with feedback is

$$f_{2f} = f_2 \times (1 + A_{MB}F) = 22 \times (1 + 1000 \times 0.012) = 286 \text{ kHz}$$

PRACTICE Problem

12.5 An amplifier has a gain of

$$A = \frac{875}{1 + j\frac{\omega}{4000}}$$

If this stage is used in a negative feedback configuration with $F = 0.03$, calculate the midband gain and bandwidth for the closed-loop amplifier. *Ans:* $G_{MB} = 32.1$ V/V, $\omega_{2f} = 109,000$ rad/s.

12.1.4 TYPES OF FEEDBACK

The feedback signal can be derived from one of two output quantities: output voltage or output current. Furthermore, the feedback signal can be subtracted from the input in two ways to form a difference signal at the input of the ideal amplifier. It is possible to subtract a feedback voltage from the input voltage by including the feedback signal in series with the input loop, which results in series feedback. It is also possible to use shunt feedback wherein

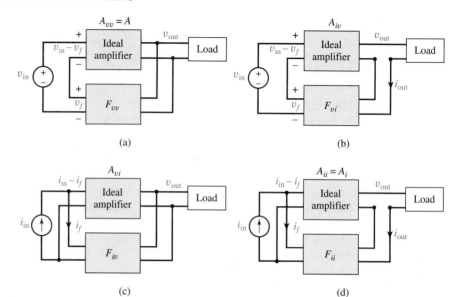

Figure 12.3
Four possible feedback types: (a) series-shunt or voltage amplifier, (b) series-series or transadmittance amplifier, (c) shunt-shunt or transimpedance amplifier, (d) shunt-series or current amplifier.

the amplifier input is shunted by a network that decreases the input current by an amount proportional to the feedback signal to form a difference current. Thus, there are four basic types of feedback: voltage-voltage or series-shunt, voltage-current or series-series, current-voltage or shunt-shunt, and current-current or shunt-series. The word before the hyphen refers to the input and the second word refers to the output of the circuit. Figure 12.3 indicates these combinations.

In general, the feedback network will be different for each amplifier configuration. The transfer function F of this network has different units for each case. For voltage-voltage feedback, F generates a feedback voltage, v_f, proportional to the amplifier output voltage, v_{out}. In this case, F is a voltage-ratio transfer function. For voltage-current feedback, F must generate a feedback voltage, v_f, in proportion to amplifier output current, i_{out}. The transfer function F is now a transimpedance. Current-voltage feedback requires F to be a transadmittance generating a feedback current, i_f, that is proportional to amplifier output voltage, v_{out}. Current-current feedback requires F to be a current-ratio transfer function developing a feedback current, i_f, in proportion to the amplifier output current, i_{out}.

It is instructive to consider the closed-loop gain expressions for the four types of amplifier. The voltage-voltage type corresponds closely to the idealized amplifier of Fig. 12.1 for which the gain expression has been developed. The low-frequency gain is

$$G = \frac{A}{1 + AF_{vv}} \qquad \textbf{(12.16)}$$

where A is the open-loop or ideal amplifier voltage gain. The feedback factor is $F_{vv} = v_f/v_{\text{out}}$. This is called a voltage feedback amplifier.

For the series-series case, we must relate the amplifier output current to the amplifier input voltage. The gain is then a transadmittance, which we shall designate G_{iv}. If F_{vi} is defined by

$$F_{vi} = \frac{v_f}{i_{\text{out}}}$$

then the transadmittance of the amplifier is

$$G_{iv} = \frac{i_{\text{out}}}{v_{\text{in}}} = \frac{A_{iv}}{1 + A_{iv}F_{vi}} \qquad \textbf{(12.17)}$$

where A_{iv} is the transadmittance of the ideal amplifier. Note that the amplifier type, transadmittance in this case, is opposite to the feedback type, transimpedance in this case.

Equation (12.17) has the same form as Eqs. (12.1) and (12.16); however, the output quantity associated with the former equation is current rather than voltage. Noting this difference, we might extend to the transadmittance amplifier some previous conclusions that apply to the voltage amplifier. The arguments for improved stability, improved bandwidth, and reduction of noise with feedback are valid for the transadmittance amplifier, but current is the output quantity to which the improvements apply rather than voltage. Equations developed earlier for frequency response, sensitivity, and noise performance are easily extended to the transadmittance amplifier with this change in output variable and with appropriate units defining A and F.

It is often useful to know the voltage gain of the transadmittance amplifier. This quantity can be found by expressing the output voltage as

$$v_{\text{out}} = A v_{ia} = A(v_{\text{in}} - v_f)$$

where A is the voltage gain of the ideal amplifier. The feedback signal is

$$v_f = F_{vi} i_{\text{out}} = F_{vi} A_{iv} v_{ia}$$

Combining the two preceding equations results in

$$G = \frac{A}{1 + A_{iv} F_{vi}} \qquad\qquad \textbf{(12.18)}$$

The voltage gain is reduced by the same factor as the transadmittance when feedback is applied.

Equation (12.18) does not correspond to Eq. (12.1) since the gain of the numerator is a voltage gain, whereas the denominator contains a transadmittance. We cannot then apply the stability, bandwidth, and noise results to the output voltage, even though these results apply to the output current as previously mentioned. Generally, the output voltage will be less sensitive to parameter changes when feedback is applied, although it is easy to visualize examples wherein the sensitivity of output voltage is only slightly affected by feedback while the output current becomes very stable. Suppose an amplifier drives a load resistance that is highly temperature dependent. The output current may be stabilized with current feedback, yet the output voltage amplitude would vary with temperature change as the load resistance changed.

A similar situation occurs in relation to bandwidth improvement. The output current bandwidth of a transadmittance amplifier is

$$\omega_2(1 + A_{iv} F_{vi})$$

where ω_2 is the output current bandwidth of the ideal amplifier. The output voltage bandwidth may not exhibit the same bandwidth unless the output voltage is proportional to output current for all frequencies up to the upper corner frequency. In general $v_{\text{out}} = Z_L i_{\text{out}}$, where Z_L is the load impedance. Often the reactive component of Z_L (due perhaps to shunt capacitance) will significantly lower the magnitude of load impedance at frequencies below $\omega_2(1 + A_{iv} F_{vi})$. Hence voltage bandwidth can be less than current bandwidth.

The overall transimpedance of the shunt-shunt feedback amplifier is found to be

$$G_{vi} = \frac{v_{\text{out}}}{i_{\text{in}}} = \frac{A_{vi}}{1 + A_{vi} F_{iv}} \qquad\qquad \textbf{(12.19)}$$

where A_{vi} is the transimpedance of the ideal amplifier. The transimpedance amplifier has a feedback factor F_{iv}, which represents a transadmittance.

The shunt-series or current amplifier has an overall current gain given by

$$G_i = \frac{i_{\text{out}}}{i_{\text{in}}} = \frac{A_i}{1 + A_i F_{ii}} \qquad \text{(12.20)}$$

where A_i is the current gain of the ideal amplifier.

In summary, the four types of amplifier each obey the closed-loop gain expression of Eq. (12.1). This gain may correspond to a voltage gain, a current gain, a transadmittance, or a transimpedance, depending on the configuration. All equations relating to bandwidth, stability, and distortion apply to the output and input quantities defined by the gain equation for the amplifier in question.

One point that is rather significant in discussing the ideal feedback amplifiers of Fig. 12.3 is that we have implied certain impedance levels in our derivations that should be considered more fully. The feedback network of the voltage amplifier shown in Fig. 12.3(a) must have negligible output impedance compared to the ideal amplifier input impedance and must also have a high input impedance compared to the resistance of the load, R_L. If such were not the case, loading effects would cause the gain equation to be inaccurate. In Fig. 12.3(b) the input impedance to the feedback network must be small compared to R_L, and the output impedance must be small compared to the input impedance of the ideal amplifier. In parts (c) and (d) the output impedances of the feedback network should be very large. The input impedance of the feedback network in Fig. 12.3(c) should be large compared to R_L, whereas that of part (d) should have a low value. Departures from these ideal conditions can be treated as shown in the following section, but they lead to somewhat more complex equations.

12.1.5 EFFECT OF FEEDBACK ON IMPEDANCE LEVEL

The input and output impedances of an amplifier can be changed greatly when feedback is applied. Occasionally feedback is used to obtain some desired impedance condition, although in many other instances the effect of feedback on impedance level may be of secondary interest.

Input Impedance: In general we can say that the input impedance of an amplifier will be increased with series feedback and decreased with shunt feedback. We can demonstrate the basis for these conclusions by referring to the amplifiers of Fig. 12.4.

The basic amplifier is now assumed to consist of an ideal amplifier with finite input and output resistances, R_{ia} and R_{oa}, respectively. Before feedback is applied, the input resistance is R_{ia} for the basic amplifier. When series feedback is applied, the input current

Figure 12.4
Feedback amplifiers: (a) series input, (b) shunt input.

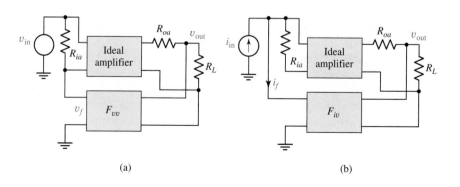

(a)   (b)

$i_{\rm in}$ is given by

$$i_{\rm in} = \frac{v_{\rm in} - v_f}{R_{ia}} = \frac{v_{\rm in} - F_{vv}v_{\rm out}}{R_{ia}} = \frac{v_{\rm in} - GF_{vv}v_{\rm in}}{R_{ia}}$$

Using the fact that

$$G = \frac{A}{1 + AF_{vv}}$$

we write

$$i_{\rm in} = \frac{v_{\rm in}\left(1 - \frac{AF_{vv}}{1+AF_{vv}}\right)}{R_{ia}}$$

Manipulating this equation leads to

$$i_{\rm in} = \frac{v_{\rm in}}{R_{ia}(1 + AF_{vv})}$$

The input impedance to the series feedback amplifier is

$$R_{\rm in-}f = \frac{v_{\rm in}}{i_{\rm in}} = R_{ia}(1 + AF_{vv}) \tag{12.21}$$

Normally, the factor $(1 + AF_{vv})$ is much greater than unity, which indicates that the input impedance increases significantly with series feedback. The same result applies to the transadmittance amplifier wherein series-series feedback is applied.

For shunt feedback at the input, we can write

$$v_{\rm in} = (i_{\rm in} - i_f)R_{ia}$$

$$= (i_{\rm in} - G_{vi}F_{iv}i_{\rm in})R_{ia} = \frac{i_{\rm in}R_{ia}}{(1 + A_{vi}F_{iv})}$$

The input impedance to the shunt feedback amplifier is

$$R_{\rm in-}f = \frac{v_{\rm in}}{i_{\rm in}} = \frac{R_{ia}}{(1 + A_{vi}F_{iv})} \tag{12.22}$$

This impedance is normally much smaller than the impedance of the basic amplifier without feedback. The same result applies to the shunt-series or current amplifier.

Output Impedance: The output impedance of an amplifier will decrease for voltage feedback and increase for current feedback. Consider the stages of Fig. 12.5. The output

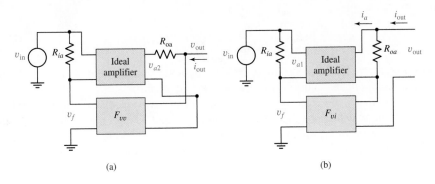

(a)   (b)

Figure 12.5
Feedback amplifiers: (a) voltage feedback, (b) current feedback.

impedance can be calculated by finding the Thévenin equivalent resistance at the output terminals. To do this we assume that the input source is shorted and that an applied voltage, v_{out}, appears at the output terminals, resulting in a current i_{out} as shown. For the voltage-voltage feedback amplifier of Fig. 12.5(a), we can write

$$i_{out} = \frac{v_{out} - v_{a2}}{R_{oa}}$$

assuming that no current is drawn by the input of the feedback network. The voltage v_{a2}, appearing to the left of R_{oa}, is simply $A(v_{in} - F_{vv}v_{out})$, but since $v_{in} = 0$ in this case, we can write

$$v_{a2} = -AF_{vv}v_{out}$$

The current i_{out} is then

$$i_{out} = \frac{v_{out} + AF_{vv}v_{out}}{R_{oa}}$$

and the output impedance is

$$R_{out-f} = \frac{v_{out}}{i_{out}} = \frac{R_{oa}}{1 + AF_{vv}} \tag{12.23}$$

Closing the voltage feedback loop reduces the output impedance of the amplifier by the factor $(1 + AF_{vv})$. The same result applies to the shunt-shunt or transimpedance amplifier.

For the current feedback stage of Fig. 12.5(b), which is a transadmittance amplifier, we can write

$$i_{out} = \frac{v_{out}}{R_{oa}} + i_a$$

assuming that negligible voltage develops across the input to the feedback network. The current i_a is

$$i_a = -A_{iv}v_{a1} = -A_{iv}F_{vi}i_{out}$$

The output current is now expressed as

$$i_{out} = \frac{v_{out}}{R_{oa}} - A_{iv}F_{vi}i_{out}$$

and

$$R_{out-f} = \frac{v_{out}}{i_{out}} = R_{oa}(1 + A_{iv}F_{vi}) \tag{12.24}$$

Closing the current feedback loop increases the output impedance by the factor $(1 + A_{iv}F_{vi})$. The same result applies to the shunt-series or current amplifier.

If high output impedance is desirable, current or series feedback at the output may be used, whereas voltage or shunt feedback at the output is appropriate when low output impedance is a major requirement. Higher input impedance levels result from voltage or series feedback at the input, whereas current or shunt feedback at the input reduces the input impedance.

EXAMPLE 12.3

The amplifying element used in a voltage feedback configuration is

$$A = \frac{1000}{1 + j\frac{f}{22\text{ kHz}}}$$

The input resistance to this stage is 10 kΩ and the output resistance is 1 kΩ. If the feedback factor is $F = 0.012$ V/V, calculate the midband values of input and output impedances of the feedback amplifier. Assume that the input and output impedances of F do not affect the result.

SOLUTION The input impedance to the feedback amplifier will be a factor of $1 + A_{MB}F = 13$ times the impedance of the amplifying element. The result is

$$R_{\text{in}-f} = R_{ia}(1 + A_{MB}F) = 10 \times 13 = 130 \text{ k}\Omega$$

The output impedance with feedback is

$$R_{\text{out}-f} = \frac{R_{oa}}{1 + A_{MB}F} = \frac{1}{13} = 77 \ \Omega$$

PRACTICE Problems

12.6 Rework Example 12.3 if the input resistance to this stage is changed to 1 kΩ and the output resistance is changed to 10 kΩ.
Ans: $R_{\text{in}-f} = 13$ kΩ, $R_{\text{out}-f} = 770 \ \Omega$.
12.7 Rework Example 12.3 if the amplifier is a shunt-series configuration with $A_i = A$ and $F_{ii} = F$.
Ans: $R_{\text{in}-f} = 770 \ \Omega$, $R_{\text{out}-f} = 13$ kΩ.

12.1.6 AC AND DC FEEDBACK

DC feedback is used to stabilize dc gain or the operating point with respect to changes in temperature or changes in device parameters. AC feedback is used to stabilize ac gain, extend the bandwidth, decrease distortion, or control impedance levels. Often an amplifier will use both dc and ac feedback, but either can be used separately. When both are used, the ac and dc feedback factors may differ.

12.2 The Practical Voltage Feedback Amplifier

IMPORTANT Concepts

1. A common configuration for feedback inserts the feedback voltage into the input loop of the amplifier. This configuration results in an amplifier input signal that is the difference between the applied input signal and the feedback voltage.
2. The practical feedback network leads to impedance loading that must be considered in gain calculations. Generator resistance and load resistance must also be considered.
3. Equations developed for the ideal amplifier can be applied to the practical amplifier after modifying the value of forward amplifier gain.

The theory developed in the previous section applies to the ideal feedback amplifier with an ideal feedback network and an ideal difference circuit. Although finite impedances can occur in the ideal case, a given impedance must be either small enough or large enough to be negligible, depending on the location of this impedance in the circuit. Some practical amplifiers approximate the ideal case closely enough to be analyzed with the ideal theory. Other amplifiers require special considerations to analyze completely. We will consider only the voltage amplifier in this section.

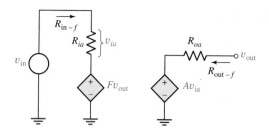

Figure 12.6
The ideal voltage feedback
configuration.

12.2.1 AN IDEAL VOLTAGE AMPLIFIER CONFIGURATION

The ideal feedback configuration shown in Fig. 12.6 assumes that the feedback network has infinite input impedance and zero output impedance. The only feedback element of this ideal circuit is the dependent source that produces the voltage $F v_{\text{out}}$.

By applying the feedback voltage in series with the input loop, a voltage difference across the amplifier terminals is created without using a special difference network. The voltage across the amplifier terminals equals $v_{\text{in}} - F v_{\text{out}}$.

For this circuit, the following equations apply:

$$G = \frac{A}{1 + AF} \tag{12.1}$$

$$R_{\text{in}-f} = R_{ia}(1 + AF) \tag{12.21}$$

$$R_{\text{out}-f} = \frac{R_{oa}}{1 + AF} \tag{12.23}$$

$$f_{2f} = f_2(1 + AF) \tag{12.25}$$

We derived these equations in the previous section.

12.2.2 USING A PRACTICAL FEEDBACK NETWORK

In practice, the feedback network will typically have finite input and output impedances. The input voltage source will also have some source impedance that appears in the input loop, and the output will drive a load resistance. Thus, when these elements are included in the feedback circuit, the above equations are no longer accurate. There are, however, some relatively simple modifications that can be made to allow the use of a similar set of equations.

The actual circuit can be represented by the schematic of Fig. 12.7. The resistor R_g represents the signal generator resistance, $R_{F\text{in}}$ is the input resistance, and $R_{F\text{out}}$ is the output resistance of the feedback network. The value of $R_{F\text{in}}$ will depend not only on the

Figure 12.7
A practical voltage feedback
configuration.

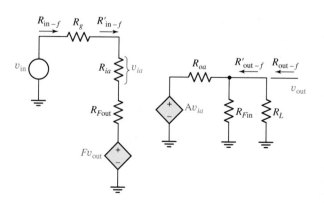

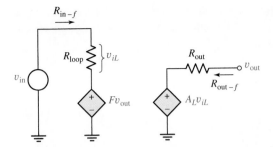

Figure 12.8
A modified configuration.

feedback network, but also on the circuit to which the output of this network is connected. In many cases, however, this value is determined primarily by the feedback network.

Since we know how to analyze the ideal feedback configuration, we direct our efforts toward modifying the configuration of Fig. 12.7 to obtain that of Fig. 12.6. Of course, the modifications must apply acceptable network manipulations to be valid.

In order to make this modification, the output impedance of the feedback network is associated with the input loop impedance of the feedback amplifier, as is the finite generator resistance. The input impedance to the feedback network is associated with the output loop of the amplifier, which leads to the configuration of Fig. 12.8.

The sum of the resistances R_g, R_{ia}, and R_{Fout} are equal to the input loop resistance, called R_{loop}. The total output resistance is given by the parallel combination of R_{oa}, R_L, and R_{Fin} and is called R_{out}. The voltage v_{iL} is the voltage across the input loop resistance. This voltage also equals $v_{in} - Fv_{out}$, which is the expression for amplifier input voltage of an ideal feedback amplifier.

The forward amplifier model now has an input impedance equal to the loop resistance, R_{loop}. The output resistance equals R_{out}, and the forward amplifier gain A_L, the loaded gain, is given by

$$A_L = \frac{v_{out}}{v_{in}} \tag{12.26}$$

This value of A_L is the gain of the amplifier when loading due to the generator resistance, the load resistance, and the input and output impedances of the feedback network are included. The ideal gain A is the ideal amplifier gain with no loading considered.

A consideration of the modified configuration of Fig. 12.8 indicates that it is identical to that of the ideal feedback amplifier. It then follows that Eqs. (12.1), (12.21), (12.23), and (12.25) can be modified to give

$$G = \frac{A_L}{1 + A_L F} \tag{12.27}$$

$$R_{in-f} = R_{loop}(1 + A_L F) \tag{12.28}$$

$$R_{out-f} = \frac{R_{out}}{1 + A_L F} \tag{12.29}$$

$$f_{2f} = f_2(1 + A_L F) \tag{12.30}$$

Note that if we wanted to find the input impedance seen by the input signal generator that includes a generator resistance R_g, we simply subtract R_g from the value found by Eq. (12.28). Thus,

$$R'_{in-f} = R_{in-f} - R_g \tag{12.31}$$

If we want to find the output impedance as seen by the load resistance, we simply note that a parallel value of R_L must be removed from the value given by Eq. (12.29). Since

$$\frac{R'_{\text{out}-f}R_L}{R'_{\text{out}-f} + R_L} = R_{\text{out}-f}$$

the value of $R'_{\text{out}-f}$ can be found to be

$$R'_{\text{out}-f} = \frac{R_{\text{out}-f}R_L}{R_L - R_{\text{out}-f}} \tag{12.32}$$

It should be emphasized that the gain A_L must be evaluated with the resistances R_g, R_L, $R_{F\text{in}}$, and $R_{F\text{out}}$ in place. Although this value will generally be found from an analysis of the circuit, it can also be calculated as

$$A_L = A \times \frac{R_{ia}}{R_g + R_{ia} + R_{F\text{out}}} \times \frac{R_L R_{F\text{in}}/(R_L + R_{F\text{in}})}{R_L R_{F\text{in}}/(R_L + R_{F\text{in}}) + R_{oa}} \tag{12.33}$$

The second term represents the input loading factor, and the third term represents the output loading factor.

From a practical standpoint there are two methods of evaluating the loaded gain, A_L. The first method removes the resistors R_g, R_L, $R_{F\text{out}}$, and $R_{F\text{in}}$, as well as the dependent source, Fv_{out}, before calculating the forward gain. The resulting value is the unloaded amplifier gain A. Equation (12.33) can then be used to calculate A_L.

A more practical method is to represent the feedback network with an input impedance that loads the amplifier output and an output impedance that appears in the input loop along with the dependent voltage source, Fv_{out}. This dependent feedback source is then assumed to be zero, removing the feedback voltage, and the overall gain is calculated with the resistances $R_{F\text{out}}$ and $R_{F\text{in}}$ remaining in the circuit. The resulting gain is A_L. We will demonstrate the first method in the following example.

EXAMPLE 12.4

A voltage feedback amplifier is constructed using the two modules of Fig. 12.9, where

$$A = \frac{600}{1 + j\frac{f}{20,000}}$$

A signal generator with a 0.5-kΩ resistance drives the voltage feedback amplifier. A 3-kΩ load is to be driven. For this feedback amplifier, calculate

Figure 12.9
Amplifier and feedback modules.

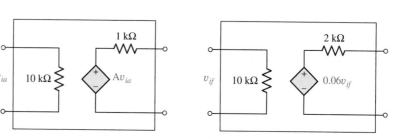

(a) The midband voltage gain

(b) The bandwidth

(c) The input impedance seen by the signal generator

(d) The output impedance seen by the load resistance

SOLUTION The feedback amplifier with generator and load resistors is shown in Fig. 12.10. The first step in calculating the needed values is to find the forward loaded gain of the amplifier. Equation (12.33) can be used to evaluate A_L as

$$A_L = A \times \frac{R_{ia}}{R_g + R_{ia} + R_{Fout}} \times \frac{R_L R_{Fin}/(R_L + R_{Fin})}{R_L R_{Fin}/(R_L + R_{Fin}) + R_{oa}}$$

$$= 600 \times \frac{10}{0.5 + 10 + 2} \times \frac{2.3}{2.3 + 1} = 335 \text{ V/V}$$

The value of F is 0.06 V/V. The factor $1 + A_L F$ is now found to be 21.1.

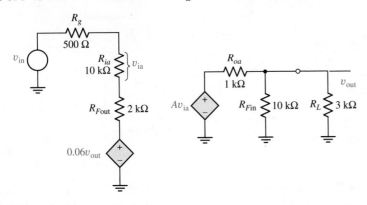

Figure 12.10
Feedback configuration for Example 12.4.

(a) The midband gain is

$$G_{MB} = \frac{A_L}{1 + A_L F} = \frac{335}{1 + 335 \times 0.06} = 15.9 \text{ V/V}$$

(b) The bandwidth is increased by a factor of 21.1 to

$$f_{2f} = 20 \times 21.1 = 422 \text{ kHz}$$

(c) The total impedance of the input loop before feedback is 12.5 kΩ. After feedback the input loop impedance is $21.1 \times 12.5 = 263.8$ kΩ. The impedance seen by the input source is less than this value by R_g or 500 Ω. Thus, the value of R'_{in-f} is 263.3 kΩ.

(d) The output impedance including the load resistance, before feedback, is

$$R_{out} = R_{oa} \parallel R_{Fin} \parallel R_L = 1 \parallel 10 \parallel 3 = 698 \ \Omega$$

With feedback this value is reduced to

$$R_{out-f} = \frac{R_{out}}{1 + A_L F} = \frac{698}{21.1} = 33 \ \Omega$$

The value of output impedance seen by the load resistor is

$$R'_{out-f} = \frac{R_{out-f} R_L}{R_L - R_{out-f}} = \frac{33 \times 3000}{3000 - 33} = 33.4 \ \Omega$$

Very few actual circuits allow the method used in Example 12.4 to be used directly. The second method is generally more useful in practical circuits and will be demonstrated in the next example.

EXAMPLE 12.5

The amplifier of Fig. 12.11 indicates a three-stage network with voltage-series feedback. We will assume that R_B, V_Z, and R_Z are selected to produce suitable dc biasing but do not affect the incremental gain due to loading. We are to find the midband voltage gain, the input resistance, and the output resistance of the feedback amplifier.

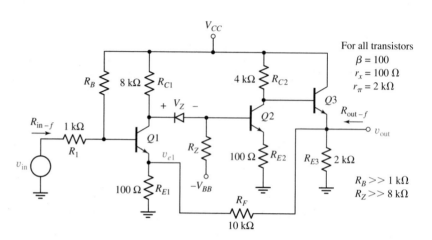

Figure 12.11
A voltage feedback amplifier.

SOLUTION A Thévenin equivalent circuit of the feedback network is taken at the output of this network, as shown in Fig. 12.12. We can now redraw the circuit as shown in Fig. 12.13 to correspond more closely to the voltage feedback amplifier. From Fig. 12.12, we see that R_{Fout} consists of a 100-Ω resistor in parallel with the 10-kΩ resistor R_F, resulting in $R_{Fout} = 99\ \Omega$. This resistance is designated R_{Th} in the figure. The resistance R_{Fin} consists of a 10-kΩ resistor in series with the output impedance at the emitter of the first transistor. Since this value will be small compared to 10 kΩ, it also is negligible, and $R_{Fin} \approx R_F = 10$ kΩ.

The amplifier of Fig. 12.13 can be drawn to correspond to the ideal stages discussed earlier. Figure 12.14 shows the appropriate identifications.

In order to evaluate the amplifier parameters, we need only evaluate the quantities A_L, F, R_{loop}, and R_{out}. The input loop impedance to the amplifier R_{loop} is

$$R_{loop} = R_1 + r_x + r_\pi + (\beta + 1)R_{Th} = 13.1\text{ k}\Omega$$

Figure 12.12
Thévenin equivalent looking into the output terminals of the feedback network.

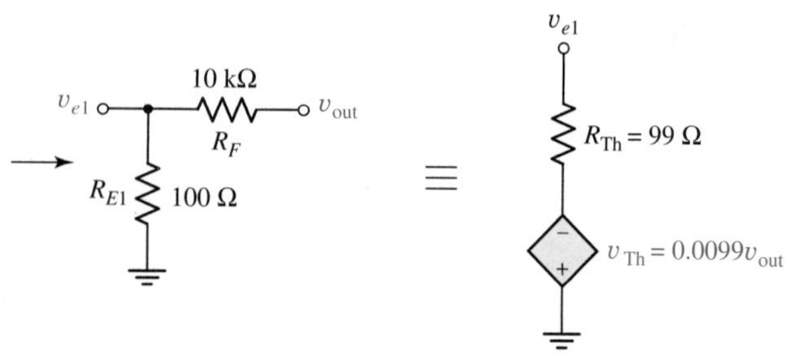

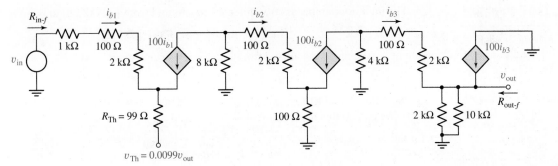

Figure 12.13
Equivalent circuit for amplifier of Fig. 12.11.

The impedance R_{out} is given by

$$R_{\text{out}} = R_{E3} \parallel R_F \parallel \frac{r_\pi + r_x + R_{C2}}{\beta + 1} = 58 \ \Omega$$

The loaded voltage gain is

$$A_L = A_{v1} A_{v2} A_{v3}$$

where

$$A_{v1} = -\frac{\beta(R_{C1} \parallel [r_x + r_\pi + (\beta + 1)R_{E2}])}{R_1 + r_x + r_\pi + (\beta + 1)R_{\text{Th}}} = -37 \ \text{V/V}$$

$$A_{v2} = -\frac{\beta(R_{C2} \parallel [r_x + r_\pi + (\beta + 1)(R_{E3} \parallel R_F)])}{r_x + r_\pi + (\beta + 1)R_{E2}} = -32 \ \text{V/V}$$

and

$$A_{v3} = \frac{(\beta + 1)(R_{E3} \parallel R_F)}{r_x + r_\pi + (\beta + 1)R_{E3} \parallel R_F} = 0.99 \ \text{V/V}$$

The gain A_L is then

$$A_L = 1170 \ \text{V/V}$$

The feedback factor is

$$F = 0.0099 \ \text{V/V}$$

The closed-loop gain is

$$G = \frac{A_L}{1 + A_L F} = 93 \ \text{V/V}$$

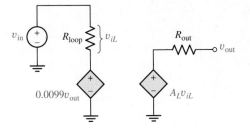

Figure 12.14
Circuit modified to be in ideal configuration.

PRACTICE **Problem**

12.8 Rework Example 12.4 if the midband gain is changed to 2000 V/V with all other parameters remaining at the values given in the example.
Ans: $G_{MB} = 16.4$ V/V, $f_{2f} = 1.36$ MHz, $R'_{in-f} = 849.5$ kΩ, $R'_{out-f} = 10.3$ Ω.

The input loop impedance with feedback is found from Eq. (12.28) to be

$$R_{in-f} = R_{loop}(1 + A_L F) = 165 \text{ k}\Omega$$

Output impedance is calculated from Eq. (12.29), which gives

$$R_{out-f} = R_{out}/(1 + A_L F) = 4.7 \ \Omega$$

It is often useful to approximate the closed-loop gain without making detailed calculations. We can do this by noting that $G \approx 1/F$ and

$$F = \frac{R_{E1}}{R_{E1} + R_F} = 0.0099 \text{ V/V}$$

giving

$$G \approx 1/0.0099 = 101 \text{ V/V}$$

An error of 8% results from applying this approximation.

The transadmittance, transimpedance, and current amplifiers can be handled in a similar manner to that of the voltage amplifier. Rather than develop the required theory at this time, we defer these considerations to the student.

12.3 **Stability of Feedback Systems**

IMPORTANT **Concepts**

1. The phase shift of an amplifier and the feedback network can change negative feedback at low frequencies to positive feedback at higher frequencies.
2. Positive feedback leads to instability; that is, an output signal can be present even when an input signal is not applied.
3. The magnitude and phase responses of the return ratio can be plotted to predict the stability of a feedback amplifier.

The feedback stages considered in the preceding sections have been idealized to the extent that the potential for instability or oscillations appears to be nonexistent. In neglecting high-frequency reactive effects we have chosen to ignore an important source of problems in feedback design. We must now consider methods of treating the stability problem in feedback amplifier design. Whereas Chapter 11 treated the specific case of stability for op amp designs, this section will deal with stability in a more general way.

A feedback amplifier applies negative feedback, and an oscillator uses positive feedback. Before considering the means by which unwanted positive feedback is introduced in an amplifier, let us consider the meaning of positive feedback. Figure 12.15 shows a positive

Figure 12.15
A positive feedback circuit.

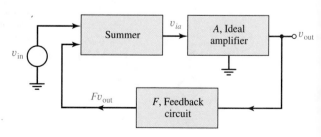

feedback circuit that might exemplify an oscillator. In this instance, the output signal is used to reinforce the input signal by summing the input and a fraction of the output.

The expression for overall gain is found to be

$$G = \frac{A}{1 - AF} \qquad (12.34)$$

If the return ratio AF approaches unity, the overall gain approaches infinity, which implies that an output signal can exist with no corresponding input. For sinusoidal oscillators the term AF is designed to approach unity at one particular frequency at the same time that the phase shift of AF is zero or multiples of $360°$. The circuit then produces a sinusoid at this frequency with no external ac input.

Returning to the feedback amplifier of Fig. 12.1 using a difference circuit, we should examine the gain expression in more detail. Overall gain is

$$G = \frac{A}{1 + AF} \qquad (12.35)$$

but this expression assumes that A and F are constant with frequency. If the return ratio, AF, happens to be frequency dependent, it is possible that positive feedback might occur. Let us assume that there exists some frequency for which the phase shift of the product AF is $180°$, and at the same time the magnitude equals or exceeds unity. Positive feedback is then present, and oscillations at that frequency can result. Whereas it is possible that unwanted oscillations might occur with only two active stages, it is highly probable that three or more stages will result in oscillations unless special precautions are taken. The reactive elements that are often negligible in multistage design can become significant when a feedback loop is added. Depletion region capacitance, stray capacitance, stray inductance, and other reactive elements are potential sources of instability problems.

12.3.1 PREDICTION OF STABILITY

Stability for a negative feedback amplifier can be predicted in several ways, given that the return ratio or AF can be expressed as a function of frequency. Perhaps the most straightforward method available to indicate stability is to plot frequency responses of both magnitude and phase of AF. A typical plot of magnitude and phase of AF as a function of frequency is shown in Fig. 12.16.

A critical frequency occurs when the phase of AF has reached $-180°$, which leads to positive feedback at this frequency rather than the desired negative feedback. This result is due to the additional $180°$ phase shift of the difference circuit. This frequency will be designated ω_{180}.

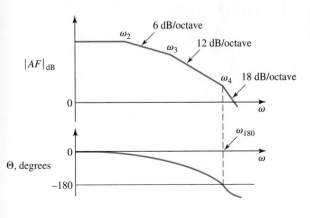

Figure 12.16
Magnitude and phase responses of return ratio AF.

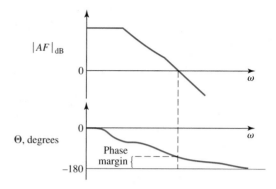

Figure 12.17
Definition of phase margin.

If the magnitude of the return ratio is less than unity at ω_{180}, the amplifier will not oscillate even though positive feedback exists. If $|AF|$ equals or exceeds unity, the circuit will oscillate. A sinusoidal output will result when $|AF|$ equals or is only slightly larger than unity, while the waveform will contain more distortion for larger values of $|AF|$. When $|AF|$ is unity at ω_{180}, the system is said to be *marginally stable*.

For most practical feedback amplifiers, stability can be predicted simply by finding the frequency at which the phase of AF equals $\pm 180°$, ω_{180}, and then calculating the magnitude of AF at this frequency.

Another measure of stability is called the phase margin, which is defined in terms of the phase shift of AF that exists when the magnitude of AF has dropped to unity as shown in Fig. 12.17. The number of degrees less than $-180°$ at this frequency is referred to as the phase margin. Most amplifier circuits target a phase margin exceeding $45°$. If θ is the phase shift of AF, the phase margin can be calculated from

$$pm = 180 + \theta \tag{12.36}$$

An unstable amplifier can be made stable by adding the proper circuit elements within the loop, which is referred to as compensation. Chapter 11 demonstrated a method of dominant pole compensation in connection with the op amp circuit. There are other possible compensation methods that are more efficient in specific applications.

EXAMPLE 12.6

The negative feedback amplifier of Fig. 12.18 has an open-loop gain of

$$A = \frac{2000}{\left(1 + j\frac{\omega}{10^7}\right)\left(1 + j\frac{\omega}{3 \times 10^7}\right)}$$

The feedback factor is

$$F = \frac{10^{-2}}{1 + j\frac{\omega}{5 \times 10^7}}$$

Figure 12.18
Amplifier for Example 12.6.

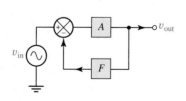

(a) At what frequency will this circuit be unstable?

(b) To what value must the midband gain, A_{MB}, be lowered to make the circuit marginally stable, that is, to make $|A_{MB}F| = 1$ at the potential oscillation frequency?

(c) If A_{MB} is lowered to 800 V/V, what is the phase margin of the system?

SOLUTION The return ratio can be written as

$$AF = \frac{20}{\left(1 + j\frac{\omega}{10^7}\right)\left(1 + j\frac{\omega}{3 \times 10^7}\right)\left(1 + j\frac{\omega}{5 \times 10^7}\right)}$$

Multiplying the factors in the denominator allows it to be written as

$$-j\,\omega^3\, 6.667 \times 10^{-23} + j\omega\, 1.533 \times 10^{-7} - \omega^2\, 6 \times 10^{-15} + 1$$

This denominator can have a 180° phase shift only if the imaginary part of the denominator equals zero, that is, if

$$\omega^3\, 6.667 \times 10^{-23} = \omega\, 1.533 \times 10^{-7}$$

(a) Solving for the frequency that gives this result leads to

$$\omega_{180} = 4.796 \times 10^7 \text{ rad/s}$$

Substituting this frequency into the expression for return ratio gives a value of $|AF| = 1.56$. Since this value is greater than unity, the circuit will be unstable at a frequency of 4.796×10^7 rad/s.

(b) The circuit will be marginally stable if the return ratio magnitude is lowered from 1.56 to 1.0, which can be done by lowering the midband gain of the amplifier to

$$A_{MB} = \frac{2000}{1.56} = 1282 \text{ V/V}$$

(c) If the midband gain is lowered to 800 V/V, the return ratio becomes

$$AF = \frac{8}{\left(1 + j\frac{\omega}{10^7}\right)\left(1 + j\frac{\omega}{3 \times 10^7}\right)\left(1 + j\frac{\omega}{5 \times 10^7}\right)}$$

The denominator of this return ratio is found to have a magnitude of 8 at a frequency of $\omega = 3.81 \times 10^7$ by iterative means. At this frequency, $|AF| = 1$, and the phase of AF is

$$\Theta = -\tan^{-1} 3.81 - \tan^{-1}\frac{3.81}{3} - \tan^{-1}\frac{3.81}{5} = -164.4°$$

The phase margin is then $180 - 164.4 = 15.6°$.

PRACTICE Problem

12.9 To what value must the midband gain of the amplifier in Example 12.6 be lowered to create a phase margin of 45°? *Ans:* 394 V/V.

DISCUSSION OF THE DEMONSTRATION PROBLEM

The feedback amplifier for the demonstration problem is repeated here and should now be recognized as a voltage feedback amplifier. The output voltage is in parallel with the input of the feedback network, consisting of R_F and R_{S1}. The output of the feedback network appears in series with the input loop.

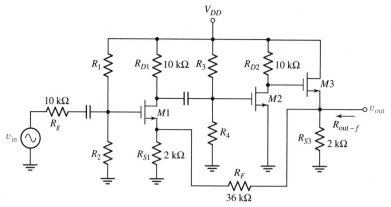

Feedback amplifier for the
Demonstration Problem

At midband frequencies, the equivalent circuit of Fig. 12.19 applies. After taking a Thévenin equivalent circuit at the output of the feedback network and noting the loading of R_F on the amplifier output, this equivalent circuit appears in Fig. 12.20. A Thévenin equivalent circuit of the signal generator and biasing resistors of the first stage is also shown.

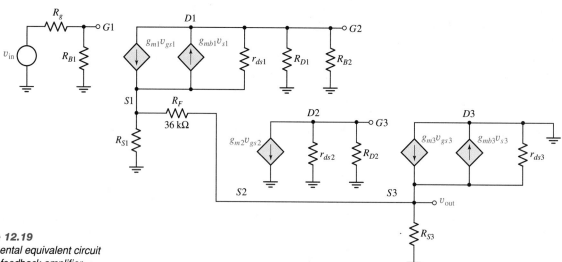

Figure 12.19
Incremental equivalent circuit
for the feedback amplifier.

In order to calculate the desired parameters of the feedback amplifier, we must first calculate the loaded voltage gain of the amplifier when the feedback voltage is removed. This voltage is removed by shorting the source of value $0.0526v_{\text{out}}$ to ground.

The input loop impedance is

$$R_{\text{loop}} = 9.1 + R_{gs1} + R_{F\text{out}}$$

where R_{gs1} is the impedance from gate to source of $M1$. Since this value is approximated as ∞, this loop impedance is also ∞.

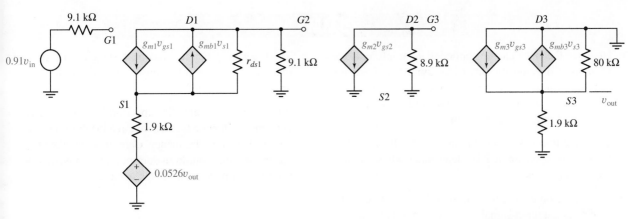

Figure 12.20
Modified equivalent circuit.

The output impedance of the source follower has been calculated in Chapter 9, and from Eq. (9.22) this value is

$$R_{\text{out}} = \frac{1}{g_m + g_{mb} + g_{ds} + G_{S3}}$$

where G_{S3} is the reciprocal of R_{S3}. For the values used in the circuit, $R_{\text{out}} = 256\ \Omega$.

The loaded voltage gain from the gate of $M1$ to output is found from the equivalent circuit of Fig. 12.20 to be

$$A_1 A_2 A_3 = (-3.59) \times (-26.7) \times 0.768 = 73.6\ \text{V/V}$$

The in-circuit gain from gate to drain of $M1$ is found by taking a Thévenin equivalent of the two current sources and r_{ds}. This value is -3.59 V/V. The gain of $M2$ is simply equal to $-g_m(r_{ds}\ \|\ R_{D2}) = -26.7$ V/V. The gain of the source follower is found from Eq. (9.24) to be 0.768 V/V.

The feedback factor is $F = 0.0526$ V/V. The midband gain from the gate of $M1$ to output with feedback is now calculated from Eq. (12.27) and is

$$G = \frac{A_L}{1 + A_L F} = \frac{73.6}{1 + 73.6 \times 0.0526} = 15.1\ \text{V/V}$$

The input voltage is attenuated by a factor of 0.91 as a result of the biasing resistors loading R_g. The overall gain calculated from the gate of $M1$ to output must be multiplied by this factor to get the actual gain, which results in

$$G_{MB} = 0.91 \times 15.1 = 13.7\ \text{V/V}$$

The output impedance with feedback is found from Eq. (12.29) to be

$$R_{\text{out}-f} = \frac{R_{\text{out}}}{1 + A_L F} = \frac{256}{4.87} = 52.6\ \Omega$$

The input impedance requires some thought to calculate. The impedance from gate to source of $M1$ is infinite. Thus, the signal generator sees only R_B, the biasing resistance. This value is 100 kΩ.

SUMMARY

➤ Negative feedback can be applied to an amplifier stage to improve the gain stability of the amplifier. Feedback can also improve the bandwidth, the signal-to-noise ratio, and the distortion within the amplifier at the expense of reduced overall gain.

➤ There are four basic configurations of feedback that can be used. A given application determines which is most appropriate to use.

➤ Depending on the configuration of the input circuit that combines the input signal with the signal that is fed back, input impedance can be increased or decreased using feedback. Output impedance can also be increased or decreased, depending on the type of configuration used to generate the signal that is fed back.

➤ The theory for the ideal feedback stage can be applied to a practical stage if loading on the gain of the amplifier is included.

➤ It is possible for parasitic capacitors or other reactive elements to change the negative feedback to positive feedback at some frequency. Care must be taken to ensure that this condition does not occur. Oscillation can result if positive feedback occurs.

PROBLEMS

SECTION 12.1 THE IDEAL FEEDBACK AMPLIFIER

12.1 The open-loop gain of the ideal amplifier of Fig. 12.1 is 1000. Calculate the closed-loop gain for (a) $F = 0.1$, (b) $F = 0.01$, and (c) $F = 0.001$. Compare the closed-loop gain in each case to the value of $1/F$. What conditions on AF allow the closed-loop gain to be approximated by $1/F$?

12.2 The open-loop gain of the ideal amplifier of Fig. 12.1 is 1000. If the closed-loop gain is to be 50, find the value of F.

SECTION 12.1.1 GAIN STABILITY

12.3 If $x = 101y + 10y^2$, calculate $S_{x,y}$ when $y = 2$.

12.4 Calculate $S_{G,T}$ for the amplifier of Example 12.1 at room temperature (300°K).

12.5 Can Eqs. (12.3) and (12.5) be used to calculate large changes in G—for example, changes of 30%? Explain.

☆ **12.6** Due to temperature changes, the midband gain of the amplifying element changes from $A = 20,000$ V/V to $A = 30,000$ V/V. What will the minimum and maximum values be for the midband gain of the amplifier, $G_{MB} = v_{out}/v_{in}$, over the same temperature range? Express results to 5-place accuracy.

Figure P12.6

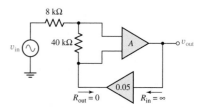

SECTION 12.1.2 SIGNAL-TO-NOISE RATIO

12.7 A 1000-Hz signal contains second and third harmonic distortion. Calculate the *SNR* if $v = 10 \cos 2000\pi t + 0.2 \cos 4000\pi t + 0.1 \cos 6000\pi t$.

☆ **12.8** A signal $v_{in} = 0.1 \sin 2000\pi t$ is presented to an amplifier with open-loop gain of $A = 200$. Non-linear distortion is introduced by the output stage,

resulting in an output signal of $v_{out} = 20 \sin 2000\pi t + 0.3 \cos 4000\pi t$. If a loop is now closed around the amplifier with $F = 0.1$ and a noiseless preamplifier is added to cause $v_{out} = 20 \sin 2000\pi t + B \cos 4000\pi t$, find the magnitude B. Compare the *SNR* of the open-loop case to the closed-loop case.

SECTION 12.1.3 BANDWIDTH IMPROVEMENT

12.9 An open-loop amplifier exhibits a gain of 1000 and a bandwidth of $\omega_2 = 10^7$ rad/s. Calculate the gain and bandwidth if the loop is closed with $F = 1/20$. Assume a -6 dB/octave gain falloff.

☆ **12.10** Sketch the frequency response of the gain of the feedback amplifier of Problem 12.9 if F is frequency dependent and given by

$$F = \frac{0.05}{1 + j\omega/10^7}$$

12.11 For each op amp of the figure

$$A = \frac{200,000}{1 + j\frac{f}{10}}$$

Figure P12.11

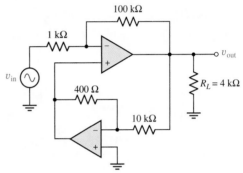

(a) Calculate the midband voltage gain of the amplifier, $G_{MB} = v_{out}/v_{in}$.

(b) Calculate the bandwidth of the amplifier and state any reasonable assumptions made.

SECTION 12.1.4 TYPES OF FEEDBACK

12.12 Identify the types of feedback of the four stages shown.

Figure P12.12

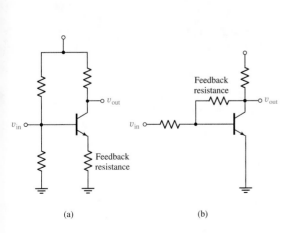

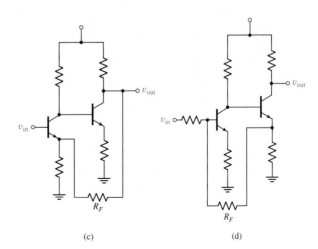

(a)　　　　　(b)　　　　　(c)　　　　　(d)

SECTION 12.1.5 EFFECT OF FEEDBACK ON IMPEDANCE LEVEL

12.13 In Fig. 12.4(a), $R_{ia} = 1 \text{ k}\Omega$, $R_{oa} = 2 \text{ k}\Omega$, $F_{vv} = 0.05$, and $A = 100$. Calculate the input and output impedance of the overall amplifier.

12.14 Repeat Problem 12.13 for the amplifier of Fig. 12.4(b). In this circuit $R_{ia} = 1 \text{ k}\Omega$, $R_{oa} = 2 \text{ k}\Omega$, $F_{iv} = 5 \times 10^{-5} \ \Omega^{-1}$, and $A_{vi} = 10^6 \ \Omega$.

12.15 Repeat Problem 12.13 for the amplifier of Fig. 12.5(b). In this circuit $R_{ia} = 1 \text{ k}\Omega$, $R_{oa} = 2 \text{ k}\Omega$, $F_{vi} = 2000 \ \Omega$, and $A_{iv} = 3 \times 10^{-3}$ A/V.

☆ **12.16** If $A = 10,000$ V/V for the amplifier in the figure,

(a) Calculate R'_{in-f} for $R_g = 0$.

(b) Calculate R'_{in-f} for $R_g = 2 \text{ k}\Omega$.

(c) Explain why R'_{in-f} depends on R_g.

Figure P12.16

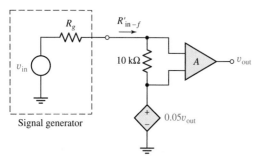

Signal generator

12.17 Calculate the midband voltage gain for the circuit of Problem 12.16(b). If A is now changed to 100,000 V/V what is the midband voltage gain of the circuit?

12.18 If $A = 100,000$ V/V in the circuit of Problem 12.16 calculate R'_{in-f} when $R_g = 2 \text{ k}\Omega$.

12.19 If $A = 10,000$ V/V in the circuit shown,

(a) Calculate R'_{out-f} for $R_L = \infty$.

(b) Calculate R'_{out-f} for $R_L = 1 \text{ k}\Omega$.

(c) Explain why R'_{out-f} depends on R_L.

Figure P12.19

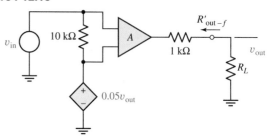

SECTION 12.2 THE PRACTICAL VOLTAGE FEEDBACK AMPLIFIER

12.20 Assume that the amplifier shown has ac coupling between stages (not shown) and that all stages are properly biased. If $r_x = 150 \ \Omega$, $r_\pi = 1 \text{ k}\Omega$, and $\beta = 120$ for all transistors, evaluate the voltage gain, input impedance, and output impedance of the amplifier.

12.21 Repeat Problem 12.20 if R_E is changed to 2 kΩ.

12.22 Repeat Problem 12.20 if R_E is changed to 100 Ω.

Figure P12.20

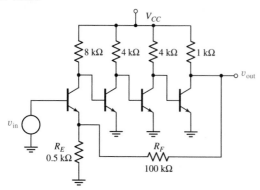

☆ **12.23** The amplifier stage has an unloaded gain of $A = -300$ V/V, an input impedance of 10 kΩ, and an output impedance of 1 kΩ. When placed in the feedback amplifier shown, calculate

(a) The voltage gain, $G = v_{out}/v_{in}$

(b) The input impedance seen by the 4-kΩ source resistance, R'_{in-f}

(c) The output impedance seen by the 4-kΩ load resistance, R'_{out-f}

Figure P12.23

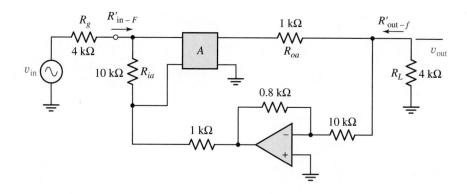

12.24 Assume that the gain-bandwidth product of the first op amp stage in the figure is $GBW = 4 \times 10^6$ Hz and the second op amp stage has a $GBW = 4 \times 10^7$. Calculate

(a) The midband voltage gain, $G = v_{out}/v_{in}$

(b) The upper 3-dB frequency of the voltage gain in Hz

(c) The output impedance R_{out-f}

Figure P12.24

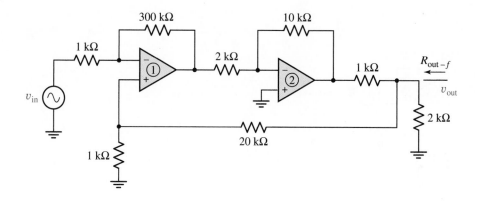

☆ **12.25** Assume that $R_{oa} = 0$ and

$$A = \frac{24{,}000}{1 + j\frac{f}{1000}}$$

for the circuit shown.

(a) Calculate the midband value of voltage gain for the amplifier.

(b) Calculate the bandwidth of the amplifier.

(c) Calculate R'_{in-f}, the input impedance seen by the signal generator.

(d) Write the expression for voltage gain as a function of frequency.

Figure P12.25

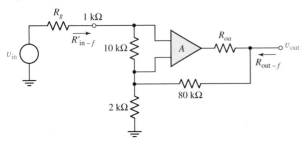

12.26 If R_{oa} is changed to 4 kΩ in the circuit of Problem 12.25,

(a) Calculate the midband value of voltage gain.

(b) Calculate R_{out-f}.

12.27 Approximate the midband voltage gain of the feedback amplifier in the figure to within 5% accuracy.

Figure P12.27

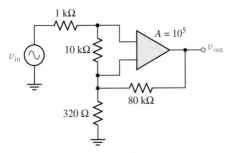

12.28 The amplifying module is used with the feedback module to create the voltage feedback amplifier shown.

(a) Calculate the loaded gain, A_L, of the amplifier.

(b) Calculate the midband gain of the amplifier, $G_{MB} = v_{out}/v_{in}$.

(c) Calculate R_A.

(d) If the bandwidth of the amplifying module is 1 MHz and falls off at the rate of −6 dB/oct above this frequency, calculate the bandwidth of the amplifier.

Figure P12.28

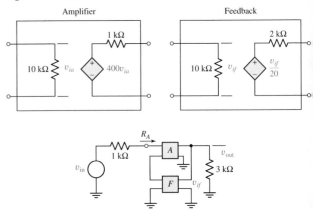

D 12.29 In the feedback amplifier of Problem 12.28, modify the value of $v_{if}/20$ in the feedback module to lead to a midband voltage gain of $G_{MB} = 34$ V/V. Calculate the bandwidth for the amplifier.

12.30 The gain A in the figure is given by

$$A = \frac{1000}{1 + j\frac{f}{10,000}}$$

(a) Calculate the midband voltage gain of the amplifier, $G_{MB} = v_{out}/v_{in}$.

(b) Calculate the bandwidth of the amplifier.

(c) Calculate R_A.

(d) Calculate R_B.

Figure P12.30

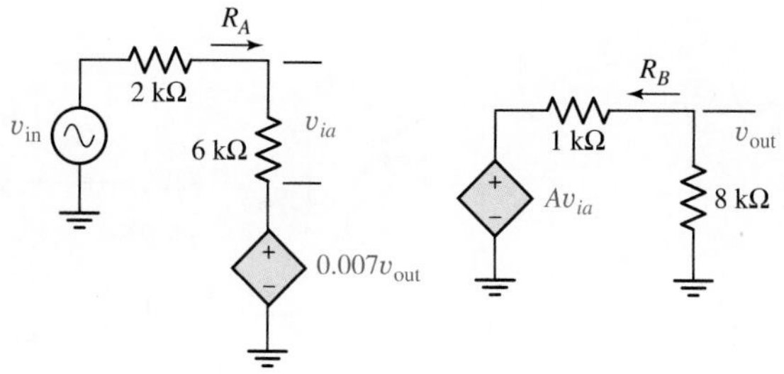

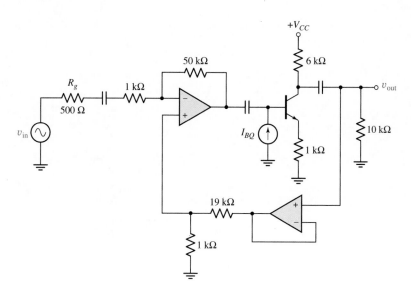

12.31 Assume a gain of

$$A = \frac{300,000}{1 + j\frac{f}{10}}$$

for both op amps shown.

(a) Calculate the midband voltage gain of the amplifier, $G_{MB} = v_{out}/v_{in}$.

(b) Calculate the bandwidth of the amplifier and state any reasonable assumptions made.

Figure P12.31

12.32 Assume amplifier stages with infinite input and zero output resistances in the figure.

(a) Calculate the midband voltage gain of the amplifier,
$G_{MB} = v_{out}/v_{in}$.

(b) Calculate R'_{in-f}.

(c) Calculate R'_{out-f}.

(d) If the bandwidth of the first amplifier stage is 10 kHz, the bandwidth of the second amplifier stage is 1 MHz, and the bandwidth of the emitter follower is 50 MHz, what bandwidth does the closed-loop amplifier have?

Figure P12.32

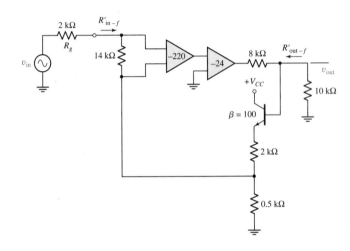

12.33 Assume that $\beta = 100$ and $V_{BE(on)} = 0.7$ V for all BJTs in the amplifier shown.

(a) Find the midband voltage gain of the amplifier,
$G_{MB} = v_{out}/v_{in}$.

(b) Calculate R'_{in-f}.

Figure P12.33

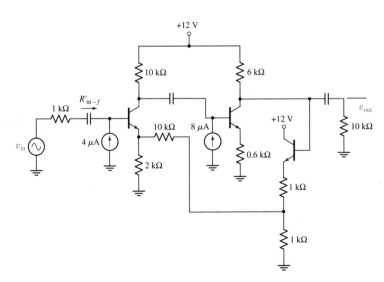

☆ **12.34** The BJT in the amplifier shown has a current gain bandwidth product of 1 GHz, and the GBW of the op amp is 3×10^6 Hz.

(a) Find the loaded gain of the amplifier, A_L.

(b) Find the midband voltage gain of the amplifier, $G_{MB} = v_{out}/v_{in}$.

(c) Find the upper corner frequency of the amplifier.

(d) Find the input impedance of the amplifier seen by the signal generator, noting that the generator impedance is 1 kΩ.

(e) Find the output impedance seen by the 6-kΩ load resistance.

Figure P12.34

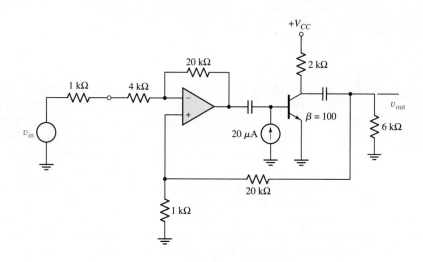

SECTION 12.3 STABILITY OF FEEDBACK SYSTEMS

☆ **12.35**

(a) If the gain of the ideal amplifier shown is

$$A = \frac{1000}{(1 + j\omega/10^6)(1 + j\omega/10^6)}$$

can the circuit become unstable?

(b) If a 0.01-μF capacitor is added in parallel to the 1-kΩ feedback resistor, can the circuit shown in the figure become unstable? If so, find the lowest frequency at which instability can occur.

12.36 Evaluate the product AF for the circuit of Problem 12.35 when a 0.01-μF capacitor is shunted across the 1-kΩ feedback resistor. Plot the magnitude and phase of AF as a function of frequency. Plot both quantities on the same graph and identify the points of instability.

12.37 Repeat Problem 12.35(a) given that the gain A is

$$A = \frac{10,000}{(1 + j\omega/10^7)(1 + j\omega/2 \times 10^7)(1 + j\omega/4 \times 10^7)}$$

Figure P12.35

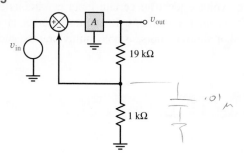

13 | *Large-Signal Circuits*

Electronic circuits can be divided into two categories: large-signal circuits and small-signal circuits. Although a great deal of space could be devoted to formal definitions of these categories, we will offer a simple guideline. Small-signal circuits limit output voltages or currents to values such that these output quantities are linear functions of an input value. It has been shown in earlier chapters that the electronic devices in amplifiers are constrained to operate at all times in a region called the active region. This circuit produces an output signal voltage or current that is some multiple times the applied input signal voltage or current. The multiplying factor is called the gain of the amplifier. If an input signal is increased to a larger value, the output signal may begin to distort and no longer be related linearly to the input signal. If the distortion becomes too great, the circuit is not operating entirely in the small-signal region. It may now be in a transition region between small-signal and large-signal operation. The circuit may still operate entirely within the active region, but the output signal experiences too much distortion to be considered a small-signal or linear circuit.

Large-signal circuits are nonlinear circuits. The output voltage or current may be related linearly to an input variable in one range of output values, but the relationship may change as a different range of output values is reached. Generally, the output variable will operate in the active region for some portion of time and in another region such as the cutoff region for another portion of time. In many cases, the output variable will move into three different regions of operation: active, cutoff, and saturation for the BJT or triode for the MOSFET. A large-signal circuit may not leave the active region as the input variable changes, but it will closely approach the cutoff or saturation region. The equations relating input variable to output variable will be radically different for each of these regions of operation.

An example of a large-signal circuit is a logic gate used in a digital computer. The kind of circuit used for logic gates can also be referred to as a switching circuit. The output voltage will exist at one voltage level for certain input conditions and switch to another voltage level for different input conditions. Examples of other large-signal circuits are ramp generators, one-shot multivibrators, astable multivibrators, and digital logic circuits.

DEMONSTRATION PROBLEM

A light-sensor has a 2-kΩ output impedance and generates a triangular waveform as shown when a light source is detected. Design a circuit that produces a rectangular output pulse of duration 300 μs and magnitude of at least 4 V when the triangular waveform is applied.

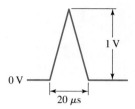

Waveform for Demonstration Problem.

This problem requires the design of a trigger circuit and a monostable multivibrator. The trigger circuit will accept the output of the light sensor and initiate a pulse of fixed duration by the multivibrator.

This chapter will consider the basic switching circuit and then proceed to a discussion of the one-shot and astable multivibrators that find application in both computing and noncomputing systems. Chapter 14 will discuss digital logic circuits that are used mainly in computer systems.

13.1 Switches

IMPORTANT Concepts

1. A simple switch can be used in various configurations to generate a high voltage level when in one state and a low voltage level when in the opposite state.
2. Nonideal switches are those that exhibit a small resistance rather than zero resistance when the switch is closed and a finite resistance rather than infinite resistance when the switch is open.
3. The BJT, the MOSFET, and the diode can be used to emulate a nonideal switch. These switches can be set to the open or closed state by some applied control voltage.

A major effort in the electronics field is spent in developing devices that emulate the behavior of a voltage-controlled switch. In such a device, an input voltage or current change opens or closes an output switch. An open switch means that no current flows between a pair of terminals, and a closed switch means that current flows freely between the pair of terminals with little resistance. Because this is an important area of electronics, we will look at this type of large-signal circuit in some detail.

Switching circuits are very important to several areas of electrical engineering. The most prominent of these areas is that of the digital computer. The electronics of these important systems are composed of nothing but switching circuits. Personal computers and larger computers are so significant in society that this area alone could justify a study of switching circuits. However, there are other areas that utilize this kind of circuit also. A few of these are the areas of electronic instruments, communication systems, and automotive electronics.

Two important elements in the realization of switches are the BJT and the MOSFET. Some of the factors involved in selecting a device relate to how closely the device emulates the switch, how small the device can be made, and how many supporting components are required.

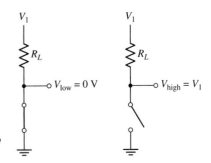

Figure 13.1
An ideal switch used to generate
two voltage levels.

13.1.1 A SWITCHING CIRCUIT

In digital circuits, an electronic switch is often used to generate two different output voltage levels: one when the switch is closed, the other when the switch is open. Figure 13.1 shows one configuration to generate these two levels using an ideal manual switch. When the switch is closed, the output terminal is connected directly to ground, which forces an output voltage of 0 V. When the switch is opened, the voltage is pulled toward V_1 through the load resistor. If R_L is small compared to the resistance that is being driven, the output voltage will equal approximately V_1. For the present time, we will assume that the resistance being driven is an open circuit or infinite resistance.

Figure 13.2 shows a nonideal switch that has some low resistance in the closed or *on* condition rather than a short circuit. In the open or *off* condition, the resistance of this switch is large but finite rather than infinite.

The characteristics of an ideal switch in both conditions are shown in Fig. 13.3(a). When the switch is off, no current flows, but an applied voltage will appear across the switch. This relationship is represented by the horizontal line that corresponds to $I = 0$. The on condition of the ideal switch is represented by the vertical line, which allows no voltage across the closed switch but will allow as much current as available.

Figure 13.3(b) shows the modification of characteristics for a nonideal switch. The off condition results in a large resistance, R_{so}, appearing across the switch instead of an open circuit. The slope of this line on the V-I characteristics is equal to $1/R_{so}$. When the nonideal switch is closed, the resistance is a small but nonzero value, designated R_{sc}. The slope of this line on the V-I characteristics is $1/R_{sc}$.

The expressions for the low and high voltages of the switching circuit of Fig. 13.2 can be derived using voltage division. When the switch is closed, presenting a resistance of R_{sc}, the low voltage level is

$$V_{\text{low}} = \frac{R_{sc}}{R_{sc} + R_L} V_1 \tag{13.1}$$

Figure 13.2
A nonideal switch.

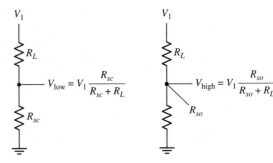

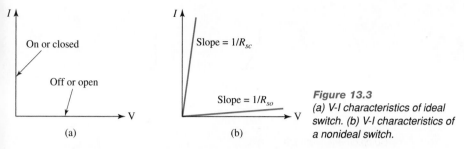

Figure 13.3
(a) V-I characteristics of ideal switch. (b) V-I characteristics of a nonideal switch.

When the switch is open, presenting a resistance of R_{so}, the high voltage level is

$$V_{high} = \frac{R_{so}}{R_{so} + R_L} V_1 \tag{13.2}$$

In an electronic circuit, a transistor will act as the switch. This switch will be turned on and off by a voltage applied to one of the terminals. Thus, it is voltage-controlled and can be caused to switch much more rapidly than can a manually controlled switch. Unfortunately, the transistor will have a finite but small on resistance and a large finite off resistance. Knowing these values allows the calculation of the two output voltage levels.

The bipolar junction transistor or BJT is commonly used in this type of switching arrangement to generate two output voltage levels. The input signal determines which state the transistor switch is in and, consequently, the voltage level of the output.

EXAMPLE **13.1**

For the switching circuit shown in Fig. 13.4,

(a) Calculate the low and high levels of V_{out} if the switch resistances are $R_{sc} = 25\ \Omega$ and $R_{so} = 400\ k\Omega$.

(b) Find the minimum value of R_{so} to ensure that V_{high} reaches a voltage of 11.98 V.

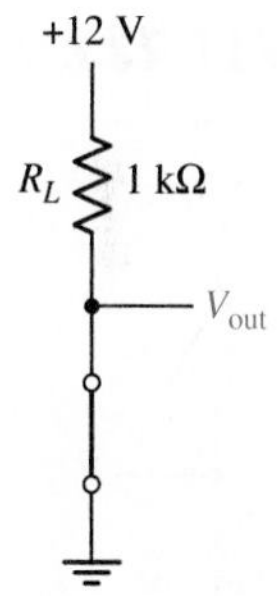

SOLUTION The voltage levels can be calculated from Eqs. (13.1) and (13.2). The low level is

$$V_{low} = \frac{R_{sc}}{R_{sc} + R_L} V_1 = \frac{25}{1025} 12 = 0.29\ V$$

The high-level voltage is

$$V_{high} = \frac{R_{so}}{R_{so} + R_L} V_1 = \frac{400}{401} 12 = 11.97\ V$$

To ensure that V_{high} reaches 11.98 V, the open resistance of the switch must be increased. This value can be found by solving the equation

$$11.98 = \frac{R_{so}}{R_{so} + 1} 12$$

for R_{so}, which results in a value of $R_{so} = 599\ k\Omega$.

+12 V

$R_L \gtrless 1\ k\Omega$

V_{out}

Figure 13.4
Circuit for Example 13.1.

13.1.2 COMPLEMENTARY SWITCHES

A second method of generating two different voltage levels replaces the resistor with a second voltage-controlled switch, as shown in Fig. 13.5. This configuration is called a complementary arrangement of switches. One switch will always be set to the open position when the other is closed and vice versa.

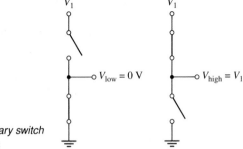

Figure 13.5
Complementary switch arrangement.

For ideal switches it is easy to see that when the top switch is open and the lower switch is closed, the output voltage will be 0 V. When the lower switch opens and the upper switch closes, the output voltage will be equal to V_1.

For nonideal switches, these same results will be approximated, depending on the values of on resistance and off resistance of the switches. Figure 13.6 indicates the two possible states for the practical case. The low and high voltage levels can be calculated easily since the two switches form a voltage divider. These values are

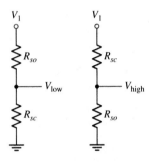

$$V_{\text{low}} = \frac{R_{sc}}{R_{sc} + R_{so}} V_1 \tag{13.3}$$

and

$$V_{\text{high}} = \frac{R_{so}}{R_{sc} + R_{so}} V_1 \tag{13.4}$$

Figure 13.6
Complementary switch equivalent circuits.

EXAMPLE 13.2

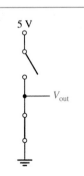

For the complementary switching circuit shown in Fig. 13.7, calculate the low and high values of V_{out} if the two switches are identical with resistances of $R_{sc} = 6$ kΩ and $R_{so} = 300$ MΩ.

SOLUTION These voltages are calculated from

$$V_{\text{low}} = \frac{R_{sc}}{R_{sc} + R_{so}} V_1 = \frac{6}{300,006} 5 = 0.0001 \text{ V}$$

and

$$V_{\text{high}} = \frac{R_{so}}{R_{sc} + R_{so}} V_1 = \frac{300,000}{300,006} 5 = 4.9999 \text{ V}$$

Figure 13.7
Circuit for Example 13.2.

13.1.3 SEMICONDUCTOR SWITCHES

The two most popular types of semiconductor switches used today are the BJT and the MOSFET. The BJT is often used in a configuration that emulates the single switch with a resistive load of Fig. 13.1. The MOSFET is often used in a configuration that emulates the complementary switching behavior of the two switches in Fig. 13.5, called a complementary MOSFET or CMOS circuit. These switching circuits form the basis of the digital circuits that are used to construct computers. One of the output voltage levels of a digital or switching circuit is associated with the binary number 1 and the other level is associated with binary zero.

At the present time, CMOS circuits are by far the most commonly used in microcomputers. This is primarily due to the huge number of MOS devices that can be fabricated on a single chip, which lowers the cost of the chip, and the continually improving performance achieved with MOSFET circuits.

The next section will consider only the very general behavior of the BJT and the MOSFET used in switching applications.

PRACTICAL Considerations

The BJT typically has a much lower on resistance than the MOSFET. The on resistance of a small BJT might be 50 Ω, and the corresponding value for a small MOSFET might be 5 kΩ. Thus the BJT can emulate a closed switch more closely than does the MOSFET.

On the other hand, the MOSFET has a much larger off resistance, which allows this device to emulate an open switch more closely than does the BJT. A small BJT might have an off resistance in the MΩ range, whereas the MOSFET off resistance is in the hundreds of MΩ range.

PRACTICE Problems

13.1 Rework Example 13.1 if $R_L = 1.2\,\text{k}\Omega$, $R_{sc} = 50\,\Omega$, and $R_{so} = 300\,\text{k}\Omega$.
Ans: (a) $V_{\text{low}} = 0.48$ V, $V_{\text{high}} = 11.95$ V;
(b) $R_{so} = 719\,\text{k}\Omega$.

13.2 Rework Example 13.2 if $R_{sc} = 8\,\text{k}\Omega$ and $R_{so} = 100\,\text{M}\Omega$.
Ans: $V_{\text{low}} = 0.0004$ V, $V_{\text{high}} = 4.9996$ V.

13.2 Using Semiconductor Devices in Switching Circuits

IMPORTANT Concepts

1. A BJT can simulate a voltage-controlled switch by presenting a low impedance between collector and emitter when saturated and a high impedance when shut off. The base-to-emitter voltage or base current controls the state of the BJT.

2. The MOSFET can also simulate a voltage-controlled switch by presenting a low impedance between drain and source when conducting and a high impedance when shut off. The gate-to-source voltage controls the state of the MOSFET.

3. The low impedance of the conducting BJT is typically in the tens of ohms range, whereas the low impedance of the MOSFET is in the range of a few kΩ. In the high-impedance states, the MOSFET exhibits a greater value than that for the BJT. This value is generally in the hundreds of MΩ range.

This section will consider the application of the BJT, the MOSFET, and the silicon diode to switching functions.

13.2.1 THE BJT SWITCH

As we have seen in Chapter 7, the BJT consists of three different semiconductor regions: the base, the emitter, and the collector as represented by the symbol for an *npn* device shown in Fig. 13.8. Each of the three distinct regions is electrically connected to a terminal, which allows connection of the regions to other circuit components. The emitter terminal is identified by an arrowhead that indicates the only possible direction of significant current flow for the *npn* device. The base terminal is the control for the device. If the base-emitter

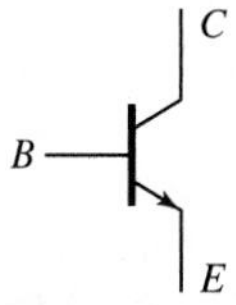

Figure 13.8
The symbol for an npn *transistor.*

voltage is such that no current flows into the base terminal, no current will flow from collector to emitter. If the base-emitter voltage is such that sufficient current is directed into the base terminal, a large current will flow from collector to emitter. These conditions allow the BJT to function as a controllable switch. The switch appears between the collector and emitter terminals. This switch is open if the base-emitter junction is not forward biased and is closed when this junction is forward biased with significant current flowing into the base.

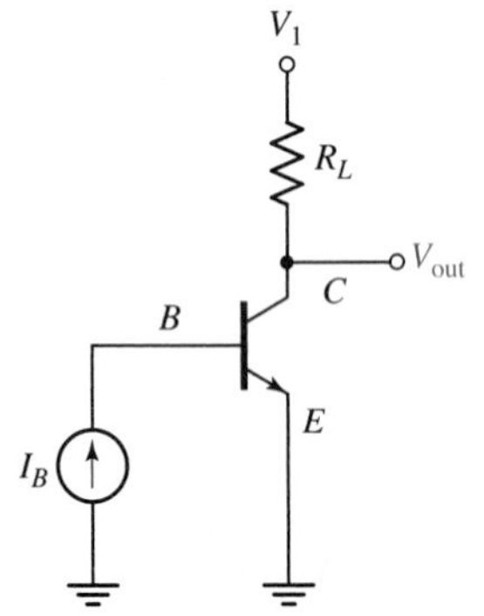

The BJT can operate in three different regions: the cutoff region, the active region, or the saturation region. For many switching applications, the device is either in the cutoff or saturation regions for most of the time, with only short durations in the active region as a transition between the other two regions is made. Figure 13.9 shows an *npn* transistor in a common switching circuit configuration.

The basic equivalent circuits for the cutoff and saturation regions for the *npn* BJT are shown in Fig. 13.10. When the BJT is off, the path between collector and emitter approximates an open switch. Although the resistance between collector and emitter is not infinite, it is typically in the range of several megohms.

When the device is on or saturated, the path between emitter and collector appears as a small resistance, R_{sat}, typically in the range of tens of ohms. This circuit is similar to that of a closed switch. In many applications, the resistance can be approximated by a short circuit with little loss of accuracy. The voltage drop across a forward-biased diode, such as the base-emitter junction, is approximated by a voltage source of value 0.6–0.7 V. Often a second voltage source called $V_{CE(\text{sat})}$ is added between collector and emitter to the on circuit of Fig. 13.10(b). This source may have a value of 0.2–0.3 V.

The characteristic curves were discussed in more detail in Chapter 7, but in that chapter, the active region was emphasized. The two curves corresponding to the on and off states in Fig. 13.10(c) are very similar to those of a switch. These curves should be compared to those of the ideal switch shown in Fig. 13.3.

Although the voltage applied to the base-emitter junction controls the state of the device, an indication of this state can be determined from the base current. If the base current is zero, the BJT is in the cutoff region and no current flows into or out of the collector or emitter junctions. Thus there is a very high resistance between the collector and emitter terminals. The cutoff region emulates an open switch that is indicated by zero base current.

To saturate the transistor or turn the switch on, the base-emitter voltage must be increased until the forward base current equals or exceeds a value of

$$I_{B\text{sat}} \approx \frac{V_1}{\beta R_L} \tag{13.5}$$

where β is the active region current gain from base to collector of the BJT. This parameter is determined in construction and is typically in the range of 100 to 300 for small devices. If the value of β is unknown, a low value of 100 can be used in calculating the value of base current needed to put the BJT into the saturation region. Although driving the base with

Figure 13.9
A switching circuit using an npn *BJT.*

Figure 13.10
(a) Cutoff equivalent circuit.
(b) Saturation equivalent circuit.
(c) Characteristic curve for BJT.

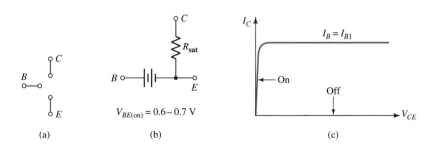

up to 10 times the needed current does not damage the transistor, it can slow the speed of switching to the cutoff region when base current is removed.

The input current source of Fig. 13.9 can be replaced by a voltage source and a series resistance to create the minimum base current given by Eq. (13.5). If the input voltage of the circuit shown in Fig. 13.11 is zero volts or negative, the transistor is in the cutoff region. When the input voltage is changed to a positive level, say V_{in}, the base current becomes equal to

$$I_B = \frac{V_{in} - 0.7}{R_B} \qquad (13.6)$$

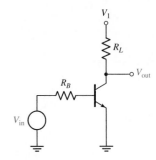

Figure 13.11
A simple BJT switch.

The voltage from base to emitter of an *npn* silicon transistor when current flows into the base is approximately 0.7 V. The voltage across the base resistor is then $V_{in} - 0.7$, as indicated in Eq. (13.6).

EXAMPLE **13.3**

If the power-supply voltage, V_1, in the circuit of Fig. 13.11 is 12 V, the load resistor is 4 kΩ, and the transistor has a β of 220, calculate a suitable value of R_B. Assume that the high level of the input voltage is 6 V and the low level is 0 V.

SOLUTION The minimum value of base current required to saturate the transistor is calculated as

$$I_{Bsat} = \frac{V_1}{\beta R_L} = \frac{12}{220 \times 4} = 13.6 \ \mu A$$

To ensure saturation, we will drive the base with 20 μA. When the input voltage switches to its high level of 6 V, the voltage across R_B equals $6 - 0.7 = 5.3$ V. The value of R_B required to allow 20 μA is

$$R_B = \frac{5.3}{20} = 265 \ k\Omega$$

This circuit will now simulate a voltage-controlled switch with a 0-V input level leading to an output voltage of 12 V. When the input voltage changes to 6 V, the transistor is driven from the cutoff region, through the active region, to the saturation region. The output voltage drops to a value near ground.

PRACTICE Problems

13.3 Recalculate R_B in Example 13.3 if $\beta = 140$.
Ans: 165 kΩ.
13.4 Recalculate R_B in Example 13.3 if $R_L = 6$ kΩ.
Ans: 390 kΩ.

13.2.2 THE MOSFET SWITCH

The MOS device consists of four regions called the gate, the source, the drain, and the substrate or body. The substrate and the source are generally tied together for digital circuits. We often ignore the substrate region and use only three terminals to represent the device in many circuits. When the source and substrate are not tied together, the effect of the voltage difference between these two terminals must be considered. This device can be fabricated as an *n*-channel (nMOS) device or as a *p*-channel (pMOS) device. The symbols for both devices are shown in Fig. 13.12.

The resistance from drain to source is controlled by the voltage applied between gate and source terminals. For the nMOS device, a zero or low value of V_{GS} causes a very large resistance to exist between source and drain, approximating an open switch. When V_{GS} exceeds some relatively small voltage called the threshold voltage, the resistance between

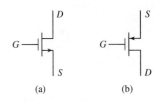

Figure 13.12
Symbols for the MOSFET: (a) the nMOSFET, (b) the pMOSFET.

Figure 13.13
MOSFET equivalent circuits:
(a) off state, (b) on state,
(c) characteristic curve for the
nMOSFET.

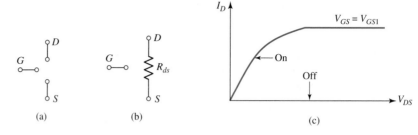

source and drain becomes much smaller. For the pMOS device, a negative value of V_{GS} must be applied to lead to a low resistance between source and drain terminals.

The MOS device approximates a switch better than the BJT does in the off state. It does not approximate a switch as closely as the BJT does in the on state. The equivalent circuits for the MOS in the two states are shown in Fig. 13.13.

The MOSFET equivalent circuit in the off state is identical to that of the BJT equivalent. In neither case is the impedance between terminals infinite, but it is quite high. It may be in the range of tens of $M\Omega$ for the BJT and a factor of 10 to 100 times higher for the MOS device. The resistance in the on state, R_{ds}, may be a few hundred to a few thousand ohms for the MOSFET. This region of operation is called the triode region.

The characteristic curves of Fig. 13.13(c) can be seen to approximate the switch characteristics of Fig. 13.3. The slope of the curve when the device is on is not nearly as steep as the ideal switch, nor is it as steep as that of the BJT in the on state.

The nMOS device is forced into the cutoff region when the magnitude of the gate-to-source voltage is less than the threshold voltage of the device. This value is determined in the construction process and may be 0.8 V for the nMOS device. For the pMOS device, the device is in the cutoff region when the gate-to-source voltage is more positive than the threshold voltage, which may be -0.8 V. When the gate-to-source voltage for the nMOS device exceeds the threshold voltage, current can flow from drain to source.

When the gate-to-drain and the gate-to-source voltages both exceed the threshold voltage, the nMOS device moves into the on state in the triode region. For the pMOS device, the gate-to-source voltage and the gate-to-drain voltage must both be more negative than the threshold voltage to put the device in the on state.

The basic configuration of the MOSFET switch corresponds to the complementary arrangement shown in Fig. 13.5. A pMOS device is used for the upper switch and an nMOS device is used for the lower switch, as shown in Fig. 13.14(a). We will assume that the voltage $V_1 = 5$ V in Fig. 13.14. When the input voltage is at 0 V, the nMOS device is in

Figure 13.14
(a) A complementary MOSFET
(CMOS) switch. (b) Model for
$V_{in} = 0$ V. (c) Model for
$V_{in} = 5$ V.

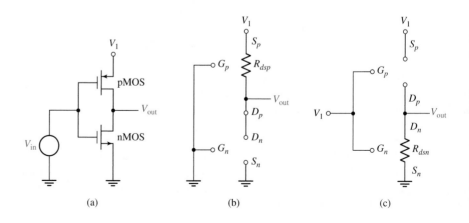

the off state since the gate-to-source voltage is 0 V. The pMOS device has −5 V from gate to source and, if this device is in the on state, the output voltage is at 5 V. An output voltage of 5 V will lead to a gate-to-drain voltage for the pMOS of −5 V, which verifies that the pMOS device is in the on state in the triode region. This situation is shown in Fig. 13.14(b).

If the input voltage changes to a higher voltage, say 5 V, the nMOS device will turn on and the pMOS device will turn off. This leads to the situation depicted in Fig. 13.14(c).

EXAMPLE **13.4**

If the power-supply voltage in the circuit of Fig. 13.14 is 5 V, $R_{dsp}(\text{on}) = R_{dsn}(\text{on}) = 6$ kΩ, and $R_{dsp}(\text{off}) = R_{dsn}(\text{off}) = 300$ MΩ, calculate the low and high voltage levels of V_{out}. Assume that the input swings from a low level of 0 V to a high level of 5 V.

SOLUTION The input voltage equals V_{GSn} while $V_{GSp} = V_{\text{in}} - 5$. When V_{in} is at 0 V, the nMOSFET is off and the pMOSFET is on. When V_{in} changes to 5 V, the pMOSFET shuts off and the nMOSFET turns on. The two voltage levels can be calculated as shown in Example 13.2, which uses the same values of on and off resistance given for the MOS devices. The results of that calculation show that $V_{\text{low}} = 0.0001$ V and $V_{\text{high}} = 4.9999$ V. When V_{in} is at the low level, V_{out} will be at its high level. When V_{in} switches to 5 V, the output voltage switches to its low level. We note that the low and high levels of the output voltage approximate the values 0 V and 5 V, respectively. Thus, the output voltage of the CMOS switch swings from the reference voltage to the power-supply voltage, referred to as a rail-to-rail swing.

The CMOS switch is also called an inverter, since a low input voltage level leads to a high output level and a high input voltage level leads to a low output level. This circuit forms the basis for several important CMOS logic circuits to be considered in the next chapter.

The on resistances of the nMOS and pMOS devices are often unequal, as are the off resistances. Fortunately, the output swing will still be rail to rail for reasonable resistance values. The designer can control these resistances to some degree by controlling the dimensions of the MOSFET channels.

PRACTICAL **Considerations**

The relatively high value of on resistance for the MOSFET makes it somewhat difficult to use this device in a switching arrangement with a fixed resistor load, especially in IC design where resistor values are limited to tens of kΩ.

The complementary arrangement used for CMOS circuits is ideal for MOSFET switching circuits. This arrangement offers rail-to-rail output voltage swings and requires no resistors. Although two devices are required to create a voltage switch, the chip real estate occupied by these two devices is minimal. Even though a single BJT and a load resistor can also emulate a switch, the real estate required on the chip is several factors greater than that used for the CMOS switch.

13.2.3 THE SEMICONDUCTOR DIODE

The diode is also a voltage-controlled switch, but it exhibits a significant difference from the transistor switches previously discussed. This difference is that the voltage used to control the state of the switch is the same voltage applied to the switch. This differs from the BJT,

PRACTICE **Problems**

13.5 Rework Example 13.4 if $R_{dsp}(\text{on}) = 2R_{dsn}(\text{on}) = 10$ kΩ and $R_{dsp}(\text{off}) = R_{dsn}(\text{off}) = 200$ MΩ.
Ans: $V_{\text{low}} = 0.000012$ V, $V_{\text{high}} = 4.99975$ V.

13.6 Rework Example 13.4 if $R_{dsp}(\text{on}) = 3R_{dsn}(\text{on}) = 18$ kΩ and $R_{dsp}(\text{off}) = 3R_{dsn}(\text{off}) = 500$ MΩ.
Ans: $V_{\text{low}} = 0.00006$ V, $V_{\text{high}} = 4.99946$ V.

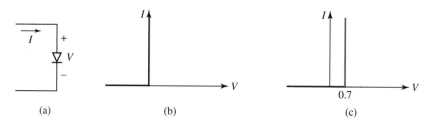

Figure 13.15
The ideal diode: (a) symbol,
(b) ideal diode characteristics,
(c) more practical
characteristics.

for example, in that the voltage applied to the base terminal determines whether the switch is on or off and the voltage at the collector terminal switches from a high to a low value. The symbol for the diode and the characteristics for an ideal diode are shown in Fig. 13.15, (a) and (b).

When current is applied to the ideal diode in the forward direction as indicated by the arrow of the symbol, no voltage drop appears. If a current source attempts to force current in the negative direction, the diode will block this current regardless of the voltage that may drop across this device. This device becomes an open circuit when reverse biased and is a short circuit when forward biased. A better approximation for the silicon diode is shown in Fig. 13.15(c). For these characteristics, the diode is an open circuit when the voltage across it is less than 0.7 V. For appreciable forward currents, a constant 0.7 V drops across the device.

The circuit of Fig. 13.16 shows one simple configuration for the diode. With a positive voltage applied to the circuit as in Fig. 3.16(a), the diode is forward biased and current will flow in the forward direction. Assuming we use the ideal diode model of Fig. 3.15(b), there will be no voltage drop across the diode. The result is an output voltage that equals the input voltage.

When the input voltage is negative, as shown in Fig. 13.16(b), no current is allowed to flow through the resistor; thus, there is no voltage drop across this element. This situation results in zero output voltage, assuming ideal diode characteristics.

Figure 13.16
The diode switch.

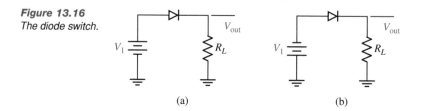

13.2.4 SWITCHING EXAMPLES

This subsection considers several examples that relate to large-signal circuit operation. Before discussing these examples, we refer to Eq. (2.33) as an important equation relating to the charging or discharging of a capacitor. In the circuit of Fig. 13.17, the switch is in the first position long enough to charge the capacitor to an initial voltage, v_i. The switch then changes position, and the capacitor begins to charge toward a target voltage, v_t. The voltage across the capacitor changes exponentially between the initial voltage and the target voltage. The equation describing this voltage is

$$v(t) = v_i + (v_t - v_i)\left(1 - e^{-t/\tau}\right) \tag{13.7}$$

We will demonstrate the use of this equation in the following examples.

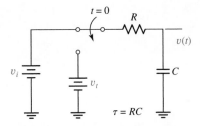

Figure 13.17
A single time constant circuit.

EXAMPLE 13.5

Assume that the BJT in Fig. 13.18 is a perfect switch with infinite off resistance and zero on resistance between collector and emitter. Assume that $\beta = 200$, $V_{BE(on)} = 0.7$ V, and the initial output voltage is 5 V. Sketch the resulting output waveform.

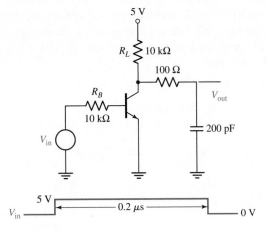

Figure 13.18
A BJT switching circuit.

SOLUTION When the input voltage is at 0 V, the BJT is in the off state. The equivalent collector-emitter circuit is shown in Fig. 13.19(a). When V_{in} switches to 5 V, Eq. (13.5) must be used to determine the value of base current required for saturation. Note that when the transistor switches on, the 100-Ω resistor instantaneously has 5 V across it since the capacitor is at 5 V. Thus, the load resistance seen by the transistor is 10 kΩ in parallel with 100 Ω, which is approximately 100 Ω. To calculate the current required for saturation, we use 100 Ω for the collector load, which gives

$$I_{Bsat} = \frac{V_1}{\beta R_L} = \frac{5}{200 \times 100} = 0.25 \text{ mA}$$

The actual base current is calculated from Eq. (13.6) as

$$I_B = \frac{V_{in} - V_{BE(on)}}{R_B} = \frac{5 - 0.7}{10} = 0.43 \text{ mA}$$

Since this exceeds the required base saturation current, the BJT is definitely saturated. The equivalent circuit of Fig. 13.19(b) applies to this situation. When the BJT switches on, the capacitor will discharge toward 0 V through the 100-Ω resistor. In order to use Eq. (13.7), the time constant, the target voltage and the initial voltage must be evaluated. The time constant is $\tau = 100 \times 2 \times 10^{-10} = 2 \times 10^{-8} = 20$ ns. The initial voltage is 5 V and the target voltage is 0 V. Using these values in Eq. (13.7) gives

$$V_{out} = 5 + (0 - 5)(1 - e^{-t/20 \text{ ns}}) = 5e^{-t/20 \text{ ns}}$$

This waveform is a simple decaying exponential.

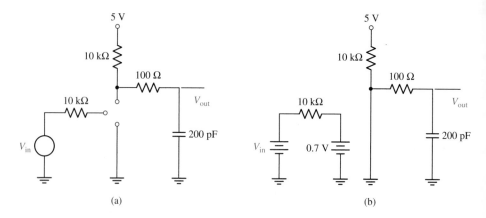

Figure 13.19
Equivalent circuits: (a) off state,
(b) on state.

The input voltage keeps the BJT in the on state for 0.2 μs, which amounts to 10 time constants. The output voltage will reach 0 V long before the input voltage switches back to 0 V. When this occurs, the BJT switches to the off state and the equivalent circuit of Fig. 13.19(a) again applies. Equation (13.7) is now used with values of $\tau = 10,100 \times 2 \times 10^{-10} =$ 2.02 μs, $v_i = 0$ V, and $v_t = 5$ V. The result is

$$V_{out} = 0 + (5 - 0)\left(1 - e^{-t/2.02\ \mu s}\right) = 5\left(1 - e^{-t/2.02\ \mu s}\right)$$

The charging time constant is much larger than the discharging time constant due to the difference in resistances for the two cases. The output waveform is sketched in Fig. 13.20.

Figure 13.20
Output waveform for circuit of
Fig. 13.18.

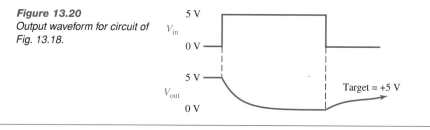

EXAMPLE 13.6

Repeat Example 13.5 with the modifications to the circuit shown in Fig. 13.21. Assume the diode requires a forward voltage of 0.7 V to conduct.

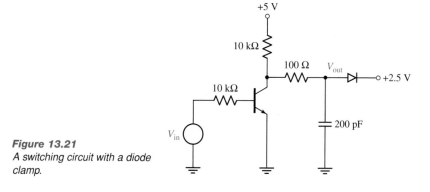

Figure 13.21
A switching circuit with a diode
clamp.

SOLUTION This circuit is modified slightly from that of Fig. 13.19 by the addition of a diode and a 2.5-V source. In this case, the capacitor voltage cannot rise above a voltage of

3.2 V. When this occurs, the diode has a forward voltage of 0.7 V and conducts current that is supplied by the 10-kΩ load resistance. This arrangement is called a diode clamp and is similar to the clipping circuit discussed in Chapter 5.

When the input voltage is at its low level, the transistor is off and the output voltage is 3.2 V, which represents the initial capacitor voltage. When the input switches to the high level, the transistor again saturates and the capacitor discharges toward 0 V. The time constant is the same as that in Example 13.5, that is, 20 ns. The only difference is that the initial voltage is 3.2 V instead of 5 V. Note that as the capacitor discharges from 3.2 V toward ground, the diode becomes an open circuit.

After the capacitor has discharged and the input voltage switches back to the low level, the capacitor begins to charge. Again, the charging time constant equals that of Example 13.5, that is, 2.02 μs. The target voltage is 5 V, but the final voltage in this case is 3.2 V rather than 5 V. The resulting waveform is shown in Fig. 13.22.

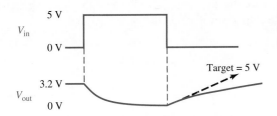

Figure 13.22
Output waveform for Example 13.6.

EXAMPLE **13.7**

When the input waveform of the CMOS circuit of Fig. 13.23 is at 0 V, the pMOS device is on and the nMOS device is off. When the input voltage switches to 5 V, the nMOS device turns on and the pMOS device shuts off. Assume that the off resistance from drain to source of each device is infinite. When on, the drain to source resistances are $R_{dsn} = 4$ kΩ and $R_{dsp} = 8$ kΩ. Sketch the output waveform.

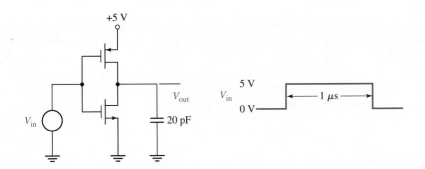

Figure 13.23
Circuit of Example 13.7.

SOLUTION If the input signal has been at the low level for a short time before changing levels, the capacitor has been charged to the 5-V level, which becomes the initial capacitor voltage. When the input waveform rises to 5 V, the p device becomes an open circuit and the n-device resistance becomes 4 kΩ. The equivalent circuit is shown in Fig. 13.24(a). The capacitor will discharge toward ground with a time constant of $\tau = 4 \times 10^3 \times 20 \times 10^{-12} = 80$ ns. The initial voltage is 5 V and the target voltage is 0 V. The output voltage will easily reach the 0-V level by the end of the input pulse width.

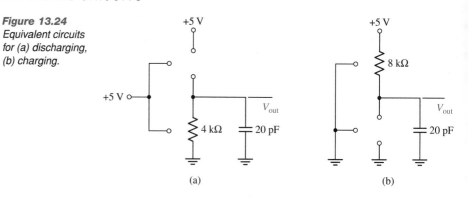

Figure 13.24
Equivalent circuits for (a) discharging, (b) charging.

(a)　(b)

13.7 A capacitor charges from 0 V toward 12 V with a time constant of 1 μs. How long does it take to charge from 0 V to 8 V?
Ans: 1.1 μs.

13.8 How long does the capacitor of Practice Problem 13.7 take to move from 2 V to 6 V?
Ans: 0.511 μs.

13.9 How long does it take V_{out} to move from 5 V to 0.1 V in Example 13.7?
Ans: 313 ns.

13.10 How long does it take V_{out} to move from 0 V to 4.9 V in Example 13.7?
Ans: 626 ns.

When the input switches back to 0 V, the *n* device shuts off and the *p* device turns on. The capacitor charges toward 5 V with an initial voltage of 0 V and a time constant of 160 ns. This time constant is twice the discharge value because the charging resistance is twice the discharge value. The output waveform is shown in Fig. 13.25.

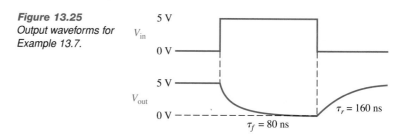

Figure 13.25
Output waveforms for Example 13.7.

13.3 Multivibrator Circuits

IMPORTANT Concepts

1. The monostable multivibrator generates an output pulse of predetermined duration in response to a short trigger pulse at the input. This duration is set by the values of certain timing components in the circuit.

2. The astable multivibrator produces a rectangular waveform at the output and requires no input signal. The frequency of the output signal is determined by the values of certain timing components in the circuit.

The family of circuits known as multivibrators is a large one, which incorporates within its subclasses many different circuit configurations. Circuit designers involved with switching circuits or waveform generation will often utilize this class of circuits.

The *multivibrator* is a regenerative circuit, meaning that within the circuit there is a positive feedback loop. The voltage gain around this loop is very high when the devices in the loop are all in their active regions, perhaps reaching a value of several thousand. The multivibrator has more than one stable or quasi-stable state in which it can exist. In these states, at least one device in the loop is not in its active region. The loop gain is very low, approaching zero for this situation.

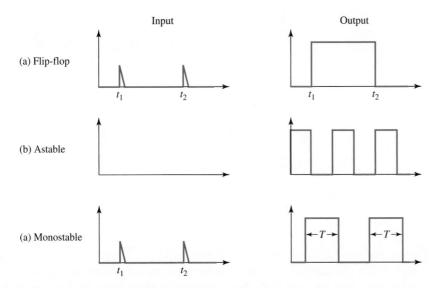

Figure 13.26
Multivibrator waveforms.

The three popular types of multivibrator are the *bistable*, the *astable*, and the *monostable*. Over the years, other names have developed that apply to these three types of circuit. One important subclass within the bistable classification is the flip-flop that is so important in computer systems. The astable is often referred to as a free-running multivibrator or as a clock circuit, and the monostable is called a one-shot circuit. The basic difference among the three types of multivibrator is the number of stable states associated with each type. The bistable has two stable states, the monostable has one stable state (and one quasi-stable state), and the astable has no stable states (two quasi-stable states). The waveforms of Fig. 13.26 show typical input and output signals that might be expected from the three types of multivibrators.

The output of the flip-flop changes from one stable state to another upon application of an input signal. It will remain in this state until returned to its original state by a second input signal. The astable requires no input while the output continually switches between the two quasi-stable states. The length of time spent at each voltage level is determined by certain components of the circuit. The output of the one-shot remains at a specific level while in the stable state until an input signal initiates the quasi-stable state. The output voltage switches to another level in this state and remains at that level until the output reverts back to the original value. The time that the circuit remains in the quasi-stable state is called the period of the one-shot and is determined by circuit components. In the bistable and monostable multivibrators, the output signal levels are independent of the amplitudes of the input signals, assuming these amplitudes are large enough to initiate switching.

The bistable is commonly used in computer systems. Both the monostable and the astable circuits are used in computing and noncomputing systems. These two devices will now be considered in some detail.

13.3.1 THE MONOSTABLE MULTIVIBRATOR

Figure 13.27 shows a simple one-shot multivibrator circuit. In the stable state, $Q1$ is off while $Q2$ is on. The resistor R_B is selected to saturate transistor $Q2$. The two resistors R_1 and R_2 form a voltage divider between the collector voltage of $Q2$, which will be approximately zero, and the supply voltage, $-V_{BB}$. The base of $Q1$ will be negative while the emitter is at 0 V; hence $Q1$ is off. The output voltage taken at the collector of $Q2$ is zero, and this condition corresponds to the one stable state of the multivibrator.

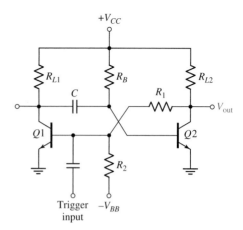

Figure 13.27
A transistor monostable multivibrator.

If an external positive trigger pulse of sufficient amplitude is applied to the base of $Q1$, it will start to turn this transistor on. At this point regeneration occurs; that is, the applied signal is amplified and returned to the base of $Q1$, as a much larger positive signal than originally applied. The original signal is amplified and inverted by $Q1$. The negative-going signal at the collector of $Q1$ is coupled through C to the base of $Q2$. The signal is amplified and inverted by $Q2$ and coupled back, as a positive-going signal, to the base of $Q1$ to drive this device into saturation. The circuit is now in the quasi-stable state. Base drive to keep $Q1$ saturated is furnished through R_{L2} and R_1 as long as $Q2$ remains off.

As the capacitor connected to the base of $Q2$ charges toward V_{CC} through R_B, it will ultimately reach a slightly positive value to turn $Q2$ on again. When $Q2$ turns on, the collector of this device drops and removes base drive to $Q1$. For the very short time that both transistors are in their active region, regeneration again occurs to drive the multivibrator back to the stable state.

For all but the most critical application, the equivalent circuits of Fig. 13.10 can be used to represent the transistor in the multivibrator. In fact, we will assume that $R_{sat} = 0$ and $V_{BE(on)} = 0$ for the saturated transistor. After considering an analysis based on these simple models, we will then extend the results to include nonideal effects. The equivalent circuit of Fig. 13.28 applies to the one-shot before the trigger pulse is applied. The capacitor has a voltage of V_{CC} across it with the polarity shown. When the positive trigger pulse is applied, transistor $Q1$ will saturate and $Q2$ will be turned off. Neglecting transistor switching times, the equivalent circuit is as shown in Fig. 13.29 instantaneously after switching.

Figure 13.28
Equivalent circuit for the stable state of the one-shot.

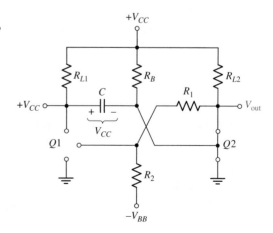

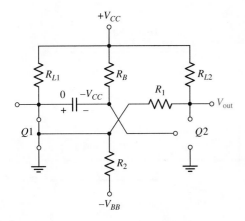

Figure 13.29
Equivalent circuit of one-shot instantaneously after application of trigger pulse.

The charge on C cannot change instantaneously; therefore, as the collector of $Q1$ switches from $+V_{CC}$ to ground, the base of $Q2$ is forced by the capacitor to switch from ground to $-V_{CC}$. Then $Q2$ shuts off, causing base drive to be supplied through R_{L2} and R_1 to $Q1$. The capacitor will charge from $-V_{CC}$ toward a voltage of $+V_{CC}$ through R_B, but when V_{B2} reaches zero volts $Q2$ will turn on. This forces $Q1$ off and is the point in time at which the circuit reverts to the original state. The length of time that $Q2$ remains in the off state is referred to as the period of the one-shot. The period can be found by calculating the time required for the capacitor to charge from $-V_{CC}$ to zero volts. The charging circuit for the capacitor is shown in Fig. 13.30.

From Eq. (13.7), the voltage on the capacitor can be expressed as

$$v_c = v_i + (v_t - v_i)\left(1 - e^{-t/R_B C}\right) \tag{13.8}$$

where the initial voltage is $-V_{CC}$ and the target voltage is $+V_{CC}$. Thus,

$$v_c = -V_{CC} + 2V_{CC}\left(1 - e^{-t/R_B C}\right)$$

Setting $v_c = 0$ and solving for the period T gives

$$T = R_B C \ln 2 = 0.69 R_B C \tag{13.9}$$

The output voltage during the period T is calculated as

$$V_{C2} = \frac{R_1}{R_1 + R_{L2}} V_{CC} \tag{13.10}$$

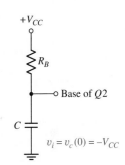

Figure 13.30
Charging circuit of timing capacitor.

At the end of the period, the capacitor has zero volts appearing across it. As $Q2$ switches on, this effectively ties the rightmost side of the capacitor to ground. The other side charges through R_{L1} to $+V_{CC}$, since the collector of $Q1$ now presents an open circuit to the capacitor. The collector voltage of $Q1$ then charges to $+V_{CC}$ governed by the recovery time constant $R_{L1}C$. From three to five of these time constants are allowed for recovery time, that is, before another trigger pulse is applied to the circuit. Waveforms at various points in the circuit are shown in Fig. 13.31.

Nonideal Effects: In the preceding analysis, very simple equivalent circuits were used, with $V_{BE(on)}$, R_{sat}, and $V_{CE(sat)}$ taken to be zero. These circuits do not reflect several aspects of transistor behavior that occasionally must be considered. The following items can cause

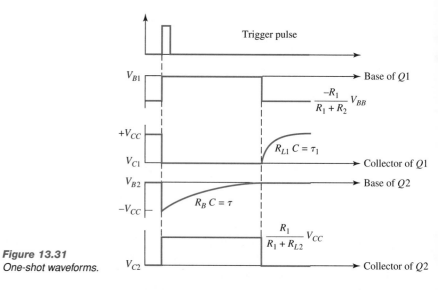

Figure 13.31
One-shot waveforms.

slight modifications in the design of the one-shot circuit:

1. $V_{BE(\text{on})}$: nonzero values of base-to-emitter voltage drop when the transistor is on.
2. $V_{CE(\text{sat})}$: nonzero values of collector-to-emitter saturation voltage. R_{sat} is generally negligible.
3. r_x: nonzero values of ohmic base resistance.
4. Switching times of transistors.
5. Breakdown voltage of reverse-biased base-emitter diode.

The presence of $V_{BE(\text{on})}$ requires that V_{B2} charge to a value of $V_{BE(\text{on})}$, rather than to zero volts, before $Q2$ will turn on to end the period. The finite saturation voltage affects the output voltage, since the lower level will now be $V_{CE(\text{sat})}$ rather than zero volts. Furthermore, the initial charge on the capacitor, instantaneously upon switching, will be $V_{CC} - V_{BE(\text{on})} - V_{CE(\text{sat})}$. The waveforms of Fig. 13.32 reflect these changes. The presence of $V_{CE(\text{sat})}$ and

Figure 13.32
Accurate one-shot waveforms.

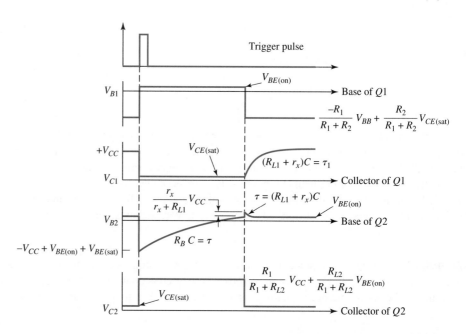

$V_{BE(on)}$ modify both collector and base waveforms slightly. These quantities become more important as the power-supply voltage becomes smaller.

It is left as an exercise for the student to verify that the expression for the period becomes

$$T = R_B C \ \ln \frac{2V_{CC} - V_{BE(on)} - V_{CE(sat)}}{V_{CC} - V_{BE(on)}} \qquad (13.11)$$

The discontinuity in the voltage V_{B2} at the end of the period occurs as $Q1$ switches off, giving the equivalent circuit shown in Fig. 13.33. The capacitor appears as a short circuit, and V_{B2} has an instantaneous value of

$$V_{CC} \frac{r_x}{R_{L1} + r_x} + V_{BE(on)} \qquad (13.12)$$

which decays exponentially to $V_{BE(on)}$.

In modern transistors made by integrated circuit processes, the reverse breakdown voltage from base to emitter may be relatively small. It is not unusual to find a device that has a reverse breakdown voltage of 3 V. When the one-shot initially switches to the state of $Q1$ on and $Q2$ off, the base voltage of $Q2$ is at its most negative value. If this value exceeds the breakdown voltage, the one-shot behavior is modified from the ideal theory. To avoid this undesired behavior, a diode can be inserted in series with the base or the emitter lead to absorb the excessive reverse bias on $Q2$.

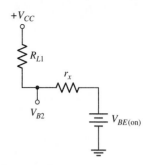

Figure 13.33
Equivalent circuit instantaneously after Q1 shuts off.

The power supply $-V_{BB}$ can be eliminated, since a silicon transistor requires 0.7 V from base to emitter in order to conduct significant current. It is unnecessary to apply a negative voltage to the base of $Q1$ to turn it off. Instead, the resistance R_2 can be eliminated or returned to ground and selected to cause a voltage of perhaps 0.3 V at the base of $Q1$ when $Q2$ is on, which will ensure that $Q1$ will be off until triggered by the input source. Generally, R_2 is eliminated and both load resistors are chosen to be equal.

Design Procedure: There are many degrees of freedom in designing a one-shot, but the following design procedure can be used in most cases.

1. Select V_{CC} and R_L.

2. Select R_B and C.

3. Select R_1 and R_2.

The selection of V_{CC} and R_L is based on the required output voltage swing and output impedance requirements. The value of R_B is selected to give $Q2$ the necessary base current, and C is selected to lead to the proper period T. It can easily be seen that Eq. (13.8) applies to this circuit also. Finally, R_1 and R_2 (if used) are selected to give $Q1$ the proper reverse bias when $Q1$ is off and to establish the necessary base drive when $Q1$ is on. A design example will demonstrate this procedure.

EXAMPLE 13.8

Design a one-shot multivibrator with a period of 1 ms and an output impedance of 5 kΩ or less. The voltage swing should be at least 9 V. Assume that $\beta_{(min)} = 50$, $V_{BE(on)} = 0$, and $V_{CE(sat)} = 0$.

SOLUTION

1. Select V_{CC} and R_L.
To ensure a 9-V swing, V_{CC} will be chosen as 10 V. A value of $R_L = 5$ kΩ will result in an output impedance of approximately 5 kΩ. The circuit configuration is shown in Fig. 13.34. Note that R_2 has been eliminated as well as the second dc power supply.

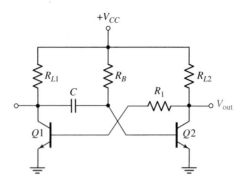

Figure 13.34
The one-shot circuit of Example 13.8.

2. **Select R_B and C.**
Since R_B must cause saturation of $Q2$, we will pick the current through R_B to be $1.5I_{B(\text{sat})}$, giving a 50% safety factor. The collector current for saturation is

$$I_{C(\text{sat})} = \frac{V_{CC}}{R_L} = \frac{10}{5\ \text{k}\Omega} = 2\ \text{mA}$$

therefore

$$I_{B(\text{sat})} = \frac{I_{C(\text{sat})}}{\beta_{(\text{min})}} = \frac{2\ \text{mA}}{50} = 40\ \mu\text{A}$$

We will allow 60 μA to flow through R_B. Since

$$I_B = \frac{V_{CC} - V_B}{R_B} = \frac{V_{CC} - 0}{R_B}$$

then

$$R_B = \frac{10}{60\ \mu\text{A}} = 167\ \text{k}\Omega$$

The period T must be 1 ms, so

$$C = \frac{T}{0.69 R_B} = \frac{10^{-3}}{0.69 \times 167 \times 10^3} = 0.0087\ \mu\text{F}$$

3. **Select R_1.**
This resistor is chosen in a manner similar to that of R_B. The current through R_1 when $Q2$ is off must be large enough to saturate $Q1$. Assuming that a base current of 60 μA is required (same as I_{B2}), we can write

$$I_{B1} = I_1 = 60\ \mu\text{A}$$

where I_1 is the current through R_1. By inspection

$$I_1 = \frac{V_{CC}}{R_{L2} + R_1} = \frac{10}{5 + R_1}$$

which gives

$$R_1 = 162\ \text{k}\Omega$$

Before we finish the design we must check to see whether the requirement of a 9-V swing will be met. The lower level of the output voltage has been found to be

$$V_{\text{min}} = 0\ \text{V}$$

The upper level is

$$V_{max} = 10 \, \frac{162}{162 + 5} = 9.7 \text{ V}$$

That is, when $Q2$ is shut off, the collector voltage swings to 9.7 V. The total net swing is

$$V_{swing} = V_{max} - V_{min} = 9.7 \text{ V} - 0 \text{ V} = 9.7 \text{ V}$$

The list of values is as follows:

$$V_{CC} = 10 \text{ V}$$

$$R_{L1} = R_{L2} = 5 \text{ k}\Omega$$

$$R_B = 167 \text{ k}\Omega$$

$$C = 0.0087 \, \mu\text{F}$$

$$R_1 = 162 \text{ k}\Omega$$

A Spice simulation of the circuit of Example 13.8 follows, along with a plot of the output voltages. A step voltage varying from 0 V to 5 V, applied through a 10-nF capacitor to the base of $Q1$, triggers the circuit. It could also be triggered by a negative-going pulse applied to the base of $Q2$.

```
EX13-8.CIR

R1 2 1 5K
R2 3 1 5K
R3 5 1 167K
R4 3 4 162K
C1 2 5 8700PF
C2 4 6 10NF
V1 1 0 10
V2 6 0 PWL(0 0 2U 5 2M 5)
Q1 2 4 0 N1
Q2 3 5 0 N1
.MODEL N1 NPN (BF=50)
.OP
.TRAN 1U 2M
.PROBE
.END
```

The output waveform of the one-shot is shown in Fig. 13.35.

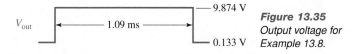

V_{out} ← 1.09 ms → — 9.874 V — 0.133 V

Figure 13.35
Output voltage for
Example 13.8.

Retriggerable and Nonretriggerable One-Shots

There are two types of one-shot that find heavy application in the digital area. One is called a retriggerable monostable and the other is the nonretriggerable type. In the nonretriggerable circuit a second trigger pulse, occurring during the period, is ignored by the one-shot. No effect on the

13.11 In the one-shot circuit of Fig. 13.27, $R_B = 100\,\text{k}\Omega$ and $C = 0.002\,\mu\text{F}$. If $V_{CC} = 6\,\text{V}$, calculate the periods for the two cases of $V_{BE(\text{on})} = 0\,\text{V}$ and $V_{BE(\text{on})} = 0.7\,\text{V}$. Assume that $V_{CE(\text{sat})} = 0\,\text{V}$. *Ans:* 138 μs, 151 μs.

13.12 Calculate the period of the one-shot of Fig. 13.27 for $R_B = 100\,\text{k}\Omega$, $C = 0.002\,\mu\text{F}$, $V_{CC} = 8\,\text{V}$, and $V_{BE(\text{on})} = V_{CE(\text{sat})} = 0\,\text{V}$. Assume that R_B is connected to a voltage of 14 V rather than to V_{CC}. *Ans:* 90 μs.

period is noticeable, and the circuit cannot be triggered again until the period and recovery time are completed. The discrete circuit monostable discussed earlier in this section is nonretriggerable.

A retriggerable monostable references the conclusion of the period to the input trigger pulse. For example, if the circuit is triggered to start the period and a second trigger pulse is applied at $t = 0.6T$, the output of the one-shot will last $1.6T$ rather than T. When a trigger pulse occurs, the period will extend a full period from that point even if an earlier period has not been completed.

13.3.2 THE ASTABLE MULTIVIBRATOR

Figure 13.36 shows a conventional astable multivibrator. It can be seen that both $Q1$ and $Q2$ can be saturated simultaneously with the base drive furnished by R_1 and R_2. Of course, this is not the mode of operation desired for the astable. The normal mode of operation is present when only one transistor is on at a given time. When this transistor switches off, the other transistor is forced on. We will see later that special circuits can guarantee the existence of this mode, but for the present time let us assume that $Q1$ is off and $Q2$ is on. Let us further assume that at time $t = 0$, $Q1$ switches on. It is at this point that we will start the analysis.

Figure 13.36
Conventional astable multivibrator.

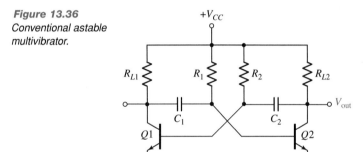

Prior to the turning on of $Q1$, the capacitor C_1 has a voltage of V_{CC} across it, as shown in Fig. 13.37. Just as in the case of the one-shot, the transition of V_{C1} from V_{CC} to zero volts causes the base voltage of $Q2$ to swing from zero to $-V_{CC}$ to preserve the charge on C_1. When $Q2$ is turned off, V_{C2} rises to $+V_{CC}$. Device $Q2$ remains off until C_1 charges to zero volts. This time is

$$t_1 = 0.69R_1C_1 \tag{13.13}$$

Figure 13.37
Instantaneous voltage on C_1 as $Q1$ turns on.

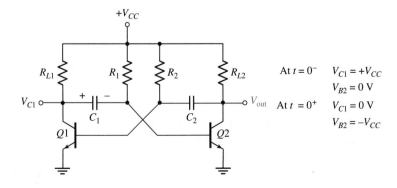

At $t = 0^-$ $V_{C1} = +V_{CC}$
            $V_{B2} = 0\,\text{V}$
At $t = 0^+$ $V_{C1} = 0\,\text{V}$
            $V_{B2} = -V_{CC}$

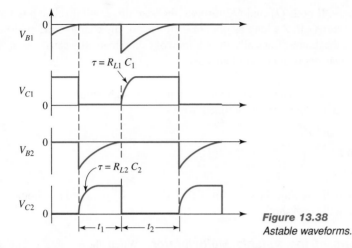

Figure 13.38
Astable waveforms.

At the end of t_1, $Q2$ turns on, dropping V_{C2} from $+V_{CC}$ to zero volts. This forces $Q1$ off until C_2 can charge from $-V_{CC}$ to zero volts. Transistor $Q1$ will be off for a time

$$t_2 = 0.69 R_2 C_2 \qquad (13.14)$$

At the end of time t_2, $Q1$ again turns on and starts the cycle once more. The cycle is seen to continue indefinitely once initiated. The total period of the astable is

$$T = t_1 + t_2 = 0.69(R_1 C_1 + R_2 C_2) \qquad (13.15)$$

Note that there is a recovery time associated with both collectors for the astable circuit. The waveforms are shown in Fig. 13.38 with exaggerated recovery times.

Methods of Starting the Astable: If the astable circuit is constructed as shown in Fig. 13.36, the application of the power-supply voltage does not guarantee that the circuit will oscillate. When the circuit does not oscillate, removal and reapplication of the power-supply voltage will usually start the oscillation. There are applications of the astable, such as in digital computer systems, where sure-starting is critical. Figure 13.39 illustrates a sure-starting astable. When V_{CC} is first applied, both sides of C will be at V_{CC} volts, since

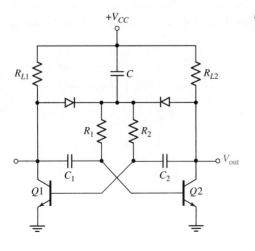

Figure 13.39
Sure-starting astable.

C has no charge. If both $Q1$ and $Q2$ turn on, the base drive will discharge C until one of the transistors turns off. As long as one transistor is off, the charge on C will be restored through one of the diodes. Basically, the addition of C automatically causes removal of base drive if the transistors turn on simultaneously.

Design Procedure: The procedure for designing an astable multivibrator is simpler than that for a one-shot. The steps are outlined here:

1. Choose V_{CC} and R_L.

2. Pick R_1 and R_2.

3. Pick C_1 and C_2.

The extension to the nonideal case, considering $V_{CE(\text{sat})}$, $V_{BE(\text{on})}$, and r_x, should be obvious after examining the one-shot case.

Modifications of the Astable Multivibrator: When the astable changes state, a reverse-bias voltage of $-V_{CC}$ appears at the base of the transistor that turns off. In some instances, the reverse breakdown voltage of the emitter-base junction is less than $-V_{CC}$. Although the breakdown is nondestructive if the circuit limits the base current to a reasonable level, the current flow through this junction can very drastically affect the period of the astable (or one-shot). Just as in the one-shot circuit, this problem can be overcome by adding a diode in series with each emitter. The reverse-bias voltage now appears across both the diode and the emitter-base junction. The total breakdown voltage is equal to the sum of the breakdown voltage of the diode and that of the junction.

There are applications of the astable or one-shot wherein it is desirable to vary the period in accordance with an input voltage. A popular approach to this problem is shown in Fig. 13.40. By connecting R_1 and R_2 to a variable voltage, the target voltage for the capacitors can be controlled. The total period becomes

$$T = (R_1 C_1 + R_2 C_2) \ln \left(1 + \frac{V_{CC}}{v_t}\right) \tag{13.16}$$

Since $\ln(1 + x) \approx x$ for small values of x, Eq. (13.16) can be written as

$$T = (R_1 C_1 + R_2 C_2) \frac{V_{CC}}{v_t} \tag{13.17}$$

when $v_t \gg V_{CC}$. For this case, the period varies inversely with the control voltage. The lower limit on v_t for the circuit of Fig. 13.40 is the voltage that will cause enough base

PRACTICE Problems

13.13 Select values of C_1 and C_2 in Fig. 13.36 to create values of $t_1 = 50\ \mu s$ and $t_2 = 60\ \mu s$. Assume that $R_1 = R_2 = 80\ k\Omega$. *Ans:* $C_1 = 906$ pF, $C_2 = 1090$ pF.

13.14 Repeat Practice Problem 13.13 if $V_{CC} = 5$ V and $V_{BE(\text{on})} = 0.7$ V. Assume that $V_{CE(\text{sat})} = 0.2$ V. *Hint:* Equation (13.11) can be applied here. *Ans:* $C_1 = 834$ pF, $C_2 = 1000$ pF.

13.15 Using the astable of Practice Problem 13.13 with $V_{CC} = 10$ V, calculate the voltage to which the timing resistors must be connected to result in periods of $t_1 = 25\ \mu s$ and $t_2 = 30\ \mu s$. *Ans:* 24.3 V.

Figure 13.40
Variable-period astable.

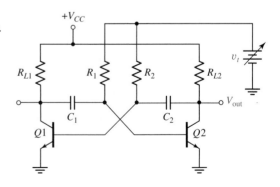

current through R_1 and R_2 to saturate the transistors. For values of v_t that are not much larger than V_{CC}, Eq. (13.16) can be applied to find the change in period.

PRACTICAL Considerations

The simple BJT multivibrators can be used at quite high frequencies of operation. Carefully designed astable circuits can reach into the 100-MHz range. One-shot periods in the nanosecond range are also possible.

Although the next device to be considered, the 555 timer, can be applied with fewer design considerations, this device is limited in high-frequency operation. One-shot periods in the microsecond range and astable frequencies in the tens of kHz are more reasonable for the 555 circuits.

13.4 The 555 Timer

IMPORTANT Concepts

1. External components can be connected to the 555 timer chip to create a monostable multivibrator. A different configuration results in an astable multivibrator.
2. A modified configuration must be used to achieve a 50% duty cycle in the astable output waveform.

In the last few decades, very few IC chips have been as popular as the 555 timer. This chip is used in a variety of timing applications. One important use of the 555 is that of a monostable multivibrator; another is that of an astable multivibrator. This section will consider the 555 monostable circuit.

13.4.1 THE 555 ONE-SHOT

A block diagram of the 555 is shown in Fig. 13.41. The three equal values of resistance establish a reference voltage of $V_{CC}/3$ at the top of the lower resistance and a voltage of $2V_{CC}/3$ at the top of the center resistor. The trigger comparator compares the voltage level of the trigger input to the reference voltage $V_{CC}/3$. As long as the input voltage is greater than $V_{CC}/3$, the comparator output is not asserted. When the input signal drops below $V_{CC}/3$, the comparator output is asserted and drives the set-reset or S-R flip-flop to the $Q = 0$ state.

If the trigger input voltage exceeds $V_{CC}/3$ and the threshold input voltage is below $2V_{CC}/3$, neither comparator is asserted. When the threshold input voltage increases above $2V_{CC}/3$, this comparator output is asserted and sets the flip-flop to the $Q = 1$ state. If both comparators are asserted simultaneously, requiring that the trigger voltage is less than $V_{CC}/3$ and the threshold voltage is greater than $2V_{CC}/3$, the reset input overrides the set and forces the flip-flop to the $Q = 0$ state. This is called a *reset-overrides-set flip-flop* for obvious reasons.

The output buffer inverts the flip-flop output to cause a high-level output when the flip-flop is reset and a low output when the flip-flop is set. When the flip-flop is set, it provides base drive current to the discharge transistor, which allows this transistor to saturate when

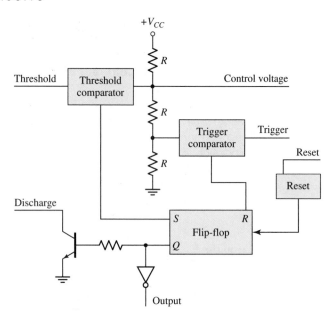

Figure 13.41
The 555 timer.

the collector circuit is completed. There is also a separate reset input to the 555 circuit. Asserting this input also saturates the discharge transistor and causes the output to be at the low level.

The circuit of Fig. 13.42 shows the connections required to convert the 555 timer into a one-shot. The small circles on the trigger and reset inputs indicate that these inputs must be driven low to assert the corresponding circuits. The circle on the discharge output indicates that this output is driven low when the discharge transistor is asserted or turned on. The $0.01\text{-}\mu F$ capacitor is connected to the control voltage input to hold this voltage constant even if a transient voltage appears on the power supply. The normal state of the flip-flop is the set state that saturates the discharge transistor, which places a short circuit across the capacitor holding the capacitor voltage to 0 V. Since the threshold input is tied to the capacitor, the threshold comparator is not asserted. The trigger input is held at a high level until the output pulse is to be initiated.

When the trigger input drops below $V_{CC}/3$ for a short duration, the flip-flop changes state. The output is now at the high level and the discharge transistor is turned off, becoming an open circuit. The capacitor can now charge through R_A toward the voltage V_{CC}. As the capacitor charges, the threshold voltage reaches a value of $2V_{CC}/3$, causing the flip-flop to

Figure 13.42
The 555 one-shot circuit.

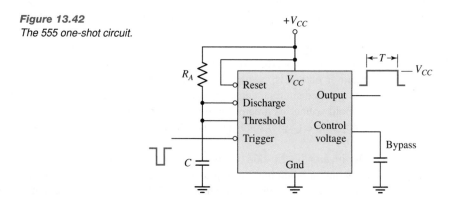

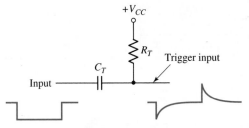

Figure 13.43
Triggering the 555 one-shot.

13.16 Select values for the one-shot of Fig. 13.42 to create a 1-ms period.
Ans: $R_B C = 0.909$ ms.
13.17 Design a trigger circuit for the one-shot of Practice Problem 13.16.

return to the set state. This ends the period, dropping the output to 0 V and again saturating the output transistor. The period T is the time taken for the capacitor to charge from 0 V to a voltage of $2V_{CC}/3$ with a target voltage of V_{CC}. This time can be calculated from the general equation for a charging capacitor given earlier; that is,

$$v(t) = v_i + (v_t - v_i)(1 - e^{-t/\tau}) \tag{13.7}$$

For the case under consideration $v_i = 0$, $v_t = V_{CC}$, $v(T) = 2V_{CC}/3$, and $\tau = R_A C$. The period T is then found by solving Eq. (13.7), with these values giving

$$T = R_A C \ln 3 = 1.1 R_A C \tag{13.18}$$

The trigger pulse width must be less than T for Eq. (13.18) to be valid. If the trigger input remains low after the period is over, the output remains high because the reset of the flip-flop overrides the set. When the input trigger pulse is longer than T, an RC differentiator should be used to ensure that the pulse reaching the trigger terminal is short enough. The resistor connects from V_{CC} to the trigger input, and the capacitor is inserted between the input line and this point as shown in Fig. 13.43. The trigger input is then held positive until the leading edge of the trigger pulse arrives. This transition is coupled through the capacitor to initiate the period. The trigger input then charges toward V_{CC} with a time constant largely determined by the $R_T C_T$ product of the trigger circuit, since the input impedance to the trigger comparator is very high. This $R_T C_T$ product should be considerably smaller than the one-shot period.

The period T can be varied from a minimum of approximately 10 μs to a maximum of several minutes by varying the elements R_A and C. Values of $R_A = 10$ kΩ and $C = 0.001$ μF lead to $T = 11$ μs, whereas $R_A = 10$ MΩ and $C = 100$ μF result in $T = 1100$ s. The 555 timer is popular for lower frequency applications, but it cannot be used when smaller rise and fall times are required.

The 555 timer is designed to operate with a power-supply voltage ranging from 5 V to 18 V. The magnitude of the output voltage varies directly with the power supply, allowing the 555 timer to be used in a variety of applications.

13.4.2 THE 555 ASTABLE

The 555 timer can also be used for an astable with the connections shown in Fig. 13.44. The capacitor voltage can swing between $V_{CC}/3$ and $2V_{CC}/3$. As the capacitor voltage rises toward $2V_{CC}/3$, it does so with a target voltage of V_{CC} and a time constant of $C(R_A + R_B)$. When it reaches $2V_{CC}/3$, the threshold comparator causes the flip-flop to change state (see Fig. 13.41). The discharge transistor saturates, and the capacitor voltage heads toward ground with a time constant of $C R_B$. When the voltage drops to $V_{CC}/3$, the trigger

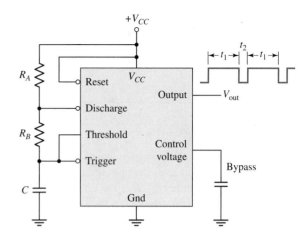

Figure 13.44
The 555 astable circuit.

comparator changes the state of the flip-flop and shuts off the discharge transistor. The capacitor again charges toward V_{CC}. The output and capacitor voltage waveforms of the 555 are shown in Fig. 13.45.

The duration of the positive portion of the waveform is called t_1 and is calculated from Eq. (13.7) with $v_i = V_{CC}/3$, $v_t = V_{CC}$, and $\tau = (R_A + R_B)C$. We then get

$$v_c = \frac{2V_{CC}}{3} = \frac{V_{CC}}{3} + \left[V_{CC} - \frac{V_{CC}}{3}\right]\left[1 - e^{-t_1/\tau}\right]$$

which can be solved for t_1 to yield

$$t_1 = C(R_A + R_B)\ln 2 = 0.69C(R_A + R_B) \tag{13.19}$$

Likewise, the time t_2 can be found to be

$$t_2 = 0.69CR_B \tag{13.20}$$

The duty cycle of the astable is defined as the ratio of the duration of the positive portion of the period to the total period, or

$$\text{Duty cycle} = \frac{t_1}{t_1 + t_2} = \frac{R_A + R_B}{R_A + 2R_B} \tag{13.21}$$

Figure 13.45
Astable waveforms.

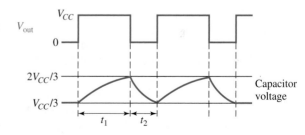

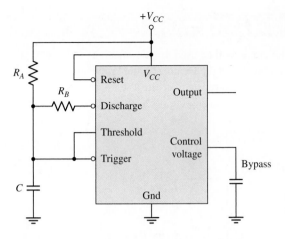

Figure 13.46
An astable circuit capable
of a 50% duty cycle.

If the value of R_B becomes much greater than R_A, the duty cycle will approach a minimum value of 50%.

Some manufacturers give an alternative definition of duty cycle as the ratio of the time the output transistor is on (low voltage) to the total period. This definition leads to

$$\text{ON duty cycle} = \frac{t_2}{t_1 + t_2} = \frac{R_B}{R_A + 2R_B} \tag{13.22}$$

Another configuration used to obtain a 50% or even smaller duty cycle, with a smaller spread of timing resistor values, is shown in Fig. 13.46. The value of t_1 is found by noting that the capacitor charges through R_A only, rather than $R_A + R_B$. This value is then

$$t_1 = 0.69CR_A \tag{13.23}$$

During t_2 the capacitor discharges from $2V_{CC}/3$ to $V_{CC}/3$. The target voltage and discharge resistance can be calculated from the circuit of Fig. 13.47. (Note that the discharge transistor is saturated.)

The Thévenin equivalent voltage, which is also the target voltage, is

$$V_{\text{Th}} = \frac{R_B V_{CC}}{R_A + R_B}$$

and the discharge resistance is

$$R_{\text{Th}} = \frac{R_A R_B}{R_A + R_B}$$

Using Eq. (13.7) with $v_i = 2V_{CC}/3$, $v_t = V_{CC}R_B/(R_A + R_B)$, $\tau = R_{\text{Th}}C$, and solving for the time for the capacitor voltage to reach $V_{CC}/3$ results in

$$t_2 = R_{\text{Th}}C \ \ln \frac{2R_A - R_B}{R_A - 2R_B} \tag{13.24}$$

The resistor R_B can be adjusted to result in a t_2 that equals t_1. In using Eq. (13.24), we must limit the value of R_B to lead to a target voltage of less than $V_{CC}/3$ or the circuit will not oscillate. Thus there is an upper limit on R_B of $R_A/2$.

PRACTICE Problems

13.18 Select values for the astable of Fig. 13.44 to result in $t_1 = 80\ \mu\text{s}$ and $t_2 = 52\ \mu\text{s}$. Use $C = 3$ nF. *Ans:* $R_B = 25$ kΩ, $R_A = 13.5$ kΩ.

13.19 Select values for the astable of Fig. 13.46 to result in equal half-periods of 50 μs. Use $C = 4$ nF. *Ans:* $R_A = 18.1$ kΩ, $R_B = 7.68$ kΩ.

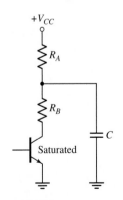

Figure 13.47
Equivalent discharge circuit of
the astable of Fig. 13.46.

In order to produce a fixed-length pulse, a one-shot circuit will be used. In this situation, a 555 timer, connected as a one-shot, can satisfy the requirements of the Demonstration Problem.

The triangular output of the light sensor must first be squared up to create a sharper trigger pulse. One method that can produce a square pulse while inverting the polarity of the trigger signal is indicated in Fig. 13.48.

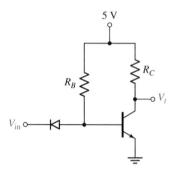

Figure 13.48
An inverting trigger circuit.

We will assume that $\beta = 150$, $V_{BE(on)} = 0.7$ V, $V_{CE(sat)} = 0.2$ V, and a forward voltage drop of 0.6 V for the diode. We recognize that the output of this trigger circuit must drive the trigger input of a 555 timer.

The 555 circuit has a high trigger-input impedance, listed in the MΩ range in the specifications. We will choose a much smaller collector resistance of $R_C = 4$ kΩ to avoid loading problems. Next, R_B will be calculated to ensure saturation of the transistor when the input voltage goes positive. When the input signal is at zero volts, the diode conducts and robs base current from the transistor. The transistor will be off, conducting no collector current. If R_B is much larger than the 2-kΩ output impedance of the sensor, this output impedance will be negligible.

When the input voltage goes positive, the anode of the diode and the base of the transistor remain 0.6 V more positive than the input voltage. When the input voltage increases above 0.1 V, the current through R_B begins to flow into the base of the transistor. As V_{in} increases further, the diode becomes reverse biased, and the current through R_B flows entirely into the base of the transistor.

To guarantee saturation, the base current is set to $1.5 I_{B(sat)}$. The collector saturation current is found to be

$$I_{C(sat)} = \frac{V_{CC} - V_{CE(sat)}}{R_C} = \frac{4.8}{4} = 1.2 \text{ mA}$$

The base current required for saturation is then

$$I_{B(sat)} = \frac{I_{C(sat)}}{\beta} = \frac{1.2}{150} = 8 \text{ }\mu\text{A}$$

We will choose R_B to produce 12 μA of base current. The value of R_B is then

$$R_B = \frac{V_{CC} - V_{BE(on)}}{I_B} = \frac{4.3}{12} = 358 \text{ k}\Omega$$

When V_{in} goes positive, V_t switches from 5 V to 0.2 V. This signal is used to trigger the following one-shot. The complete circuit is shown in Fig. 13.49.

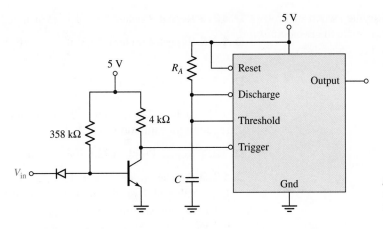

Figure 13.49
*One-shot circuit for
Demonstration Problem.*

We must select the RC time constant to satisfy the requirement of $T = 0.3$ ms. The relation between the pulse duration or period of the one-shot and RC is expressed by Eq. (13.18) as

$$T = 1.1 R_A C \tag{13.18}$$

Choosing $C = 10$ nF gives a value of

$$R_A = \frac{T}{1.1 \times C} = \frac{3 \times 10^{-4}}{1.1 \times 10^{-8}} = 27.3 \text{ k}\Omega$$

SUMMARY

➤ When used in digital circuits, the BJT and the MOSFET emulate electronically controlled switches.

➤ In such applications, these devices are either in the cutoff region, conducting no current, or in the high-conductance region, which is the saturation region for the BJT and the triode region for the MOSFET.

➤ The one-shot multivibrator is used in applications that require an output pulse with a fixed duration, initiated by an input pulse.

➤ The astable multivibrator generates a rectangular wave at the output with a period and duty cycle that are determined by element values.

➤ The 555 timer can be used for an astable or a one-shot multivibrator.

PROBLEMS

SECTION 13.1 SWITCHES

13.1 A switch has an open or off resistance of 10 MΩ and a closed or on resistance of 100 Ω. Calculate the two voltage levels of V_{out} for the circuit shown. Assume that $R_L = 5$ kΩ.

Figure P13.1

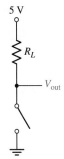

13.2 If two switches are identical to the switch of Problem 13.1 and are used as shown, calculate the two output voltage levels. Assume that $S1$ and $S2$ must always be in opposite or complementary states.

Figure P13.2

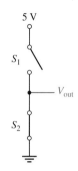

5 V

S_1

V_{out}

S_2

D 13.3 The circuit of Fig. 13.11 has a 4-kΩ load resistance and a power supply of $V_1 = 20$ V. If $V_{BE(\text{on})} = 0.6$ V and $\beta = 90$ for the transistor, calculate the maximum value of R_B to drive the transistor to saturation if $V_{\text{in}} = 6$ V.

D 13.4 Repeat Problem 13.3 if V_1 is changed to 10 V.

D 13.5 Repeat Problem 13.3 if $V_1 = 10$ V and $R_L = 6$ kΩ.

D 13.6 In the circuit of Fig. 13.11, the transistor has values of $V_{BE(\text{on})} = 0.6$ V and $\beta = 90$. If $R_B = 40$ kΩ, $V_{\text{in}} = 6$ V, and $V_1 = 20$ V, what is the minimum value of R_L required for saturation?

D 13.7 Repeat Problem 13.6 if $\beta = 130$.

D 13.8 For the circuit of Fig. 13.14(a), $V_{\text{in}} = V_1 = 5$ V. If the value of R_{dsp} for the pMOS device is 100 MΩ, what is the maximum value of R_{dsn} to ensure that $V_{\text{out}} \leq 0.2$ V?

D 13.9 Repeat Problem 13.8 to ensure that $V_{\text{out}} \leq 0.1$ V.

SECTION 13.2 USING SEMICONDUCTOR DEVICES IN SWITCHING CIRCUITS

13.10 Rework Example 13.5 if R_L is changed to 5 kΩ. How is the fall time of the output voltage affected by R_L?

13.11 Rework Example 13.7 if the on resistances are $R_{dsn} = 8$ kΩ and $R_{dsp} = 8$ kΩ.

13.12 In the triode region, the p device has a value of $R_{dsp} = 10$ kΩ and the n-device triode resistance is $R_{dsn} = 6$ kΩ. In cutoff, these resistances become infinite. Prior to switching, the circuit has reached a steady state.

(a) After the input goes positive from 0 V to 5 V, how long does it take V_{out} to reach 1 V?

(b) After the input goes negative from 5 V to 0 V, how long does it take V_{out} to reach 3.5 V?

Figure P13.12

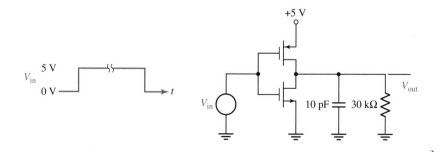

13.13 Assume that the off resistances of both devices are infinite and the on resistances are $R_{dsp} = 8$ kΩ and $R_{dsn} = 4$ kΩ. The circuit is in the steady state before either input is switched. Plot the output voltage waveform, showing times and voltage levels at switching times.

Figure P13.13

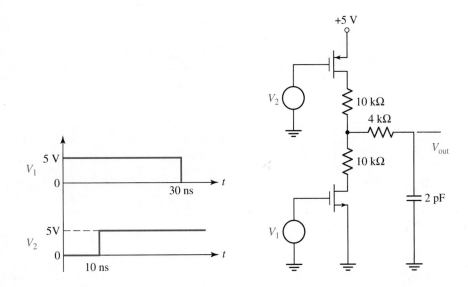

13.14 The p device has an on resistance of $R_{dsp} = 8$ kΩ and an off resistance of 100 MΩ. For the n device, the on resistance is 6 kΩ and the off resistance is 180 MΩ.

(a) Sketch the output voltage.
(b) Calculate the output voltage just prior to switching.
(c) Calculate the output voltage at $t = 200$ ns.
(d) Calculate the output voltage at $t = 1$ s.

Figure P13.14

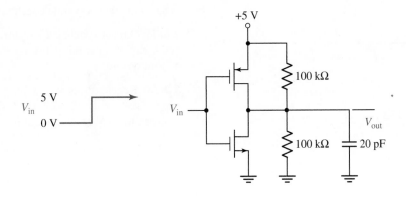

13.15 The input voltage was at 0 V until steady state was achieved in the circuit. The base drive is sufficient to saturate the transistor when $V_{in} = 6$ V.

(a) Plot the output voltage as a function of time.

(b) What is V_{out} just prior to the time that V_{in} returns to 0 V?

(c) What is V_{out} at $t = 30$ ns?

Figure P13.15

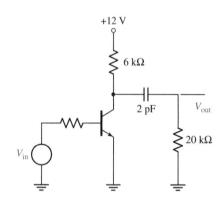
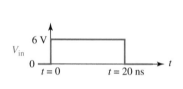

SECTION 13.3.1 THE MONOSTABLE MULTIVIBRATOR

13.16 Assume that $\beta = 100$, $V_{BE(on)} = 0.7$ V, and $V_{CE(sat)} = 0.2$ V. The diode requires 0.7 V to conduct current in the forward direction.

(a) Instantaneously after switching to $Q1$ on and $Q2$ off, what is the voltage V_{B2}?

(b) What is the target voltage for V_{B2}?

(c) Calculate the period of the one-shot.

Figure P13.16

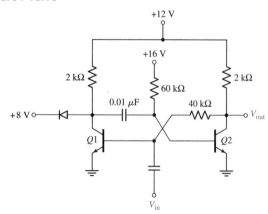

D 13.17 The one-shot multivibrator of Fig. 13.27 uses matched transistors with $\beta = 60$, $V_{BE(on)} = 0.7$ V, and $V_{CE(sat)} = 0.2$ V. The following element values are employed: $V_{CC} = 12$ V, $V_{BB} = 6$ V, $R_{L1} = R_{L2} = 3$ kΩ, and $R_B = 150$ kΩ. Choose values of C, R_1, and R_2 to lead to a 1-ms period. The output voltage swing should be at least 11 V.

D 13.18 Design a one-shot circuit using a single 8-V power supply and matched transistors with $\beta = 200$, $V_{CE(sat)} = 0.2$ V, and $V_{BE(on)} = 0.7$ V. The period should be 50 μs. The output impedance should be 1 kΩ or less. What is the recovery time constant of the circuit?

D 13.19 Repeat Problem 13.18 for a 12-V power supply and a required period of 100 μs. Do a Spice simulation of your design.

D 13.20 Repeat Problem 13.18 given that the power-supply voltage is 12 V, but the timing resistor R_B is returned to 24 V. Do a Spice simulation of your design.

D 13.21 Design a one-shot circuit using the transistors and power supply of Problem 13.18. The timing resistor should be returned to the wiper of a potentiometer, as shown in the circuit, and should result in a period that can be varied from 20 μs to 40 μs.

D 13.22 Assume matched transistors with $V_{CE(sat)} = 0.2$ V, and $V_{BE(on)} = 0.7$ V. Calculate V_1 to result in a period of $T = 150$ μs.

Figure P13.21

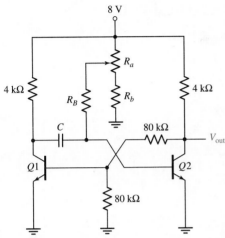

Figure P13.22

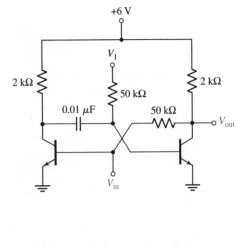

SECTION 13.3.2 THE ASTABLE MULTIVIBRATOR

13.23 Assume that $V_{BE(on)} = V_{CE(sat)} = 0$ V. If t_1 is defined as the time that $Q1$ is on and t_2 is defined as the time $Q2$ is on, calculate the voltages V_1 and V_2 to result in $t_1 = 0.2$ ms and $t_2 = 1$ ms.

13.26 Repeat Problem 13.25, given that $R_1 = R_2 = 400$ kΩ and v_t varies from 60 V to 80 V. Use both Eqs. (13.16) and (13.17) to calculate T and plot the results on the same graph.

D 13.27 The astable circuit of Fig. 13.40 uses transistors with $V_{BE(on)} = 0.6$ V and $\beta_{min} = 120$. Given that $v_t = 30$ V and $V_{CC} = 12$ V, select reasonable element values to give a symmetrical total period of 1 ms with an output impedance of 3 kΩ or less. Simulate your design with Spice.

Figure P13.23

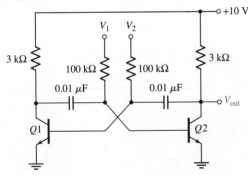

D 13.24 Using the transistors and power supply of Problem 13.18, design an astable multivibrator with $t_1 = 40$ μs and $t_2 = 60$ μs. The output impedance should be 4 kΩ or less.

13.25 In the astable circuit of Fig. 13.40, $V_{CC} = 12$ V, $R_{L1} = R_{L2} = 4$ kΩ, $R_1 = R_2 = 80$ kΩ, and $C_1 = C_2 = 0.01$ μF. Plot the total period as a function of v_t as this value varies from 6 V to 24 V.

SECTION 13.4.1 THE 555 ONE-SHOT

13.28 If the one-shot is triggered at $t = 0$, calculate the period T.

Figure P13.28

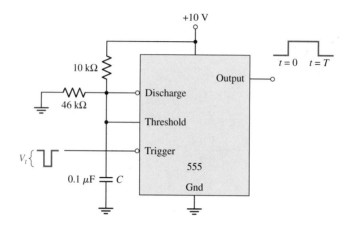

13.29 What is the minimum value of V_t required to trigger the one-shot of Problem 13.28? Select the capacitor C to lead to a 10-ms period.

D 13.30 Select R_A and C in Fig. 13.42 to give the one-shot a period of 2.4 ms.

D 13.31 Select R_A and C in Fig. 13.42 to give the one-shot a period of 120 μs.

D 13.32 A trigger pulse of period 2.4 ms must initiate a 1.1-ms period for the one-shot of Fig. 13.42. Select proper values for the differentiator circuit of Fig. 13.43.

D 13.33 A pulse of 0 V to 5 V with a period of 8 ms should initiate a 2-ms pulse on its trailing edge. Design a 555 one-shot and trigger circuit to accomplish this.

SECTION 13.4.2 THE 555 ASTABLE

13.34 Calculate t_1 and t_2 for the circuit shown.

Figure P13.34

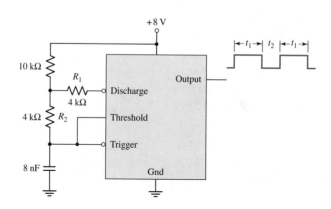

13.35 Calculate t_1 and t_2 for the circuit of Problem 13.34 if R_1 is changed to 10 kΩ.

13.36 Calculate t_1 and t_2 for the circuit of Problem 13.34 if R_2 is changed to 1.5 kΩ.

D 13.37 Select R_A, C, and R_B in the astable circuit of Fig. 13.46 to result in a total period of 1 ms with a 50% duty cycle.

D 13.38 Select R_A, R_B, and C for the 555 astable of Fig. 13.44 to result in $t_1 = 200\,\mu s$ and $t_2 = 120\,\mu s$.

D 13.39 Repeat Problem 13.38 for $t_1 = 22$ ms and $t_2 = 16$ ms.

D 13.40 Select R_A, R_B, and C for the astable of Fig. 13.46 to lead to $t_1 = t_2 = 80$ ms.

14 | *Basic CMOS Logic Circuits*

Essentially every personal computer now made is based on CMOS or closely related logic circuits. Computer chips are fabricated with hundreds of millions of MOSFET devices on each chip. These devices now operate at high switching speeds and with very little power consumption. For these reasons CMOS logic circuits have become the logic family of choice for computers and other digital systems. Consequently, BJT logic circuits such as transistor-transistor logic (TTL) and emitter-coupled logic (ECL) are not discussed in this textbook, even though they are still used in some applications.

DEMONSTRATION PROBLEM

Implement the logic expression

$$X = \bar{A}\bar{B} + CD$$

using NOR/NAND logic. Draw the device-level schematic for this realization. Implement this same expression minimizing the number of devices used.

It is necessary to understand how to implement NOR and NAND gates using CMOS logic to complete the first part of this demonstration problem. The second implementation requires a knowledge of minimization techniques as applied to CMOS logic circuits.

14.1 The CMOS Inverter

IMPORTANT Concepts

1. The simple CMOS inverter forms the basis of CMOS logic gates such as the NAND gate and the NOR gate.
2. A mathematical analysis of the inverter can be used in applications requiring more accuracy than the simple model.

The CMOS switch discussed in Chapter 13 is often called an inverter. This circuit is shown in Fig. 14.1. The CMOS inverter is extremely important because it forms the basis of most gates in the CMOS logic family. The operation of this building block should be understood before proceeding to a study of the more complex logic circuits. In addition, the inverter can be used as an amplifier of analog signals with proper biasing. This application will not be considered in this chapter.

In the inverter, the drains of the p device and the n device are connected together as are the gates. We assume that the input logic signal swings from zero volts to V_{DD} volts; that is, it is a rail-to-rail signal. We will first analyze the low-frequency behavior of the inverter before considering the transient behavior. For low-frequency behavior, the capacitors associated with the MOS devices can be considered to be open circuits.

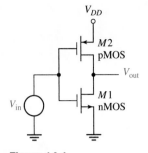

Figure 14.1
The basic CMOS inverter.

14.1.1 LOW-FREQUENCY BEHAVIOR OF THE INVERTER

An initial overview of the inverter will aid the more accurate analysis presented in succeeding paragraphs. The output voltage is taken at the junction of the two drain terminals. We note that the drain currents in both devices will always be equal in magnitude when no load is connected to the output. If either one of the two devices is in the off state, no current flows in the output circuit.

When the input signal is at zero volts, the n device will be off, presenting a very high impedance between drain and source of this device. The p device has V_{DD}, perhaps 5 V, connected to its source while the gate is at zero volts, which results in a value of $V_{GSp} = -5$ V. The p device will be in the conducting region. A finite resistance will be presented by the p device between its drain and source terminals. Without having any further information, the output voltage can be found to be at 5 V. This conclusion can be reached easily from the equivalent circuit of Fig. 14.2.

The output circuit appears as a voltage divider, with R_{dsp} having some finite value since this device is in the conducting region. As we shall later see, the p device is in its triode region for this input condition. The value of R_{dsn} can be taken as infinite. The output voltage is then

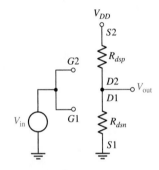

Figure 14.2
Equivalent output circuit.

$$V_{out} = V_{DD}\frac{R_{dsn}}{R_{dsp} + R_{dsn}} = V_{DD} \tag{14.1}$$

When the input voltage changes to V_{DD}, the situation is reversed in that the n device turns on and enters its triode region while the p device turns off. This situation leads to a value of R_{dsn} being finite and a value of R_{dsp} that is essentially infinite. Using the voltage division circuit of Fig. 14.2 with these resistance values results in an output voltage of

$$V_{out} = V_{DD}\frac{R_{dsn}}{R_{dsp} + R_{dsn}} = 0 \tag{14.2}$$

We can now return to the question of which region the conducting transistor is in for these input voltages. When the input is at zero volts, the output is at V_{DD} volts; thus the drain of the conducting p device is also at V_{DD} volts. Both source and drain voltages of the p device are equal and of such a value that both ends of the channel have electrons induced as a result of these voltages. The conducting or p device is then in the triode region. The same reasoning can be applied to the n device when $V_{in} = V_{DD}$ to conclude that this device is also in the triode region.

14.1.2 GRAPHICAL ANALYSIS OF THE INVERTER

One method used to analyze the inverter is similar to the load line approach used in earlier chapters. The inverter is partitioned into two sections. One section is the n device, with its source connected to ground. The other section consists of the p device in series with the power supply. These two sections are shown in Fig. 14.3.

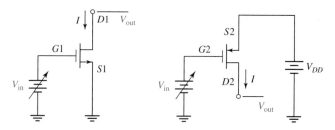

Figure 14.3
A partitioned inverter.

The characteristic curves of the n device are shown in Fig. 14.4(a), assuming a 5-V power supply and threshold voltages of $V_{Tn} = -V_{Tp} = 1$ V. We note that the drain-to-source voltage of this device is also the output voltage of the inverter, and the drain current is the current of the output loop that includes the other device and the power supply. Furthermore, the input voltage equals the gate-to-source voltage of the n device.

Also shown in Fig. 14.4(b) is the set of curves for the p device and the power supply. These curves are plotted for the output voltage of the inverter versus the current leaving the drain of the p device. It is assumed that the constants K_p and K_n as defined below are identical for both devices.

These constants are

$$K_n = \frac{\mu_n W_n C_{ox}}{2L_n} \tag{14.3}$$

and

$$K_p = \frac{\mu_p W_p C_{ox}}{2L_p} \tag{14.4}$$

Note that the curves for the p device and the power supply are generated from $V_{out} = V_{DD} + V_{dsp}$, noting that V_{dsp} is always negative in this configuration. The current I is the current that leaves the drain of the p device. The parameter that is varied to generate the set of curves is the gate-to-source voltage, which equals $V_{in} - V_{DD}$ for the p device. Since

Figure 14.4

Characteristic curves for (a) the n-device, (b) the p-device and the power supply.

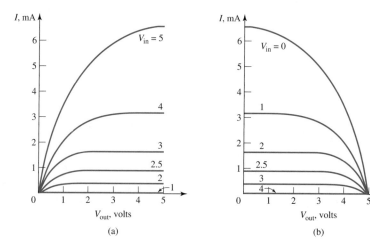

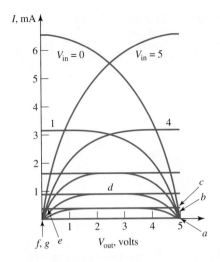

Figure 14.5
The output current versus output
voltage of the inverter.

V_{DD} is constant, the curves are plotted as V_{in} varies. The parameter λ is assumed to be zero for both devices.

The two sets of characteristics can then be plotted on the same axes in terms of V_{out} versus I as done in Fig. 14.5. For any given value of V_{in}, the n device must operate along the corresponding curve, as also must the p device. The only possible operating point for the inverter is the point where the two curves intersect. Using this information allows a *transfer characteristic* of V_{out} as a function of V_{in} to be plotted.

When $V_{in} = 0$, the inverter operates at point a in Fig. 14.5, which corresponds to $I = 0$ and $V_{out} = V_{DD} = 5$ V. As V_{in} increases to 1 V, the operating point remains at point a. As V_{in} increases above the threshold voltage, for example to 1.5 V, the operating point moves to point b. The p device is in the triode region while the n device is in the active region. The drop across the p device is quite small, leading to an output voltage that is near 5 V. As the input voltage moves to 2 V, the operating point moves to point c. The p device is still in the triode region with a small voltage drop, while the n device is in the active region and has a large voltage drop. Finally, when the input voltage reaches 2.5 V, the gate-to-source voltage across each device has a 2.5-V magnitude. Both devices are in the active region and the operating point is at point d, which indicates an output voltage of 2.5 V also. As the input voltage increases to 3 V, then 4 V and 5 V, the operating point moves to points e, f, and g.

This information can be presented in terms of V_{in} versus V_{out} or versus I as shown in Fig. 14.6. The plot of V_{out} versus V_{in} is called the transfer characteristics of the inverter. The change in V_{out} with V_{in} increases greatly as both devices are in the active region. When only one device is in the active region while the other is in the triode region, the change in V_{out} as V_{in} varies is quite small.

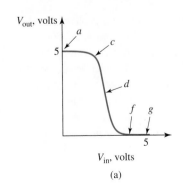

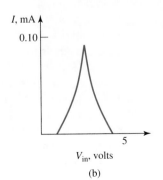

Figure 14.6
(a) V_{out} versus V_{in}.
(b) I versus V_{in}.

Table 14.1 Conditions of Devices as V_{in} Varies

V_{in}, V	nMOS	pMOS	V_{out}, V
0	cutoff	triode	5
1	cutoff	triode	5
2	active	triode	4.9
2.5	active	active	2.5*
3	triode	active	0.1
4	triode	cutoff	0
5	triode	cutoff	0

*Output may fall between 1.5 and 3.5 V.

The progression of states as the input voltage ranges from zero to V_{DD} is summarized in Table 14.1.

At $V_{in} = 2.5$ V, the graph of Fig. 14.5 shows that the output is indeterminate. The curves for the *n* device and *p* device intersect over a range of values rather than as a single point. This results from the assumption that $\lambda = 0$. In the practical case, λ will be nonzero and the curves will both have a finite slope. In this case, the curves will intersect only at a single point and this intersection will be near 2.5 V. Figure 14.6(a) corresponds to this practical situation.

In logic circuits, the major concern is often with the extreme values of V_{out}, that is, 0 V and 5 V, due to the fact that the input signal will exhibit a rail-to-rail swing and the output will move quickly between points *a* and *g* of Fig. 14.6(a).

The graph of current versus V_{in} in Fig. 14.6(b) shows that no current flows through the inverter when the input is at 0 V or 5 V. Only when the inverter is changing states does a finite value of current flow.

PRACTICAL Considerations

A personal computer (PC) includes many inverters or inverter-like gates. When power is applied to the PC but it is not performing any operations, very few circuits other than clock-related circuits are switching. The total average current drawn from the dc power supply is quite low. The power delivered to the PC from the power supply is equal to the product of power-supply voltage and average current drawn from the supply and is also relatively low.

As the PC begins performing various operations, more switching takes place. For CMOS circuits, more inverters and gates are drawing a transition current similar to that shown in Fig. 14.6(b). The average current increases as well as the product of average current and power-supply voltage. Many laptops experience heating as the power dissipation increases to higher levels.

One method of *thermal management* for a laptop consists of controlling the clock speed to lower the average current drawn. When excessive heating is detected, the thermal circuit signals the clock to cut its frequency in half. Doing so decreases the average current correspondingly and brings the dissipation toward an acceptable level. If the dissipation continues at an excessive level, the thermal circuit causes the clock frequency to cut the frequency in half a second time. Until the temperature begins to decrease, the thermal circuit controls the clock frequency. When the temperature decreases to some predetermined value, the clock frequency is restored to its nominal value. This method of thermal management allows the computer to continue functioning as overheating is approached, although at a slower rate, rather than shutting down.

The graphs of Figs. 14.5 and 14.6 are useful in understanding the behavior of the inverter and in developing a mathematical description of this circuit.

14.1.3 MATHEMATICAL ANALYSIS OF THE INVERTER

The current equations for the n device and p device in all three regions are expressed as

$$I_n = I_p = 0 \quad \text{cutoff} \tag{14.5}$$

$$I_{Dn} = 2K_n\left[(V_{GS} - V_{Tn})V_{DS} - \frac{V_{DS}^2}{2}\right] \quad \text{triode} \tag{14.6}$$

$$I_{Dp} = -2K_p\left[(V_{GS} - V_{Tp})V_{DS} - \frac{V_{DS}^2}{2}\right] \quad \text{triode} \tag{14.7}$$

$$I_{Dn} = K_n[V_{GS} - V_{Tn}]^2 \quad \text{active} \tag{14.8}$$

$$I_{Dp} = -K_p[V_{GS} - V_{Tp}]^2 \quad \text{active} \tag{14.9}$$

[handwritten margin notes: $V_{GS} = V_{DD} - V_{in}$, $V_{DS} = V_{out} - V_{DD}$ *]*

The negative signs on the p device drain currents merely signify that drain current flows out of the drain terminal. The active region equations also neglect λ.

The analysis of the inverter is rather simple when one of the two devices is in the off state. When this is true, the entire power-supply voltage will drop across the off device. For example, when the input voltage is less than V_{Tn}, the n device is off and the power-supply voltage drops across this device, resulting in $V_{out} = V_{DD}$. When the input voltage is increased so that the gate-to-source voltage of the p device causes this device to turn off, the power-supply voltage drops across the p device, resulting in $V_{out} = 0$ V. This drop would occur for input voltages above $V_{DD} + V_{Tp}$ where V_{Tp} has a negative value. If the threshold voltages are equal in magnitude at 1 V, the n device will be off when V_{in} is less than 1 V and the p device will be off when V_{in} is greater than 4 V.

As V_{in} increases above the threshold voltage of the n device, this device enters the active region and the p device remains in the triode region. As V_{in} continues to increase, the p device will enter its active region. The steepest region of the transfer characteristics occurs when both devices are in their active regions.

One particular point of interest occurs when the output voltage equals one-half of the power-supply voltage. When the constants K_n and K_p are identical for both devices as well as the magnitudes of threshold voltage and the constants λ, an input voltage of $V_{DD}/2$ will result in an output voltage of $V_{DD}/2$. This situation requires a p device that has a wider channel area than the n device by a factor of μ_n/μ_p, where μ is the mobility. This conclusion results from solving the equation

$$K_n = \frac{\mu_n W_n C_{ox}}{2L_n} = K_p = \frac{\mu_p W_p C_{ox}}{2L_p}$$

If $L_n = L_p$, which is sometimes required for digital ICs, this equation can be solved to find that

$$\frac{W_p}{W_n} = \frac{\mu_n}{\mu_p}$$

to result in $K_n = K_p$.

If the two devices do not have identical values of K, the approximate input voltage that leads to equal drops across both devices can be found by neglecting λ and equating the active region currents of the two devices. Doing so leads to $I_{Dn} = -I_{Dp}$ or from Eqs. (14.8) and (14.9)

$$K_n(V_{in} - V_{Tn})^2 = K_p(V_{in} - V_{DD} - V_{Tp})^2 \tag{14.10}$$

Taking the square root of both sides of Eq. (14.10) results in

$$\sqrt{K_n/K_p}\,(V_{in} - V_{Tn}) = \pm(V_{in} - V_{DD} - V_{Tp}) \tag{14.11}$$

Since the left side of Eq. (14.11) will be positive for the range of V_{in} being considered, the negative sign must be associated with the right side of the equation, which will lead to a positive value for the right side of the equation. Solving for V_{in} then yields

$$V_{in} = \frac{V_{DD} + V_{Tp} + \sqrt{K_n/K_p}\,V_{Tn}}{1 + \sqrt{K_n/K_p}} \tag{14.12}$$

As an example of the use of this equation, we will assume equal-sized devices with $K_n = 3K_p$ and $V_{Tn} = -V_{Tp} = 1$ V. For a 5-V power supply, the value of V_{in} to give $V_{out} = 2.5$ V is found from Eq. (14.12) to be 2.1 V. The overall transfer characteristic for this case is shown in Fig. 14.7 along with two other ratios of K-values. Although different-sized devices shift the transfer characteristics, one significant fact remains constant in logic applications: The output voltage for $V_{in} = 0$ V is $V_{out} = V_{DD}$, and for $V_{in} = V_{DD}$ the output is $V_{out} = 0$ V. In other words, the output voltage swings rail to rail regardless of the ratio of K_n to K_p.

There are variations of CMOS design wherein the output swing is not rail to rail and is maximized or tailored to meet some specification by choosing a particular ratio of K_n to K_p. These designs are called *ratioed* designs and are not as straightforward as the design of CMOS circuits.

It is possible to calculate output voltages for various input values or to calculate required input voltages for a specified output voltage. Although a simulation program would generally be used for this purpose, we will consider these calculations in more detail to foster a better understanding of MOS behavior.

Output Voltage for Specified Input Voltage: If the parameters K_p, K_n, V_{Tp}, and V_{Tn} are known, along with the input voltage, it is not always apparent in which region each device operates. It is appropriate to make an educated guess or assumption as to the correct operating region and then base the calculations on this assumption. If the calculation of V_{out} verifies that the operating regions were those assumed, the answer is correct. If the value of V_{out} calculated results in one or both devices not existing in the assumed regions, the result has no validity.

Figure 14.7
Transfer characteristics for different device sizes.

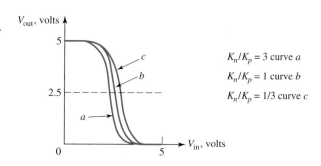

$K_n/K_p = 3$ curve a
$K_n/K_p = 1$ curve b
$K_n/K_p = 1/3$ curve c

Equation (14.12) is a good starting point for this procedure. The value of V_{in} calculated from Eq. (14.12) indicates the voltage needed to put both devices in their active regions. If the actual value of V_{in} applied to the inverter input is higher than the value calculated from Eq. (14.12) but the p device is still on, we may assume that the p device is in the active region while the n device is in the triode region. This fact is also reflected in Table 14.1. An applied value of V_{in} lower than that calculated by Eq. (14.12), along with an input voltage that exceeds the threshold voltage of the n device, leads to an assumption that the n device is in its active region while the p device is in the triode region. An example will demonstrate these concepts.

EXAMPLE 14.1

For the inverter shown in Fig. 14.8, $K_n = 80\ \mu\text{A/V}^2$ and $K_p = 60\ \mu\text{A/V}^2$. The threshold voltages are equal to 1 V in magnitude. If λ is negligible for both devices, find the output voltage for an input voltage of 2.5 V.

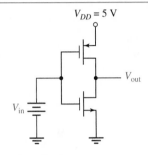

$V_{DD} = 5\ \text{V}$

SOLUTION Using Eq. (14.12), we find that the value of V_{in} that puts both devices into their active regions is 2.39 V. Since the applied value of V_{in} is slightly higher than this value, Table 14.1 suggests that the n device is in the triode region and the p device is in the active region. The current leaving the drain of the p device can then be expressed as

$$-I_{Dp} = 60[V_{\text{in}} - 5 + 1]^2 = 60[-1.5]^2 = 135\ \mu\text{A}$$

Figure 14.8
A CMOS inverter.

This current must equal the drain current of the n device calculated from the triode region equation; that is,

$$135 = 80\big[2(V_{\text{in}} - 1)V_{\text{out}} - V_{\text{out}}^2\big] = 80\big[3V_{\text{out}} - V_{\text{out}}^2\big]$$

Using the quadratic equation to solve for V_{out} gives two results: 0.75 V and 2.25 V. If we check the first result, we see that the n device would be in the triode region for this output voltage since $V_{GD} = V_{GS} - V_{DS} = 2.5 - 0.75 = 1.75\ \text{V}$, which exceeds the threshold voltage. The p device would be in the active region for this output voltage; thus the output voltage calculated is accepted as the correct result. Note that when the result of 2.25 V is checked, the n device would not be in the triode region.

As V_{in} is raised to a value of 4 V, the p device cuts off and the output voltage reaches 0 V.

Input Voltage for Specified Output Voltage: This problem is somewhat more complex when λ is taken as zero for both devices. The reason for this is the ambiguity of output voltage when both devices are in the active region. The equivalent circuit of the output is shown in Fig. 14.9, with both devices represented by current sources. The characteristic curves for this situation are shown in Fig. 14.10. We note that the output voltage may fall anywhere between V_{DSPn}, the drain-source pinchoff voltage of the n device, and $V_{DD} - |V_{DSPp}|$ for the given value of V_{in}, due to the zero slope of the curves as V_{out} changes. For real devices, with finite values of λ, the n device curve slopes upward and the p device curve slopes downward. A given value of V_{in} leads to only one possible point of intersection.

A suggested method of calculating the input voltage for a required output voltage is to again start the procedure with a solution of Eq. (14.12) for V_{in}. Knowing this value of V_{in} allows V_{DSPn} and $V_{DD} - |V_{DSPp}|$ to be found. If the desired output voltage falls within this range, the required value of V_{in} has been calculated.

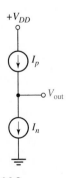

$+V_{DD}$

I_p

V_{out}

I_n

Figure 14.9
Active region output model with $\lambda_n = \lambda_p = 0$.

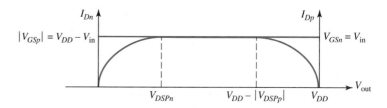

Figure 14.10
Characteristic curves for both devices.

If the desired value of V_{out} falls outside this range, the assumption of operating regions is then obvious. An example will demonstrate this approach.

EXAMPLE 14.2

For the inverter shown in Fig. 14.8, $K_n = 80\ \mu\text{A/V}^2$ and $K_p = 60\ \mu\text{A/V}^2$. The threshold voltages are equal to 1 V in magnitude. If λ is negligible for both devices, find the input voltage, V_{in}, that leads to an output voltage of 3.1 V. Repeat this calculation if an output voltage of 4.5 V is desired.

SOLUTION The calculation of $V_{\text{in}} = 2.39$ V to put both devices in the active region was done in Example 14.1. This value allows us to calculate

$$V_{DSPn} = V_{GSn} - V_{Tn} = 2.39 - 1 = 1.39\ \text{V}$$

and

$$V_{DD} - |V_{DSPp}| = V_{DD} - |V_{GSp} - V_{Tn}| = 5 - |-2.61 + 1| = 3.39\ \text{V}$$

Since the desired output voltage is 3.1 V and falls within the range from 1.39 V to 3.39 V, the required value of V_{in} is 2.39 V.

For the desired output voltage of 4.5 V, we see that this value could only be obtained if the p device is in the triode region and the n device is in the active region. The p device current leaving the drain is calculated from Eq. (14.7) as

$$-I_{Dp} = 60[2(V_{\text{in}} - 5 + 1)(-0.5) - 0.5^2]$$
$$= 60[2(4 - V_{\text{in}})0.5 - 0.25] = 60[3.75 - V_{\text{in}}]$$

Equating this value to the active region drain current of the n device gives

$$60[3.75 - V_{\text{in}}] = 80[V_{\text{in}} - 1]^2$$

Solving this equation results in two possible input voltages: 2.11 V and 0.86 V. Only the first answer leads to operating regions consistent with the assumption; thus, the answer is $V_{\text{in}} = 2.11$ V.

Calculation of the drain currents for each device for this input voltage leads to agreement as well, with a value of

$$-I_{Dp} = I_{Dn} = 98\ \mu\text{A}$$

As we mentioned at the beginning of this section, the CMOS inverter is the basic building block of CMOS gates and other logic circuits. It is important to understand the operation of the inverter in order to extend its use to the synthesis of more complex logic circuits.

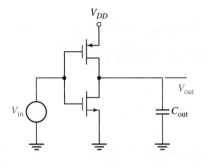

Figure 14.11
An inverter with parasitic capacitance.

14.1.4 INVERTER SWITCHING SPEED

The high-frequency switching limitations of the inverter are imposed by the parasitic capacitances of the devices. For the inverter of Fig. 14.11, the output circuit determines the speed of switching. The output capacitance consists of the drain-to-substrate capacitance of both devices plus the gate-to-drain capacitance of both devices. The junction between drain and substrate for each device will be reverse biased; thus as the output voltage changes, so also will the drain-to-substrate capacitance. The result is nonlinear capacitances that could lead to a difficult problem to analyze. However, since the output voltage is a rail-to-rail swing, it is not difficult to find linearized values for these capacitors as the output switches between V_{DD} and 0 V.

In addition to the nonlinear capacitance as the output switches, the resistance seen by the capacitance also varies nonlinearly. Typically, simulations are done to find accurate switching times. The circuits of Fig. 14.12 allow the switching times to be approximated.

The values of R_{dsp} and R_{dsn} are average resistance values in the triode regions of the respective devices. A simple method of evaluating the triode resistance is to assume that it equals the value near $V_{DS} = 0$. For the nMOS device, this value is given by Eq. (6.23)

$$R_{\text{lin}} = \frac{1}{\frac{\mu W}{L} C_{\text{ox}} (V_{GS} - V_T)} \tag{14.13}$$

The value of V_{GS} when the nMOS device moves to the triode region is $V_{GSn} = V_{DD}$.

Knowing the capacitances and resistances allows the time constants to be calculated as

$$\tau_{\text{neg}} = R_{dsn} C_{\text{outneg}} \quad \text{and} \quad \tau_{\text{pos}} = R_{dsp} C_{\text{outpos}}$$

where C_{outneg} and C_{outpos} are average capacitance values as the output swings negative and positive, respectively.

A typical value of R_{lin} for a 0.5-μ device is 4 kΩ, whereas the capacitance might be 100 fF. The resulting time constant is 0.4 ns.

PRACTICE Problems

14.1 Repeat Example 14.1 for an input voltage of 2.1 V. *Ans:* $V_{\text{out}} = 4.51$V.
14.2 Repeat Example 14.2 for output voltages of 1.3 V and 0.3 V. *Ans:* $V_{\text{in}} = 2.4$V, 2.84V.
14.3 If $\mu C_{\text{ox}} W/L = 100$ μA/V^2 and $V_T = 0.8$V, calculate R_{lin} when $V_{GS} = 5$V. *Ans:* 2.38 kΩ.

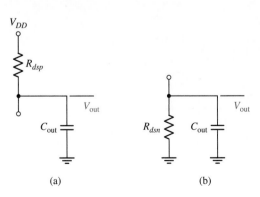

Figure 14.12
Approximate switching circuits for the inverter.

14.2 CMOS Logic Gates

IMPORTANT Concepts

1. CMOS circuits can be used to implement NOR and NAND gates.

2. These gates use a configuration based on that of the inverter.

We will now proceed to a discussion of CMOS logic circuits based on the inverter configuration. This discussion assumes that the reader has some familiarity with logic circuits, Boolean algebra, and Karnaugh or K-maps.

14.2.1 THE NOR GATE

The truth table for a NOR gate is

A	B	X
0	0	1
0	1	0
1	0	0
1	1	0

where A and B are inputs and X is the output.

When either input is a logic 1, the output will be 0. The only combination of inputs that results in an output of 1 is $A = 0$ and $B = 0$. The logic expression that describes the NOR gate is

$$X = \overline{A + B} \tag{14.14}$$

We will assume that a positive logic convention is used in the following discussion; that is, the high voltage level corresponds to logic 1. The two-input NOR gate can be implemented in CMOS, as shown in Fig. 14.13. The general logic symbol for the NOR gate is also indicated in this figure. The small circle at the output represents an inversion following the OR gate to create the NOR gate.

As we have seen, all CMOS logic gates require a pMOS device and an nMOS device for each input. Each complementary pair is connected in a manner similar to the inverter, except that there may be additional series or parallel devices in the drain-source circuits.

Figure 14.13
(a) A two-input CMOS NOR gate. (b) General logic symbol for the NOR gate.

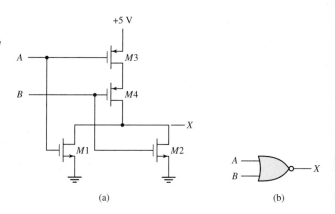

(a)   (b)

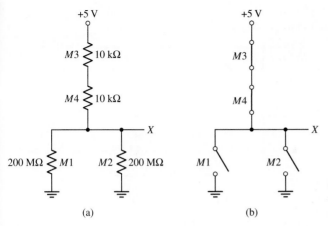

Figure 14.14
(a) Equivalent circuit for
A = B = 0. (b) Approximate
equivalent circuit.

When both inputs are at 0 V or logic 0, both nMOS devices are off and both pMOS devices are on. This input condition leads to 0-V gate-to-source voltages for both $M1$ and $M2$ and -5-V gate-to-source voltages for $M3$ and $M4$. The output voltage will be very near 5 V or logic 1, as shown in Fig. 14.14(a). This figure assumes an on resistance for the p devices of 10 kΩ and an off resistance for the n devices of 200 MΩ. The actual high-level output voltage in Fig. 14.14(a) can be calculated to be

$$V_{\text{high}} = \frac{10^8}{10^8 + 2 \times 10^4} \, 5 = 4.999 \approx 5 \text{ V}$$

Since the output voltage is so close to 5 V, the approximate equivalent circuit of Fig. 14.14(b) can be used in many situations. When one or both inputs switch to logic 1 or 5 V, at least one nMOS device turns on and at least one pMOS device turns off. The approximate equivalent circuits of Fig. 14.15 apply to this case.

In either case, a very high resistance exists between the power-supply voltage and output, and a very low resistance exists between output and ground, which forces the output to be at 0 V or logic 0. Based on this discussion, the gate circuit of Fig. 14.13(a) is seen to satisfy the conditions of the truth table for the NOR gate.

The NOR gate can be extended to a higher number of inputs by adding a complementary pair of devices for each additional input. The additional pMOS device is placed in series

PRACTICE **Problems**

14.4 The inputs A and $\bar{B}$ drive a two-input NOR gate. Inputs C and B drive a second NOR gate. The outputs of these gates drive a third two-input NOR gate to create a variable X. Express X as a logic function.
Ans: $X = AC + AB + \bar{B}C$.
14.5 How many MOS devices are required for an eight-input NOR gate?
Ans: 16 devices.

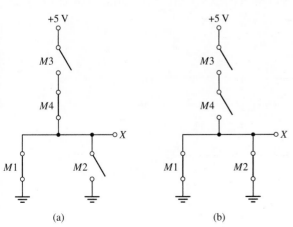

Figure 14.15
(a) Equivalent circuit for A = 1,
B = 0. (b) Equivalent circuit for
A = B = 1.

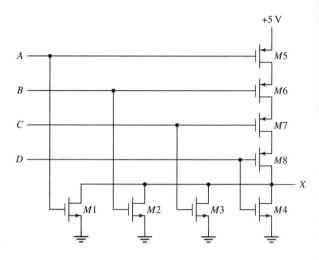

Figure 14.16
A four-input CMOS NOR gate.

with the other pMOS devices, and the additional nMOS device is placed in parallel with the other nMOS devices. Figure 14.16 shows a four-input NOR gate.

14.2.2 THE NAND GATE

A two-input NAND gate satisfies the following truth table.

A	B	X
0	0	1
0	1	1
1	0	1
1	1	0

The logical expression for this gate is

$$X = \overline{AB} \qquad\qquad (14.15)$$

For a positive logic system in which a high voltage level is defined as a logic 1, the NAND gate output is at the high level for all but one input combination. If both inputs are high, the output should be low. The CMOS gate of Fig. 14.17 satisfies this truth table. Also shown in the figure is the logic symbol for the NAND, which is equivalent to an AND gate followed by an inversion.

Figure 14.17
(a) A two-input CMOS NAND gate. (b) General logic symbol for the NAND gate.

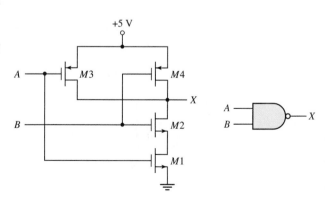

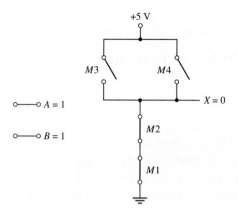

Figure 14.18
Approximate equivalent circuit for A = B = 1.

For the combination of inputs $A = B = 1$, both $M1$ and $M2$ will be turned on and both $M3$ and $M4$ will be off. A very high impedance exists between the power supply and the output terminal, while a low impedance exists between the output terminal and ground. An approximation to this output condition appears in Fig. 14.18. For any other input condition, the impedance from power supply to output will be low, while the impedance from output to ground will be high. These conditions lead to a high output voltage, as seen in Fig. 14.19.

If the high impedance level of the MOS devices is 200 MΩ rather than infinite and the low impedance level is 10 kΩ rather than zero, the output voltages for the four input combinations are found to be

A	B	X, volts
0	0	4.9999
0	1	4.9998
1	0	4.9998
1	1	0.0010

These voltages are so close to 0 and 5 V that the output can again be considered rail to rail.

PRACTICE Problems

14.6 The inputs A and $\bar{B}$ drive a two-input NAND gate. Inputs C and B drive a second NAND gate. The outputs of these gates drive a third two-input NAND gate to create a variable X. Express X as a logic function.
Ans: $X = A\bar{B} + BC$.
14.7 How many MOS devices are required for a six-input NAND gate?
Ans: 12 devices.

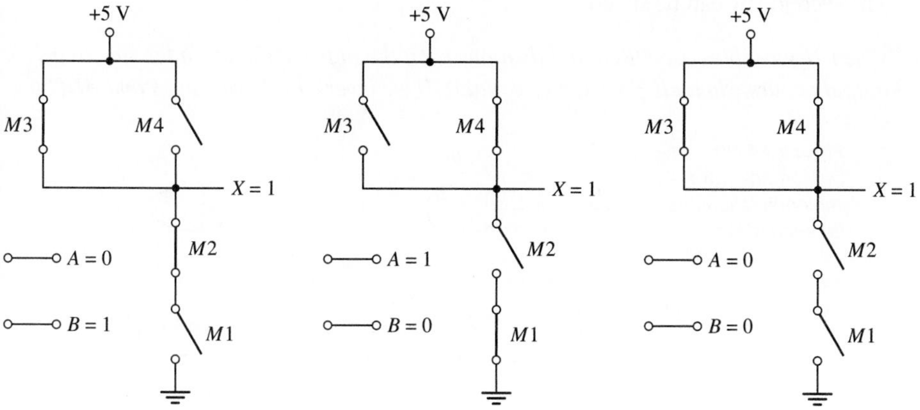

Figure 14.19
Conditions leading to X = 1 for the NAND gate.

14.3 Logic Function Realization

IMPORTANT Concepts

1. NOR gates and NAND gates can be used to realize any combinational logic function.
2. Whereas an inverter can be used to convert a NOR gate to an OR gate or to convert a NAND gate to an AND gate, DeMorgan's laws can be used to implement logic expressions in terms of NOR-NAND logic.
3. It is possible to realize simple logic functions while minimizing the number of devices used. These circuits are not restricted to the use of NOR-NAND logic only.

Most logic circuits can be divided into two types: *combinational* or *sequential*. Combinational circuits have outputs that depend only on the input values. These circuits do not depend on time or on past inputs. Sequential circuits have output values that may depend on input values, on time, and on past input values. Sequential circuits are often complex circuits that employ combinational circuits as building blocks. This section will consider the realization of combinational circuits using CMOS devices.

Before we consider logic function realization, we will review two important rules that will assist us in designing logic circuits. These rules are referred to as DeMorgan's laws.

14.3.1 DEMORGAN'S LAWS

These two laws allow us to convert between forms of logic equation and also convert between types of gate used. DeMorgan's laws are given by

$$\overline{ABC \cdots N} = \bar{A} + \bar{B} + \bar{C} + \cdots + \bar{N} \tag{14.16}$$

and

$$\overline{A + B + C + \cdots + N} = \bar{A}\bar{B}\bar{C} \cdots \bar{N} \tag{14.17}$$

We can state Eq. (14.16) in words as follows:

When N variables are ANDed and then inverted, the only combination leading to a 0 output occurs when all N variables equal 1. If we invert the N variables and OR the result, the only combination leading to a 0 output again occurs when all N variables equal 1.

The second law can be stated as:

When N variables are ORed and then inverted, the only combination leading to a 1 output occurs when all N variables equal 0. If we invert the N variables and AND the

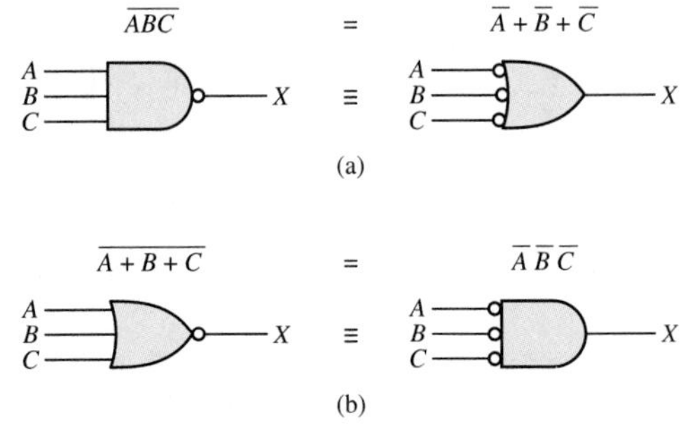

Figure 14.20
DeMorgan's laws implementations: (a) first law, (b) second law.

(a)

(b)

result, the only combination leading to a 1 output again occurs when all N variables equal 0.

These word statements essentially prove the results without the necessity of using truth tables. A physical implementation of the two laws again leads to significant practical results. Figure 14.20 shows these equivalencies for three input variables.

In terms of gates we see that a NAND gate is equivalent to an OR gate with inverted inputs and a NOR gate is equivalent to an AND gate with inverted inputs. As a memory aid we note that any gate, AND or OR, followed by an inverter is equivalent to the opposite type gate, OR or AND, respectively, with inverted inputs.

Alternate forms of DeMorgan's laws are found by complementing both sides of Eqs. (14.16) and (14.17), yielding

$$ABC \cdots N = \overline{\bar{A} + \bar{B} + \bar{C} + \cdots + \bar{N}} \tag{14.18}$$

and

$$A + B + C + \cdots + N = \overline{\bar{A}\bar{B}\bar{C} \cdots \bar{N}} \tag{14.19}$$

Application of these two laws along with the Boolean identities allows a designer more flexibility to implement a given expression in terms of the gates available.

PRACTICE Problems

14.8 Write $X = \bar{L} + M + \bar{N}$ in terms of a NAND expression. *Ans:* $X = \overline{L\bar{M}N}$.
14.9 Write $Y = A\bar{B}\bar{C}\bar{D}$ in terms of a NOR expression. *Ans:* $Y = \overline{\bar{A} + B + C + D}$.

14.3.2 NOR-NAND REALIZATION OF LOGIC FUNCTIONS

DeMorgan's laws can be used to convert a general combinational logic function to a form that can be implemented with NOR gates or NAND gates. Let us demonstrate the key points in this method. Suppose we have a function

$$X = A(B + CD)$$

that we must implement with NOR gates and inverters. The form of the function we should obtain is one that can be realized directly with NOR gates. An expression such as

$$Y = \overline{\overline{(R + S)} + \overline{(P + Q)}}$$

results from two stages of NOR gates, as shown in Fig. 14.21. We can approach this form for the function X by applying DeMorgan's laws to obtain

$$X = \overline{\bar{A} + \overline{B + CD}} = \overline{\bar{A} + \bar{B}\,\overline{CD}} = \overline{\bar{A} + \bar{B}(\bar{C} + \bar{D})}$$
$$= \overline{\bar{A} + \bar{B}\bar{C} + \bar{B}\bar{D}} = \overline{\bar{A} + \overline{B + C} + \overline{B + D}}$$

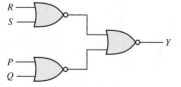

Figure 14.21
Two-stage gating.

The final expression can be realized as shown in Fig. 14.22.

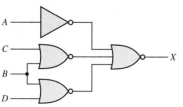

Figure 14.22
Implementation of
$X = A(B + CD)$ *using NOR gates.*

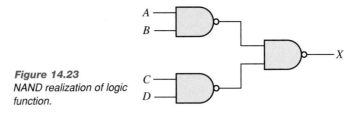

Figure 14.23
NAND realization of logic function.

Let us consider the implementation in CMOS of the function

$$X = AB + CD$$

We first use DeMorgan's law to write

$$AB + CD = \overline{\overline{AB} \cdot \overline{CD}}$$

This form of the logic function represents a two-level NAND relationship that can be implemented as shown in Fig. 14.23. Each NAND gate can then be realized by the CMOS NAND discussed earlier, resulting in the circuit of Fig. 14.24.

The key to CMOS realization is to manipulate the logic function to the NAND or NOR forms and then use the CMOS implementations with NAND or NOR gates.

Figure 14.24
CMOS NAND gate realization of logic function.

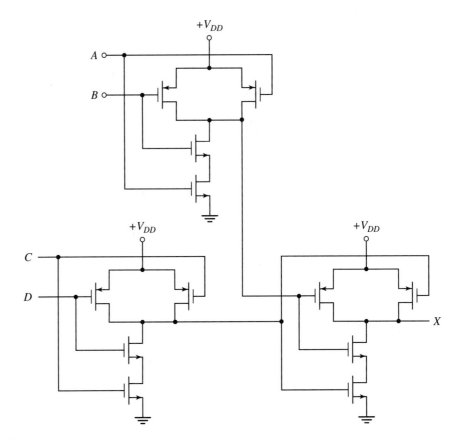

EXAMPLE 14.3

The logic expression $X = A\bar{B}\bar{C} + A\bar{B}C + \bar{A}BC$ is to be implemented with CMOS logic gates.

(a) Implement this function directly.

(b) Reduce the expression with DeMorgan's laws and implement.

(c) Minimize the expression using a K-map, then use DeMorgan's laws to implement.

(d) Compare the total number of transistors needed with each approach.

SOLUTION (a) The direct implementation of this logic expression requires three three-input AND gates and one three-input OR gate. In order to implement these in CMOS, the AND gates must be NAND gates followed by an inversion. The OR gate will be a NOR gate followed by an inverter. Figure 14.25 shows this implementation of X. Each of the four three-input gates requires six transistors, and each of the four inverters requires two transistors each. The total number of transistors for this implementation is 32.

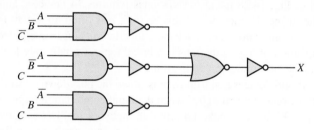

Figure 14.25
Direct implementation of X.

(b) We can apply DeMorgan's law to write

$$X = A\bar{B}\bar{C} + A\bar{B}C + \bar{A}BC = \overline{\overline{A\bar{B}\bar{C}} \cdot \overline{A\bar{B}C} \cdot \overline{\bar{A}BC}}$$

This expression can be realized with four three-input NAND gates as in Fig. 14.26. Each NAND gate employs six transistors, leading to a total of 24 transistors.

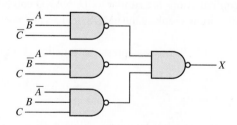

Figure 14.26
NAND gate implementation.

(c) The K-map of Fig. 14.27 allows the expression to be reduced. The reduced expression can be written as

$$X = A\bar{B} + \bar{A}BC$$

Again applying DeMorgan's law allows this expression to be written in NAND form as

$$X = \overline{\overline{A\bar{B}} \cdot \overline{\bar{A}BC}}$$

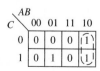

Figure 14.27
K-map for the logic expression
X = ABC̄ + AB̄C + ĀBC.

PRACTICE Problems

14.10 Use DeMorgan's laws to reduce
$Y = PQR + \bar{P}RS + \bar{Q}RS$
to simplest form.
Ans: PQR + RS.

14.11 Express $X =$
$\overline{AB} + CD$ in terms of single variables or their complements.
Ans: $AB\bar{C} + AB\bar{D}$.

The implementation of this expression appears in Fig. 14.28 and requires two two-input NANDs and one three-input NAND. A total of 14 transistors implements this expression.

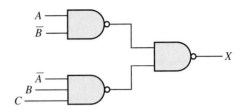

Figure 14.28
A reduced implementation of X.

(d) The direct implementation requires 32 devices, while the NAND form of this expression decreases this number to 24 devices. If the expression is first reduced and then implemented in NAND form, the number of transistors required is only 14.

14.3.3 REALIZING LOGIC FUNCTIONS WITH MINIMUM TRANSISTOR COUNT

The reduced implementation of Fig. 14.28 using NAND gates requires 14 devices. This number can be reduced further using the methods of this section. It is often important to minimize the number of devices required in order to pack more functionality on an integrated circuit chip. Although minimization of logic functions is a broad topic, we will only look at one basic idea for CMOS device minimization at this time.

In CMOS logic gates, the nMOS devices make up what is called the pull-down network or PDN. Its function is to pull the output down to 0 V when the input combination requires a logic 0 at the output. The pMOS devices make up a pull-up network or PUN to force the output to V_{DD} when the input combination dictates a logic 1 at the output. When the PUN is asserted or in the low-impedance state, we note that the PDN is not asserted and when the PDN is asserted, the PUN is not asserted. Thus, the PDN is the complement of the PUN, meaning it is in the opposite state to the PUN. The PUN should be asserted and in the low-impedance state when the output is $X = 1$, whereas the PDN is asserted when the output is $X = 0$.

Understanding this behavior of CMOS gates allows us to extend beyond the realization of simple NOR and NAND gates. In order to realize a logic function X we can use the configurations of Fig. 14.29. These configurations are similar to the NAND and NOR gate configurations, except that pMOS logic circuits replace the pMOS devices and nMOS logic circuits replace the nMOS devices.

For the NAND configuration of Fig. 14.29(a), we express the function X in terms of two smaller functions as

$$X = \overline{F_1 \cdot F_2} \tag{14.20}$$

The functions F_1 and F_2 can be complex logic functions. When both F_1 and F_2 are asserted by the input combination, the PDN goes to the low-impedance state and the PUN goes to the high-impedance state. This pulls the output to logic 0 as dictated by the expression for X.

For the NOR configuration of Fig. 14.29(b), the function X is manipulated to the form

$$X = \overline{F_1 + F_2} \tag{14.21}$$

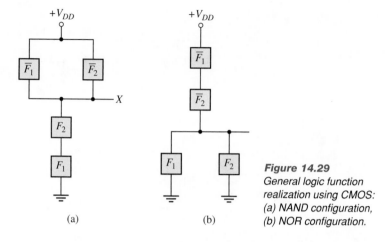

Figure 14.29
*General logic function
realization using CMOS:
(a) NAND configuration,
(b) NOR configuration.*

When either F_1 or F_2 is asserted by the input combinations, the PDN goes to the low-impedance state and the PUN goes to the high-impedance state. This pulls the output to logic 0, as it should.

An example will demonstrate the use of this method.

EXAMPLE 14.4

Implement the logic expression $X = AB + CD$ in CMOS, using the minimum number of devices.

SOLUTION Using DeMorgan's law, we write this function as

$$X = \overline{\overline{AB} \cdot \overline{CD}}$$

which corresponds to the form of Eq. (14.20) where

$$F_1 = \overline{AB} \quad \text{and} \quad F_2 = \overline{CD}$$

We note that when the logic conditions to satisfy both F_1 and F_2 occur, the PDN network must be asserted and in the low-impedance state.

Using DeMorgan's law on the two logic expressions $\overline{AB}$ and $\overline{CD}$ allows X to be expressed as

$$X = \overline{(\bar{A} + \bar{B}) \cdot (\bar{C} + \bar{D})}$$

This expression can be implemented in the NAND configuration of Fig. 14.29(a). The PDN is designed to be asserted when $F_1 = (\bar{A} + \bar{B})$ is asserted and $F_2 = (\bar{C} + \bar{D})$ is asserted. The PDN is shown in Fig. 14.30.

The PUN is chosen to be complementary to the PDN in every way. Parallel elements in the PDN become series elements in the PUN and vice versa. The same is true of larger subnetworks. The final function realization is depicted in Fig. 14.31. We note that the parallel devices in the PDN driven by $\bar{A}$ and $\bar{B}$ appear as series devices in the PUN. This is also true for the devices driven by $\bar{C}$ and $\bar{D}$. In the PDN, the subnetwork driven by $\bar{A}$ and $\bar{B}$ is in series with the subnetwork driven by $\bar{C}$ and $\bar{D}$. The PUN includes the subnetwork driven by $\bar{A}$ and $\bar{B}$ in parallel with the subnetwork driven by $\bar{C}$ and $\bar{D}$.

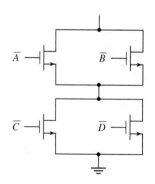

Figure 14.30
*NAND configuration for PDN
realization.*

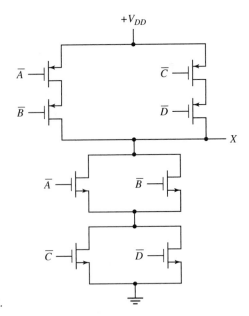

Figure 14.31
NAND configuration for X.

This realization requires only eight devices, compared to the 12 used in the circuit of Fig. 14.24.

Before leaving the topic of CMOS logic functions, we will consider the effect of using the distributive law in minimizing the number of devices. A function such as

$$Y = AB\bar{D} + B\bar{C}D$$

can be implemented in several ways. We can use DeMorgan's law to write Y as

$$Y = \overline{\overline{AB\bar{D}} \cdot \overline{B\bar{C}D}} = \overline{(\bar{A} + \bar{B} + D)(\bar{B} + C + \bar{D})}$$

We could now realize Y in two-stage logic, using two three-input NAND gates followed by a two-input NAND gate similar to the realization of Fig. 14.23. This would require a total of 16 MOS transistors. Following the method of Example 14.4, we could implement the logic function with 12 devices as indicated in Fig. 14.32.

It is possible to reduce the number of devices further by using the distributive law to write Y in the form

$$Y = B(A\bar{D} + \bar{C}D)$$

Again applying DeMorgan's law, we can write

$$Y = \overline{\bar{B} + \overline{(A\bar{D} + \bar{C}D)}}$$

The circuit of Fig. 14.33 realizes this expression with 10 MOS transistors. The distributive law can often save a small number of transistors and should be considered in applications that require minimization.

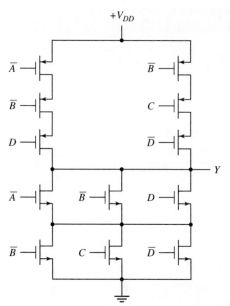

Figure 14.32
Realization of Y.

PRACTICAL Considerations

Since present-day personal computers use chips with several million devices, the methods discussed in this chapter are presented to teach the basic ideas of CMOS logic design. In addition, an understanding of these basic ideas leads to an understanding of the automated methods that must be applied to these ultra-large-scale integrated circuits.

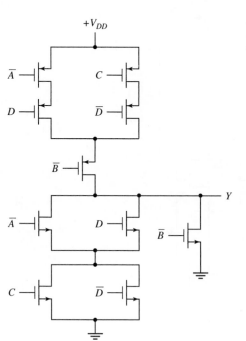

Figure 14.33
Implementation of Y *based on distributive law.*

PRACTICE Problem

14.12 What is the minimum number of devices required to implement $X = AB + \bar{B}C$ in CMOS? *Ans:* Eight devices.

DISCUSSION OF THE DEMONSTRATION PROBLEM

We first want to implement the expression

$$X = \bar{A}\bar{B} + CD$$

using NOR/NAND logic. We use DeMorgan's laws to manipulate this expression to give

$$X = \overline{A + B} + CD = \overline{(A + B) \cdot \overline{CD}}$$

The implementation of this expression is given in Fig. 14.34.

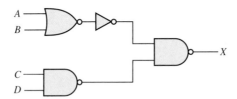

Figure 14.34
Implementation of the Demonstration Problem.

The schematic for this implementation in CMOS circuits is shown in Fig. 14.35. This implementation uses 14 devices.

Figure 14.35
CMOS schematic for Demonstration Problem.

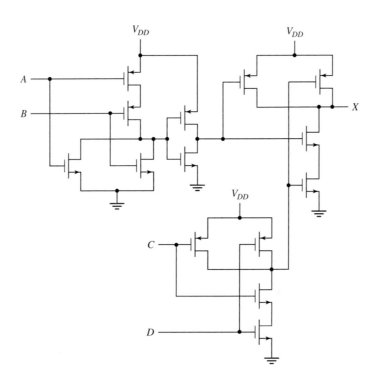

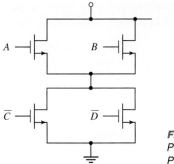

Figure 14.36
PDN for the Demonstration Problem.

To minimize the realization of this function, we could choose either of the forms shown in Fig. 14.29. We will choose the configuration of Fig. 14.29(a), which results in an output of

$$X = \overline{F_1 \cdot F_2}$$

We identify the two functions as

$$F_1 = A + B \qquad \text{and} \qquad F_2 = \overline{CD}$$

We now design the PDN to be asserted when both of these functions are asserted. The result is a PDN that presents a low-impedance path to ground when both expressions $A + B$ and $\overline{CD} = \bar{C} + \bar{D}$ are asserted. The network of Fig. 14.36 will implement this PDN. The PUN is formed as the complementary circuit to the PDN, resulting in the final circuit shown in Fig. 14.37.

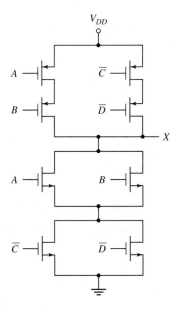

Figure 14.37
Minimized realization of the Demonstration Problem.

We note that this realization uses only eight devices rather than the 12 devices used in the NOR/NAND realization.

SUMMARY

➤ CMOS logic is universally used in personal computers because of the high density of devices that can be fabricated on a chip, the low power requirements, and the good frequency performance of these small MOS devices.

➤ The CMOS inverter forms the basis of the CMOS NOR gates and NAND gates.

➤ Logic functions can be implemented using NOR-NAND logic. DeMorgan's laws are useful in converting a general logic function to NOR-NAND form.

➤ Minimization techniques can reduce the transistor count to a minimum number in the realization of logic expressions.

PROBLEMS

SECTION 14.1 THE CMOS INVERTER

14.1 Rework Example 14.1 if K_n is changed to 45 μA/V^2.

14.2 Rework Example 14.1 if K_p is changed to 100 μA/V^2.

14.3 A CMOS inverter is powered by a 5-V supply. Given that $K_n = \mu W C_{ox}/2L = 30\ \mu$A/V^2 for the n device and $K_p = \mu W C_{ox}/2L = 40\ \mu$A/V^2 for the p device. If $V_{Tn} = -V_{Tp} = 0.8$ V, find

(a) V_{in} to give $V_{out} = 2.5$ V

(b) V_{in} to give $V_{out} = 0.5$ V

14.4 Repeat Problem 14.3 if K_n is changed to 50 μA/V^2.

D 14.5 For the inverter of Problem 14.3, find the value to which K_p must be changed to cause an output voltage of 2.5 V when the input voltage is 2.44 V.

D 14.6 For the inverter of Problem 14.3, find the value to which K_p must be changed to cause an output voltage of 2.5 V when the input voltage is 2.58 V.

SECTION 14.2 CMOS LOGIC GATES

14.7 Draw the schematic for a three-input CMOS AND gate.

14.8 Draw the schematic for a three-input CMOS OR gate.

D 14.9 Implement the function

$$X = A\bar{B} + C$$

using only CMOS NOR gates. Show the finished schematic. Assume that input variables and complements are available to drive the system.

D 14.10 Implement the function

$$X = A\bar{B} + C$$

using only CMOS NAND gates. Show the finished schematic. Assume that input variables and complements are available to drive the system.

D 14.11 Implement the function

$$X = A\bar{B} + ABC$$

using only CMOS NOR gates. Show the finished schematic. Assume that input variables and complements are available to drive the system.

D 14.12 Implement the function

$$X = A\bar{B} + ABC$$

using only CMOS NAND gates. Show the finished schematic. Assume that input variables and complements are available to drive the system.

D 14.13 Implement the function

$$X = A\bar{B}C + \bar{A}\bar{B}C + AB\bar{C}$$

using only CMOS NOR gates. Show the finished schematic. Assume that input variables and complements are available to drive the system.

D 14.14 Implement the function

$$X = A\bar{B}C + \bar{A}\bar{B}C + AB\bar{C}$$

using only CMOS NAND gates. Show the finished schematic. Assume that input variables and complements are available to drive the system.

☆ **D 14.15** Implement the function

$$X = A\bar{B}CD + AB\bar{C}\bar{D} + \bar{A}BCD + \bar{A}BC\bar{D}$$

using only CMOS NOR gates. Show the finished schematic. Assume that input variables and complements are available to drive the system.

☆ **D 14.16** Implement the function

$$X = A\bar{B}CD + AB\bar{C}\bar{D} + \bar{A}BCD + \bar{A}BC\bar{D}$$

using only CMOS NAND gates. Show the finished schematic. Assume that input variables and complements are available to drive the system.

SECTION 14.3 LOGIC FUNCTION REALIZATION

14.17 For Problems 14.17 through 14.24, evaluate the logic function realized by the circuit shown.

Figure P14.17

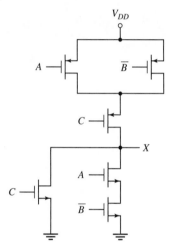

14.18

Figure P14.18

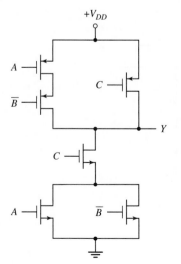

14.19

Figure P14.19

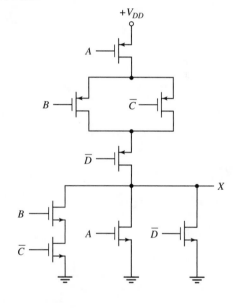

☆ **14.20**

Figure P14.20

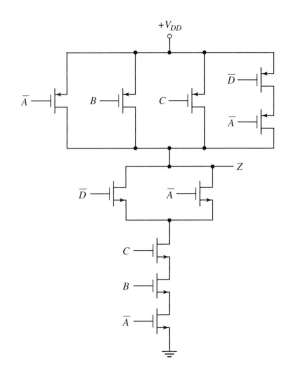

☆ **14.21**

Figure P14.21

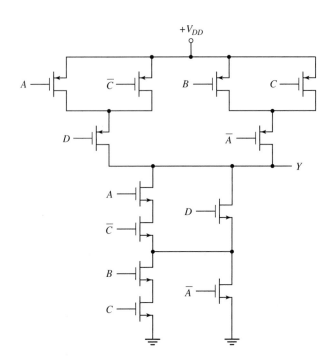

☆ **14.22**

Figure P14.22

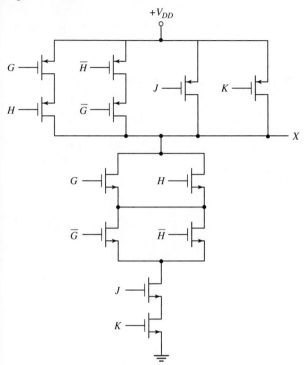

☆ **14.23**

Figure P14.23

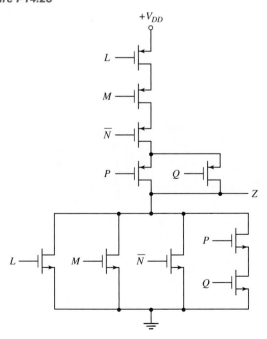

☆ **14.24**

Figure P14.24

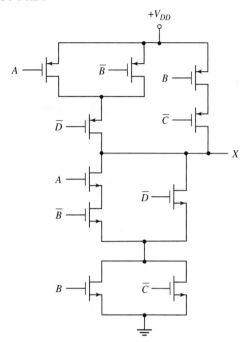

D 14.25 Using as few CMOS NOR gates and inverters as possible, implement the logic expression

$$Z = AB + \bar{B}C$$

Show the implementation in terms of logic gates. Assume that input variables and complements are available to drive the system.

D 14.26 Implement the function

$$X = A\bar{B} + C$$

using the fewest devices possible. Show the finished schematic. Assume that input variables and complements are available to drive the system.

☆ **D 14.27** Implement the function

$$X = A\bar{B} + ABC$$

using the fewest devices possible. Show the finished schematic. Assume that input variables and complements are available to drive the system.

☆ **D 14.28** Implement the function

$$X = A\bar{B}C + \bar{A}\bar{B}C + AB\bar{C}$$

using the fewest devices possible. Show the finished schematic. Assume that input variables and complements are available to drive the system.

☆ **D 14.29** Implement the function

$$X = A\bar{B}CD + AB\bar{C}\bar{D} + \bar{A}BCD + \bar{A}BC\bar{D}$$

using the fewest devices possible. Show the finished schematic. Assume that input variables and complements are available to drive the system.

☆ **D 14.30** Implement the function

$$Y = (AB + \bar{B}C)(CD + A\bar{D})$$

using the fewest devices possible. Show the finished schematic. Assume that input variables and complements are available to drive the system.

☆ **D 14.31** Implement the function

$$X = \bar{A}\bar{B} + \bar{B}\bar{D} + \bar{A}\bar{C} + \bar{C}D + ABC$$

using the fewest devices possible. Show the finished schematic. Assume that input variables and complements are available to drive the system.

Basic Equations in Amplifier Design

The following material summarizes several key equations used in the design and analysis of various amplifying stages.

A1.1 THE COMMON-SOURCE MOSFET STAGE

The common-source stage is shown in Fig. A1.1.

Figure A1.1
A common-source MOSFET stage.

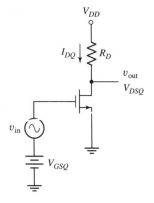

DC Equations

$$I_{DQ} = \frac{\mu C_{\text{ox}} W}{2L} \left[V_{GSQ} - V_T \right]^2 \left[1 + \lambda (V_{DSQ} - V_{DSP}) \right] \qquad (1)$$

$$V_{DSQ} = V_{DD} - R_D I_{DQ} \qquad (2)$$

where

Quantity	Definition
I_{DQ}	Quiescent drain current
V_{GSQ}	Quiescent gate-to-source voltage
V_{DSQ}	Quiescent drain-to-source voltage
V_{DSP}	Drain-to-source pinchoff voltage
μ	Mobility of induced carriers in channel
C_{ox}	Capacitance per unit area of the gate oxide
W	Width of channel
L	Length of channel from drain to source
λ	Channel length modulation parameter

Parameter Estimation

$$g_m = \sqrt{2\mu C_{\mathrm{ox}}(W/L)I_{DQ}} \tag{3}$$

$$r_{ds} = \frac{1}{\lambda I_{DP}} \approx \frac{1}{\lambda I_{DQ}} \tag{4}$$

Incremental or AC Equations

$$R_{\mathrm{out}} = R_D \parallel r_{ds} \tag{5}$$

$$A_{MB} = -g_m R_{\mathrm{out}} \tag{6}$$

A1.2 THE COMMON-EMITTER STAGE

Figure A1.2 represents a simple common-emitter BJT stage.

Figure A1.2
A common-emitter BJT stage.

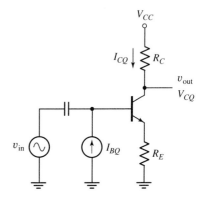

DC Equations

$$V_{BE(\mathrm{on})} \approx 0.7V \tag{7}$$

$$I_{CQ} = \beta I_{BQ} \tag{8}$$

$$V_{CQ} = V_{CC} - R_C I_{CQ} \tag{9}$$

where

Quantity	Definition
I_{CQ}	Quiescent collector current
I_{BQ}	Quiescent base current
V_{CQ}	Quiescent collector voltage
$V_{BE(\mathrm{on})}$	Quiescent base-to-emitter voltage
β	Current gain from base to collector

Parameter Estimation

$$r_e = \frac{kT}{q I_{EQ}} \tag{10}$$

or

$$r_e = \frac{26}{I_E(\text{mA})} \qquad (11)$$

Incremental or AC Equations

When R_E is present,

$$R_{\text{out}} \approx R_C \qquad (12)$$

$$A_{MB} = \frac{-\beta R_C}{(\beta + 1)(r_e + R_E)} \approx \frac{-R_C}{r_e + R_E} \qquad (13)$$

A1.3 THE BJT DIFFERENTIAL STAGE

Figure A1.3 shows a simple BJT differential stage with matched transistors.

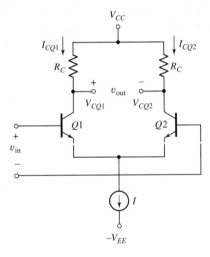

Figure A1.3
A BJT differential stage.

DC Equations

$$I_{CQ1} = I_{CQ2} = \frac{I}{2} \qquad (14)$$

$$V_{CQ1} = V_{CQ2} = V_{CC} - R_C \frac{I}{2} \qquad (15)$$

$$v_{\text{out}Q} = V_{CQ1} - V_{CQ2} = 0 \qquad (16)$$

Parameter Estimation

$$r_e = \frac{kT}{q I_{EQ}} = \frac{2kT}{qI} \qquad (17)$$

or

$$r_e = \frac{26}{I_E(\text{mA})} = \frac{52}{I(\text{mA})} \qquad (18)$$

$$g_m = \frac{\alpha}{r_e} \approx \frac{1}{r_e} \qquad (19)$$

Incremental or AC Equations

When R_C is small,

$$R_{\text{out}} \approx R_C \tag{20}$$

$$A_{MB} = \frac{v_{\text{out}}}{v_{\text{in}}} = \frac{-\beta R_C}{(\beta + 1)r_e} = \frac{-\alpha R_C}{r_e} = -g_m R_C \tag{21}$$

A1.4 THE MOSFET DIFFERENTIAL STAGE WITH CURRENT MIRROR LOAD

Figure A1.4 shows a MOSFET differential stage with matched transistors and a current mirror load.

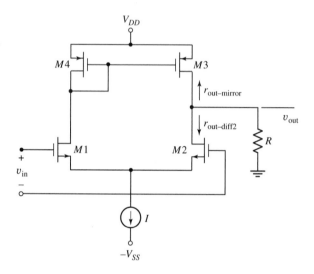

Figure A1.4
A MOSFET differential stage with current mirror load.

DC Equations

$$I_{DQ1} = I_{DQ2} = \frac{I}{2} \tag{22}$$

$$V_{SQ1} = V_{SQ2} = -\left[V_T + \sqrt{LI/\mu C_{\text{ox}}W}\right] \tag{23}$$

Parameter Estimation

$$g_m = \sqrt{2\mu C_{\text{ox}}(W/L)I_D} = \sqrt{\mu C_{\text{ox}}(W/L)I} \tag{24}$$

Incremental or AC Equations

$$R_{\text{out}} = r_{\text{out-mirror}} \parallel r_{\text{out-diff2}} \parallel R \tag{25}$$

$$A_{MB} = \frac{v_{\text{out}}}{v_{\text{in}}} = g_m R_{\text{out}} \tag{26}$$

If $R_{out} \approx R$, then

$$A_{MB} = \frac{v_{out}}{v_{in}} = g_m R \tag{27}$$

A1.5 OP AMP STAGES
The Noninverting Stage

Figures A1.5 and A1.6 show a noninverting op amp stage and an inverting op amp stage. The noninverting op amp stage satisfies the following relationships:

Figure A1.5
A noninverting op amp stage.

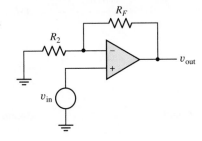

Figure A1.6
An inverting op amp stage.

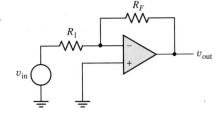

$$A_{MBni} = 1 + \frac{R_F}{R_2} \tag{28}$$

$$f_{2ni} = \frac{GBW_{ni}}{A_{MBni}} = \frac{GBW_{oa}}{A_{MBni}}$$

$$= \frac{GBW_{oa}}{1 + \frac{R_F}{R_2}} = GBW_{oa} \frac{R_2}{R_2 + R_F} \tag{29}$$

$$GBW_{ni} = GBW_{oa} \tag{30}$$

where

Quantity	Definition
A_{MBni}	Midband voltage gain of noninverting stage
f_{2ni}	Upper 3-dB frequency of noninverting stage
GBW_{oa}	Gain-bandwidth product of op amp
GBW_{ni}	Gain-bandwidth product of noninverting stage

The Inverting Stage

The inverting op amp stage satisfies the following relationships:

$$A_{MBi} = -\frac{R_F}{R_1} \tag{31}$$

$$f_{2i} = \frac{GBW_i}{A_{MBi}} = \frac{GBW_{oa} \frac{R_F}{R_1+R_F}}{A_{MBi}}$$

$$= \frac{GBW_{oa} \frac{R_F}{R_1+R_F}}{\frac{R_F}{R_1}} = GBW_{oa} \frac{R_1}{R_1 + R_F} \tag{32}$$

$$GBW_i = GBW_{oa} \frac{R_F}{R_1 + R_F} \tag{33}$$

where

Quantity	Definition
A_{MBi}	Midband voltage gain of inverting stage
f_{2i}	Upper 3-dB frequency of inverting stage
GBW_{oa}	Gain-bandwidth product of op amp
GBW_i	Gain-bandwidth product of inverting stage

Answers to Selected Problems

CHAPTER 2

2.1 (a) $A = 600$ V/V; (b) $I_L = 1.5$ mA

2.3 $A = 507$ V/V

2.6 $A_o = 1008$ V/V; $A_{dB} = 60.1$ dB

2.9 Limited by low drive to 30 gates

2.12 $n = 3.58$

2.15 5093 Hz

2.18 $\Delta f = 31$ kHz

2.22 $R_{Th} = 99\ \Omega$; $E_{Th} = 7.88\ \sin\ \omega t$

2.25 $I_B = 3.54\ \mu A$

2.28 $R_{in} = 1.148$ kΩ

2.30 $v_{out} = v_{Th} = 1.961(v_{in} - v_{out})$; $R_{out} = 331\ \Omega$; (b) $A_v = 0.662$ V/V

2.32 $A_v = -1499$ V/V

2.36 $t_1 = 25.5\ \mu s$

2.38 $t_1 = 34.7\ \mu s$

2.41 (a) $\tau = 6.43$ ms; (b) $t = 12.9$ ms

CHAPTER 3

3.2 (a) $v_{out} = -0.6\ \sin\ \omega t$ V; (b) $A_v = -30$ V/V; (c) $V_{out} = 6\ V$

3.6 $A_{MB} = 121$ V/V

3.10 $A(\omega) = \dfrac{A_{MB}}{\left(1 - j\frac{f_1}{f}\right)\left(1 - j\frac{f_1'}{f}\right)}$; where $A_{MB} = 121$ V/V

$f_1 = 2.61$ Hz; $f_1' = 12.2$ Hz

3.14 $f_{high} = 20{,}000$ Hz

3.18 $f_{low} = 40$ Hz; $f_{high} = 42{,}000$ Hz

3.21 $A = \dfrac{64}{\left(1-j\frac{10}{f}\right)\left(1-j\frac{2.6}{f}\right)}$

3.25 $A = \dfrac{-10.9}{1-j\frac{28.9}{f}}$

3.28 $C_{c1} = 0.289\ \mu F;\ C_{c2} = 0.159\ \mu F$

$A = \dfrac{-16}{\left(1-j\frac{20}{f}\right)\left(1-j\frac{50}{f}\right)}$

3.31 $C_{c1} = 0.161\ \mu F$

3.35 $C_{c2} = 0.52\ \mu F$

3.38 $X = 4.2$

3.40 $A = 8.39$ V/V

3.43 $C_{c1} = 1\ \mu F;\ C_{c2} = C_{c3} = 1.14\ \mu F;\ C_{c4} = 1.59\ \mu F;\ f_{\text{low}} = 46$ Hz

3.46 $A_{MB} = 4571$ V/V; $f_2 = 206$ kHz

CHAPTER 4

4.5 2-V source in series with a $\frac{2}{3}$ Ω resistor

4.8 (a) $v_{\text{out}} = 10\cos\omega t$; (b) $v_{ab} = 3.33\cos\omega t$; (c) $v_{\text{out}} = 1.67\cos\omega t$

4.10 $V_{\text{out}} = 10.5$ V

4.12 $R_{\text{in}} \approx 1\ k\Omega$; $R_{\text{out}} \approx 0$; $A_{MBi} = -100$ V/V; $f_{2i} = 39.6$ kHz

4.15 Upper stage: $A_{MB} = -0.5$ V/V; Lower stage: $A_{MB} = 1$ V/V

4.18 (a) $A_{MBi} = -3$ V/V; (b) Yes; (c) $f_{2i} = 800$ kHz; (d) Yes

4.21 $GBW_{oa} = 1.82$ MHz

4.25 $A_{MBni} = 10.67$ V/V; $f_{2ni} = 125$ kHz

4.28 (a) $A_{MBo} = 45$ V/V; (b) $f_{2o} = 224$ kHz

(c) $A = \dfrac{45}{(1 + jf/720,000)(1 + jf/244,000)}$

(d) $R_1 = 75.7$ kΩ; $R_2 = 21.2$ kΩ; (e) $f_{2o} = 269$ kHz

4.33 $GBW_{oa-\min} = 2.61$ MHz

CHAPTER 5

5.1 $p = N_A = 2 \times 10^{13}/\text{cm}^3$; $n = 1.125 \times 10^7/\text{cm}^3$

5.3 $\sigma = 1.54 \times 10^{-3}$ S/cm

5.7 $N_D = n = 4.8 \times 10^{16}/\text{cm}^3$

5.10 $n_p(0) = 1.84 \times 10^{10}/\text{cm}^3$

5.12 $p_n = 2.25 \times 10^5/\text{cm}^3$; $n_p = 2.25 \times 10^2/\text{cm}^3$

5.15 $I = 1.44$ mA

5.18 $C_{dl}(12V) = 3.8$ pF

5.22 $V = 2.13$ V; $I = 19.3$ mA

5.26 Ideal diode $v_{\text{outp}} = 50$ V; for the 0.6 V drop, $V_{\text{outp}} = 49.4$ V. The percentage difference is much smaller, and the 10-kΩ resistance leads to less attenuation.

5.32 $i_{\text{ac}} = 0.108$ mA $\sin \omega t$

5.34 (a) For $V = -3$, $C = 10.5$ pF; (b) For $V = -16$, $C = 6.49$ pF

5.38 v_{out} has a peak value of 9.4 V. $V_{\text{dc}} \approx 2.99$ V

5.42 $V_{\text{dc}} = 52.3$ V

CHAPTER 6

6.3 $R_{\text{lin}} = 2.08$ kΩ

6.5 (a) Active; (b) Active; (c) Triode

6.8 From Eq. (6.25), $I_D = 1.28$ mA for both cases. At 4 V, $I_D = 1.36$ mA. At 8 V, $I_D = 1.48$ mA.

6.10 $g_m = 8.37$ mA/V

6.12 r_{ds}(at $V_{GS} = 2$) $= 69.4$ kΩ; r_{ds}(at $V_{GS} = 4$) $= 9.77$ kΩ

6.15 $A_{MB} = -6.28$ V/V

6.18 $V_{\text{max}} = +8$ V; $V_{\text{min}} = 1.945$ V

6.21 $A_{MB} = -5.19$ V/V

6.24 (a) $V_{DQ} = 16.22$ V, $V_{SQ} \approx 0.76$ V;
(b) $v_{\text{out}} = -0.0337 \cos \omega t$; (c) $f_{\text{low}} = 398$ Hz

6.27 $A_{MB} = -13.9$ V/V

6.31 $R_1 = 434$ kΩ

6.35 $A_{MB} = -7.68$ V/V

CHAPTER 7

7.2 $I = 0.154$ mA; $I = 337$ mA

7.5 $V_{\text{out}} \approx 0V$

7.8 $V_{\text{out}} = -11.25 \sin \omega t$ only when this expression is positive.

7.11 $V_{\text{out}} = 16.8 - 0.3 \sin \omega t$. $A_{MB} = -30$ V/V; gain is reduced, proportional to R_C.

7.15 $v_{\text{out}} = -2.5 \sin \omega t$, $V_{CEQ} = 10$ V

7.18 (a) $R_B = 970$ kΩ; (b) $R_B = 2.43$ MΩ; (c) $R_B = 1.94$ MΩ

7.22 $R_1 = 191$ kΩ, $V_{p-p} = 6$ V

7.25 $A_{MB} = 5.74$ V/V

7.28 $A_{MB} = -33.3$ V/V, $V_{CQ} = 12$ V

7.30 $A_{MB} = -60.4$ V/V, $R_{\text{in}} = 3.313$ kΩ, $R_{\text{out}} = 2$ kΩ

7.33 $A_{MB1} = 0.988$ V/V, $A_{MB2} = 0.663$ V/V

7.36 $C = 16.6$ pF

7.40 $f_2 = 13$ kHz

7.44 $A_{MB} = -58.5$ V/V, $f_2 = 150$ kHz

7.48 For 7.4, $C_D = 1592$ pF, $D = 1.63$, $C_M = 1010$ pF. For 7.45, $C_D = 159.2$ pF, $C_M = 1010$ pF, $D = 7.34$.

CHAPTER 9

9.1 $I_D = 0.215$ mA

9.4 $I_{DP} = 265$ μA and $V_{DSP} = 3.18$ V. These values cannot be reached in this circuit since $V_G = V_D$.

9.6 (a) $I_{in} = 157$ μA; (b) $V_{GS} = V_{DS} = 2.56$ V; (c) Assume that $V_{T2} = 0.889$ V; (d) $I/I_{in} = 1.14$

9.9 $W_1 = 1764$ μ

9.11 $f_2 = 484$ MHz

9.15 $C_L = 131 - 49 = 82$ fF; $f_{2o} = 2.29$ MHz

9.18 $V_{GS} = 1.7$ V, $A_{MB} = 102.8$ V/V

9.23 $f_2 = 61.6$ MHz

9.25 $A_{MB} = -33.1$ V/V

CHAPTER 10

10.1 $R = 4.31$ kΩ

10.4 $+3.75\%$, $r_{out} = 54$ kΩ

10.6 $\frac{I_o}{I_{in}} = 3.40$

10.8 $r_{scm} = 50$ kΩ, $r_{wcm} = 600$ kΩ

10.11 $A_{MB} = -1824$ V/V

10.14 (a) $A_{MB} = -914$ V/V; (b) $A_{MB} = -1474$ V/V

10.17 $A_{MD} = -315$ V/V

10.19 $A_{MB} = -681$ V/V

10.21 $f_2 = 237$ kHz

CHAPTER 11

11.1 $A_{CM} = 0.1$ V/V, $A_D = 98.15$ V/V

11.3 $A_{CM} = 0.0579$ V/V

11.5 $A_D = -110.3$ V/V

11.7 (a) $I_{E1} = 0.571$ mA, $I_{E2} = 0.429$ mA; (b) 6.78 V (*peak*), $A_D = -90.3$ V/V

11.9 $A_D = 27$ V/V

11.12 $A_{MB} = 29.3$ V/V

11.14 $R = 8$ kΩ

11.19 (a) $A_D = 3.1$ V/V; (b) $A_D = 2.68$ V/V

11.21 $A_{MB} = 3917$ V/V

11.24 $v_{\text{out}} = 1.7 \sin \omega t$ V

11.27 For stability, $P_1' = 55$ rad/s

11.32 $R_3 = 50$ kΩ

CHAPTER 12

12.1 (a) $G = 9.9$ V/V; (b) $G = 90.9$ V/V; (c) $G = 500$ V/V
   (a) $1/F = 10$; (b) $1/F = 100$; (c) $1/F = 1000$

12.3 $S_{x,y} = 1.17$

12.7 $SNR = 44.7$

12.9 $G_{MB} = 19.6$ V/V; $\omega_{2f} = 510$ Mrad/s

12.12 (a) Series-series/transadmittance; (b) Shunt-shunt/transimpedance;
   (c) Series-shunt/voltage; (d) Shunt-series/current

12.15 $R_{\text{in}-f} = 7$ kΩ, $R_{\text{out}-f} = 14$ kΩ

12.18 $R_{\text{in}-f}' = 50$ MΩ

12.21 $A_{MB} = 51$ V/V

12.24 (a) $G_{MB} = 20.6$ V/V; (b) $f_{2f} = 626$ kHz; (c) $R_{\text{out}-f} = 13.7$ Ω

12.27 $G_{MB} = 251$ V/V

12.31 (a) $G_{MB} = 17.2$ V/V; (b) $f_{2f} = 633$ kHz

12.35 (a) No; (b) $\omega = 1.1$ Mrad/s

CHAPTER 13

13.3 $R_B = 97$ kΩ

13.6 $R_L = 1.65$ kΩ

13.10 Fall time is unaffected by R_L. $\tau_f = 20$ ns, $\tau_r = 1.02$ μs

13.13 At $t = 0$, capacitor discharges from 5 V toward 2.19 V with $\tau = 23.8$ ns. At
   $t = 10$ ns, the capacitor has reached 3.15 V. The capacitor now discharges toward
   ground with $\tau = 28$ ns. At $t = 30$ ns, the voltage remains at 1.54 V.

13.16 (a) $V_{B2} = -7.8$ V; (b) $v_t = 16$ V; (c) $T = 265$ μs

13.22 $V_1 = 17.3$ V

13.25 $(v_t\ V, T\ \text{ms}) = (6, 1.76),\ (12, 1.11),\ (18, 0.82),\ (24, 0.65)$

13.29 $V_{t-\min} = 6.67$ V, $C = 0.73$ μs

13.34 $t_1 = 77.3$ μs, $t_2 = 114$ μs

13.37 $C = 0.145$ μF, $R_B = 4280$ Ω

CHAPTER 14

14.1 $V_{out} = 2.75$ V

14.3 (a) $V_{in} = 2.62$ V; (b) $V_{in} = 2.99$ V

14.6 $K_p = 36.2 \ \mu A/V^2$

14.9

Figure P14.9

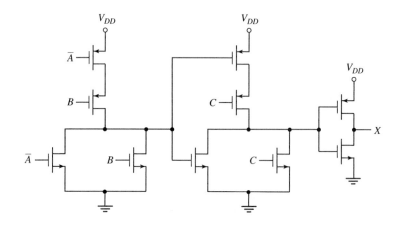

14.12

Figure P14.12

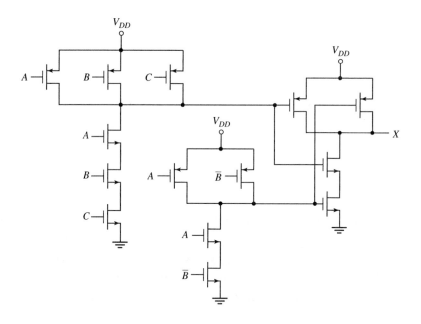

14.17 $X = \bar{A}\bar{C} + B\bar{C}$

14.20 $Z = A + \bar{B} + \bar{C}$

14.22 $X = \bar{G}\bar{H} + GH + \bar{J} + \bar{K}$

14.25

Figure P14.25

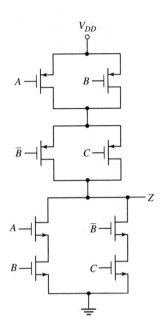

14.31

Figure P14.31

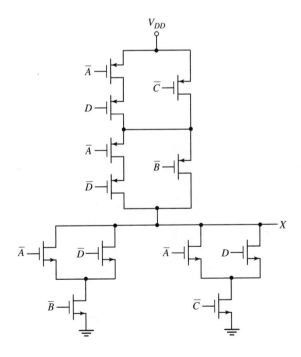

Index

485